# CAMBRIDGE LIBRARY COLLECTION

*Books of enduring scholarly value*


## Physical Sciences

From ancient times, humans have tried to understand the workings of the world around them. The roots of modern physical science go back to the very earliest mechanical devices such as levers and rollers, the mixing of paints and dyes, and the importance of the heavenly bodies in early religious observance and navigation. The physical sciences as we know them today began to emerge as independent academic subjects during the early modern period, in the work of Newton and other 'natural philosophers', and numerous sub-disciplines developed during the centuries that followed. This part of the Cambridge Library Collection is devoted to landmark publications in this area which will be of interest to historians of science concerned with individual scientists, particular discoveries, and advances in scientific method, or with the establishment and development of scientific institutions around the world.

## The Life of William Thomson, Baron Kelvin of Largs

The mathematician and physicist William Thomson, 1st Baron Kelvin, (1824–1907) was one of Britain's most influential scientists, famous for his work on the first and second laws of thermodynamics and for devising the Kelvin scale of absolute temperature. Silvanus P. Thompson (1851–1916) began this biography with the co-operation of Kelvin in 1906, but the project was interrupted by Kelvin's death the following year. Thompson, himself a respected physics lecturer and scientific writer, decided that a more comprehensive biography would be needed and spent several years reading through Kelvin's papers in order to complete these two volumes, published in 1910. Volume 2, beginning in 1871, covers not only Kelvin's mature research, but also more personal aspects of his life, including his love of music and sailing, his experiments with compasses and navigation, and the relationship between his scientific discoveries and his religious beliefs.

Cambridge University Press has long been a pioneer in the reissuing of out-of-print titles from its own backlist, producing digital reprints of books that are still sought after by scholars and students but could not be reprinted economically using traditional technology. The Cambridge Library Collection extends this activity to a wider range of books which are still of importance to researchers and professionals, either for the source material they contain, or as landmarks in the history of their academic discipline.

Drawing from the world-renowned collections in the Cambridge University Library, and guided by the advice of experts in each subject area, Cambridge University Press is using state-of-the-art scanning machines in its own Printing House to capture the content of each book selected for inclusion. The files are processed to give a consistently clear, crisp image, and the books finished to the high quality standard for which the Press is recognised around the world. The latest print-on-demand technology ensures that the books will remain available indefinitely, and that orders for single or multiple copies can quickly be supplied.

The Cambridge Library Collection will bring back to life books of enduring scholarly value (including out-of-copyright works originally issued by other publishers) across a wide range of disciplines in the humanities and social sciences and in science and technology.

# The Life of William Thomson, Baron Kelvin of Largs

Volume 2

Silvanus Phillips Thompson

CAMBRIDGE UNIVERSITY PRESS

Cambridge, New York, Melbourne, Madrid, Cape Town, Singapore,
São Paolo, Delhi, Dubai, Tokyo, Mexico City

Published in the United States of America by Cambridge University Press, New York

www.cambridge.org
Information on this title: www.cambridge.org/9781108027182



This edition first published 1910
This digitally printed version 2011

ISBN 978-1-108-02718-2 Paperback

# LIFE OF LORD KELVIN

MACMILLAN AND CO., LIMITED
LONDON · BOMBAY · CALCUTTA
MELBOURNE

THE MACMILLAN COMPANY
NEW YORK · BOSTON · CHICAGO
ATLANTA · SAN FRANCISCO

THE MACMILLAN CO. OF CANADA, LTD.
TORONTO

*Lord Kelvin and His Compass.*

*From a photograph by Annan Glasgow. 1902.*

# THE LIFE

OF

# WILLIAM THOMSON

## BARON KELVIN OF LARGS

BY

SILVANUS P. THOMPSON

IN TWO VOLUMES

VOL. II

MACMILLAN AND CO., LIMITED
ST. MARTIN'S STREET, LONDON
1910

# CONTENTS

## CHAPTER XV

### THE "LALLA ROOKH," THE BRITISH ASSOCIATION, AND THE "HOOPER"



## CHAPTER XVI

### IN THE 'SEVENTIES

## CHAPTER XVII

### NAVIGATION: THE COMPASS AND THE SOUNDING MACHINE



## CHAPTER XVIII

### GYROSTATICS AND WAVE MOTION



## CHAPTER XIX

### IN THE 'EIGHTIES

## CHAPTER XX

### THE BALTIMORE LECTURES



## CHAPTER XXI

### GATHERING UP THE THREADS



## CHAPTER XXII

### THE PEERAGE

## CHAPTER XXIII

### THE JUBILEE: RETIREMENT



## CHAPTER XXIV

### THE GREAT COMPREHENSIVE THEORY

## CHAPTER XXV

### VIEWS AND OPINIONS



## CHAPTER XXVI

### THE CLOSING YEARS



## APPENDICES

# LIST OF PLATES

# CHAPTER XV

## THE *LALLA ROOKH*, THE BRITISH ASSOCIATION, AND THE *HOOPER*

BY the purchase of the *Lalla Rookh*, a smart sailing-yacht of 126 tons, Sir William Thomson became acquainted with navigation in a new phase. All his life he had been fond of sailing; but by the possession of this craft he acquired at first hand a most intimate knowledge of seamanship and of its needs. For many years the cruises of the *Lalla Rookh* occupied a considerable part of the six months between the sessions of the University. When the end of October 1870 compelled him to lay her up in the Gareloch for the winter, he left her with regret. He looked keenly forward to the first of May when he should be able to join his ship. He was now planning an expedition to the Canaries, to be followed by an extensive cruise in the Hebrides with a party of scientific friends in the coming autumn, and it became necessary to fit out the yacht with furniture and bedding. To this end he took counsel with Mrs. Tait, resulting in a lively and characteristic correspondence :—

GLASGOW COLLEGE, *March* 29, 1871.

DEAR MRS. TAIT—The question, cotton or linen, for the *Lalla Rookh's* berths has, after anxious consideration and consultation with naval experts, been decided in favour of linen. The cotton fabric seems to be too hygrometric to be suitable for sea-going places.

Will Glasgow do as well as Belfast for getting such a number (or area) as is required?

The area for mattress is approximately rectangular, and 3 ft. 9 i. by 7 ft. In fixing on the size of sheet I would wish to avoid an error which seems to have originated in the Levant prior to 725 B.C. (Isaiah xxviii. v. 20, second clause [1] of the v.), and which is still deplorably prevalent at sea.

I think I ought to have in all 12 pr., and therefore (as the acct. enclosed shows 4 pr. to be already provided) 8 pr., with the proper proportion of pillow slips, would be enough. The other things which I want are, so far as I can judge: 5 dozen towels, equal and similar to those provided by you for the N.P.L. [Natural Philosophy Laboratory.] 6 large bath sheets of similar material. Sometimes bath sheets are made thicker (apparently with the idea of maintaining a constant proportion of thickness to length or breadth), which is a mistake.

3½ dozen damask table napkins—"double damask," I understand from T′, has been decided.

10 tablecloths. I forgot to measure the table yesterday when I was at Greenock to see the *L. R.*, now fresh coppered and almost ready to be launched, but the dimensions will be sent to you by Captain Flarty. I think the best quality of damask should be taken for the tablecloths, as drops from the skylight, accidents through want of steadiness of platform, etc., etc., require the strongest resistance against shabbiness of appearance that the material can give.

I should also have a proportionate quantity of glass-

[1] [The verse in question runs—"For the bed is shorter than that a man can stretch himself on it; and the covering narrower than that he can wrap himself in it."]

towels, cook's cloths, and dusters, which, with what I have already, should be enough to serve for several weeks away from port.

Whatever of the above is to be had best from Belfast, will you order it for me? For the rest, any hints you can give will be gratefully received.

I am not unconscious (but as much as possible the reverse) that I am asking a very great benefit, and taking advantage to the utmost of the promise you gave me to help me, when I write so troublesome a list of wants. But you must allow me absolutely to restrict your kindness to ordering the things for me, and directing that the hemming and marking be done by the people who supply them, and who certainly will, if required, find persons ready to undertake those works.

The Committee on Ships of War will continue its sittings during May, and I am afraid much of June, partly in London and partly at sea with the Channel Fleet. So I must give up Teneriffe for this year, which I do with great regret. Will you tell Guthrie[1] that I hope for another visit from him before May, as we got scarcely anything of the books done last time, there was so much time wasted on tops, etc. Could he not come from Ap. 18 to 27, which would include an opening cruise of the *L. R.* to Arran, Friday till Monday?—Yours always truly, WILLIAM THOMSON.

The project of an autumn cruise with a party of scientific friends is set out in a letter to Helmholtz, terminated by a postscript from Professor Tait:—

GLASGOW COLLEGE, *March* 30/71.

DEAR HELMHOLTZ—I hope you will be able to come to the meeting of the British Association at Edinburgh in the first week of August. After it is over (and I wish it were over now, as I have the misfortune to be president-

[1] Peter Guthrie Tait, his friend and colleague.

elect) I want you to come and have a cruise for a few weeks among the Hebrides and West Highlands in a schooner of $128 \times 10^6$ grammes, which will be my only summer quarters besides the new College here. I hope Tait will come too, but he has a great aversion to being afloat, and without the inducement of your company he would scarcely be persuadable. I would also ask Clerk Maxwell and Huxley and Tyndall, which would reach nearly to the capacity of the *Lalla Rookh.* Will you let me have a line when your plans are fixed?

Many thanks for your last letter. I hope the remaining anxieties of the campaign in respect to your son soon ceased, and that he has got through unhurt. I say nothing just now in reply to what you said about the sympathies of England.—Believe me, yours always truly,

WILLIAM THOMSON.

On the last page of this letter Professor Tait wrote :—

DEAR PROF. HELMHOLTZ—As Thomson has sent this *through* me, doubtless for some great moral purpose, I beg to add that I have *no* aversion to being afloat, but that I prefer to spend my few holidays in active physical work, such as the game of golf.—Yours truly,

P. G. TAIT.

GLASGOW COLLEGE, *April* 9, 1871.

DEAR MRS. TAIT—Many thanks for your kind letter. I do not know the dimensions of the pillows, and could not well get them till Wednesday, as they are in store at Gourock. I think it would be safe to make the pillow-slips of the same size as for land pillows, which, I suppose, are something less than 3 ft. 9 in. long. If you think so you might let them be made accordingly. But in any case I shall have the dimensions of the actual pillows despatched by post from Gourock, addressed to you, on Wednesday. For the sheets I think $2\frac{3}{4}$ yards might be rather short for Guthrie when he comes on a cruise with me. At sea it is desirable to have rather more of the sheet to turn over at the top than in beds less exposed to acceleration. Three yards would be a safe length. Two yards will be a very good breadth, sufficient even for sleeping through several tacks.

I am very sorry to hear that Guthrie has been so ill. I cannot think it was good for him to allow Dr. Crum Brown to pull out a tooth. We regretted very much his not being here to meet Maxwell, Joule, etc.

Do not let him shirk the August cruise with Helmholtz, Huxley, Tyndall, and Maxwell, who I hope will all come.

If the boat race is to be at all, it is right that Cambridge should win, and they seem to have pulled splendidly last time. [April 1, 1871.]

I forgot that you had asked about the tablecloths. I am in a difficulty about them. I understood from Guthrie that the breadth determined the length, each being made one and indivisible in certain absolutely fixed proportions. I think the length 5 f. 4 i. must be when the table is at its shortest. But it is capable of prolongation, and I believe about 4 can sit on each side. The breadth you have is accurate. I shall write to you on Tuesday giving the maximum length.—Believe me, yours very sincerely, W. THOMSON.

WESTERN CLUB, GLASGOW,
*Tuesday evening* [*April* 11].

DEAR MRS. TAIT—The *L. R.* table is, I find, of invariable length, and the values of the constants which you have are correct. Those of the T-cloths which you proposed are therefore no doubt perfectly right.

The pillow-slip question is more difficult. An expert who has been employed on board told me that the pillows are presumably of the full breadths of the berths. If so there will be several different sizes. Captain Flarty will send you the length and breadth of each from Gourock as soon as possible, by to-morrow evening's post, I hope.

I write in the greatest haste, as I am just going to sit

(in the chair) at a meeting to promote united-non-sectarian-compulsory education on the same model as the Irish national education, which Roman Catholics, Presbyterians, and Church-people would not give fair play to in Ireland.

. . . . . . .

The gathering referred to was a great public meeting in the City Hall. From Sir William's speech, as chairman, as reported in the *Glasgow Herald*, the following extracts are taken. After reading apologies for absence from influential representatives of the Free Church and the United Presbyterian Church, he said:—

He believed the feeling was very strong in these Churches generally in favour of the views to be advocated that night. What the meeting really desired was something very much analogous to that which had been given to England in the bill for national education which had now become law in that part of the United Kingdom. There were certain blots, undoubtedly, in the Scottish Education Bill, for what reason he knew not. It seemed to be supposed that Scotland required a more denominational, a more sectarian system of education than England. No mistake could be greater than this. Scotland, of all countries in Christendom, was the one most prepared, most ready to accept a united non-sectarian national system of education. Scotland was prepared to make this a thoroughly religious system. We were not here in Scotland to have a godless education. It would not be a godless education that would be supported by the Free Church, the United Presbyterian Church, and by the Established Church for the people of Scotland. What was desired was, in the first place, education in these elements of knowledge and art which were necessary for any religious education whatever. What was desired was to make religious education possible in the first place by the universal teaching of the arts of reading and writing, and to make this part of the national education compulsory. It seemed that opinion in England was strongly divided with reference to the question "compulsory or non-compulsory"; but, on the other hand, it seemed that in Scotland there was a very strong feeling indeed among the people that compulsory education was desirable. If the people of Scotland desired compulsory education, he felt very confident indeed that they had only to say so, and it would be provided for them by Parliament. It was quite clear, however, that if education was to be compulsory, the national system of education in connection with which compulsory statutes were founded must be unsectarian. There was one other point upon which Scotland desired something different from that which had ever been provided in England, and that was a degree of elasticity in the national system, in virtue of which it should not be confined merely to reading, writing, and arithmetic. Scotland did not desire schools from which the use of the globes should be excluded. A school without maps would not be satisfactory in any parish of Scotland, nor would a school be satisfactory without music.

Later, in reference to one of the motions proposed, he said: The motion before the house did not propose to exclude the Bible from the schools. The Bible was truly and avowedly national, and he desired to ask the meeting to say that the motion — which demanded a provision that no religious catechism, or formulary which is distinctive of any particular denomination, should be taught in the schools—did not apply to the Bible.

At the conclusion the chairman said he wished to call attention to the danger that hung over Scotland at present. Was Scotland, he asked, to be made the stepping-stone from the system of mild denominationalism in England, to utter and destructive denominationalism in Ireland? Unless they resisted strenuously the efforts to carry the denominationalism proposed in this bill, Scotland would bitterly rue her part in such a matter.

[*About April* 23, 1871.]

DEAR MRS. TAIT

. . .

The *L. R.*, unfortunately, was not ready for my proposed cruise at this time owing to the weather, which made communication with the shore at Gourock difficult. I trust she will be ready by Friday, but ready or not I sail on Friday southwards, and hope the E. wind will not stop till I reach Land's End. If the linen is not ready to cross from Belfast on Thursday it might come leisurely to the Admiralty, where I shall be on Wed. week, ditto fortnight, do. etc.

Excuse great haste, and believe me, yours always truly, W. THOMSON.

[*P.S.*]—Tell Guthrie I have no time to answer his letter to-day as I am overdue to go out to the Rouken.

On May 5 he was at the Admiralty, but left that evening for Plymouth to sail in Plymouth Sound.

*L. R.*, DARTMOUTH, *May* 8 [1871].

DEAR MRS. TAIT—Since writing the above I have arrived in London. I made my first attempt to get a quiet forenoon of work yesterday in the *L. R.* (it being now Tuesday the 9th) which, you may tell Guthrie, was very promising, although it only resulted in the miserable fiasco of twelve letters all "business," and of the most trivial but inevitable kind. However, I hope for better things. The only interruption in the course of three hours was a great trawler fouling me. The cutter and gig had both gone to shore, one for water and the other to land J. T. B. and his father, who went to take a walk and leave me quiet. The captain, steward, and cook were shoving her (the trawler) off when I came on deck on hearing the noise, and soon after we got her clear. "I hope there's nothing broke, sir?" "No" (replied Captain F.). "I am glad to hear it, sir," were the last words from the trawler. I intended to write and give you a history of the voyage from the Clyde to Penzance, and how thoroughly enjoyed it was by J. T. B. and David King, the latter faintly denying that he would have enjoyed it still more if she had been on the slip at Greenock all the time. J. T. B. was more reticent, but I believe felt as deeply. I should also, if I had achieved my project of writing to you on board, have given you many details of a trip to the Eddystone and a voyage from Plymouth to Dartmouth against strong east wind. D. K. being replaced by J. T. B.'s father. The latter remarked that the best thing about yachting was going on shore, an opinion in which I by no means concur. But all such matters rapidly lose importance, and the "log" that is not written during the voyage is never written at all. The one unforgettable thing is the linen and the marking of the kitchen towels, than which nothing could be better. I never attributed the marking to Guthrie, but only the address on the parcel from Edinburgh.

Your most kind letter about the B.A. reached me here (London to-day). The five reasons are, each separately, irresistible. I shall certainly stay at 17 Drummond Place if I am able. I shall write as soon as I know. Tell Guthrie I am here (London) three days of every week [1] (address Athenæum Club, Pall Mall). Therefore he may give high praise to the *L. R.* if I get, as I intend, two good working days weekly. Tell him also that I met Dr. Lyon Playfair just now, who told me that he had quite lately seen Dr. N. Arnott, and that the latter intends to give £1000 to each of the Scotch Univ^ies^, but that he has taken a crick, and though Mrs. N. A. had strongly urged Ed. as well as Glasgow (on account of the work done in the *P. L.*[2] there) he was stiff about beginning with Glasgow for a trial. He remarked that I had not called the last time I was in London. I hope Guthrie's cold is better. Tell him that a good cruise in the *L. R.* will be requisite to brace him against these recurring attacks, which seem to be partially (if not wholly) due to overdoing the links. Will you not bring him with you to London when you come? Even that would do him good, and if you would both come from a Friday till Wednesday to the *L. R.* the cure would be complete.

*Monday Morning* [*May* 15, 1871]
TRAIN, WEYMOUTH TO LONDON.

DEAR MRS. TAIT—On receiving your most kind letter I wrote immediately to my sister [3] to ask if I might accept your invitation. I do not mean that I put it exactly in that way, but I pointed out forcibly how much

[1] (Four days this week to-day for Admiral Halstead's fleet, and Sir Joseph Whitworth's ordnance.)—Wed., Thurs., Frid. (from 12 till 5 in the Admiralty, except when it is 11 to 5 on account of extra tediousness of witnesses, or 9 till 6 Shoeburyness expedition as last week).

[2] [Physical Laboratory.] [3] [Mrs. King, then resident in Edinburgh.]

better she would be without me, and I said that you had very kindly invited me to stay at 17 D. P. [Drummond Place]. I received her answer just before setting out on a short tour on the Continent, from which I am now returning, and which has prevented me from earlier writing to you. I think that as I promised last September to stay with her, and she has Dr. Gladstone only, and says she would be greatly disappointed if I did not come, I am not free to do otherwise. I could easily prove this is a great advantage to you and Guthrie, but your letter disarms me, and I can only say that unless it were to be very different from all my visits to Greenhill Gardens and previous ones to D. P., it would have been one of the few pleasures that remain pleasures to me, to have the prospect of being at D. P. during the impending meeting.

Often when kept in Glasgow by affairs or by his laboratory work, Sir William Thomson would retreat for the week-ends to his yacht to gain quiet and rest. If he had no relations or friends on board he would take his secretary with him, that he might get on with work. Rising early he would take a plunge, before breakfast, in the sea, swimming round the yacht, and in spite of his lame leg climb with agility on board by the rope. When there were no observations or soundings to take he would sit for hours with green book and pencil in hand working at calculations and meditating over his problems; or he would pace the deck smoking a quiet cigar. Often he would work on far into the night. He was a daring navigator, and would sail far into the season when other yachts were laid up, sometimes in darkness and in severe weather. Once when he was sailing in the teeth of a gale his assistant John Tatlock, who often was with him as amanuensis, heard Captain Flarty saying half-aloud in Sir William's presence: "You will not rest till you have your boat at the bottom." He took no notice. He never seemed to tire. With all the sailors he was extremely popular; their only grievance was that he would sometimes pop up on deck in the small hours of the morning to make sure that the watch was at his post and awake. In all the operations of sailing he took the keenest interest, and became a most expert navigator. Happy though he was to be thus alone, he was still happier if he could secure for a few days' cruise his brother or some member of their related families, nephews or nieces, many of whom retain the most joyous recollections of the days spent on board the *Lalla Rookh.*

Two short extracts from letters to Miss Jessie Crum show the use he made of his yacht:—

*May* 15.—*Train, Weymouth to London. Monday morning.*—I received your letter on Friday afternoon just as I was leaving the Admiralty, and read it in the train on my way to Southampton to the *L. R.*

*May* 17, *Wedy.*—*Athenæum.*— . . . Soon after daybreak (last) Saturday I sailed for Cherbourg . . . to Portland on Sunday morning. Lord Dufferin is ordered by the Queen to Balmoral for 3 weeks, and the Committee is therefore adjourned until after the 10th of June. I sail for Lisbon to-night.

From Lisbon he sent Miss Crum a long letter about his doings there. The *Lalla Rookh* had sailed from Portland to the bar of Lisbon in 6 days 23 hours. On his return he wrote to Helmholtz:—

THE ATHENÆUM, *June* 14, 1871.

MY DEAR HELMHOLTZ—I have only this morning, on returning to London from a cruise to Lisbon and back in the *Lalla Rookh*, received your letter of the 11th May. I am very sorry you will not be able to be at the meeting of the Association in Edinburgh—and many others will be sorry also. But I am glad that you will come and sail with me in the West Highlands, and I shall take care to have the *Lalla Rookh* in a convenient position (probably in the Clyde or possibly Oban), at whatever time suits you. I asked Huxley and Tyndall to come for a cruise immediately after the meeting, but, unfortunately, neither of them could accept, and I shall therefore most probably remain chiefly at the College in Glasgow after the meeting until your arrival. You must arrange to spend as much as possible of your holiday in Scotland, and if you wish to mix a little work with it as you did before in Arran, you will find writing not impossible in the *Lalla Rookh*.

I congratulate you and Mme. Helmholtz most sincerely on the safe return of your son from the war.—Believe me, yours very truly, WILLIAM THOMSON.

On June 24th he writes from the India Office, where he has been attending a meeting of the Examiners for the India Telegraph Service, telling of the progress of his Admiralty work. He is just going down to Portsmouth with William Froude to sail in the *Lalla Rookh* to Torquay. He has been staying in London with Dr. Gladstone;[1] he has also given a party, of which he tells Miss Crum,

[1] Dr. John Hall Gladstone, F.R.S., who had married the eldest daughter of Dr. David King.

bringing together several of his old comrades in the C.U.M.S.:—

> Blow, and Shedden and his wife completed the number, six in all. Pollock played the hautboy, Blow accompanied on the piano, and Blow played on the violin, unaccompanied. He did not play very much, as he had been playing in the Crystal Palace (last day of Handel Festival) "Israel in Egypt," and was tired, and only got up to London for 7.30 dinner. I had succeeded in getting the room James Bottomley recommended (as the one in which his chemical monthly dinners had taken place) and all went off very well. It was a strange reunion, like a return from the other world—Shedden, Blow, Pollock, and myself, who had not been all together since the end of 1846, when Pollock, then a new-comer to Cambridge, quickly began to be intimate with Blow, Shedden, and me, just before Blow and Shedden were leaving Cambridge. I have often looked forward to such a reunion merely as an occasion when the music would have been a happy enjoyment. We had a visit from Blow at the Langham Hotel, but could not get any opportunity for music. It can never again be what it was, and it is too full of sadness for the present.

On July 1, he writes from Cowes that in the previous week he had sailed on Monday to Torquay, thence up to Southampton. On Wednesday he had run up to London to stay the night with Pollock at Hampstead. Lord and Lady Dufferin have come down to Cowes to yacht with him. "I have," he adds, "however, really found the *L. R.* the quietest and best place attainable for work." Work meant here the preparation of his Inaugural Address as President of the British Association, to be held on August 2nd, at Edinburgh.

To his brother-in-law, Alexander Crum, he wrote :—

*L. R.*, HURST ROAD, SOLENT, *July* 11.

My B. A. address destroys everything now. I cannot write a word of it, but it effectually prevents me from writing or doing anything else. . . . Helmholtz is coming from Germany for the sole purpose of a cruise about the 12th or 15th [of August]. . . . Shedden and his wife, W. Young, and a young man (Roberts) from the *Nautical Almanac* office, who has been calculating tides for me (as Brit. Ass. Committee) for four or five years, came down with me on Saturday for a few days' cruise.

*July* 18.—*Tuesday* (*Athenæum*).— . . . landed at Portsmouth this morning on my way to London.

I have made some slight beginnings of actual writing for the Address, and have a great mass of matter, greater than I shall find space for, to bring in. My difficulty will be to get proper arrangement and condensation, and I feel as if it must necessarily be a very unsatisfactory thing at best. I had George King with me from Saturday till to-day. . . . George began the day by reading a number of chapters of 1st Corinthians, and spent a great part of the remainder in writing for me,[1] towards the Address. I have taken some of the proceeds to the printers to-day, and hope to give some more instalments this week.

I dine with Huxley alone to-day to talk over Association and other matters for the sake chiefly of my Address.

I shall be here daily till Saturday, but am staying at Pollock's, Hampstead. On Sat. I go to the *L. R.* for quiet.

Sir William Thomson's Presidency of the British Association, at Edinburgh, on August 2, 1871, was

[1] [Mr. George King remembers how Sir William paced up and down the deck, dictating a few words at intervals, very slowly, making many corrections, while the yacht lay becalmed off Bournemouth.]

an event of great importance. His Address was awaited with expectancy, for he was to be introduced by none other than Huxley, with whom he had crossed swords with knightly courtesy indeed, but with deadly earnest, in the matter of Geological Time; and he was known to be opposed to some of the developments of the doctrines of Evolution that for a decade had been revolutionising men's minds as to the origin of things. Nor were the expectations of the assembled men of science disappointed; for the Address, though somewhat lengthy and discursive, proved of surpassing interest. The assembly was a brilliant one. Huxley, the retiring President, was accompanied on the platform by the Emperor Dom Pedro of Brazil, and by a crowd of most distinguished savants, British and foreign, also by a number of the leading Professors of the Scottish Universities. On rising to vacate the chair, he expressed cordial thanks to the officers and members for the support given to him, and congratulated the Association on the good work accomplished during the past year. Then turning toward the President-elect, he introduced him with exquisite courtesy in the words already quoted on p. 550 above.

Sir William Thomson's address began with a reference to the origin of the British Association and the aims of its founders, in particular Brewster and Herschel, the latter of whom had passed away but two months before. He also referred to the recent death of De Morgan; to the work of the

Meteorological Observatory at Kew since its establishment by the Association in 1842; and to the need of national laboratories for research. Our Government, he declared, fatally neglected the advancement of science. Glancing at the Reports on different branches of science, which had formed a conspicuous feature of the Association's past work, he particularised Cayley's Report of 1857 on Theoretical Dynamics, and Sabine's Report of 1838 on Terrestrial Magnetism, as having been of utmost service to scientific men, as well as of practical utility. He suggested the establishment of a British Year-book of Science as a need of the time. Then, turning to recent advances in particular branches, he pointed out that many of them owed their origin to protracted drudgery. "Accurate and minute measurement," he said, "seems to the non-scientific imagination a less lofty and dignified work than looking for something new. But nearly all the grandest discoveries of science have been but the rewards of accurate measurement and patient, long-continued labour in the minute sifting of numerical results." He instanced, as cases in point, the discovery of the theory of gravitation by Newton, that of specific inductive capacity by Faraday, that of thermodynamic law by Joule, and that of the continuity of the gaseous and liquid states by Andrews. Then he turned to the labours of Gauss and Weber, who had founded the absolute system of measurement of magnetism and electricity, and Weber's resulting discovery that the ratio of the electromagnetic and electrostatic units is a velocity. Maxwell he eulogised for his discovery that this velocity is physically related to the velocity of light. This led him to reflect how much science even in its most lofty speculations, gains in return for benefits conferred by its application to promote the social and material welfare of man. "Those," he declared, "who perilled and lost their money in the original Atlantic Telegraph were impelled and supported by a sense of the grandeur of their enterprise, and of the world-wide benefits which must flow from its success; they were at the same time not unmoved by the beauty of the scientific problem directly presented to them; but they little thought that it was to be through their work that the scientific world was to be instructed in a long-neglected and discredited fundamental discovery of Faraday's." Next, dealing with the kinetic theory of gases, which he described as the greatest achievement yet made in the molecular theory of matter, he particularly praised Clausius for having thus given the foundation for estimates of the absolute dimensions of atoms, and of their rates of diffusion. Maxwell had completed the dynamical explanation of the known properties of gases by bringing in viscosity and thermal conductivity. No such comprehensive molecular theory had ever been imagined before the nineteenth century; but Sir William Thomson was not satisfied. Definite and complete as it seemed, it was yet but a part of a still more comprehensive theory in which all physical science would be represented with every property of matter shown in dynamical relation to the whole. But there could be no permanent satisfaction to the mind in explaining heat, light, elasticity, diffusion, electricity, and magnetism by statistics of great numbers of atoms, if all the while the properties of the atom itself are assumed. "When the theory, of which we have the first instalment in Clausius' and Maxwell's work, is complete, we are but brought face to face with a superlatively grand question, What is the inner mechanism of the atom?" This at once led to a sketch of the arguments by which he himself, in independence of Loschmidt and of Johnstone Stoney, had arrived at ideas about the size of atoms. He scorned to enter into any questions of priority in this affair. "Questions of personal priority, however interesting they may be to the persons concerned, sink into insignificance in the prospects of any gain into the secrets of nature." The atom must henceforth not be regarded as a mystic point endowed with inertia and attraction, nor as infinitely small and infinitely hard. It must be regarded as "a piece of matter with shape, motion, and laws of action, intelligible subjects of scientific investigation." The prismatic analysis of light here came in to reveal new facts as to atomic constitution. The observational and experimental foundations were the discovery by Fraunhofer of the coincidence of certain dark solar spectrum lines with bright lines in flames; the rigorous test of this by Miller; the identification of the D-lines as belonging to sodium; the discovery of Foucault (see p. 224) that the voltaic arc can emit the D-rays on its own account and at the same time absorb them when they come from another quarter; the teachings of Stokes (see p. 300) as to the physical significance of the spectrum lines, and the inherent isochronism of the vibrations of an atom; the inferences from the dark lines as to the chemistry of the sun; the prodigious and wearing toil of Kirchhoff, and of Ångström, of Plücker, and of Hittorf, in preparing spectrum maps and in identification of spectra under various physical conditions. The chemists, following Bunsen, discovered new metals; biologists applied spectrum analysis to animal and vegetable substances; and the astronomers, led by Huggins, carried spectroscopic research to the stars and comets. Well might the lecturer point out that "scientific wealth tends to accumulation according to the law of compound interest." Solar and stellar chemistry had garnered great results. Rarely before in the history of science had enthusiastic perseverance, directed by penetrative genius, produced within ten years so brilliant a succession of discoveries. We were now to have a solar and stellar physics: for Miller, Huggins, and Maxwell had shown that the spectroscope afforded a means of measuring the relative velocity with which a star approaches to or recedes from the earth, and had found that not one of them had so great a velocity as 315 kilometres per second to or from the earth, a most momentous result in respect to cosmical dynamics. Then came a brief review of the nebular hypothesis of the solar system—a hypothesis invented before the discovery of thermo-dynamics, otherwise the nebulæ would not have been supposed to be fiery. Helmholtz's supposition of 1854, that mutual gravitation between the parts of the original nebula might have generated the heat of the sun, had been extended by his own further suggestion that gravitation might account for all the heat, light, and motions in the universe; while recent spectroscopic observation had shown that Tait's theory of comets, in which the head of the comet is regarded as a group of meteoric stones, furnished at least a probable explanation of that feature of their constitution. Astronomy and cosmical physics, therefore, well illustrated the truth that the essence of science consists in inferring, from phenomena which have come under actual observation, the conditions that were antecedent, and in anticipating future evolutions. Even naturalists of the present day were not appalled or paralysed by the prodigious difficulties of acting up to this ideal. They were now struggling, boldly and laboriously, to pass out of the mere "Natural History stage," and to bring Zoology within the range of Natural Philosophy. But science brought a vast mass of inductive evidence against the hypothesis of spontaneous generation, to confute the idea that dead matter might have run together or crystallized or fermented into organic cells or germs or protoplasm. "Careful enough scrutiny has in every case up to the present day discovered life as antecedent to life. Dead matter cannot become living without coming under the influence of matter previously alive." "This," said Sir William, "seems to me as sure a teaching of science as the law of gravitation." "I confess to being deeply impressed by the evidence put before us by Professor Huxley; and I am ready to adopt, as an article of scientific faith, true through all space and through all time, that life proceeds from life, and from nothing but life." The passage which followed startled even the most advanced thinkers present. "How, then, did life originate on the Earth? Tracing the physical history of the Earth backwards on strict dynamical principles, we are brought to a red-hot melted globe on which no life could exist. Hence, when the Earth was first fit for life there was no living thing on it. There were rocks, solid and disintegrated, water, air all round, warmed and illuminated by a brilliant sun, ready to become a garden. Did grass and trees and flowers spring into existence, in all the fulness of ripe beauty, by a fiat of Creative Power? or did vegetation, growing up from seed sown, spread and multiply over the whole Earth? Science is bound, by the everlasting law of honour, to face fearlessly every problem which can fairly be presented to it. If a probable solution, consistent with the ordinary course of nature, can be found, we must not invoke an abnormal act of Creative Power. . . . When a volcanic island springs up from the sea, and after a few years is found clothed with vegetation, we do not hesitate to assume that seed has been wafted to it through the air, or floated to it on rafts. Is it not possible, and, if possible, is it not probable, that the beginning of vegetable life on the Earth is to be similarly explained? Every year thousands, probably millions, of fragments of solid matter fall upon the Earth. Whence came these fragments? What is the previous history of any one of them? Was it created in the beginning of time an amorphous mass? This idea is so unacceptable that, tacitly or explicitly, all men discard it. It is often assumed that all, and it is certain that some, meteoric stones are fragments which have been broken off from greater masses and launched free into space. . . . Should the time when this Earth comes into collision with another body, comparable in dimensions with itself, be when it is clothed as at present with vegetation, many great and small fragments, carrying seed and living plants and animals, would undoubtedly be scattered through space. Hence and because we all confidently believe that there are at present, and have been from time immemorial, many worlds of life besides our own, we must regard it as probable in the highest degree that there are countless seed-bearing meteoric stones moving about through space. If at the present instant no life existed upon this Earth, one such stone falling upon it might, by what we blindly call *natural* causes, lead to its becoming covered with vegetation. I am fully conscious of the many scientific objections which may be urged against this hypothesis, but I believe them all to be answerable. . . . The hypothesis that [some][1] life [has actually] originated on this Earth through moss-grown fragments from the ruins of another world may seem wild and visionary; all I maintain is that it is not unscientific [and cannot rightly be said to be impossible]." A brief peroration touched the then burning question of Evolution *versus* Design. "From the Earth stocked with such vegetation as it could receive meteorically, to the Earth teeming with all the endless variety of plants and animals which now inhabit it, the step is prodigious; yet, according to the doctrine of continuity, most ably laid before the Association by a predecessor in this chair, Mr. Grove, all creatures now living on earth have proceeded by orderly evolution[2] from some such origin." He then quoted from the conclusion of Darwin's great work on *The Origin of Species*, a couple of sentences about the numerous forms of life—plants, birds, insects, worms—different, interdependent, yet "all produced by laws acting around us," and about the "grandeur in this view of life

[1] The words in brackets were added by Lord Kelvin himself when he reprinted the address in 1894 in vol. ii. of his *Popular Lectures and Addresses.*

[2] Professor Huxley, in a later discourse, gently brushed aside the importance of Thomson's suggestion in the following words: "I think it will be admitted that the germs brought to us by meteorites, if any, were not ova of elephants, nor of crocodiles; not cocoa-nuts, nor acorns; not even eggs of shell-fish or corals, but only those of the lowest forms of animal and vegetable life. Therefore, since it is proved that from a very remote epoch of geological time the earth has been peopled by a continual succession of the higher forms of animals and plants, these either must have been created or they have arisen by evolution. And in respect of certain groups of animals, the well-established facts of palæontology leave no rational doubt that they arose by the latter method."

with its several powers having been originally breathed by the Creator into a few forms or into one, from which endless forms, most beautiful and most wonderful, have been and are being evolved." Then he continued: "With the feeling expressed in these two sentences I most cordially sympathise. I have omitted two sentences which come between them, describing briefly the hypothesis of 'the origin of species by natural selection,' because I have always felt that this hypothesis does not contain the true theory of evolution, if evolution there has been, in biology. Sir John Herschel, in expressing a favourable judgment on the hypothesis of zoological evolution, with, however, some reservation in respect to the origin of man, objected to the doctrine of natural selection that it was too like the Laputan method of making books, and that it did not sufficiently take into account a continually guiding and controlling intelligence. This seems to me a most valuable and instructive criticism. I feel profoundly convinced that the argument of design has been greatly too much lost sight of in recent zoological speculation. Reaction against frivolities of teleology such as are to be found, not rarely, in the notes of learned commentators on Paley's *Natural Theology*, has, I believe, had a temporary effect in turning attention from the solid and irrefragable argument so well put forward in that excellent old book. But overpoweringly strong proofs of intelligent and benevolent design lie all around us; and if ever perplexities, whether metaphysical or scientific, turn us away from them for a time, they come back upon us with irresistible force, showing to us, through nature, the influence of a free will, and teaching us that all living beings depend on one ever-acting Creator and Ruler."

Received with great applause, this address evoked many perplexities in its hearers. It was known that Sir William did not accept the doctrine of natural selection; and many of the orthodox Scottish clergy, who looked to him for some pronouncement, were aghast to find him appealing to the principle of continuity, and to discover that he was an evolutionist who, if he put back the origin of life on this earth to some distant globe or planet whence it had been meteorically introduced, would by an equal logical necessity put it back from such globe or planet to one yet more distant, and so on *ad infinitum*; and they were disposed to regard him as a greater sinner against the then popular theology than even Darwin himself. Others seemed to regard the hypothesis of the meteoric introduction of life as a huge scientific joke.[1] Maxwell made it the subject of one of his rhyming *jeux d'esprit*, which was sung at the *Red Lion* dinner. For two successive weeks *Punch* poked good-humoured fun at him in verse. The issue of August 12, 1871, contained a poem by Tom Taylor, entitled: "The Truth after Thomson, as versed by a Modern Athenian," a really clever summary of the address, from which we cull the following sample:—

[1] *Vide*, for example, *St. Paul's Magazine*, Sept. 1871.

But say, whence in those meteors life began,
From whose collision came the germs of man?
Still hangs the veil across the searcher's track,
We have but thrust the myst'ry one stage back.
Below the earth the elephant we've found,
Below him of the tortoise touched the ground;
But what the tortoise bears? Dig as we will,
Beneath us lies a deep unsounded still:
Sink we with DARWIN, with ARGYLL aspire,
Betwixt angelic or ascidian sire,
Though ne'er so high we soar, or deep we go,
The infinite's above us and below:
Beyond the creeds and fancies of the hour,
Looms, fixed and awful, A Creative Power.

In several successive years at the Association meetings Sir William reiterated his view. At Plymouth in 1877, when a certain meteorite (or model of it) was shown, he was keen to explain how, though the stone presented marks of fusion on the surface, the interior might have remained quite cool, so that if there had been in some deep crevice of it a bit of moss it would not have been burned; or if there had been lurking there a Colorado beetle it might have survived to become the father of a numerous progeny. Whereupon the witty Dr. Samuel Haughton remarked that he would not much mind the father-beetle coming in the crevice of a meteoric stone if only it had had the foresight to leave the old mother beetle at home!

The following letter of Feb. 11, 1882, shows that Sir William persisted in his views.

*11th Feby.* '82.

DEAR DUKE OF ARGYLL—I am much interested to see that independently you have come to the same conclusion regarding the source of all our terrestrial energy as I had been forced to come to a long time ago. You will see the thing referred to on page 22 of the enclosed address. It is more fully developed in an article under the title "On Mechanical antecedents of Motion, Heat, and Light," which is published in the British Association *Report* for 1854.

As to the extract from *The Times*, which I return, the writer does not seem to have noticed that while saying that ardent faith in the existence of numerous inhabited worlds throughout space, such as Sir David Brewster had expressed,' was more sentimental than scientific, I had myself expressed a very strong conviction, not only that there is life in other worlds than this, but that some of the life in this world is in all probability of meteoric origin; and that I returned to the subject again and again in the British Association Meeting at York, and obtained the appointment of a Committee to investigate meteoric dust, chiefly with a view to ascertaining whether any of it contains either traces or actual specimens of life. . . .—Believe me, yours very truly,

WILLIAM THOMSON.

Sir William Thomson also took part in the proceedings of the sectional meetings of the Association, and in presenting the Report of the Committee on Tidal Observations, added an extempore statement as to the determination of the amount of tide in the solid body of the globe, which he pronounced to be far more rigid than a globe of glass of the same size would be.

The Association over, Sir William Thomson hastened to the quiet of his yacht. During calm days he made some extremely interesting observations on the sets of capillary ripples which are originated in water streaming past a fixed narrow obstacle, such as a fishing line. These he described, with the theory of them, in letters to Tait, dated Aug. 16 and 23. They are reprinted in Appendix G of the *Baltimore Lectures*, 1904. On the 24th he was joined by Helmholtz, who came from Germany too late for the meetings. Helmholtz's letters to his wife give so graphic a picture of his Scottish friends and their activities, that a few extracts must be given. The extracts are taken from letters ranging from August 20 to September 14:—

St. Andrews has a splendid bay, with fine sands which slope sharply up to the green links. The town itself is built on stony cliffs. There is a lively society of sea-side visitors, elegant ladies and children, and gentlemen in sporting costumes, who play golf. This is a kind of ball-game, which is played on the green sward with great vehemence by every male visitor, and by some of the ladies:—a sort of ball game in which the ball lies on the ground and is continuously struck by special clubs until it is driven, with the fewest possible blows, into a hole, marked by a flag, about an English mile distant. The entire round over which each party wanders amounts to about ten English miles. They drive the ball enormously far at each blow. Mr. Tait knows of nothing else here but golfing. I had to go out with him; my first strokes came off—after that I hit either the ground or the air. Tait is a peculiar sort of savage; lives here, as he says, only for his muscles, and it was not till to-day, Sunday, when he dared not play, and did not go to church either, that he could be brought to talk of rational matters. The Browns are also here, and he (Crum Brown) will accompany me to-morrow to Sir William. At dinner we had a chemist, Andrews, from Belfast, with his wife and daughter, and to-day Professor Huxley, the famous evolutionary zoologist, all pleasant and interesting people. From Sir William we had yesterday two telegrams and two letters, to-day two telegrams with changing directions. The yacht squadron will sail earlier, and the latest instructions are that we go to-morrow evening to Glasgow to sleep in Thomson's house at the College, and on Tuesday join the yacht squadron at Inveraray on Loch Fyne. W. Thomson must be now just as much absorbed in yachting as Mr. Tait in golfing.

(INVERARAY, *Aug.* 24, 1871.)—I came yesterday with Professor Crum Brown, who luckily stuck to me till we reached the *Lalla Rookh,* in order to witness here the festivities of the clans-folk belonging to the Duke of Argyll at the reception of their future chieftainess, the Princess Louise. On Sunday we had dinner with Crum Brown, with whom is staying a great mathematician from London, Sylvester, in aspect extremely Jewish, but otherwise an important and presentable person. After dinner we had to leave the ladies and retreat to the smoking-room ; Tait would not allow anything else, but we got on well. Mr. Sylvester has been treated by Mr. Gladstone about as badly as could have happened at the hands of a Prussian Cultus-minister—or even worse ; and there was great indignation about it expressed by the company. As to their attendance at worship, they all excused themselves, as also did the ladies, on account of the rain. On Monday afternoon I travelled with Prof. Crum Brown to Glasgow. In Glasgow we slept in College, where a nephew of W. Thomson did the honours. The interior of the house was not yet finished, neither carpeted nor painted, full of old furniture not yet put into place, and it produced an indescribably sad impression, as if no one cared about it, in contrast to the old house which Lady Thomson had managed. In one corner of the dining-room hung an exceedingly fine and expressive portrait of her, and below it the couch where she used to lie, and her coverlet. I was very sad and could scarce restrain my tears. It is very sad when men lose their wives, and their life is left desolate. . . . There are about forty yachts assembled here, slender and elegantly built ships, and some of them tolerably large. Thomson's belongs to the larger sort, is a two-master, and is quite commodious. At the moment, besides Professor Crum Brown and myself, there are, on the yacht, Thomson's two sisters-in-law, another relation Houldsworth, and a London physicist Gladstone. My cabin is just about so large that I can stand upright in it beside the narrow bed: the rest of the space is less lofty, yet it contains wash-table, dressing-table, and three drawers, so that I can arrange my things well. For washing the space is rather small, particularly when the ship rolls and one cannot stand firm. To-day we began the morning by running on deck wrapped in a plaid and sprang straight from bed into the water. After that an abundant breakfast was very pleasant. Then came visits to the other yachts, and so the day has up to now passed very pleasantly in spite of the rain.

(GLASGOW COLLEGE, *Sunday evening*, *Aug.* 27.)—Thursday was still worse: we went to lunch on shore although the waves were already so high that the yachts began to be unsafe at anchor. We saw some Highland sports and dances. . . . Yesterday morning there was less wind, but sun and rain alternately. The morning was passed in preparations for departure, which was accomplished about one o'clock. Thomson and his men manœuvred the ship very cleverly, and the afternoon was passed with tolerably good weather, while we sailed back slowly along Loch Fyne. But then the wind caught us, and we went at a surprising speed the last two-thirds of our course to Greenock, the port for Glasgow. This evening we are to go with two nieces of Thomson's to Largs; Monday to Belfast.

On board the yacht they studied the theory of waves, "which," says Helmholtz, "he (Thomson) loved to treat as a kind of race between us." When Thomson had to go ashore at Inveraray for some hours, as he left he said: "Now, mind,

Helmholtz, you're not to work at waves while I'm away."

On Aug. 31st Sir William wrote to his sister, Mrs. King, from the yacht in Bangor Bay, County Down:—

I am just going to land along with Prof. Helmholtz, and Dr. Andrews, who came down last night and slept in the *L. R.*, to see a regatta to-day and accompany us to Clandeboye, Lord Dufferin's. We shall be at Clandeboye till after dinner to-morrow night, and then sail for Skye. Post Office, Portree, and, care of Professor Blackburn, Roshven, Fort William, are the best addresses. . . . We dined with James on Thursday after Helmholtz had an opportunity of seeing Dr. Andrews in his laboratory. . . . On Friday morning a party of twelve came down (Dr. and Mrs. Andrews and two daughters, Prof. Everett, and James and his family, and Mary Bottomley) making seventeen in all. . . . Late in the evening, a wonderfully beautiful moonlight night, Dr. A., J. T. B., Helmholtz, and I, drove down to Cultra and got on board the *L. R.* about midnight. We went on shore to breakfast with W. B. at Cultra this morning, and had a fine sailing-day for the regatta since.

*Aug.* 31, BELFAST.—We arrived off Holywood about one o'clock this afternoon. We do not leave till Sunday night about midnight, Lord Dufferin having asked Prof. Helmholtz and me to come to his house on Saturday to stay over the Sunday.

After a very pleasant visit to Clandeboye they sailed from Belfast on Sunday night, but had very bad weather, which prostrated them all—"even our Admiral," says Helmholtz. The party consisted of Sir William, his brother, his brother-in-law, two nephews, and the Geheimrath. They visited Oban, Loch Etive, and Tobermory. Thence to Roshven, whence Helmholtz wrote on Sept. 9th :—

W. Th. was very eager to arrive here, where his colleague Mr. Blackburn, Prof. of Mathematics in Glasgow, has a lonely property, a very lovely spot on a bay between the loneliest mountains. The Atlantic showed itself this time very friendly, and we came quickly here, so that in the afternoon we could take an excursion with the family and dined with them. . . . I expect that in the next day or so we shall abruptly begin our return, for Sir W. is very undecided as to the north side of Skye. . . . Mrs. B. has a remarkable talent for painting animals. She fashions all her doings and household ways to suit her professional tastes. . . . It was all very friendly and unconstrained. W. Thomson presumed so far on the freedom of his surroundings that he always carried his mathematical note-book about with him, and as soon as anything occurred to him, in the midst of the company, he would begin to calculate, which was treated with a certain awe by the party. How would it be if I accustomed the Berliners to the same proceedings? But the greatest *naïveté* of all was when on the Friday he had invited all the party to the yacht, and then as soon as the ship was on her way, and every one was settled on deck as securely as might be in view of the rolling, he vanished into the cabin to make calculations there, while the company were left to entertain each other so long as they were in the vein; naturally they were not exactly very lively. I allowed myself to seek amusement in balancing myself up and down on the deck, in wavering grace, and occasionally setting cataracts of sea-water to run off my waterproof.

After cruising in the Sound of Skye they returned through the Sound of Mull, where, being becalmed, they made experiments on the velocity of propagation of the smallest ripples

SIR WILLIAM THOMSON'S YACHT *LALLA ROOKH* (126 TONS).

that can be formed on water, and so back to Glasgow.

*L. R.*, LARGS BAY, *Oct.* 29, '71.

DEAR HELMHOLTZ—I have too long omitted to write to Du Bois Reymond in acknowledgment of the notice he sent me of my having been elected to the Berlin Academy. I received it on my way through Glasgow to the *L. R.* after the British Association, and left it in the house, which is now all in confusion, being handed over to painters and paperhangers. It may be some time yet before I can find the official intimation, and as I am anxious not to delay writing to Du Bois Reymond, you would oblige me much by telling me what is the proper designation of the Academy? Imperial? Royal? Berlin Academy of Sciences, I presume; also what is the designation of my own appointment—corresponding member? foreign member?

I hope you found all well at home when you arrived, and that all "went well" in respect to the marriage. I suppose you are now fairly launched on your University "Semester." Our "session" commences to-morrow week, and by this day week the *Lalla Rookh* will be at her winter moorings in the Gareloch. I have lived on board ever since you left (not merely because my house has been uninhabitable), but except two trips to Loch Fyne and two to Arran I have been chiefly between Largs and Greenock, and working hard at my reprint etc. of *Electrostatics and Magnetism*, which I am anxious to get launched before Christmas. It has been "on the stocks" for about five years.

You should look at Cauchy and Poisson on Waves, the *Concours de* 1815, when you have time. The *point* lies in the evaluation of the function

$$\int_0^\infty \cos mx \cdot \cos(t\sqrt{gm})dm$$

(for the case of motion in two dimensions); considered as a function of $x$ it is a fluctuating function of a very curious character. We must have it tabulated by the British

Association's Function-calculating Committee. Cauchy makes the thing very clear. Poisson I don't know so well yet. Both would be greatly improved by diagrams showing the forms of the waves and the laws of variation at different depths, etc. I was under a misapprehension when I spoke to you lately on the subject. I thought that a single disturbance at a point or along an infinite straight line, such as is produced by dipping a solid into water and not raising it out, but leaving it at rest, could not cause oscillations. What it does really is to cause a positive swell to spread out in each direction, followed by a series of undulations, negative and positive, finer and finer, and at any one place of the water, becoming finer and finer in length from crest to crest ultimately in proportion to $\frac{1}{t^2}$. After ten or twenty waves have passed a point at distance $x$ from the place of disturbance, the wave length (in the case of motion in two dimensions) is very approximately

$$\frac{4\pi x^2}{gt^2}, \text{ or } 2\pi x \frac{x}{\frac{1}{2}gt^2},$$

where $x$ must be a large multiple of the diameter of the disturbing body, but a small fraction of $\frac{1}{2}gt^2$.

Did you meet Strutt[1] when you visited his family in England? I hear that he would have been the new professor in Cambridge if Maxwell had not accepted—Believe me, yours always truly,

WILLIAM THOMSON.

On Nov. 2, still cruising off Largs, he wrote to Professor Andrews that he was awaiting Napier to make trials of his pressure-log, after which the yacht was to sail to winter quarters in the Gareloch.

At the end of the cruising season he wrote to Dr. J. Hall Gladstone:—

[1] Lord Rayleigh.

*Lalla Rookh*,
Gareloch, *Nov.* 4, 1871.

My dear Gladstone—You have heard from my sister that I am to be in London this day week. Even should it not be convenient to you to let me stay with you this time, I hope to have the pleasure of seeing you in the course of the few days that I shall be in London. I do not, however, wish to delay so long answering about the Tidal Committee in reply to Mr. Unwin's letter. The present Committee of the British Association on Tides is a new one, which was appointed about four years ago, and has been continued from year to year since that time, with grants of money for calculating results of observations such as those given by tide-gauges, and generally for promoting the investigation of tides. . . .

The Committee will be glad to receive the curves of the Calcutta Tide-gauge, and to apply the method of reduction which we have been following if we find that it can be done with advantage. . . .

I am now on the point of "flitting," as we say in Scotland, from my summer quarters on board the *Lalla Rookh* to the College. I am alone with one man on board waiting for my train, the others having just sailed away in the "cutter" and "gig" for Greenock to leave the boats there for the winter, and to find places, chiefly no doubt in foreign going ships, for themselves. . . .—Believe me, yours always truly,

William Thomson.

The business in London was a petition for the prolongation of the patent for the mirror galvanometer. Sir John Karslake, Q.C., was counsel for the petitioner; Mr. Archibald for the Crown. Six weeks later Sir William wrote to his assistant, Mr. Leitch, who was in charge of the recorder at Suez:—

*Dec.* 14, 1871.

My dear Leitch— . . . Ten days ago the Privy Council gave me a prolongation for 8 years of my

1858 patent. My formal petition for the prolongation was made last summer, and the three cis-Indian and the three ultra-Indian Companies all lodged objections. They, however, withdrew their objections before the petition was heard, and promoted rather than opposed my case. I also got assistance from Sir C. Lampson, who was deputy-chairman of the Atlantic Telegraph Company, and from Mr. Saward, their secretary. Also Mr. Willoughby Smith, Sir James Anderson, Sir Daniel Gooch, Captain Sherard Osborne, Mr. Pender, and other influential people in the companies were favourable. . . .
—Yours truly, W. THOMSON.

Further details are given in a letter to Miss Jessie Crum, then abroad :—

GLASGOW UNIVERSITY, *Dec.* 12, 1871.

DEAR JESSIE—I have been hearing of you all in several indirect ways, the last of which was Mary's letter to Dr. Rainy, which he brought to me one day. I hope you are getting on well, and feeling comfortable in your villa. I should be much obliged by a letter from either you or Mary, when you have time to write. You must look upon this simply as a begging letter. I cannot give anything in return for what I have been asking, as the things I have been kept incessantly busy with are dull and uninteresting, except so far as getting through little by little what must be done is interesting.

I was in London again from Saturday last till Wednesday about my petition for prolongation of my 1858 patent. I had been warned by Grove (who was my counsel until he was promoted to be a judge) to expect nothing, and to consider that even a prolongation for one year would be a good result. The Privy Council gave 8 years. The case altogether went off very well. The judges early intimated that they did not require any more evidence as to the "merits of the invention," and they showed a liberal spirit in respect to accounts, etc.

Varley had prepared an admirable apparatus for illustrating the action of my mirror instrument, and showed it in action to the judges, which had a very good effect. The Telegraph Companies (8 now in all) with whom I have come to agreement are all very pleasant and friendly, and the new instrument is making its way eastwards (now as far as Suez, and going off to-day to Aden and Bombay). Until the time when I was coming home from Brest, when we were at Barra House, there was nothing settled. As soon as anything should be settled, it went into unsettlement, with another prospect of a lawsuit, again up till that time. I well remember the warm congratulations and sympathy we had when we hurried home from Kissingen the year before, and things seemed to be settled in London. Then I went off again, and all the winter we were in Edinburgh it was a subject of anxiety to my dearest Margaret. It was not till the August following that I could tell her it was all settled. Since that time those things have gone as prosperously in every respect as possible; but she only knew the perturbations and toils, from some of which she suffered greatly by over-fatigue going to Valencia in 1858. Near the end of April, when very good accounts of the new instrument came from St. Pierre, and the Indian Companies were all wanting to have it, she said, "It is just the fruit of your labours."

I must stop now, and go on with my book on Electricity, which is chiefly compiled from things written more than twenty years ago, and some which I wrote in Edinburgh the last winter we were there. Macmillan is pressing me to get it out by Christmas, if possible, and I am at it every moment of spare time.

With love to your mother and Mary, I am, yours always affectionately, WILLIAM THOMSON.

In January 1872 Sir William was busy over the proofs of his reprint of papers on *Electrostatics and Magnetism*, which had been on hand for four years. In February he was in London with Dr. Gladstone;

then went to Edinburgh to work with Tait at proofs of the smaller *Elements of Natural Philosophy*, for the Oxford Press (see p. 472).

On March 29 he wrote of his doings to Miss Jessie Crum: new cable schemes, trials of telegraph instruments old and new, correspondence "with my old friend De Sauty, and several others of the old Atlantic people, who are all much taken up with the recorder, and (under instructions from Sir James Anderson) doing their best to get it to work well." He is proposing a short spring cruise before session ends, and then to sail to Gibraltar to see the recorder working there. He has a prospect, after the British Association is over, at the end of August, of going to Quebec with Dr. Norman Macleod, but the project was cut short by the death of Dr. Macleod in June. Two of his nephews will be required as lieutenants in the new Atlantic cable scheme. "There is quite an epidemic amongst the laboratory students of desire to become *telegraph engineers*."

Then comes a commercial shadow across the path.

GLASGOW UNIVERSITY, 3 *April*, 1872.

DEAR JENKIN—I am sorry to hear what you tell me. I have no confidence in B——, and would require any statements as to the use of the mirror to be very carefully sifted before we can admit them. It would be necessary for him actually to have used the mirror on the cable, and also at a time found inconsistent with my claims, before we could admit any weight to the objection to our rights. Find out, if possible, taking whatever law advice is necessary, to what extent experimental use of an invention in that way, confessedly mine, can invalidate my claim. If he only experimented with it on the cable, and did not use it for practical working on the line, I do not believe his objection will be valid. Try, however, if possible, should the case look bad against us, to make a compromise, as the companies no doubt admit the moral right. Of course we know that directors can't be generous with their shareholders' money, but the proper mixture of generosity and worldly wisdom, escaping litigation, and procuring us as allies and assistants to their signalling arrangements, may commend itself to them. We have another string to our bow in the recorder. For all their lines it must cut out the mirror, and that speedily. But be cautious in using or showing this string. If we can get our terms for the mirror consented to, we can make more use of our recorder rights than if we put them forward now. In the course of six months, I believe, I could give thorough good recorders for their lines. You may feel confident as to this, and use it as you think best.—Yours truly,

WILLIAM THOMSON.

By April 11 he is able to send word to Leitch that Sir James Anderson now considers the recorder to be *the instrument* for all their cable stations.

On April 28 he writes again, from London, he has been suddenly called up on business of the "Great Western Telegraph Co."; that he intends to go to J. T. Bottomley's marriage at Belfast; and that on Friday he hopes to be at rest on his yacht in the Gareloch, ready to put to sea. He is wishing to sail for Gibraltar as soon as possible, that he may be free to go later to Bermuda. "On Friday I got the last MSS. of the book out of hands."

The Great Western Telegraph Company was a project to lay a cable *via* Madeira and the Bermudas to Boston and to the West Indies; but later by arrangement with the earlier companies the project was altered, though the cable for this work had been manufactured and the ship *Hooper* specially designed for laying it; and it became merged in the Western and Brazilian Telegraph Co., which laid cables in five sections from Para to Rio Janeiro, touching at Pernambuco. Eventually this and other South American cables were taken over by the London Platino-Brazilian Telegraph Company.

About June 1st Sir William wrote to Helmholtz:—

50 GROSVENOR PLACE, LONDON, S.W.

DEAR HELMHOLTZ—I am going to Scotland to-night, and return to London about the middle of next week, to spend two days in this house (of Mr. Spottiswoode, President of the London Mathematical Society).

On Saturday the 21st I hope to sail from Torquay for Gibraltar, and to call at London on my way back, visiting the telegraph stations at both places, my recorder being now in constant use there.

There is now a great telegraph project in the course of execution to lay cables from England to Bermuda, and then to New York and St. Thomas. The manufacture of the cables has commenced, and Fleeming Jenkin and I being engineers to the Company are obliged one or other of us to be very frequently in London. We have a great deal of electric testing to do—insulation, electrostatic capacity, and resistance of the copper conductor, also testing the strength of the iron wire and of the finished cable. The laying will not be commenced till this time next year. I am living chiefly on board the *Lalla Rookh*, off the south of England, and coming up to

London when necessary. I can only get mathematical work done in the yacht, as elsewhere there are too many interruptions.

A few days ago I despatched the very last of my volume of *Electrostatics and Magnetism* to the printers, except the preface, and I am now getting to work on Vol. II. of the *Natural Philosophy*, and the reprint of Vol. I.

I hope you have been well, and your family all well, since we parted at the "Albert Quay." Is your new laboratory finished or making satisfactory progress? I hope it will turn out in all respects satisfactory to you.—Believe me, yours very truly,

WILLIAM THOMSON.

By this time the new company was fairly afloat, and the partners had to keep a staff of electricians at work, some at Millwall, others at Mitcham under David T. King, to superintend the manufacture. Sir William had to spend two or three days each week at the works. He has a way of turning up at the Millwall works on a surprise visit, arriving once at 2 A.M. in a dripping mackintosh, with a black bag in his hand, "for all the world like a tea-traveller," as one of the assistants writes. He is living the rest of his days on his yacht, cruising round Torquay, or taking his friends — Dr. Gladstone, Mr. Varley, and Dr. Siemens—trips to Sheerness or Margate. He varied these amusements by reading to the London Mathematical Society a paper on the reduction of Polynomial Quadratics, which he had worked out in the quietude of his yacht.

On June 24th he wrote to Lord Rayleigh, from

Torquay, respecting certain paragraphs of "the book":—

I am on the wing for Gibraltar (and other telegraph stations—Lisbon, Brest—if time before the B. A., Brighton, Aug. 14, permits). I hope to despatch from Gibraltar all I have to say in the way of additions or amendments to the first two or three sheets of Vol. I., so that the reprint may go on forthwith. Meantime, or as soon as possible, amendments or suggestions for early parts or any part of the volume sent to Tait will be thankfully received.

Then he sails for Gibraltar one Sunday morning from Gravesend. But just as they weigh anchor the "Thames Mission" boat comes up, and Sir William orders Captain Flarty to stop the yacht while the minister conducts service for them on board. By June 24th he has got to Torquay, and has taken aboard the new recorder for Gibraltar, and some new instruments for sounding and for measuring speed at sea. While he is away affairs at home do not flag. White is pushing on with improved recorders; and Donald MacFarlane, writing him to report progress of the laboratory work in the new building of the University, says: "I have taken possession of the spare room above the staircase (without leave), and in one corner of it I have stowed all the packing-boxes which were always in the way."

Returning to England, August 1st, he writes in the train, from London to Torquay, to his sister-in-law a detailed account of his tour:—

I have had a very pleasant and satisfactory cruise, and

made useful as well as interesting visits to the three telegraph stations, Gibraltar, Lisbon, and Porthcurno (though only three hours at the last in consequence of a letter, reinforced by a telegram, summoning me to attend a meeting of the "Great Western Tel." Board in London yesterday). At Gibraltar my old enemy, but now very good friend, De Sauty, who was at the other end of the cable in 1858, has managed admirably with the recorder, and has entirely given up the mirror in all the work of the station. I found him as agreeable and obliging as possible in every way. We were almost constantly at work in the telegraph office from the Sunday[1] morning, when I arrived, till the Saturday morning, when I sailed for Lisbon. . . .

The rest of the letter is full, moreover, of lively details about the monkeys on the Rock of Gibraltar that came early in the morning to visit the telegraph station there; of his trip towards Algiers in the *Lalla Rookh* with De Sauty on board; and of his voyage home *via* Lisbon. To-morrow he will sail from Torquay to Cowes for the R. Y. S. Regatta.

Brighton was the scene of the British Association meeting of 1872, and Sir William went there for three days to introduce his successor, Dr. Carpenter, into the presidential chair, and to read two papers—one on the Identification of Lights at Sea, the other on the Use of Steel Wire for Deep Sea Sounding. In the latter he narrates how in the Bay of Biscay he has corrected the charts, using

[1] "Particularly the Sunday, which at all the stations of submarine lines is the great day for testings and adjustments, lawful on the ground of necessity and mercy. About five o'clock on the Sunday the cable has generally done its week's work, and is nearly at rest till about eleven on Monday forenoon; but for three weeks together it has been never once clear, which is about as bad as Mr. Pickwick's cab horse."

a lead sinker of only thirty pounds at the end of a three-mile line of thin pianoforte wire.

At the Mathematical and Physical Section, in proposing a vote of thanks to the president, who had referred to Professor Zöllner's electric theory of comet's tails, he told how some time since, at a workmen's philosophical institute at Millwall, an intelligent man produced a glass tube which cracked when an iron wire was laid along its inside. The workmen were puzzled by the fact, but at last agreed that it must be electrical! The same merit lay at the bottom of Zöllner's theory, namely, *omne ignotum pro electrico.*

More cruising about the Clyde completes the holiday, and in September he is back at the University.

In October 1872 Sir William Thomson was elected to one of the two life Fellowships at Peterhouse, founded for men distinguished in Science or Letters; the eminent Greek scholar, Richard Shilleto, having been elected to the other in 1867.

There was now big work in hand over the manufacture of the new cable, and the building of the cable-ship for laying it. He seeks advice from his engineering brother:—

GREAT WESTERN TEL. CO.,
103 CANON STREET, LONDON,
*Oct.* 30, 1872.

MY DEAR JAMES—Hooper's Telegraph manufacturing company have ordered for cable laying a ship 350 ft. long, 55 ft. beam, 36 ft. moulded depth; builders' measurem$^{t}$ = 4940 tons.

Jenkin and I both strongly urge a hydraulic arrangement to give power of manœuvre—that is to say, a pump and water pipes to give means of discharging water perpendicularly across the length at any one of four places, A, A′, B, B′, or at two of them simultaneously. I calculate that water discharged through an aperture of $\frac{1}{4}$ square metre (say $2\frac{1}{2}$ square feet) at a velocity of $6\frac{1}{4}$ metres per second, that is to say, $\frac{6\frac{1}{4}}{4}$ or $1\frac{9}{16}$ tons per second, would give a pressure of one ton. I would wish to be able to give at least 1 ton simultaneously at A and B, and therefore would need to be able to discharge not less than $3\frac{1}{8}$ tons per second, or 728 gallons per second, or say 44,000 gallons per minute. The head of water corresponding to the discharge velocity of $6\frac{1}{4}$ metres per sec. is $\frac{(6\frac{1}{4})^2}{2 \times 98} = 2$ metres. I should be much obliged by your telegraphing to me to above address on Friday morning your opinion as to the centrifugal pump and water-ways that would be required for this, and your opinion regarding Gwynne's pumps, of which I send you printed prospectus by same post with this. You might also write to me on Friday, addressing St. Peter's College, Cambridge. . . . The ship is to be made by Mitchell, Newcastle, and it is to be finished and round in the Thames by 26 April, subject to £100 penalty per day for delay after that date.

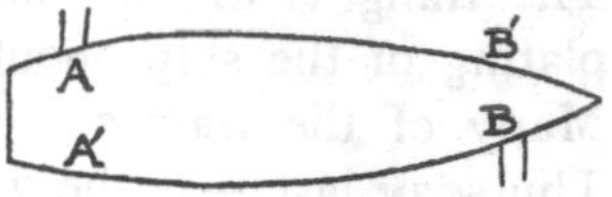

I was made a Fellow of Peterhouse under a new statute which allows men eminent in literature or science to be elected independently of marriage. I shall go back to Cambridge on Saturday on my way to Glasgow.—Your affectionate brother, W. THOMSON.

GLASGOW COLLEGE, *Nov.* 5, 1872.

DEAR JAMES—I think the hydraulic thwart ship propeller, according to the data of your telegram, will do. I spent yesterday at Newcastle with the shipbuilder (Mr.

Swan of "Mitchell and Co."), and he has found a place for it . . . [here follow ten octavo pages of details] . . . The thing is of extreme urgency, as in three weeks the plating of the ship about the stern will have commenced. Many of the frames are up already. I only heard on Thursday last that she was to be built. I wish they had told me beforehand, and I would have had a thwartship propeller in the original plans, which would have saved a good deal of money on what will have now to be spent to get it applied. Your professional charge and expenses must be charged, with the wheel and work of the ship-builders putting it in, to Hooper's Company. If we can get a practicable scheme, it is, I think, certain that the Company will adopt it.—In haste, your affec[t] brother,

W. THOMSON.

UNIVERSITY, EDINBURGH, *Monday 31st*,
[*Nov.* 18, 1872, *Post-mark*].

DEAR JAMES—Thanks for your telegram. Mitchell's are quite confident about thwartship screw below main screw shaft, 6 ft. diameter of screw. Three blades to be driven by a wire rope round grooved rim 6 ft. diameter surrounding blades of screw. You will receive in a few days from me (or from Mr. Froude) their drawings.

The engine is to be on deck, and I have a telegram from them to-day (scarcely time to have read it yet) to effect that we may have 60 lbs. pressure, and no limit to size of cylinders.—Yours, W. T.

Great haste.

Sir William was at this time living in his half-furnished residence in the professors' court at the University, his nephew, James T. Bottomley, residing with him and acting as assistant in his laboratory work and teaching. A well-known feature of his household was "Dr. Redtail," a grey parrot with red tail feathers, who had been bought in Seven Dials. Of this favourite bird many stories are told.

The best authenticated is his greeting of his master as he hurried in from the laboratory to join an invited party at lunch:—"Late again, Sir William! Late again!"

At the end of the year he has a proposal to convey to Helmholtz:—

ATHENÆUM CLUB, LONDON,
*Dec.* 2, 1872.

DEAR HELMHOLTZ—I enclose a letter of Dr. Cookson, Master of Peterhouse and Vice-Chancellor of the University of Cambridge, which he requested me to transmit to you. It is written in consequence of a suggestion I made to him when I saw him three days ago at Cambridge, that he should ask you to give the "Rede Lecture" for 1873. I hope you will accept. You would choose your own subject—anything upon which you would like to speak for an hour, or an hour and a half, to a cultivated audience. It is given annually in the Senate House of the University, and the authorities are always anxious to have a man of high distinction. So far as I know, no one not a British subject has hitherto been asked to give the lecture. You would probably, if you accept, prefer to have the lecture fully written out, and to read it to the audience. It is desirable that it should be afterwards published.

In 1866 I was asked to give the "Rede Lecture." I accepted, and chose for my subject the "Dissipation of Energy." I did not succeed in getting it written out, and it has not been published, but I hope sometime to write it out (with, no doubt, many changes and additions) and to publish it. I hope very much you will be able and willing to accept. I would make a point of being at Cambridge at the time. Dr. Cookson will be glad to hear from you as soon as may be in reply. Address, The Rev. Dr. Cookson, Master of Peterhouse, Cambridge.

I hope all goes well with you at Berlin. I should be glad to hear from you.

I am here for a few days on telegraph business, and I go

to-morrow to Cornwall to test a new cable which has been just laid from the Lizard to Bilbao. I shall be in Glasgow again by next Monday, I trust. I shall send you very soon a printed paper describing the best way I have found for managing the large tray battery, which has been doing well. I am getting a battery of eighty trays of larger size[1] than those you have, and I expect to get a very powerful electric light from it.—Believe me, yours always truly, WILLIAM THOMSON.

*P.S.*—With trays the same size as yours, I get the resistance of each cell as low as 12 of an ohm.

GLASGOW COLLEGE, *Jan.* 8, 1873.

MY DEAR HELMHOLTZ—We are very sorry that you are unable to undertake the "Rede Lecture." I cannot share your misgivings about success in interesting the audience had you been able to undertake it, but only regret that your engagements in Berlin make it impossible for you to do so.

You have heard, no doubt, before now of the sad loss we have had in the death of Rankine. I send you by this post a copy of the *Glasgow Herald* (Dec. 28), containing an article on his life and scientific work by Tait; also a copy of the same newspaper for Dec. 26, containing two articles, all of which I think will interest you. We lost Archibald Smith,[2] too, in the same week, whose name you may know from the great work he has done for navigation in respect to correcting the compass error in iron ships. He was a very old and excellent friend of mine. He has been a hard-working Chancery barrister almost ever since he took his degree at Cambridge as "Senior Wrangler" in 1836, or else he must, with his great mathematical powers and inclination for physical science, have been one of the foremost men of science of this country.

I have urged my brother, James Thomson (who is at present Professor of Engineering in Queen's College, Belfast,

[1] The zincs 22 inches square.

[2] [See Obituary Notice by Sir W. T., *Proc. Roy. Soc.* xxii., pp. i.-xxiv.]

and has been so for many years) to apply for Rankine's vacant chair. I should feel much obliged by your writing to me a very short statement of your opinion of my brother's merits as a scientific investigator, or qualifications for a chair of Engineering. I have received such letters to-day from Andrews, Tait, and Joule, in answer to similar requests which I made to them. I expect one from Maxwell. These four, and one from you if you will write it to me, shall be laid before Mr. Bruce, the minister ("Home Secretary") who has to make the appointment, and I think should constitute sufficient evidence in support of my brother's application. I thank you very much for your corrections and remarks on our *Treatise.* Some of the former we had noticed. All will be taken advantage of. I instructed Macmillan to send you a copy of my *Electrostatics and Magnetism*, which was published just before Christmas. Wishing you and Mrs. Helmholtz "a good new year" as we say in Scotland—I remain, yours truly,

W. THOMSON.

*P.S.*—I am hard at work just now with your

$$x = \phi + \epsilon^{\phi} \cos \psi + \sigma$$
$$y = \psi + \epsilon^{\phi} \sin \psi + \tau$$

and trying to help myself by it to find the shape of a coreless cylindrical vortex couple.

In this winter of 1872-73 Sir William Thomson sent several technical communications to the newly-founded Society of Telegraph Engineers, of which he was a foundation member and vice-president. These were On a New Form of Joule's Tangent Galvanometer, On the Measurement of Electrostatic Capacity, Tests of Battery, and On a Tray Battery for the Siphon Recorder. This last invention was a form created by the necessity of providing a constant current for the electromagnet of the recorder, and consisted of a pile of lead-lined shallow wooden trays about a yard square containing zinc grids and sulphate of copper as a modification of Daniell's well-known type of cell. In the early spring he read several papers to the Edinburgh Royal Society, only the titles of which remain; also two communications in March to the Institute of Engineers in Scotland, on "Signalling through Cables" (illustrated by a model cable) and "On the Rope-dynamometer." He was also very full of the question of distinguishing lighthouse lights by flashing signals, and on signalling the letters of the Morse code by flags and by waving lights. He contributed to *Good Words* of March 1873 an article on "Lighthouses of the Future" (see p. 725 below).

The appointment in March 1873 of Professor James Thomson, LL.D., to the chair of Engineering at Glasgow, as successor to Rankine, was a great joy to his younger brother. In the summer of 1873 the James Thomsons lived in Sir William's College house, and reported to him that the day after he left for Brazil his parrot, "Doctor Redtail," had surprised the household by saying "Sir William Thomson gone to Liverpool."

GLASGOW COLLEGE, *March* 15, 1873.

DEAR HELMHOLTZ—I have delayed too long writing to thank you for your most valuable letter regarding my brother's qualifications for the chair of Engineering. It must, I am sure, have had more influence in promoting his appointment than almost any other document put into the hands of Mr. Bruce, the Home Secretary. I have now the satisfaction of being able to tell you that he has been appointed to the chair. He will remain in Belfast to finish the business of the present session there, and next November will enter on his duties in Glasgow. I hope and fully expect that he will have much more time here for original research than the comparatively inconvenient arrangement of the "Queen's University" allows him in Belfast, and he will find my laboratory a great aid.

I hope all goes well with you as to your new laboratory and school of experimental science.

Remember me kindly to your wife, and believe me, yours always truly, WILLIAM THOMSON.

[*P.S.*]—I expect a visit from Joule when my brother comes over in the course of a week or two, to be formally admitted to the chair. He is President-elect of the British Association at the meeting appointed for Bradford in Sept. next. Is there any chance of your being present? I am sorry that I shall not be able to be there as I am to be away in Brazil laying cables.

YACHT *LALLA ROOKH*,
LARGS, *May* 25, 1873.

MY DEAR ANDREWS— . . . I ought sooner to have written to thank you and Mrs. Andrews for your very kind invitation, but I waited till I could see my way as to a possible time for going across to Belfast. I have had a great deal on hand—seeing the new cable-ship *Hooper*, and sailing round in her on her first voyage from the builder's yard at Newcastle to Millwall Dock, etc., etc. I have now to get sounding apparatus, and one of my laboratory students indoctrinated in the use of it, despatched by a steamer to sail from Liverpool on the 31st for Para, and take soundings along the coast of Brazil from Para to Pernambuco. I hope about ten days hence to be able to sail across, and to look after the setting up of an eclipsing arrangement which the Harbour Commissioners have ordered for the light in Holywood bank. If I can manage to remain a night in Belfast it will be a great pleasure to me to avail myself of your invitation, should the time, which I am sorry is still necessarily uncertain, be convenient to you and Mrs. Andrews.—Believe me, yours very truly,

W. THOMSON.

On April 23 Sir William wrote to Miss Jessie Crum that on the Friday before he had set out on a three days' cruise with his nephew W. Bottomley and Dr. James Napier. They saw the Ardrossan harbour light, "an excellent distinguishing light introduced by Mr. Thomas Stevenson." On May 28 he wrote again from London, where the *Hooper* was taking in cable, that he was returning to Glasgow to sail in the yacht for Liverpool with Mary and Mr. Watson[1] to see sounding apparatus on a ship.

Preparations were now far advanced for the laying of the new cable from Pernambuco to Para. He sent word to his brother:—

(Post-mark—London, S.W., July 16, 1873.)

CABLE-SHIP *HOOPER*, *June* 15/73.

DEAR JAMES— . . . The cable-ship came out of dock yesterday, and after about two days here is to sail for Plymouth. It may be Saturday next, or more probably a few days later, that we leave finally for Brazil. Having seen the cable, and arrangements for testing all right, and the ship away from the factory, I leave her to-morrow morning, and after a day and a half in London, leave (I trust) to-morrow afternoon for Cowes, to sail thence westwards. I have a cable (the "Direct Spanish") to test at Lizard before going away in the *Hooper*, and I hope to be able to sail there, and possibly further to Porthcurno,

[1] Rev. Charles Watson, D.D., who had married Miss Mary Gray Crum. He was Free Church minister at Largs, and died 1908.

and see trials of my new automatic sender there, and, still sailing in the *L. R.*, get back to Plymouth in time. If wind does not answer I shall have to take train.—Your affec$^{te}$ brother, W. T.

On Friday, June 20th, the *Hooper* sailed from the Thames, having on board some 2500 miles of cable. On the 26th she landed the shore end at Lisbon, and proceeded westwards with the rest of the cable. "Here we are," wrote Jenkin to his wife from the *Hooper*, on June 29, "off Madeira at seven o'clock in the morning. Thomson has been sounding with his special toy ever since half-past three (1087 fathoms of water)." On July 7th Sir William wrote to his sister-in-law from the *Hooper*, then lying in Funchal Bay, that they had been there a week and would be there a week more. A few days after leaving Plymouth a fault had been found in the cable in a length of 543 miles that was coiled in one of the three tanks; and as the fault was 400 miles down the coil they had had a prodigious work in uncoiling, splicing pieces, and recoiling. The expense to Hooper's Company was some £200 per day; but it was well that the stoppage had been here, not at Cape Verde or at Pernambuco. He had been struck by the marvellous beauty of the island. "It has been impossible," he added, "to keep off Darwinism, and although Madeira gave Darwin some of his most notable and ingenious illustrations *and proofs* (!) we find at every turn something to show (if anything were needed to show) the utter futility of his philosophy."

An incident related by R. L. Stevenson in his memoir of Fleeming Jenkin, deserves mention :—

I shall not readily forget with what emotion he once told me an incident of their associated travels. On one of the mountain ledges of Madeira, Fleeming's pony bolted between Sir William and the precipice above; by strange good fortune, and thanks to the steadiness of Sir William's horse, no harm was done; but for the moment, Fleeming saw his friend hurled into the sea, and almost by his own act: it was a memory that haunted him.

A month later Thomson wrote a further account of the events of the voyage :—

*Aug.* 8, *Friday.*
PLAÇA DO COMMERCIO, RECIFE, PERNAMBUCO.

We hope to be under weigh for Para, paying out cable from the stern of the *Hooper*, before dark this evening. . . .

I have bought a parrot, green, with splendid red tips to his wing shoulders and end-wing feathers, dark blue outer wing feathers, light blue and white head, brilliant yellow breast.[1] The colouring is as rich and varied as Mrs. Bowden Fullarton's dress, and even more harmonious in general effect. . . .

Tell Mary that we have had a great deal of dot and dash practice between the *Hooper* and the *Paraense*, both by lamps at night and (with far more difficulty) by various other means in the day-time, to be ready to receive her soundings, and tell her where to go next in choosing out track for Para. We had some admirable lamp signalling several evenings at Funchal between the *Hooper* and Mr. Blandy's house, about 1½ miles distant. The Miss Blandys learned "Morse" very well and quickly, and both sent and read long *telegrams* the first evening they tried it, to the admiration of France and other old telegraphers on board.

[1] This parrot, named "Professor Papagaio," lived many years in the College House. When he died he was stuffed, and is now in the Hunterian Museum in Glasgow University.

The ladies in question were the daughters of Charles R. Blandy, Esq., one of the principal residents of Madeira, at whose villa Sir William was welcomed. The delay to the expedition lasted over a fortnight, but at last the repairs were completed. An eye-witness has recounted how, when the anchor was weighed, and the *Hooper* steamed slowly out of Funchal Bay, a figure was seen waving a floating streak of white drapery from a window of the house on the hill high above the port. "G-O-O-D-B-Y-E" was spelled out. "Eh! What's that? What's that?" said Sir William, adjusting his eye-glass the better to catch the signals. "Good-bye, good-bye, Sir William Thomson." And as the ship's hull dipped beyond the horizon the white streak still fluttered "Good-bye."

# CHAPTER XVI

## IN THE 'SEVENTIES

HOLDING his fellowship at Peterhouse, Sir William Thomson now frequented Cambridge more often; and on returning from Pernambuco he paid a visit there on his way north. He wrote to his sister, Mrs. King:—

ST. PETER'S COLLEGE, CAMBRIDGE,
*Oct.* 22, 1873.

MY DEAR ELIZABETH— . . . I am here till the 29th, when there is an important College meeting which I should have had to come back from Glasgow to attend if I had been there. Meantime I am very busy, having (in consequence of having been re-elected to a fellowship) accepted the office of "additional examiner" for the Senate-house examination of next January. Making questions and meeting with the other examiners and the moderators is my present occupation. Then in January there will be some days of hard work examining the answers. Since coming here last week I have been again rowing in an eight-oar (the first time since 1846) with the "ancient mariners," of whom Fawcett, the (blind) member for Brighton, is a chief.

David (jun$^{r}$)[1] has been doing very well indeed. He is not to go out in the *Hooper* this trip (to lay the Pernambuco-Bahia-Rio Janeiro sections, for w$^{h}$ she leaves the Thames on 3rd of Nov.), but will remain in charge at Millwall.

[1] David Thomson King, who was drowned at sea (see p. 655).

This, I think, will be better for his progress afterwards than going to sea just now would have been, as it makes him known to Mr. Heugh and others as holding a responsible position. He will probably go out on the Para-St. Thomas (a very important part of the work) next spring. I shall try to get him a short holiday soon. I shall be in London from the 30th Oct. till the 3rd Nov. to make final arrangements and see the *Hooper* off.—Your affec$^{e}$ brother, W. T.

In December 1873 Sir William read a paper to the Royal Society of Edinburgh on a new method of determining the material and thermal diffusion of fluids.

He wrote on Christmas day from Knowsley to Mrs. King:—

Yesterday I came here on a visit to Lord and Lady Derby for a few days. On Saturday or Monday I go to Mere Old Hall, near Knutsford, William Crum's place, to remain till the end of the holidays. I have to be at the Royal Society, Edinburgh, on Monday week to "read" a paper, which, however, will not, I fear, be written till after the reading. As Mrs. Johnstone told you, I shall have to be there after this winter, having been elected to be President.

My Cambridge work (as one of the examiners[1] for the "Mathematical Tripos" of 1874) will keep me very busy till the end of January, when it will be over. I have brought the examination papers here (a very large heap) for revisal, etc. About the 20th of January I shall have to go there and remain till the list showing the result is given out.

A letter to Dr. King followed:—

[1] As examiner for the Mathematical Tripos Sir William Thomson introduced various changes to give greater width of studies in the direction of Natural Philosophy. That these reforms did not please all the Cambridge mathematicians was natural; but Maxwell, who had paved the way for them, rejoiced.

CAMBRIDGE, *Jan.* 27/74.

MY DEAR DAVID—

. . . . .

I should have written sooner, if only to say so much; but that I have been absolutely overwhelmed with examination papers (answers to our printed questions) for the "Mathematical Tripos," that is to say, the Cambridge University examination for mathematical honours. The work is exceedingly interesting to me, but most laborious and wearisomely plodding. For my share of one sitting of the candidates I got 10½ lbs. of papers of written answers. I have had seven such hauls, and scarcely any of them less than 5 lbs. By the same post with this (or by to-morrow's) I shall send a specimen of our printed papers of questions $w^h$ it may interest you to see. The questions marked with roman numerals in it are mine, the "arabic" by another examiner. I shall enclose it in a number of the *Telegraphic Journal* containing a report of an "address" I was obliged to make in London on my way here. I had only (after enormous labour with Tatlock in two days) succeeded in getting enough written to occupy 4 MINUTES, and the prospect had made me feel as if I had a millstone round my neck for a fortnight before the day. So after I read the little beginning piece, the rest was a "leap in the dark" altogether. I had really not an idea of what I was going to say, so I was thankful when it was all over. I was surprised a few days later with a copy of the *Telegraphic Journal* containing the report, which had been taken (very well as I thought) by a shorthand writer. It seems to contain every word I said, with only a few errors. . . .

I would like very much to make a cruise in the Mediterranean, but next May and June I shall in all probability not be free to do so.

The Society of Telegraph Engineers, destined later to blossom into the Institution of Electrical Engineers, was then not three years old. Sir

William Thomson, as its president, in his inaugural address [1] dealt chiefly with the reflected benefits which science gains from its practical applications, and the benefit of the systems of measurement that grew up out of the requirements of the practical telegraphist. Terrestrial magnetism was still, so far as its cause was concerned, a mystery; so was that of terrestrial electricity. But telegraph engineers, by investigating the facts over the globe, could help to solve these mysteries. He regarded the Telegraph Engineers as a society for establishing harmony between theory and practice in electrical engineering, and in electrical science generally, by organized co-operation.

Within a month he gave another presidential address to the Glasgow Geological Society on the Influence of Geological Changes on the Earth's Rotation, and communicated a paper on Deep-sea Sounding. At Edinburgh he read an important paper on the Kinetic Theory of the Dissipation of Energy.

On April 10 he took a preliminary cruise of four days on the yacht with a party including Jenkin and some former students.

To Charles Abercromby Smith (now Sir Charles), of Cape Town, he wrote :—

GLASGOW UNIVERSITY, *April* 28, 1874.

MY DEAR SMITH—You know by this time that I am again a colleague of yours, as Fellow of Peterhouse. It

[1] See *The Telegraphic Journal*, vol. ii. p. 67, Jan. 15, 1874; *Soc. of Telegr. Engineers' Journal*, iii. pp. 1-15, 1874; *Pop. Lectures*, ii. p. 206.

is pleasant to be again associated with a former pupil and friend, though we are pretty nearly at two extremities of a diameter of the earth. Do you remember Tatlock? at all events he remembers often hearing about you, and of your thermo-electric experiments in the laboratory of the old College. . . . This is written in his hand. As I have so many engagements, and so much laboratory work that I am kept constantly standing and walking about, I can seldom sit down to write anything, and am obliged to do nearly everything I wish in black and white by dictation.

I examined for the mathematical tripos last January, which gave me a good deal of work from about this time last year till the beginning of February, first composing the questions, and then having all the heavy labour of examining the answers. I was at Cambridge in all at different times about five weeks, and enjoyed this very much, as it was very pleasant for me to live once again in the old College, which by the way, as you perhaps know too, has been greatly improved and beautified *at much expense*. . . . This will be delivered to you by Mr. Coles, who I believe is already known to you. He is, I believe, to disclose to you, and others who may be interested, a new form of cable which has been designed by Hooper's Telegraph Works Company for connecting the Cape with Aden and Mauritius. It is a form of cable in which I have great confidence. The hempen insulation is of the general character which both Professor Jenkin and I have long advocated as being the most suitable for a deep-sea cable, but it is a very great improvement indeed on anything of this kind that we ever either designed ourselves or have seen designed by others.

He was already making plans for the summer. On March 26 he wrote to Froude that he must be in London on 20th of April for a soiree of the Telegraph Engineers, and that he intended to sail from Falmouth on 2nd of May for Madeira. The *Lalla*

*Rookh* was ready. He left instructions to have put into his Glasgow house a new heating stove to give next winter a heat "like Madeira," and to procure plants and flowers to decorate it in the autumn, and departed almost gaily for the trip. But this time it was not cable-laying that took him to Madeira.

Soon he wrote to Mrs. King, then in Florence :—

*L. R.*, FUNCHAL BAY, MADEIRA,
*Tuesday, May* 12, 1874.

MY DEAR ELIZABETH—I believe you heard from Lizzie that I intended to sail from Falmouth for Madeira on the 2nd of May. The *Lalla Rookh* has done well—taken me to the island, 1200 sea miles from Falmouth, in $6\frac{3}{4}$ days. I anchored exactly at noon on Sunday in Funchal Bay, an hour before the *Hooper*, which I had left at Greenhithe on Friday week after testing the cable on board, and which sailed from the Thames on the day following. Yesterday I was answered *Yes* to a question which I asked very soon after the English people came out of forenoon church on Sunday. I was here for sixteen days last June and July on account of a fault in the cable. Otherwise this greatest possible blessing could not have come to me,—that is as *we* see, —but surely it is "not chance." When I came to Madeira in the *Hooper* it had never seemed to me possible that such an idea could enter my mind, or that this life *could* bring me any happiness. I thank God always that I was brought here. When I came away in July I did not think happiness possible for me, and indeed I had not begun even to wish for it. But I carried away an image and impression from which the idea came, and before I landed at Dover in October I had begun to wish for it. Hope grew stronger till yesterday, when I found that I had not hoped in vain. I cannot write more just now, but I send this because I do not wish a mail now on the point of leaving to go without bringing the good news. When you know Fanny you will be able to really congratulate me. Even now I think you will be glad for my sake. . . .—Your ever affectionate brother, WILLIAM THOMSON.

The next letter is to Helmholtz:—

YACHT *LALLA ROOKH*, FUNCHAL BAY, MADEIRA,
*June* 23, 1874.

DEAR HELMHOLTZ—I am to be married in Madeira to-morrow. I enclose a photograph, and I hope you will know the original before very long. Let me have a line addressed Athenæum Club, London, to say if you are to be at the British Association in Belfast. I do not intend to be at the meeting, but if you are to be there we might see you on your way to or from it. We think of sailing from Madeira in the *Lalla Rookh* about the middle of July, but have not made up our minds whether to make as short a passage as we can to England, or to touch at Gibraltar, Lisbon, Vigo, Corunna, on our way, or to keep a more westerly course and make a little cruise among the Azores. The future mistress of the *Lalla Rookh* promises to be a very good sailor, having already been out a good many times for a day's sail, one of them round the Desertas (about 70 miles) and always hitherto escaped sea-sickness. Still it remains to be seen whether a yacht cruise on the open Atlantic is a pleasure in direct or in inverse proportion to its duration.

My present happiness is due to a fault in the cable which kept the *Hooper* for sixteen days in Funchal Bay last summer. I hope you and Mrs. Helmholtz and your children are all well.—With kind regards, I remain, yours always truly, WILLIAM THOMSON.

He wrote the same day a similar letter to Dr. J. Hall Gladstone, adding a congratulation on his election to the Fullerian Professorship of Chemistry at the Royal Institution:—"To be Faraday's successor is indeed an honour. I am sure you will find the post most congenial to you."

The *Glasgow Herald* of July 4, 1874, contained the following announcement:—

MARRIAGES.

At the British Consular Chapel, Funchal, Madeira, on the 24th ult., Sir WILLIAM THOMSON, Professor of Natural Philosophy in the University of Glasgow, to FRANCES ANNA, daughter of CHARLES R. BLANDY, Esq., of Madeira.

To his sister Sir William wrote:—

ST. ANNA, MADEIRA, *July* 5, 1874.

MY DEAR ELIZABETH—

. . . . . . .

On the 24th we rode away in the afternoon to a place called St. Antonio de Serra, about 4 miles ride from Funchal, and 2000 feet above the sea level. We lived there in a house belonging to an uncle of Fanny's for a few days and then came across to this place. We have been taking rides and walks every day and enjoying to the utmost the beauties of Madeira. On Thursday next we return to Funchal, and remain about 10 days in Mr. Blandy's house before sailing away in the *Lalla Rookh*.—Your affec brother, W. T.

The homeward voyage in the yacht was shortened, for off Finisterre she broke her main gaff, and finished the voyage under top-sail to Cowes for repairs. Sir William and Lady Thomson paid a hurried visit to London, returning to Cowes for further cruising between engagements in town, which prevented them from going to the British Association at Belfast. Here James Thomson was to be president of the Engineering section, and to him, on August 12, Sir William wrote, from the Great Western Hotel, Paddington:—

My dear James—

. . . .

We left Cowes on Friday to come here on business. I have been overwhelmed with arrears of correspondence—reports of recent expeditions. The *Hooper* is expected home about the 18th, and I must be here for some time after that to decide what is to be done with the defective cable which the *Hooper* brings home (which was to have been laid between Cayenne and Demerara, but is brought back because defective). I don't know how long this may keep me, but it may be that for several weeks yet I must be within call of London. We return to the *Lalla Rookh* at Cowes to-morrow to remain "at home" in her until we return to London for the *Hooper*. . . .

W. Bottomley tells me you are going to refer to the eclipsing system of distinguishing lighthouses. I trust the one on Holywood Bank will be in action and giving practical proof of the plan. You can scarcely be too strong in expressions, as the NEED for distinction in REAL experience, though sailors and admiralty officials *believing honestly* that they speak from experience are quite ready to deny it.

I think you might refer to the soundings by pianoforte wire. . . .

When we get quite free from London we shall probably sail for the Clyde, weather and time permitting. We should touch at several places on the way so as to have chiefly sailing by day and resting in port by night. If the Holywood Bank eclipsing light is as I hope it will be, we should probably go into Belfast Lough to look at it on our way, and even without that inducement we might make Belfast one of our ports and take a few hours to run up and see our friends there. You, I suppose, will be back in Glasgow before that time; but when we get back to the Clyde I hope you will be able to come and have a few days' sailing with us. We shall take the earliest opportunity after getting to Scotland to go to the College, and perhaps remain there a few days; but the yacht will be our headquarters probably till about

NETHERHALL, LARGS, BUILT BY SIR WILLIAM THOMSON, 1875.

the middle of October. You must take some thorough rest after you get over the B.A. and your address. I am afraid in the meantime you will be too busy to allow you much rest. . . .—Your affectionate brother,

W. THOMSON.

They were not able to attend the British Association, but Sir William communicated two papers: one on Perturbations of the Compass at Sea, and another on his Improvements in Compasses. To Dr. King he sent a message strongly disapproving of Tyndall's famous presidential address, and of the dictum in which he had discerned in matter "the promise and potency of all terrestrial life." Thomson thought it "especially inappropriate." Later they sailed to Belfast, thence to Largs to visit friends and relations; then went to Glasgow to see "home." After that there was a final cruise to Arran before the session began.

On October 29th Sir William wrote to Mrs. King:—

We have bought a little piece of ground, Kirklands, Largs, bounded on the south and north-west by the Noddle Burn (Noddsdale Burn, properly), on the road to Barr's farm, to build a house on. We shall begin building before very long, I hope. I thought you would be interested to have the earliest news of this.

"Netherhall" was the name given to the house,[1] a commodious country mansion in the Scottish baronial style, the building of which occupied many months. Sir William, with the aid of his brother,

[1] For a full description of the building, see *The Building News*, June 27, 1890. It is there stated that the cost was £12,000.

took a large part in the planning of the building, which was original in many ways. He himself engaged a master-mason and a master-carpenter, instead of letting the contract to a builder—a costly and vexatious plan in the sequel. A dozen years later the estate was much improved by the purchase of additional ground. Netherhall was the scene in after years of many family reunions, and of extended hospitalities presided over by the gracious hostess. It was here that Sir William sought quiet hours when the yachting days were over; and to it he retired when he withdrew from his chair in 1899.

Sir William Thomson was now 51 years of age, but, save for his slight lameness, as active as in youth. For nearly thirty years he had never felt the want of money, and for some years past had enjoyed a very large professional income.[1] He was supremely happy in his domestic life. Lady Thomson had been welcomed into the circle of family relations, and directed his household with rare dignity and grace. His academic duties were lightened by the devoted assiduity of his official assistant MacFarlane, and of his demonstrator and deputy-lecturer, James T. Bottomley.

A graphic picture of Sir William as he appeared

[1] The income of the partnership of Thomson, Varley, and Jenkin was considerable, for they derived handsome profits from their inventions. Varley's patent for the signalling condenser was very profitable. Sir William, writing to Jenkin in 1881, speaks of "the quadrant electrometer, the mirror galvanometer, and the last recorder patent, which is now bringing us £3000 from the Eastern Telegraph Co., £2100 from the Eastern Extension, and £1500 from the Anglo." The partnership of Thomson and Jenkin as consulting engineers to various cable companies was also extremely profitable, and brought them each several thousands a year.

to his students has been recorded by Professor Andrew Gray, then one of the merry students who filled the ten benches of the lecture theatre, afterwards his trusted secretary and scientific assistant, and finally his successor in the chair of Natural Philosophy. By his kind permission [1] the following extracts are given from his admirable book :—

The writer will never forget the lecture-room when he first beheld it, from his place on Bench VIII., a few days after the beginning of session 1874-75. Sir William Thomson, with activity emphasised rather than otherwise by his lameness, came in with the students, passed behind the table, and, putting up his eye-glass, surveyed the apparatus set out. Then, as the students poured in, an increasing stream, the alarm weight was released by the bell-ringer, and fell slowly some four or five feet from the top of the clock to a platform below. By the time the weight had descended the students were in their places, and then, as Thomson advanced to the table, all rose to their feet, and he recited the third Collect from the Morning Service of the Church of England. It was the custom then, and it is still one better honoured in the observance than in the breach (which has become rather common), to open all the first and second classes of the day with prayer; and the selection of the prayers was left to the discretion of the professors. Next came the roll-call by the assistant; each name was called in its English or Scottish (for the clans were always well represented) form, and the answer "adsum" was returned.

The vivacity and enthusiasm of the Professor at that time was very great. The animation of his countenance as he looked at a gyrostat spinning, standing on a knife-edge or on a glass plate in front of him, and leaning over so that its centre of gravity was on one side of the point of support; the delight with which he showed that hurrying of the precessional motion caused the gyrostat to rise, and retarding the precessional motion caused the gyrostat to fall, so that the freedom to "precess" was the secret of its not falling; the immediate application of the study of the gyrostat to the explanation of the precession of the equinoxes, and illustration by a model of a terrestrial globe, arranged so

[1] *English Men of Science. Lord Kelvin, an Account of his Scientific Life and Work.* By Andrew Gray, LL.D., F.R.S., V.-P.R.S.E. London, J. M. Dent and Co., 1908.

that the centre should be a fixed point, while its axis—a material spike of brass—rolled round a horizontal circle, the centre of which represented the pole of the ecliptic, and the diameter of which subtended an angle at the centre of the globe of twice the obliquity of the ecliptic; the pleasure with which he pointed to the motion of the equinoctial points along a circle surrounding the globe on a level with its centre, and representing the plane of the ecliptic, and the smile with which he announced, when the axis had rolled once round the circle, that 26,000 years had elapsed—all these delighted his hearers, and made the lecture memorable.

Then the gyrostat, mounted with its axis vertical on trunnions on a level with the fly-wheel, and resting on a wooden frame carried about by the Professor! The delight of the students with the quiescence of the gyrostat when the frame, gyrostat and all, was carried round in the direction of the spin of the fly-wheel, and its sudden turning upside down when the frame was carried round the other way, was extreme, and when he suggested that a gyrostat might be concealed on a tray of glasses carried by a waiter their appreciation of what would happen was shown by laughter and a tumult of applause.

On one occasion, after working out part of a calculation on the long fixed blackboard on the wall behind the table, his chalk gave out, and he dropped his hand down to the long ledge which projected from the bottom of the board to find another piece. None was there, and he had to walk a step or two to obtain one. So he enjoined MacFarlane, his assistant, who was always in attendance, to have a sufficient number of pieces on the ledge in future to enable him to find one handy wherever he might need it. MacFarlane forgot the injunction, or could not obtain more chalk at the time, and the same thing happened the next day. So the command was issued, "MacFarlane, I told you to get plenty of chalk, and you haven't done it. Now have a *hundred* pieces of chalk on this ledge to-morrow; remember, a *hundred* pieces; I will count them!" MacFarlane, afraid to be caught napping again, sent that afternoon for several boxes of chalk, and carefully laid the new, shining, white sticks on the shelf, all neatly parallel, at an angle to the edge. The shelf was about sixteen feet long, so that there was one piece of chalk for every two inches, and the effect was very fine. The class the next morning was delighted, and very appreciative of MacFarlane's diligence. Thomson came in, put up his eye-glass, looked at the display, smiled sweetly, and turning to the applauding students, began his lecture.

In the higher mathematical class, to which he lectured on

Wednesdays at noon, Thomson was exceedingly interesting. There he seemed to work at the subject as he lectured; new points to be investigated continually presented themselves, and the students were encouraged to work them out in the week-long intervals between his lectures. Always the physical interpretation of results was aimed at; even intermediate steps were discussed. Thus the meaning of the mathematical processes was ever kept in view, and the men who could follow were made to think while they worked, and to regard the mathematical analysis as merely an aid, not an end in itself. "A little expenditure of chalk is a saving of brains," "the art of reading mathematical books is judicious skipping," were remarks he sometimes made, and illustrated his view of the relative importance of mathematical work when he regarded it as the handmaid of the physical thinker.

The closing lecture of the ordinary course was usually on light, and the subject which was generally the last to be taken up—for as the days lengthened in spring it was possible sometimes to obtain sunlight for the experiments—was often relegated to the last day or two of the session. So after an hour's lecture Thomson would say, "As this is the last day of the session I will go on for a little longer after those who have to leave have gone to their classes." Then he would resume after ten o'clock, and go on to eleven, when another opportunity would be given for students to leave, and the lecture would be again resumed. Messengers would be sent from his house, where he was wanted for business of different sorts, to find out what had become of him, and the answer brought would be, hour after hour, "He is still lecturing." At last he would conclude about one o'clock, and gently thank the small and devoted band who had remained to the end for their kind and prolonged attention.

In the course of his lectures Thomson continually called on his assistants for data of all kinds. In the busiest time of his life—the fifteen years from 1870 to 1885—he trusted to his assistants for the preparation of his class illustrations, and it was sometimes a little difficult to anticipate his wishes, for without careful rehearsal it is almost impossible to make sure that in an experimental lecture everything will go without a hitch. The digressions, generally most interesting and instructive, in which he frequently indulged, almost always rendered it necessary to bring some experiment before the class which had not been anticipated, and all sorts of things were kept in readiness lest they should be wanted suddenly.

He wrote shortly to Prof. Andrews :—

UNIVERSITY, GLASGOW, *Nov. 5th*, '74.

DEAR ANDREWS—We are now settled here for the session. At your convenience I shall be glad to have the thermoelectric battery and the magnet to test them. . . . Is my compression apparatus ready? I shall be very glad indeed to have it when it is so, and greatly obliged to you for the trouble you have so kindly taken for me about it. We said good-bye with much regret to the *Lalla Rookh* on Friday last, and left her to be laid up for the winter in the Gareloch. After a succession of severe gales, in one of which (the very violent one that wrecked the *Chusan* at Ardrossan) she dragged both anchors, and went ashore in the Gareloch, we had a week of fine weather and a little beautiful sailing to Arran and in the Firth of Clyde. So we were very sorry to leave her. We were fortunately not on board the night of her shipwreck, by a mere chance of an unexpected meeting of the "University Court" keeping us in Glasgow. She was got off without damage and by ourselves without assistance or expense, after being 36 hours on soft sand in a very good position.

This autumn and winter he had much correspondence with Dr. John Hopkinson about electrolytic action, contact electrification, and the residual charges of Leyden jars, which the latter was investigating in connection with the composition and optical properties of glass.

Then came terrible news. The *La Plata*, a ship chartered by Siemens Brothers to carry some 250 miles of cable to South America for the Western and Brazilian Telegraph Company, had foundered off Ushant on Nov. 29th. Amongst the sixty lives lost was that of young David Thomson King, Sir William's nephew, one of the skilled electricians of the expedition, a man of great promise. To Mrs. King, then in Rome, Sir William wrote :—

David did his duty nobly to the last. Of that I think you might be sure, even if we had known nothing but that there had been a disaster. One of the survivors told Willie Bottomley that he had seen him very near the end, hard at work passing coals along the deck. . . . The news of what happened so soon after receiving cheery letters from him has been a terrible shock to us all. The unspeakable grief it will be to you and David, and to young David's brother and sisters, is in all our minds. God only can mitigate it to you, and to Him we pray that it may be mitigated. I cannot write more now. I need not tell you that I share your grief, and we have the warmest sympathy of my dear wife.

Sir William and Lady Thomson spent Christmas at Knowsley and then went to William Crum's house near Knutsford. At the New Year they went to Edinburgh, where Sir William was to read papers on two new instruments, a tide-calculating machine and a tide-gauge, and on some curious phenomena of capillary attraction which he had lately found out.

A project for enlarging Owens College, Manchester, was then being discussed, and some objection being taken to the expansion of its chemical and other laboratories as a degeneration from the legitimate work of a University, Thomson wrote to *The Times* (of Jan. 20, 1875) a remonstrance :—

If to give experimental science and chymistry their proper position in the curriculum, and to found new professorships and provide them with all the apparatus for complete and successful study, are tests of degeneration, Oxford, and Cambridge headed by its Chancellor, and all the Scotch Universities, have been labouring strenuously for many years to earn this reproach. . . . Surely a university must be an emporium of knowledge. . . . Owens College does, as far as its means allow it, perform the true functions of a University, the cultivation and the teaching to all comers of the whole body of human knowledge. It has become what it is by a true process of natural selection. It now wants, and it deserves, national funds, supplementary private munificence, and a charter constituting it a National University.

Principal Greenwood, Professors Ward and Balfour Stewart, and Professor (now Sir Henry) Roscoe, wrote thanking him for his generous recognition of their case. The matter dragged on for many months.

His reply to Roscoe was :—

*April* 26, 1876.

MY DEAR ROSCOE—Ever since receiving your letter with papers regarding Owens College, which I read with much interest and satisfaction, I have been in a very aggravated condition of high pressure accumulating as usual till the final blow-off at the end of the session now taking place. This has prevented me from sooner writing to you to express my most cordial approval of the proposal to now move energetically in promoting the earliest possible creation of the University of Manchester. The work already done and the position already taken by Owens College seems to me to make a University of Manchester inevitable. I utterly differ from those who object to any moderate multiplication of universities throughout the British dominions such as we have in Scotland, and I believe it will be a very great benefit indeed in England if you succeed, as I hope you will, and that very soon, in obtaining a charter constituting Owens College into a national University.—With best wishes for your success, I remain, yours truly, WILLIAM THOMSON.

Two years later, a word from Sir William Thomson again revived the project, as narrated in *The Life and Experiences* of Sir H. E. Roscoe, pp. 175-176.

The fame of the Manchester Penny Science Lectures for the People reached what the Scots term "the second city of the Empire," and I was invited to give the opening lecture of a similar series, which a committee in Glasgow had established. My lecture, on "The Chemical Action of Light," was given in 1878, in the large St. Andrew's Hall, before a crowded audience. After it was over Sir William Thomson (Lord Kelvin), who had presided at the lecture, remarked to me, "Why do not you make Owens College into a University?" I said that we had often talked about it, but that we thought the time had not yet arrived for so momentous a proposal. He replied: "You are quite mistaken. The time *has* arrived; you have quite attained a University position, and you ought to make it the University of Manchester." Thinking the matter over, I came to the conclusion that he was right, and on returning home consulted my friend Dr. Ward, now Master of Peterhouse, who was then our Professor of History, on the subject. He at once fell in with the suggestion, and after some time we convinced our Principal, Dr. Greenwood, who was of a cautious disposition, that we ought to make the attempt to found a new English University.

On Feb. 3rd, 1875, Sir William delivered in the City Hall, Glasgow, a lecture to about 2000 persons on "The Tides." Only a small part of this[1] was written out; and the reports of it are brief.

[1] See *Popular Lectures*, vol. iii. pp. 191-201; also see *The Engineer*, Feb. 19, 1875.

Of the activities of this period, and of the demands upon Thomson's time, some idea may be formed from the following letter to James Napier:—

*Feby.* 16, '75.

DEAR NAPIER—The new method in my paper does take into account the fact of the centre of gravity not being in a vertical with the point of support, and the effect of this fact on the heeling error. Indeed, what I have called the heeling error is simply the effect of that fact. If you please I would explain it in the physical section of the Philosophical Society in the most unpopular manner you wish. I can also, if you please, with great ease, do the same for the tides, and shall be glad to do so if desired, but on this condition, that not a single scrap in black and white is expected from me for the *Proceedings.* I am now three years deep in unfulfilled pledges for the Geological and Philosophical Societies, and I have resolved to stop payment and put up shutters in my factory of papers for the *Proceedings* of either Society. Bringing apparatus and giving an explanation of it at the Society, or explaining theoretical subjects mathematically or non-mathematically, I shall be glad to do at any time provided I am not expected either to look at a report of it taken by any one else, or to give anything in black and white myself for publication. Every moment of time henceforth that I can give to writing for publication must be devoted to T & T′, and to the description of physical experiments bringing out new properties of matter, of which I have now a good many on hand, for the Royal Society of London.

In these months Sir William had much correspondence about Lighthouses with Dr. Hopkinson. On the same topic also, he wrote to Prof. Peirce of Harvard:—

*Ap.* 22, 1875.

DEAR PROF. PEIRCE—I send you by book packet along with this two articles on deep-sea sounding by pianoforte wires. You will see that I have been obliged to contrast our own Admiralty not very favourably with yours in respect to liberality towards suggestions for improvement. Our authorities are indeed exceedingly heavy to move in any way which involves the introduction of new ideas. I am not allowing suggestions for distinction of lighthouses by groups of eclipses of short and long, according to the Morse system, to fall asleep. A lighthouse on this plan has been in action on Holywood Bank, Belfast Lough, since the first of November, and I am told by the harbour authorities that captains, pilots, and sailors are all much pleased with it. They know it with perfect confidence from any other light afloat or ashore. It was till last November a red fixed light which could only be seen about two miles off, and was constantly liable to be mistaken for a ship's port-side light. When my eclipsing machine was applied to it the red shade was removed, and in all respects the light was left unchanged. Besides having the advantage of being unmistakable for anything else, it is now visible for a distance of five miles. Its distinguishing mark is the Morse letter U - - — (dot dot dash). The duration of the dot eclipse is half a second, the duration of the dash is three times as long. But I keep in the background the fact that, adhering simply to the letters of the Morse alphabet, we can with the greatest ease give twenty-eight distinctions, each thoroughly unmistakable for any other. This has a tendency to frighten "practical" men. I therefore to some "practical" men scarcely venture to call the distinguishing mark which I would propose for any one light a "letter," and rather say "You will know that is Sanchoty Head, a short and a long," and have it printed in the list of lights; and again "You will know that is Gay Head, a long and a short." Nobody can but answer "Yes" to this question. The moment you see a short and a long you know that it is not Gay Head, but

Sanchoty Head, or in the case of two lighthouses not very far from one another it may be judged expedient to have one of them marked by different numbers of eclipses. The Sanchoty Head, for instance, might be the letter E (one short eclipse of half a second occurring every ten seconds, or every five seconds, as may be judged expedient).

Then the Gay Head might be the letter U (- - —), or the letter D (— - -), or the letter R (- — -). (You may tell them that a short and a long is the letter N if you please; but I am told that it would perplex "practical" men to tell them so much: I don't believe it, because I don't find sailors more apt to be confused than landsmen by anything in reason. Still, as long as I speak to Lighthouse Boards and their secretaries and engineers I am quite ready to say nothing about the Morse alphabet, but when I get to sea I do not find that a sailor is at all more apt to mistake the Holywood light for anything else when I tell him "that mark means the letter U").

Is there any hope that your Lighthouse Board would give the eclipsing system a trial? For three years back I have continually pressed it on our Hydrographic Office, but I have had even more cold water thrown on this proposal than on the pianoforte wire sounding. I should not be at all sorry for either them or the Elder Brethren if the United States authorities were to anticipate them in this matter as they have in the sounding by pianoforte wire. I have had two years' struggle with the compass department of our Hydrographic Office to induce them to take up some suggestions I have brought before them for the correction of the compass in iron ships. They are most obliging in giving me information, but are utterly immovable for anything like co-operation for anything new, as they were in the matter of sounding by pianoforte wire. I now see that if anything is to be done in the matter the whole business of it will fall upon myself, and I have therefore resolved to take out a patent for a new form of compensator and appliances which I have now made. I doubt whether with all the obligingness that has been shown to me I should be more successful in getting a trial of it than I have been in getting a trial of the pianoforte wire by our people, until I can offer it to them as a patented invention. I enclose a small paper by which if you glance at the conclusion you will see that my great need is small needles. I have now a compass on my table before me with four needles, of which the longest are each two inches long. In the course of this summer I hope to get a thorough trial of it at sea with the appliances which I have designed for determining the error due to the ship's magnetism. In the course of the last two or three days it has helped me to a very startling discovery regarding the correction of the quadrantal error by soft iron correctors as hitherto practised. The best thing hitherto done in this way undoubtedly is the iron cylinders of the Liverpool Compass Committee applied to our Admiralty standard compass. The results are described in page 174 of the second volume of your reprint of papers on the magnetism of ships and the deviation of the compass. There I find the maximum deviation produced by a pair of the 12-inch cylinders with their ends 7 inches from the centre of the compass card stated to be 12° 50′. Now I find by experiment that considerably more than one-half of this error is due to magnetization induced in the corrector by the needles of the compass! Hence if the 12-inch cylinders were applied at Liverpool to correct a quadrantal deviation of maximum value 12° 50′, the error would be enormously over-corrected in the same ship in the Gulf of St. Lawrence, and very much under-corrected at the equator. An infinitesimal needle placed at the position of the centre of the Admiralty compass card in those trials would have shown about 5° of effect only instead of 12° 50′. I am experimenting just now to find as exactly as need be the amount, but I have already seen that it is certainly less than one-half that found in the reported trials. I intend to send a short communication to the Royal Society on this subject, and hope to have the pleasure of sending you a copy when I get it in print.

In April he read a paper to the London Mathematical Society, and gave a discourse to the Royal Institution on Tides; also three papers to the Edinburgh Royal Society, one on the Theory of the Spinning Top, one on the Vibrations of a Stretched String of Gyrostats, and a third (in collaboration with Dr. John Perry) on the Capillary Surface of Revolution. In June he communicated to the Royal Society two memoirs, one on the Effects of Stress on Magnetization, the other on Electrolytic Conduction in Hot Glass.

The loss of the *Schiller* on the Scilly Isles in May gave Sir William Thomson occasion to address to *The Times* of May 12th a remonstrance on the neglect to provide means for taking rapid soundings which would have enabled the ship to keep outside the 50 fathom line. He was then at Peterhouse.

The summer was filled up with cruises on the yacht at Cowes and other places; and in August he and Lady Thomson went to Bristol for the meeting of the British Association. Here he read no fewer than seven papers, three of them mathematical, two on Tides and his Tide-calculating Machine, one on Magnetism, and one on Lighthouses.

At the concluding meeting Sir William Thomson moved a vote of thanks to the President, Sir John Hawkshaw, in the following terms:—

In him they found none of the Mephistophelian cynicism which said:—

Grau, theurer Freund, ist alle Theorie,
Und grün des Lebens goldner Baum.

For Sir John Hawkshaw the golden tree of life retained its greenness in the revivifying influence of true theory. Whoever heard him describe the tunnel which is to connect England with France could not but have felt, even in such a result as that, that they had a work of the greatest possible interest for human beings, irrespective of all scientific questions—a result which would give them a different feeling from any which they could now have towards their neighbours. He wished their president success in that undertaking, and hoped in a few years they would see it finished.

On Sept. 2 he attended at Bristol a meeting called in support of the foundation in that city of a University College. To this project he accorded the warmest welcome, and expressed the hope that any of them living twenty-five years hence might see it grown into a fully-organized University.

From Bristol Sir William and Lady Thomson went to visit the Duke of Argyll at Inveraray, where Royalty was being entertained. The memoirs of the Rev. Principal Story, who was also a guest, record the following incident:—

*Sept. 27.*—The Thomsons dined this evening, and in the saloon, soon after the Queen came in, Princess Louise, in handing her something, upset a flower-glass and spilt the water, which simple incident sent H.M. into fits of laughter, which seemed increased by the spectacle of Sir William Thomson, in an access of loyalty, wiping up the water with his pocket-handkerchief and then squeezing it into his tea-cup. The Queen smiled over it till she went away. *Solvuntur risu.*

On Nov. 11 Thomson gave, in the Glasgow City Hall, a public lecture on Navigation, referred to on p. 719 below. On the walls were hung two great charts—reproduced on pages 97 and 385 of his *Popular Lectures*, vol. iii.—of his voyage in the *Lalla Rookh* to and from Madeira in 1874, marked with observations and soundings. When he referred to these as taken during the sailing of "a certain ship," on a certain eventful voyage two years ago, the audience recognising the allusion cheered enthusiastically.

In December he dealt with vortex statics at the Royal Society of Edinburgh. He wrote to Hopkinson on his Leyden jar experiments.

THE UNIVERSITY, GLASGOW,
*Dec.* 13, 1875.

DEAR HOPKINSON—A long time ago, in the old University buildings, I made just such experiments on alternate positive and negative return charges. I used to keep a jar charged positively for several hours, sometimes for several days, then negative for several minutes or several hours, then positive again for a shorter time and so on, lastly discharge and put to electrom^r^. I always got alternate positive and negative return charges, but I forgot how many alternations I succeeded in getting. I never published the result, but I might have properly done so, and I think you should publish yours. I spoke of them to a good many people (Maxwell, probably enough, among them), and showed them in my class. I think, probably, similar observations must have been made last century, and I daresay a hunt into old books and journals might find something of them, but I have never seen them mentioned in print, and I don't believe they are generally known.

You will certainly, I think, get the result however infinitesimal you make the capacity of the non-coated part. I shall be glad to send to the R. S. a statement of your results if you please. It would be perfectly well suited for the *Proceedings*. People always seem to see it by common sense. If there can be a return charge at all there must be the possibility of alternately positive and negative return charges. (If one note can soak out of Baron Munchausen's post-horn, why not the whole tune —only it ought to have come out backwards!)

The existence of return charge, of course, disproves the simple exponential formula $A\epsilon^{-\lambda t}$, but it will be curious to find if your experimental series is better than any other obvious empirical formula.—Yours very truly,

WILLIAM THOMSON.

The year 1876 was one of astonishing activity. During the spring Thomson made several communications to the Royal Society of Edinburgh on Vortex Columns, on an Application of James Thomson's Integrator to Harmonic Analysis, on Vortex Theory of Gases, and on Thermodynamic Motivity. To the Royal Society of London he gave three papers on Mechanical Integration. The physical laboratory was in full swing, with new experiments on the secular diffusion of liquids, and new forms of gyrostats. He was also actively corresponding with Stokes and other scientific friends, as the following letters show:—

ARDINCAPLE CASTLE, HELENSBURGH,
DUMBARTONSHIRE, *Jan.* 15/76.

MY DEAR ANDREWS—

. . . . . .

We are all greatly delighted in my laboratory with what you have given us—my old assistant MacFarlane is in raptures to see carbonic acid compressed to the liquid state in that always ready way. It will be a splendid lesson to my students, and I feel that henceforth they will every session know more of the meaning of liquids and gases and vapours than I have ever been able to teach them before.

I was not able before I left Glasgow to get the warming apparatus applied, but you may be sure that we shall not wait patiently long, before doing this and going through the *continuity* cycle.

I need not attempt to tell you how much I thank you and how grateful I feel for all the trouble you have taken, and how much I value the result. When I see it set up in my apparatus room I think of it as a monument which will remain to generation after generation of professors and students of Natural Philosophy in the University of Glasgow long after I am gone, and will still be valuable and I hope valued as now.—Believe me, yours always,

W. THOMSON.

On Jan. 21 he wrote to Hopkinson on a form of pendulum suspension, and asked him to come to stay with him, as on Feb. 16 Huxley was coming to Glasgow to lecture, and on 23rd Aitken was to show experiments on the rigidity of a running chain. He wrote again to Andrews:—

THE UNIVERSITY, GLASGOW,
*March* 11, 1876.

DEAR ANDREWS—I thank you very much for your warm congratulations on the Matteucci prize. To have such congratulations from you and other friends adds greatly to the gratification which such a reward for scientific work brings. . . . I have been very much engrossed with the compass and sounding machine, but still trying to get something done in the laboratory. We have begun to get some very fine results as to the effect of transverse pull on magnetization by applying and removing hydraulic pressure on the interior of a gun barrel under the influence of longitudinal magnetizing force. We find opposite effects to those of longitudinal pull, and as in the case of longi[l] pull a certain degree of magnetizing force for which the effect of pull is zero with contrary effects for weaker and stronger magnetizing forces.

THE UNIVERSITY, GLASGOW,
*March* 17/76.

MY DEAR ANDREWS—I hope you will accept the presidentship,[1] and that you and Mrs. Andrews will take up your quarters here with us for the meeting. It will be a great pleasure to us, and as all the sections meet in the University buildings, I think you will find it more convenient and less fatiguing than living at any greater distance could be. So you must make up your mind to come to us. We hope to have the Taits and the Crum Browns also with us. . . .

I have to-day been seeing an exquisite electric experiment, quite a first-rate discovery, by Dr. John Kerr. You may have seen it in the *Phil. Mag.* a few months ago. It is literally an "illumination of Faraday's lines of force." A very moderate tension (by a small electric machine) much inferior to that required to produce a spark, between two wires about $\frac{1}{4}$-inch apart, in bisulphide of carbon, instantly brightened the flame of a lamp which had been extinguished by Nicol's prisms, and the light disappeared again the instant the electric tension was reduced to zero (by touching the prime conductor). The effect was the same as that produced when a piece of glass was substituted for the liquid and pulled in the direction which had been that of the electric force, by holding a broad slip of window glass in the course of the light between the Nicol's prisms, with the bisulphide also in the course. The electric effect was annulled by bending the glass by a moderate force with one's fingers, and holding it so that the light passed through the condensed part. When held so that the light passed through the dilated part of the glass the luminous result of the electric force was augmented. I *bespoke* an

[1] Of the British Association to be held in Glasgow that year.

exhibition of the experiments for the Association. Kerr lives in Glasgow (Principal of the Free Church Normal School) and is continuing his investigations. No doubt he will have new results to show when the time of the meeting comes. He was a pupil of mine for the first two or three years of my professorship, now *thirty years* ago.

A Special Loan Collection of Scientific Apparatus was exhibited in the spring of 1876 at South Kensington; and to this Thomson contributed a remarkable show of instruments, electrometers, galvanometers, compasses, the Tide-predicter, and an array of miscellaneous instruments of precision. Conferences, too, were arranged at which lectures were given by eminent men. The remarkable series of Thomson's electrometers was described in two lectures by James T. Bottomley; while Sir William himself gave, on May 17, a discourse on Electrical Measurement, which is reprinted in his *Popular Lectures*, vol. i. pp. 423-454.

To celebrate the centenary of the Independence of the United States the American Government organized the Centennial International Exhibition of 1876 at Philadelphia. Sir William Thomson was invited to act as one of the judges in Group 25, comprising instruments of precision, research, experiment, and illustration, including telegraphy. He left Glasgow on May 20, and wrote to Helmholtz from New York:—

*S.S. Russia arriving* New York,
*May* 30/76.

My dear Helmholtz—Just before leaving England I heard with pleasure that you and Mrs. Helmholtz are coming to the meeting of the British Association in Glasgow next September, and in the hurry of coming away I was unable to write to you and ask you to stay with us in the University. I now do so to catch the mail steamer which leaves New York to-morrow.

My wife bids me say that she looks forward with pleasure to the opportunity of making the acquaintance of you and Mrs. Helmholtz, and joins me in hoping that you will be our guests during the meeting of the Association.

We are now on our way to Philadelphia, where I am to be one of the judges for scientific apparatus during June. We intend to be in Glasgow by the end of July, but "Continental Hotel, Philadelphia," would find me for a letter posted before the end of June.

We have had a very fine passage across, with just enough of rough weather to test thoroughly a new compass, which I shall show you when you come to Glasgow. It behaved perfectly well throughout, notwithstanding a great shaking from the screw (which almost prevents me from being able to write legibly).

I have also been trying soundings by pianoforte wire from the steamer going at 14 knots, and succeeded in getting the bottom with ease in 40 fathoms. But I dare say you will hear enough and more than enough of such matters when you come to Glasgow so I need not trouble you with them now.—Believe me, yours very truly,

WILLIAM THOMSON.

The members of the group of judges met on May 25 and elected Sir William as president, Professor Joseph Henry consenting to preside until his arrival. He reached Philadelphia on June 5, and stayed till June 28. He specially undertook to write reports on electrical and magnetic apparatus and telegraphy. The general report on instruments was written by Henry. Of the 385 reports of specific awards in these classes, no fewer than forty-one were written by Thomson, the most important being as follows :—

1. Edison's Automatic Telegraph; 3. Graham Bell's Telephone and Multiple Telegraph; 4. Elisha Gray's Electric Telephone and Multiple Telegraph; 5. Philps's Printing Telegraph; 6. Quadruplex Telegraph; 14. Gray's Printing Telegraph; 19. Farmer's Magneto-electric Machine; 24. Gramme's Magneto-electric Machine; 27. Siemens's Submarine Cables; 221. Lyman's Deep Sea Wave Machine; 230. Ritchie's Floating Compass; 247. Wallace's Regulator for Electric Light; 359. Jamin's Powerful Steel Magnet; 383. Edison's Electric Pen.

The Report No. 3, on Bell's Electrical Inventions, comprises: 1st, a description of his Multiple Telegraph, and 2nd, of his Electric Telephone. The latter is as follows :—

In addition to his electro-phonetic multiple telegraph, Mr. Graham Bell exhibits apparatus by which he has achieved a result of transcendent scientific interest—the transmission of spoken words by electric currents through a telegraph wire. To obtain this result, or even to make a first step towards it—the transmission of different qualities of sound, such as the vowel sounds—Mr. Bell perceived that he must produce a variation of strength of current in the telegraph wire as nearly as may be in exact proportion to the velocity of a particle of air moved by the sound; and he invented a method of doing so—a piece of iron attached to a membrane, and thus moved to and fro in the neighbourhood of an electro-magnet, which has proved perfectly successful. The battery and wire of this electro-magnet are in circuit with the telegraph wire and the wire of another electro-magnet at the receiving station. This second electro-magnet has a solid bar of iron for core, which is connected at one end, by a thick disc of iron, to an iron tube surrounding the coil and bar. The free circular end of the tube constitutes one pole of the electro-magnet, and the adjacent free end of the bar-core the other. A thin circular iron disc held pressed against the end of the tube[1] by the electro-magnetic attraction, and free to vibrate through a very small space without touching the central pole, constitutes the sounder by which the electric effect is reconverted into sound. With my ear pressed against this disc, I heard it speak distinctly several sentences, first of simple monosyllables, "To be or not to be" (marvellously distinct); afterwards sentences from a newspaper, "S.S. Cox has arrived." (I failed to hear the "S.S. Cox," but the "has arrived" I heard with perfect distinctness); then "City of New York," "Senator Morton," "The Senate has passed a resolution to print 1000 extra copies," "The Americans in London have made arrangements to celebrate the Fourth of July." I need scarcely say I was astonished and delighted; so were others, including some other judges of our group, who witnessed the experiments and verified with their own ears the electric transmission of speech. This, perhaps the greatest marvel hitherto achieved by the electric telegraph, has been obtained by appliances of quite a homespun and rudimentary character. With somewhat more advanced plans and more powerful apparatus, we may confidently expect that Mr. Bell will give us the means of making voice and spoken words audible through the electric wire to an ear hundreds of miles distant.

[1] It will be remembered that Sir William Thomson brought a pair of instruments back from America, showed them at the British Association at Glasgow, made private trials of them before and after the meeting, but got no good results, and appears to have been misled by the attachment of the iron diaphragm of the receiver (which was in this instrument screwed down) into supposing that in some way it acted differently from those shown at Philadelphia. In the subsequent litigation in this country it was successfully maintained that Thomson's discourse and his exhibition of the American instruments did not constitute an anticipation in this country of the telephone as patented here for Bell by Morgan Browne, as altered by amendment and disclaimer.

Report No. 24 on Gramme's machines gives a curious rudimentary theory of the dynamo, which it recommends as a subject for scientific inquiry.

Returning to England he wrote, on Aug. 2, a long letter to his former student, Professor John Perry, then in Japan, who had sent him the results of a research, by Professor Ayrton and himself, on Contact Electricity:—

I wish I had more time to keep up a correspondence with you and Ayrton, and other good friends of our laboratory corps, now in many parts of the world. Please tell Ayrton and Dyer this from me. You might also tell them that I have just returned from America, where I passed a month in most interesting work at Philadelphia where I was judge. . . . Lady Thomson was with me, and we had a very interesting trip in America, in the course of which we saw Niagara Falls, Toronto, Montreal, Boston, and Newport. . . . I made soundings from the *Russia* and *Scythia* going at 14 knots without reducing speed. I found it perfectly easy to haul in the wire, of which I sometimes had as much as 300 fathoms with a 22 lb. iron sinker and a pressure-gauge for measuring the depth. I found bottom in 68 fathoms quite unexpectedly in a place where 1900 fathoms was marked on the chart. . . . I have been so pulled about from the beginning of last winter up to the present time by pressing engagements that I have never yet succeeded in completing for publication our joint paper on the Capillary Surface of Revolution. I hope, however, soon to be able to overtake the work of preparing it for press.

He wrote again to Helmholtz:—

*Aug.* 9, 1876.

DEAR HELMHOLTZ—I wrote to you from New York, on my arrival there at the end of May, to say that it would give myself and my wife much pleasure if you and

Mrs. Helmholtz will stay with us in the University during the meeting of the British Association to commence on the 4th of September in Glasgow.

I hope you received my letter and have a favourable answer to give, or have already sent one which may be wandering about in America (where we were rather unfortunate in missing letters). Will you let me have a line to say if we may expect you. You will find our house and ourselves in Glasgow ready to welcome you on the 4th; but in any case don't be later than the afternoon of the 5th, as Dr. Andrews' opening address will be on the evening of that day.

Hoping to see you soon I say no more just now (except that I am at work every moment that the *Lalla Rookh* permits on precession and nutation of a rotating liquid in a rigid ellipsoidal shell). We shall have a great deal to talk over when you come.—Yours always truly,

WILLIAM THOMSON.

*P.S.*—I find that the quasi-rigidity produced by vortex motion is such that a rotating liquid enclosed in a rigid shell giving a boundary to the liquid of any non-spherical figure of revolution, moves approximately as if it were a rigid body, when the shell is turned slowly round any axis, or when any periodic motion having a great multiple of the period of the fluid's rotation, is given to the shell. I have verified or illustrated this by a large globe of thin sheet copper, made slightly oblate and filled with water. It is provided with a point to spin on, and a stem to spin it, and it spins very much like a solid top, but more steadily! Quick nutations become rapidly extinguished, and even the slower precession wears away sooner than it does in a solid top. So that the liquid top very soon comes to "sleep" spinning round a vertical axis.

At the British Association of 1876, held at Glasgow in the University buildings, Sir William was in great force. He presided over the Physical and Mathematical Section, giving an address, the first part of which was devoted to what he had seen in America, the progress of science in her universities and in her navy, the wonders of the exhibition, the telephone—"the greatest by far of all the marvels of the electric telegraph"—the administration of patent law. Then he turned to his real subject:—

A conversation which I had with Professor Newcomb one evening last June in Professor Henry's drawing-room in the Smithsonian Institution, Washington, has forced me to give all my spare thoughts ever since to Hopkins's problem of Precession and Nutation, assuming the earth a rigid spheroidal shell filled with liquid. Six weeks ago, when I landed in England, after a most interesting trip to America and back, and became painfully conscious that I must have the honour to address you here to-day, I wished to write an address, of which science in America should be the subject . . . But the stimulus of intercourse with American scientific men left no place in my mind for framing or attempting to frame a report on American science. Disturbed by Newcomb's suspicions of the earth's irregularities as a time-keeper, I could think of nothing but precession and nutation, and tides and monsoons, and settlements of the equatoreal regions, and melting of polar ice. Week after week passed before I could put down two words which I could read to you here to-day; and so I have nothing to offer for my address but a review of evidence regarding the physical conditions of the earth; its internal temperature; the fluidity or solidity of its interior substance; the rigidity, elasticity, plasticity of its external figure; and the permanence or variability of its period and axis of rotation.

After this preamble came the discourse (reprinted in vol. ii. of the *Popular Lectures*, p. 238) on the

Physical Condition of the Earth. Amongst other things it contained the conclusion, that examination of underground temperatures showed the age of the earth to be between 90 millions and 50 millions of years. Further, we might be quite sure that the earth is solid in its interior, any internal spaces occupied by lava or other liquid being small in comparison with the whole. The hypothesis of a rigid crust enclosing liquid violates dynamical astronomy as well as physics; and the tides decide against a flexible crust covering a liquid interior. But now thrice to slay the slain: suppose the earth to be a thin crust of rock resting on liquid matter, its equilibrium would be unstable! The axis of the earth, may in the slow subsidences of ocean-beds and the gradual upheaval of continents, gradually have shifted, and the poles may have changed their positions by as much as 10, 20, 30, or 40 degrees, without at any time causing any perceptible sudden disturbance of land or water. Lastly, since the date on which an eclipse of the moon was seen in Babylon in 721 B.C., the earth has slowed down in her rotation by 11½ seconds per annum; while apparently, in consequence of some settlement in the equatoreal regions, the earth, which in the twelve years from 1850 to 1862 lost some 7 seconds, began going faster again, and gained 8 seconds from 1862 to 1872.

To the ordinary business of the section Sir William Thomson contributed no fewer than twelve papers. These were: on Precessional Motion of a Liquid; on Secular Diffusion of Liquids; on a Case of Instability of Steady Motion; on the Nutation of a Solid Shell containing Liquid; on Compass Correction; Effects of Stress on Magnetization; on Contact Electricity; on a New Form of Astronomical Clock; on Deep-sea Soundings in a Ship moving at High Speed; on Naval Signalling; on a Practical Method of Tuning a Major Third; and a Physical Explanation of the Mackerel Sky. The last two have never been published; but the substance of the earlier of the two was given in 1878 in another paper,[1] and shows two points of interest. The "revolving" character of the sound heard in the slow beats of an imperfect unison proves that, in a sense, the ear can perceive phase as well as pitch in a simple tone. Secondly, when tuning a major third by the beats heard if the tuning is imperfect, the beats become much more marked if at the same time a perfectly tuned fifth is introduced. The explanation given of the mackerel sky[2] was that there are two strata of cloud one moving over the other, the slipping of one stratum over the other producing waves, in which portions of the air rose and fell. Difference of temperature was not necessary; but the essential was that one or other of the two strata should be very near the point of hygrometric saturation; for then it would be clear when it sank to its lowest point and cloudy when it rose to its highest. The

[1] "On Beats of Imperfect Harmonies," *Proc. Roy. Soc. Edin.* ix. p. 602, April 1878; reprinted in *Pop. Lectures*, ii. p. 394.

[2] See *Symons's Monthly Meteorological Magazine*, xi. p. 131, 1876.

new Astronomical Clock, which was mentioned amongst the communications to the section, had been described in 1869 to the Royal Society, and was now on view in the hall of Sir William's house at the University. It had a modified dead-beat escapement, in which the pallets were touched for but one three-hundredth part of the beat by the escapement tooth, and so arranged that the action of the escapement did not stop the train of wheels.

*Nature*, of Sept. 7, 1876, contained a delightful biography of Sir William Thomson, compiled by Professor Tait, and illustrated by Jeens's exquisite portrait engraved on steel. It concluded as follows:—

The following opinion of Sir William Thomson's merit as a worker in science has been sent us by Prof. Helmholtz :—" His peculiar merit, according to my own opinion, consists in his method of treating problems of mathematical physics. He has striven with great consistency to purify the mathematical theory from hypothetical assumptions which were not a pure expression of the facts. In this way he has done very much to destroy the old unnatural separation between experimental and mathematical physics, and to reduce the latter to a precise and pure expression of the laws of phenomena. He is an eminent mathematician, but the gift to translate real facts into mathematical equations, and *vice versa*, is by far more rare than that to find the solution of a given mathematical problem, and in this direction Sir William Thomson is most eminent and original. His electrical instruments and methods of observation, by which he has rendered amongst other things electrostatical phenomena as precisely measurable as magnetic or galvanic forces, give the most striking illustration how much can be gained for practical purposes by a clear insight into theoretical questions; and the series of his papers on thermodynamics and the experimental confirmations of several most surprising theoretical conclusions, deduced from Carnot's axiom, point in the same direction."

British science may be congratulated on the fact that in Sir William Thomson the most brilliant genius of the investigator is associated with the most lovable qualities of the man. His single-minded enthusiasm for the promotion of knowledge, his wealth of kindliness for younger men and fellow-workers, and his splendid modesty, are among the qualities for which those who know him best admire him most.

YACHT, *LALLA ROOKH*,
*Sep.* 18, 1876.

MY DEAR HELMHOLTZ—We were very sorry not to have you with us at the British Association, and to hear that your not feeling well enough to come would deprive us of this pleasure. I hope you are now quite re-established in health, and have fully enjoyed your sojourn among the mountains.

If you come so far as London to see the exhibition of Physical Apparatus in Kensington, will you not come a little farther and see us in the University or the *Lalla Rookh*.

We shall probably remain living chiefly on board the yacht till about the middle of October, soon after which we go to London and Cambridge, where I have to attend my annual College meeting at Peterhouse on the 31st of October. . . . —Believe me, yours always truly,

WILLIAM THOMSON.

In October 1876 Sir William Thomson's friend and quondam College Tutor, the Rev. H. W. Cookson, died, and with his death the Mastership of Peterhouse became vacant. Some of the Fellows desired to have Sir William Thomson as Master, while others were in favour of the Rev. James

Porter, his junior by three years, who for some time had been Tutor, and was still holding the office. At the election, which took place on October 28, 1876, in the College Chapel according to Statute, Porter received a majority of votes, and became Master. On Porter's death in 1900, Lord Kelvin was again approached with a view to his accepting the Mastership, but he definitely declined; and, on the candidature of Dr. A. W. Ward being brought to his notice, he united with his Fellows in electing him to that office.

In February 1877 Sir William lectured at Govan on Telegraph and Lighthouse Signals, advocating the teaching of the Morse alphabet in all schools, and denouncing the "Tite Barnacles" of the Admiralty and other Boards for opposing his plan of giving each lighthouse a distinctive kind of flashing light, as if sailors could not learn to distinguish long and short flashes!

On Feb. 18, 1877, he wrote to Mrs. King:—

The compass causes a quite unprecedented addition to my occupations, but it is very interesting, and as it takes me a good deal about shipping it is not like plodding at writing or "book work." Willy Bottomley is most helpful, and has got on very well with some adjustments which he has done all himself. He was down with me from Friday morning till Saturday evening in a new ship, *Balmoral Castle*, belonging to Messrs. Donald Currie & Co., which has been fitted out with three of my compasses. We went to Gareloch head on Friday evening and "swung" early next morning to adjust compasses. *I* should not have required this, but there were two other compasses on board, and it was thought right to have the ordinary certificate of the usual adjustment, both for mine and them.

He wrote to Lord Rayleigh :—

*March* 8, 1877.
THE UNIVERSITY, GLASGOW.

DEAR LORD RAYLEIGH—I see by this morning's paper that you and Lady Rayleigh are to be passengers in the *Balmoral Castle* to-morrow from Dartmouth for the Cape. There are three of my compasses on board—an azimuth compass, and two steering compasses in the wheel house.

I should be very glad if you would look at them a little and see how they behave, and much obliged if you would write me a line from Madeira or the Cape, telling me about them.

The enclosed printed papers about the compass (and a new sounding machine which has made a good beginning already in several ships) will tell you all you may care to know about my system of adjustment.

My wife joins in wishing you and Lady Rayleigh a pleasant passage, and I remain, yours very truly,

WILLIAM THOMSON.

On Feb. 22 he gave to the Glasgow Geological Society a lecture on Geological Climate, which has been dealt with in Chapter XIII. p. 536; and on April 24 he lectured to the Institution of Engineers and Shipbuilders in Scotland on Compass-adjustment in the Clyde. The compass work, described in detail in the next chapter, was assuming a commercial importance. The greater part of his time was spent in White's workshops, whither he would repair almost daily after he had finished his morning lecture and had made the round of the laboratory. The correspondence in which he was involved was very heavy for several years. He was also busy with Tait revising the *Treatise on Natural Philosophy*, a second edition of which was now called for. On March 7 appeared in the *Scotsman* his letter on vivisection, dealt with subsequently (p. 1105). In April he wrote to Hopkinson on the inductive capacity of glass, with which the latter had experimented. He encouraged him to try as many different kinds of glass as possible, and find out whether glass of the same composition always had the same electrostatic capacity.

On May 15 he wrote again to Hopkinson that he had just received splendid reports as to the Holywood Bank Lighthouse at Belfast, for which he had designed a distinctive eclipsing light. He showed these reports to the Elder Brethren of Trinity House, to whom he had that day gone to explain his compass and his sounding machine. He was to sail for Madeira in a few days in the *Lalla Rookh* from Cowes.

To Mrs. King he wrote on July 1 from Madeira, giving a narrative of the voyage:—

. . . I have been busy ever since we got away from London, at Dartmouth, Plymouth, Vigo, and at sea and here, with Tides Report for the British Association, etc., etc., and the great book (T and T′), *it* chiefly. I hoped to have a large instalment for the printers off by this mail, but the steamer has been announced suddenly to-night and goes away early in the morning, so this must wait a little longer. On the voyage, and since coming here, I have also had much to do in trials of the sounding machine leading to what I hope is to be considerable improvement.

The return voyage, from August 5 to August 7, in thick weather, afforded a triumphant proof of the advantage of taking flying soundings. Sir William saw to the fitting up, on board H.M.S. *Minotaur*, of his own compass and sounding machine before he went on to Plymouth for the British Association on August 15.

At Plymouth Sir William was ubiquitous. He read the report on the tides of Mauritius drawn up by himself and Captain Evans. He spoke on the magnetic susceptibility of iron; on Laplace's tidal equation; on the variations of barometric pressure; on the marine azimuth mirror; on his machine for taking flying soundings; on his depth-recorder; on his compass; on the Needles lighthouse, for which he advocated a distinctive flashing light. He added to the gaiety of the mathematical section by raising the question (see p. 610) whether a beetle could not live on a meteoric voyage from Mars to the earth. He and Dr. Samuel Haughton then delighted the audience by talking in their respective brogues into Graham Bell's telephone, which Mr. (now Sir William) Preece had brought over.

On October 26 Sir William urged on the Shipmasters' Society the need for lighthouse reform. He wrote on the same subject to *The Times* on November 12.

In 1877 he was elected a foreign associate of the *Institut* of France in place of von Baer.

In the previous year the Italian Society of Science, known as the *De' Quaranta*, had awarded him the Matteucci prize as "the one who has contributed the most in the world to the advancement of science by his writings and discoveries."

Down to this date electric lighting, in the modern sense, was practically unknown. Solitary arc lamps had been used for theatre-effects, for magic-lantern work, and, in rare cases, for lighthouses. Primitive dynamo-electric machines, based on Faraday's great discovery of 1831, had been built. Thomson had even reported on Holmes's machine, which was a feature of the 1862 Exhibition. He had never concerned himself, however, either with electric lamps or with electric generators. Even the advances made in 1866-67 by Wilde, Varley, Siemens, and Wheatstone had left him unconcerned. Gramme's commercial machines of 1871 had apparently not interested him till he reported on them for the Philadelphia Exhibition of 1876. Down to January 1, 1880, he had read no fewer than 360 scientific papers, and taken out 17 patents; but not one of these touched even the fringe of the question of electrical engineering, the generation of electric energy for lighting and power. Once only, in 1857 (see p. 397), had he ever spoken on the possibility of using electric motive power, and on that occasion he put it aside as impracticable. But the question was raised in a discussion at the Institution of Civil Engineers on January 22, 1878, on Siemens's recent improvements described in a paper by Messrs. Higgs and Brittle; and following up a remark of Dr. C. W. Siemens on the possibility of distant transmission, Sir William Thomson spoke. With characteristic agility of mind he leapt forward to the logical issue :—

He believed that with an exceedingly moderate amount of copper it would be possible to carry the electric energy for one hundred, or two hundred, or one thousand electric lights to a distance of several hundred miles. The economical and engineering moral of the theory appeared to be that towns henceforth would be lighted by coal burned at the pit's mouth, where it was cheapest. The carriage expense of electricity was nothing, while that of coal was sometimes the greater part of its cost. The dross at the pit's mouth (which formerly was wasted) could be used for working dynamo-engines of the most economical kind, and in that way he had no doubt that the illumination of great towns would be reduced to a small fraction of the present expense. Nothing could exceed the practical importance of the fact to which attention had been called : that no addition was required to the quantity of copper to develop the electric light at a distance. The same remarks would apply to the transmission of power. Dr. Siemens had mentioned to him in conversation that the power of the Falls of Niagara might be transmitted electrically to a distance. The idea seemed as fantastic as that of the telephone or the phonograph might have seemed thirteen months ago ; but what was chimerical then was an accomplished fact now. He thought it might be expected that, before long, towns would be illuminated at night by an electric light produced at the pit's mouth or by a distant waterfall. The power transmissible by the machines was not simply sufficient for sewing-machines and turning-lathes, but, by putting together a sufficient number, any amount of horse-power might be developed. Taking the case of the machines required to develop one thousand H.P., he believed it would be found comparable with the cost of a one thousand H.P. engine ; and he need not point out the vast economy to be obtained by the use of such a fall as that of Niagara, or the employment of waste coal at the pit's mouth. . . .

For lighthouses, the great adaptability of the electric light to furnish increase of power when wanted gave it a value which no other source of light possessed. . . .

The value of electric light was also of great importance in regard to public health. The detrimental character of the fumes of gas in public buildings was well known. There were no fumes from the electric light, and the quantity of heat generated by it was vastly less than that produced by gas or oil for the production of the same light. It had been stated by a previous speaker that the gas employed in an economical gas engine to drive a machine to produce the electric light would produce the same light with one-fifteenth part of the combustion ; and that meant not only increased economy, but a great advantage in regard to public health.

When a year later Sir William gave evidence before the Select Committee of the House of Commons on Electric Light (see p. 691), he repeated much of the above, adding that the manufactories of whole towns might be driven by a supply of electric power so transmitted. To such application of electricity as motive power there was no limit ; it might do all the work that could now be done by the most powerful steam-engine. He advocated that by legislation in the interests of the nation, and of mankind, they should remove all obstacles, such as those arising from vested interests, and should encourage inventors to the utmost. There would be no danger of terrible effects from the employment of electric power, because the currents employed would be continuous and not alternating !

Commenting on this evidence, *Nature* remarked: "This may be called a fanatical view of the electric light"!

About this time Sir William Thomson undertook to write an article on "Elasticity" for the *Encyclopædia Britannica* (ninth edition). Much of the theoretical matter had already been worked out by him and his brother James in his Royal Society memoir (p. 319, above), or was already embodied in the *Treatise* of Thomson and Tait, but the compilation of descriptive matter and the collection of numerical data involved much labour; and new experiments had to be made in the laboratory by Donald Macfarlane, and by Andrew Gray and Thomas Gray, to fill up the lacunæ. To the article was appended a reprint of the "Mathematical Theory of Elasticity," in seventeen short chapters, dealing with strains and stresses in homogeneous bodies, both those that are isotropic (that is, destitute of "grain" and alike in every direction) and those that are aeolotropic (having definite structure, different in different directions, as in wood or in crystals). Following Green, he showed that, in the general case, completely to define the elasticity of a body there were required twenty-one coefficients, six of which were the principal elasticities, and fifteen others depending on the type of strain. He classified strains, and defined the various moduluses that specify the relation between the strains and the stresses that produce them.

The importance of Lord Kelvin's many contributions to the subject may be gathered from the fact that in Professor Karl Pearson's standard work, the *History of Elasticity*, more than one hundred pages are devoted to them. The late Professor G. F. FitzGerald has given the following appreciation:—

> His treatment of the elasticity of solids has largely influenced the whole British school of elasticians, and his fecundity in devising elastic mechanisms has emphasized his fight against the supposed necessity that, in a simple solid, the rigidity is three-fifths of the compressibility. This may, no doubt, be used as a definition of a simple solid, but in that case hardly any value can be attached to a relation that is hardly ever of any service. Our knowledge of the structure of matter is far too meagre for us to deduce, *a priori*, such a result, and Lord Kelvin has always advocated the scientific method of proving by induction that, in a large number of cases, we can reduce the elastic properties of a solid to its rigidity and compressibility, and deduce its behaviour under stress from a knowledge of these two qualities.

A proof of the "Elasticity" article being sent to Tait, it was returned with the following effusion written on the last page. The last two verses are by Tait, the rest by Maxwell:—

Count up the stresses O;
Weigh well the stresses O.
For what's our life but just a strife
Where strains elicit stresses O?

To Nature blind my torpid mind
Cared not what cork or jelly meant,
Nor could expound the stresses round
The differential element.

Now, better taught, maturer thought
That state of mind reverses O,
And finds great fun in twenty-one
Elastic moduluses O.

He's blest who dares let worldly cares
And worldly people joy on all,
And learns to express six types of stress,
Each unto each orthogonal.

Vex not my ears, ye crystal spheres,
Your harmony's insipid O;
But play again that six-fold strain
My parallelepiped O!

Joy to the fair, be gems their care,
Ornated trains and tresses O;
But leave the sage his bliss, his rage,
Related strains and stresses O!

Green grew the stresses O;
T. *slew* the stresses O.
The hardest grind I e'er did find
Was reading T. on stresses O.

For the *Encyclopædia* also he afterwards wrote the article on "Heat." It begins with the sentence, "Heat is a property of matter which first became known to us by one of six different senses." He distinguished between the sense of temperature and the tactile (or muscular) sense by which we perceive the roughness or smoothness of things. The article consisted of four principal sections: calorimetry, thermometry, transference of heat, and statistical tables. The calorimetric section is distinguished by the special prominence given to Joule's work, and by a comment on the circumstance that fifty years passed before the scientific world was converted by the experiments of Davy and Rumford to the rational conclusion as to the non-materiality of heat: "a remarkable instance of the tremendous efficiency of bad logic in confounding public opinion and obstructing true philosophic thought." The section on thermometry is marked by extreme precision of language, and the emphasis laid on Regnault's work with "mercury-in-glass" and "air-in-glass" thermometers. A large amount of space is devoted to "water-steam-pressure-thermometers" and sulphurous acid thermometers, by means of which he was endeavouring to realize actual thermodynamic instruments. He recommended for practical use constant-pressure gas thermometers of nitrogen or hydrogen. Under transference of heat he referred to Fourier's *Théorie Analytique* and its solutions of heat problems, "of which it is difficult to say whether their uniquely original quality, or their transcendent interest, or their perennially important instructiveness for physical science, is most to be praised." At various points in the article the work done by himself and his assistants in the Glasgow laboratory, in the determination of constants of conductivity and of emissivity, is emphasized. He added a strange appendix: a compendium[1] of "Fourier mathematics," admirable in itself, but thrown together like mere hints to a lecturer.

Returning to May 1878, we find Thomson busy with A. M. Mayer's groups of floating magnets, of which he wrote to *Nature*, and with a Harmonic Analyser based on the mechanical integrator devised by his brother James (p. 692). On May 10 he discoursed at the Royal Institution on the Effects

[1] Originally written in 1850 in his mathematical diary, and published in the *Quarterly Journal of Mathematics* in 1856; finally reprinted in *Math. and Phys. Papers*, ii. p. 41, in 1884.

of Stress on Magnetization. On May 20 he was addressing the Edinburgh Royal Society on Columnar Vortices. In the same month he gave two papers to the Physical Society of London. At the Paris Exhibition of 1878 he was awarded a gold medal for his compass and sounding machine. To the British Association at Dublin he sent three papers: on the Tides of the Mediterranean (along with Captain Sir F. J. Evans); on the Magnetism of Ships; and on the Influence of the Straits of Dover on the Channel Tides.

In November 1878 we find him again urging in *The Times* the question of distinguishing lights for lighthouses. In January 1879 there appeared in *The Times* an anonymous article on Aids to Safe Navigation, from which it may be gathered that the sounding machine had been tried on H.M.S. *Minotaur* to the satisfaction of Admirals Beauchamp Seymour and Lord Walter Kerr. The latest improvements in the compass, the azimuth mirror, and the deflector were mentioned and praised.

Having, in February 1879, to lecture at the Royal Institution on the "sorting demon" (p. 286), of Maxwell, he stayed with his friend Sir William Siemens, busy over the evidence which each of them was shortly to give before the Select Committee on Electric Light. Sir William and Lady Thomson spent a week in June at the Siemenses' country house near Tunbridge Wells.

In his evidence before the Committee, Sir William referred to the problem of utilizing the

Falls of Niagara, and to Siemens's proposals as outlined in his presidential address to the Iron and Steel Institute in 1877. He gave an estimate of the quantity of copper suitable for the economical transmission of power by electricity to any distance, and applied the calculation to the case of Niagara. He pointed out that, under practically realizable conditions of intensity, a copper wire half an inch in diameter would suffice to take 26,250 horse-power from water-wheels driven by the Fall, and (losing only 20 per cent on the way) to yield 21,000 horse-power at a distance of 300 British statute miles; the prime cost of the copper amounting to £60,000, or less than £3 per horse-power actually yielded at the distant station.

All through the preceding months Sir William had been busy revising "Thomson and Tait," and the first part of Vol. I. of the new edition appeared in June. It was reviewed in *Nature* of July 3, 1879, by Clerk Maxwell—almost the last thing done before his lamented death. Maxwell quoted from p. 225 the statement (obviously adopted from Newton) that "matter has an innate power of resisting external influences, so that every body, as far as it can, remains at rest or moves uniformly in a straight line." "Is it a fact," asked Maxwell, to whom "this Manichæan doctrine of the innate depravity of matter" seemed superfluous, "that 'matter' has any power, either innate or acquired, of resisting external influences? Is a cup of tea to be accused of having an innate power of resisting the sweetening influence of sugar because it persistently refuses to turn sweet unless the sugar is actually put into it?" Maxwell praised, as a feature of the new edition, the greater use of the generalized equations of motion, which had too long been left neglected "in sanctuary of profound mathematics":—

> The credit of breaking up the monopoly of the great masters of the spell, and making all their charms familiar in our ears as household words, belongs in great measure to Thomson and Tait. The two northern wizards were the first who, without compunction or dread, uttered in their mother tongue the true and proper names of those intellectual-ancepts which the magicians of old were wont to invoke only by the aid of muttered symbols and inarticulate equations. And now the feeblest among us can repeat the words of power and take part in dynamical discussions which but a few years ago we should have left for our betters.

This would seem the appropriate place for a brief mention of the very remarkable series of machines for the performance of mathematical operations which Sir William Thomson designed between the years 1876 and 1879. His brother, Professor James Thomson, had conceived a mechanism[1]—called the disk-ball-cylinder integrator—which enabled any desired fraction of the motion of a revolving disc to be communicated to a cylinder pivoted above it, by the intermediation of a heavy metal ball or globe which rested on the disk and pressed against the cylinder. If the globe lay

[1] "An Integrating Machine, being a New Kinematic Principle," *Proc. Roy. Soc.* xxiv. p. 262, 1876.

actually on the centre of the disk it communicated no motion to the cylinder. If it were shifted (by a suitable fork) along a diameter of the disk, parallel to the cylinder, to a point near the outer edge of the revolving disk, it caused the cylinder to revolve actively. In intermediate positions the amount of motion so communicated was proportional to its distance from the centre. On learning of this ingenious combination, Sir William immediately perceived that it could be applied not only to such simple integrations as the quadrature of curves, but to other much more difficult problems of the calculus, the integration of products, the solution of linear differential equations, and above all to the mechanical performance of *harmonic analysis*, the process by which, when any complicated (single-valued) periodic function is given, it can be resolved into a series of constituent terms which themselves are simply periodic. The heavy arithmetical labour involved in the ordinary calculation of the integrals might thus be saved. Lord Kelvin's own account will be found in the Appendix on Continuous Calculating Machines at the end of Vol. I. Part I. of the Thomson and Tait *Treatise*. The principal publications on the subject were :—

1. On an instrument for calculating the integral of the product of two given functions. *Proc. Roy. Soc.* xxiv. p. 266, 1876.

2. Mechanical integration of the linear differential equations of the second order, with variable co-efficients. *Ib.* xxiv. p. 269.

3. Mechanical integration of the general linear differential equation of any order with variable co-efficients. *Ib.* xxiv. p. 271.

4. An application of James Thomson's integrator to harmonic

analysis of meteorological, tidal, and other phenomena, and to the integration of differential equations. *R. Soc. Edin. Proc.* ix. p. 138, 1876.

5. Harmonic Analyser. *Proc. Roy. Soc.* xxvii. p. 371, 1878.

6. On a machine for the solution of simultaneous linear equations. *Ib.* xxviii. p. 111, 1879.

7. The tide-gauge, tidal harmonic analyser, and tide predicter. *Inst. Civil Engineers Proc.* lxv. pp. 2-25, 1881.

The Harmonic Analyser for calculating the elements of tides was constructed. An example may be seen in the Science Collections at South Kensington. The machine for solving differential equations did not prove to be practical.

Clerk Maxwell died on November 5, and thus the Cavendish chair at Cambridge became vacant. Lord Rayleigh wrote to Sir William Thomson, opening the question whether he would exchange Glasgow for Cambridge. His reply was decisive :—

LONDON, *Nov.* 17, 1879.

DEAR LORD RAYLEIGH—Your letter was forwarded from Glasgow and reached me here this morning.

I feel that my destiny is fixed for Glasgow for the rest of my life. I felt strongly attracted to Cambridge at the time the new chair of Experimental Physics was founded, but resolved then to remain in Glasgow. If you could see your way to take the chair it would, I am sure, be much for the benefit of the University, and of science too, as the Cavendish Laboratory would give you means of experimenting and zealous and highly instructed assistants and volunteers, and would naturally lead you to more of experimental research than might be your lot, even with all your zeal and capacity for investigation, if you remain independent. If, however, you feel that in taking the professorship you would be taking a burden on you to any degree uncongenial or inconvenient, or tending to occupy time which you might more advantageously spend in writing or in independent research, and that you would only hold it till some of the younger men might show suitable qualifications, I would not advise you to take it. Hopkinson would, I believe, be a candidate if you do not accept the chair; and with his strong mathematical foundation, and decided inclination for physical science, and particularly for experimental work, I believe he would be very successful should he be elected.

I need not say that it would be a great satisfaction and pleasure to myself to look forward to having you at Cambridge should you decide to take the professorship.

Believe me, yours very truly,

WILLIAM THOMSON.

# CHAPTER XVII

## NAVIGATION: THE COMPASS AND THE SOUNDING MACHINE

GREAT as were the achievements of Lord Kelvin in other directions, his contributions to the science and practice of navigation were of no mean importance. Fond of the sea from his youth, he had during his experiences in the laying of ocean cables, and, later, on board his own yacht, become a master in seamanship. "I am a sailor at heart," he declared in 1892, in his inaugural address as President of the Institute of Marine Engineers. His penchant for sailing is shown by the nautical examples given as illustrations of hydrodynamical principles in § 325 (edition of 1879) of Thomson and Tait's *Natural Philosophy*, where the tendency of a body to turn its flat side across the direction of motion through a fluid is connected with the steering of a square-rigged ship, the "wearing" of a fore-and-aft rigged vessel, and the difficulty of getting a ship in a heavy gale out of the trough of the sea; concluding with the remark, that the risk of going ashore *in fulfilment of Lagrange's equations* is a frequent incident of "getting under way" while lifting anchor.

Thomson's contributions to navigation may be summarized under seven heads :—

1. His Compass.
2. His Sounding Machine.
3. His work on Lighthouse Lights.
4. His work on Tides, and his Tide Analyser, Tide-Gauge, and Tide-Predicting Machines.
5. His connection with the Admiralty on the Committee upon the Designs of Ships of War (1871), and on the Committee to review the Types of Fighting Ships (1904).
6. His mathematical investigations of Waves in general.
7. His tables for facilitating the use of Sumner's methods of finding the place of a ship at Sea.

## The Kelvin Compass

About the year 1871 Sir William Thomson was asked by his friend, the Rev. Norman Macleod, to contribute an article to his newly founded magazine *Good Words*. He chose as topic the mariner's compass, and was thus caused to direct his attention to the construction of that indispensable instrument of navigation. He thought he had a pleasant and easy task in describing an instrument which had been for six hundred years in regular use by European mariners; but when he began he found himself compelled to criticize, so grave were the defects which he found in the compasses then in use. And having noted the defects, his mind, ever on the alert as to practical needs, turned toward possible improvements; and so began the work that eventually made his name familiar to every nautical man. It was not till 1874 that his first article was sent to *Good Words*; and it was five years afterwards when his second article appeared. "When I tried," he said, "to write on the mariner's compass, I found that I did not know nearly enough about it. So I had to learn my subject. I *have* been learning it these five years." His *Good Words* articles deal, however, with Terrestrial Magnetism in general, and with the early history of the compass, and do not touch his own inventions.

Many were the defects of the old marine compasses in vogue up to 1870. They were often sluggish in their action, and apt to stick at times. In the Navy they were useless during action, as the concussion of the vessel during gun-fire put them out of service. During stormy weather, when the vessel was rolling, they were liable to oscillate in a way that made them misleading. Some of these defects arose from the weight of the moving part. From time immemorial the compass for use at sea has consisted of a pivoted magnet carrying affixed above it the circular "card." This card, five centuries ago, used to be inscribed with the eight principal "winds," divided into halves and quarters, so making up the thirty-two "points" that have been in use all the world over. Until about 1840 the magnetic needle was either an oval plate or a simple flat bar of steel drilled with a central hole to receive the cap by which it was poised on the pin, the ends of the bar being pointed. In 1842, as the result of the investigations of an Admiralty committee, the Admiralty Standard Compass was adopted. In this the card was seven and a half inches in diameter, and under it were fastened four "needles"—two on each side of the middle—each needle being a long straight bar of flat clock-spring, set with the breadth of the bar perpendicular to the card. The card with the needles weighed about 1600 grains, its weight being borne by a jewelled cap fixed in the centre of the card, and resting on a fine point of hard iridium alloy. In the merchant marine larger compasses, with needles ten, twelve, or even fifteen inches long, were favoured. Those on board the *Great Eastern* were eleven and a half inches. Sailors like large compass cards, as they are more easily read; and the cards of larger size are steadier in a heavy sea. As Thomson himself pointed out, the secret of the steadiness of a large compass lies in its period of vibration being long. On the other hand, it was supposed that the more powerful its magnetic moment in proportion to its moment of inertia, the better would be the directive action of the compass. It may be, from the point of view of getting over friction on the pin; but when the ship is rolling in the ocean, the more powerful the magnet the less steady will the compass be. The natural period of vibration of the Admiralty standard compass is about nineteen seconds, that of the ten-inch compass of the merchant ships about twenty-six seconds. The great weight of the "card" led to errors due to friction on its pivot; and the sailors used often to kick the binnacle to make the card move, or agitated it with a twiddling line. Worse than all this, the considerable length of the magnets made it difficult to correct the errors due to the magnetism of the iron of the ship. So long as ships were made of wood, and no iron bars or stays were permitted[1] near the compass, it would obey fairly the directive action of the earth. But when iron ships began to replace the "wooden walls," it was found that the magnetism of the ships themselves produced serious errors. Early in the nineteenth century Flinders had introduced the practice of "swinging" the ship, that is, turning her slowly round, while the deviations of her compass from the true direction were observed (by taking bearings of a distant object) in different positions. He also invented the use of the "Flinders bar," a rod of soft iron placed near the compass to correct for changes in the magnetism of the ship due to the vertical component of the earth's magnetism. Barlow introduced the plan of correcting the influence of the ship's own magnetism by placing disks of soft iron in suitable positions near the compass. After the mathematical investigations of Poisson and of Airy, about 1838, methods of correction by the use of permanent magnets and of soft-iron globes were introduced, and Archibald Smith had extended the mathematical investigation, and had formulated practical rules which were

[1] An instance occurred when the *Niagara*, a wooden frigate, in laying the cable of 1858, was found to have deviated seriously from her course, the many tons of iron in the cable on board having produced an unsuspected error in her compass.

adopted by the Admiralty for correcting the deviations of the compass. The magnetism of the hull of a ship which thus disturbs the compass, and prevents its pointing truly along the magnetic meridian of the place, is partly of a permanent nature, depending on the position in which the ship lies during building, and is partly of a temporary nature due to the temporary magnetization of the iron of the ship, depending on the direction in which she is pointing for the moment, whether northwards or southwards, or in some oblique position across the magnetic meridian. As the magnetism of the ship is thus constantly changing, the correction must be an automatic one, namely, that effected by masses of soft iron subject to the same influences which cause the disturbance. To correct for the permanent magnetism of the ship, permanent magnets were placed in a reversed position beneath the compass-bowl in the binnacle; and to correct for the temporary induced magnetism of the ship's hull two globes of soft iron were placed beside the compass, one on either hand, at the level of the compass card, as directed by Airy, the then Astronomer-Royal. Smaller compasses were also in use, the received opinion being that the smaller a card is the more correctly it points, but the larger a card the more accurately it is read.

In the course of his investigation Thomson discovered that the long magnets in common use had several disadvantages. By being long and powerful, they themselves induced magnetism in the soft-iron globes used as correctors, and in some cases created more error[1] than before. Only by using much larger globes at a greater distance could the long needles be corrected. Thus to correct the needles of the *Great Eastern* would have required globes weighing more than a hundred tons each! Aboard battleships the effect of gun-fire was disastrous, compasses with heavy cards being simply unusable.

Thomson revolutionized all this. The needles must be made short; yet for accuracy of observation the card must be kept large; to diminish friction on the pivot it must be made light; and for steadiness its moment of inertia must be increased by throwing as much of the weight as possible into the rim. He therefore built a gossamer structure of threads on a light aluminium rim, below which were suspended four, six, or eight slender steel needles, and bearing the "card" printed on a circle of thin paper with all the central part cut away.

But all this was not effected without many stages of experiment and trial.

While tentatively proceeding he wrote to the acting hydrographer, Captain (later Sir Frederick J.) Evans, who had in 1870 published his well-known *Elementary Manual* on the compass.

GLASGOW UNIVERSITY, *Nov.* 25, 1873.

DEAR CAPTAIN EVANS—As you have kindly excused the trouble I have already given you, I am emboldened to trouble you farther.

[1] See Archibald Smith and F. J. Evans in *Proc. Roy. Soc.*, April 18, 1861, p. 179; where the errors of the "corrected" standard compass of the *Great Eastern* are given as between 5° and 6°.

What is the whole weight of needle, compass card, and magnets, of the best Admiralty compass at present made? Has it two needles, or four, and what are their lengths? I am not sure whether 6 inches is the diameter of the card or the length of the longest of the needles. To what degree of accuracy can you depend on the readings? What is the greatest displacement at which the compass will stick from the correct position when left undisturbed? I presume this is a small fraction of one degree.

I am disposed to advocate, as better than anything at present afloat, a little toy compass on gimbals which I have seen lately, made in London, and which is sold for 18s. electro-silvered, or 21s. electro-gilt. It has one needle and a compass card of the usual form, but only about an inch diameter. I can read it to a fraction of a degree; and hitherto I have not detected any sticking away from the correct position. I have not yet made accurate experiments. Its smallness renders it perfectly easy to correct perfectly the quadrantal deviation by the Liverpool cast-iron cylinders. But I object to the movable compass card altogether, and have made a small compass with a pair of needles about half an inch long, and a fine glass pointer (after the method originally given by Joule some twenty-five years ago) which I expect will be thoroughly satisfactory. I am arranging it with soft-iron correctors for the quadrantal deviation, and an accurately graduated magnetic adjustment for the semicircular deviation, by which I shall be able to have the compass kept correct from day to day in all latitudes, without losing any of the advantage of keeping a complete log of the ship's magnetic condition. I hope to be able to bring it with me to London and show it to you next week. Meantime I would feel much obliged by your answer to my questions, if convenient, by return of post; as besides the "biographical sketch"[1] of Archibald Smith, I have in

[1] In this obituary notice of January 1874 there occurs the following characteristic note: "When the needles of a standard compass are reduced to something like half an inch in length, and not till then, will the theoretical perfection and beauty, and the great practical merit, of Airy's correction of

hand a short article on the "Mariner's Compass" for *Good Words*, which should also be finished by the end of this week.—Yours truly, WILLIAM THOMSON.

Captain Evans, R.N.

*P.S.*—Can you also tell me the dimensions, number of needles, and weight of the largest compass supplied to some of these grand modern merchant ships? From yours and Archibald Smith's paper, I understand that the standard compass supplied to the *Great Eastern* had a card 12 inches diameter! Is this correct? Will you also tell me, roughly, the period of vibration of a standard compass, in London, of your ordinary pattern? I expect that my small compass with glass indicator will be altogether free from "hunting" in the roughest weather at sea; and that the amount of heeling-over will be directly shown by its performance during the ordinary rolling of a ship at sea.

Evans's reply has not been preserved; but he had not much sympathy with Thomson's ideas. In the *Glasgow News* of March 18, 1874, we find a brief description of the first form put forth by Thomson:—

Sir William's compass consists of a pair of steel needles, on the model of a needle prepared for the galvanometer by Dr. Joule, the founder of the science of thermodynamics. The needles are each half an inch long, and are supported on a framework of aluminium and glass rods, weighing in all $1\frac{1}{2}$ grains, and hung by a single fibre of unspun silk $\frac{1}{20}$ of an inch long.

He had two days previously shown the new compass to the Royal Society of Edinburgh.

Writing a week later to Froude, he complains

the compass by soft-iron and permanent magnets (which theoretically assumes the length of the needle to be infinitely small in proportion to its distance from the nearest iron or steel) be universally recognised, and have full justice done to it in practice."

"how people in all departments of the Admiralty, except the construction department, are averse to, and how if they could they would be impregnable against, all suggestions from without." Evidently he had received scant encouragement in that quarter.

Sir William himself, in 1885, on the occasion of a lawsuit which arose as to his patents, gave the following account of his invention :—

I began to give attention to the mariner's compass with a view to its general improvement about 1871, and after working at the subject for five years, I took my first patent relating to these improvements, namely, my patent No. 1339, A.D. 1876. In 1874 I communicated a mathematical paper to the British Association, which was published in the *Philosophical Magazine* for that year, in which I showed that largeness of vibrational period is favourable for steadiness of the compass at sea in stormy weather. Up to that time no practical method of obtaining largeness of vibrational period consistently with other essential qualities for good working, particularly smallness enough of needles to allow the proper correction of the compass for the deviations produced by the iron of the ship, had been invented. After a year and a half of experimental work on land and at sea, I succeeded in devising a method of carrying into practicable effect the mathematical principles of that paper, and as a result the light compass card with small needles and radial supports of both needles and card described in my specification, No. 1339, 1876, has come to be very extensively used in the Royal Navy, and in nearly all the other navies in the world, as well as in the mercantile marine.

Thirty-five years prior to my mathematical investigation, the late Astronomer-Royal, Sir George Biddell Airy, had shown how the errors of the compass, depending on the influence experienced from the iron of the ship, may be perfectly corrected by magnets and soft iron placed in the neighbourhood of the binnacle.

Although up to the year 1876, when my improved compass card came into use, Airy's method for correction had never been thoroughly carried into effect, partial applications of it had come into general use. There was not then, as there had been twenty years previously, the question between the advocates of correctors and the advocates of no correctors. The absolute necessity for magnetic correctors, to make the compass a practically working instrument of navigation, had, by the increased quantity of iron used in the construction of modern ships, been forced upon even the most extreme advocates of the system of no correction. The only question was how the correction could be effected in the best manner for practical purposes. The chief drawback to the perfect application of the Astronomer-Royal's method had been the great size of the needles in the ordinary compass, which renders one important part of his correction—the correction of the quadrantal error for all latitudes, by masses of soft iron placed on two sides of the binnacles—practically unattainable, and which limits and sometimes partially vitiates the other part of the correction, or that which is performed by means of magnets placed in the neighbourhood of the compass. I therefore endeavoured to make a mariner's compass with much smaller needles than those previously in use; but it was only after several years of very varied trials that I succeeded in producing a compass with all the qualities which I found to be necessary for thoroughly satisfactory working at sea, in all weathers, and in every class of ship. One result at which I arrived, partly by lengthened trials at sea in my own yacht, and partly by dynamical theory, analogous to that of Froude with reference to the rolling of ships, was that steadiness of the compass at sea was to be obtained not by heaviness of needles or of compass card, or of added weights, but by longness of vibrational period of the compass, however this longness is obtained. By the term vibrational period, or the period (as it may be called for brevity) of a compass, I mean the time the compass card with its needles takes to perform a complete vibration to and fro, when deflected horizontally through any angle not exceeding 30° or 40°, and left to itself to vibrate freely. Thus, if the addition of weight to the compass card improves it in respect to steadiness at sea, it is not because of the additional friction on the bearing point that this improvement is obtained; on the contrary, dulness of the bearing point, or too much weight upon it, renders the compass less steady at sea, and, at the same time, less decided in showing changes of the ship's head than it would be were the point perfectly fine and frictionless, supposing for the moment this to be possible. It is by increasing the vibrational period that the addition of weight gives steadiness to the compass, while, on the other hand, the increase of friction on the bearing point is both injurious in respect to steadiness, and detrimental in blunting or breaking it down, and boring into the cap, and so producing sluggishness after a short time at sea. If weight were to be added to produce steadiness, the place to add it would be at the very circumference of the card. The conclusion at which I arrived was that no weight is in any case to be added beyond that which is necessary for supporting the card, and that with small enough needles to admit of the complete application of the Astronomer-Royal's principles of correction. The length of period required for steadiness at sea is to be obtained without sacrificing freedom from frictional error, by giving a large diameter to the compass card, and by throwing to its outer edge as nearly as possible the whole mass of rigid material which it must have to support it.

In the compass set forth in my specification No. 1339, 1876, these qualities are attained by supporting the outer edge of the card on a thin rim of aluminium, and its inner parts on a series of spokes or radial threads stretched from the rim to a small central boss of aluminium, spokes, as it were, of a wheel. The card itself is of thin strong paper, and all the central parts of it are cut away, leaving only enough of it to show conveniently the points and degree divisions of the compass. The central boss consists of a thin disc of aluminium, with a hole in its centre, which rests on the projecting lip of a small aluminium inverted cup mounted with a sapphire cap, which rests on a fixed iridium point. The weight of the central boss aluminium cap amounts in all to about five grains. It need not be more for a 24-inch than for a 10-inch compass. For the 10-inch compass the whole weight on the iridium point, including rim, card, silk thread, central boss, and needles, is about 180 grains. The limit to the diameter of the card depends upon the quantity of soft iron that can be introduced without inconvenient cumbrousness on the two sides of the binnacle to correct the quadrantal error.

With the small needles of my new compass, the complete practicable application of the Astronomer-Royal's principles of correction is easy and sure; that is to say, correctors can be applied so that the compass shall point correctly on all points, and these correctors can be easily and surely adjusted at sea, from time to time, so as to correct the smallest discoverable error growing up, whether through change of the ship's magnetism, or of the magnetism induced by the earth, according to the changing position of the ship.

In 1876 he endeavoured to interest the aged Astronomer-Royal in his invention, and sent him a compass, followed by the accompanying letter:—

*March* 3, 1876.

MY DEAR SIR GEORGE—I thought you would be interested in my new compass. The objects I had in view were steadiness of indication at sea, and suitableness for the application of your method of correcting in all its completeness. Your description of an accurate method of finding the amount with direction of the permanent magnetic force has not, so far as I know, hitherto been realized, even in experimental arrangements on board ship. This I propose to carry out in connection with my new compass and the apparatus you saw; it is nearly complete, though it was not attached to the rough illustrative instrument which I sent to you on Saturday.

If you had allowed the instrument to remain at the Observatory I should have called on Monday, and explained all parts of my plan. The main points are:—1st, Needles so small that the distances of their different parts from the nearest parts of the soft-iron correctors differ by amounts which may be neglected as infinitely small in comparison with the distances themselves. 2nd, Soft-iron correctors so small that the distances of their different parts from the permanent magnet correctors may be regarded as infinitely small in comparison with the whole distance. 3rd, Accurately graduated adjustment of the permanent correcting magnets.

The monstrous size of the needles used in modern great merchant steamers would require tons (I might say hundreds of tons) of soft iron, to fulfil conditions above, with sufficient accuracy to give a good practical result for your correction for the quadrantal error.

It is fortunate that the very small needles which I have introduced for the sake of conditions 1 and 2 have also a very great advantage in respect to the mechanical qualities of the compass over any make of larger size. The compass I sent to you is an illustration of what I propose for a *standard* compass. It is read with great ease to a quarter of a degree, which is a good deal closer than is required or can be usefully practised at sea.

I gain greatly in mechanical quality, of course, by doing away with the fly card, and (at all events for a standard compass) this without any sacrifice of convenience. I intend to give my compass a thorough trial at sea. Shall I let you see it when complete before trial at sea? I shall also, if you desire it, give you a complete statement of the details of observation by which I propose to find the data required for varying the adjustment of the magnets so as to keep the compass correct at all points. —Believe me, yours truly,

WILLIAM THOMSON.

Sir George Airy.

The compass was sent by the hand of the late Mr. J. Munro, one of his pupils, who tells of its reception :—

One day, I remember, Sir William Thomson desired me to take his new compass to Sir George Airy at the Royal Observatory, Greenwich Park, and ask him what he thought of it. A crude, experimental instrument, mounted on gimbals in a wooden box, it nevertheless contained the essential features of the improvement ; and after I presented it to Sir George, he looked intently at it for some time, apparently in deep thought, and simply said, " It won't do." When I returned to Sir William, and told him of this verdict, he ejaculated, with a trace of contempt, " So much for the Astronomer-Royal's opinion ! "

One objection urged by the Admiralty officials against his compass was that the permanent magnets were not fixed, but required to be adjusted in position. He pointed out in reply that the supposed permanent magnetism of the ship, which they were designed to correct, was itself subject to change. And he told how, on H.M.S. *Thunderer*, in August 1877, when H.R.H. the Prince of Wales (now our King) visited her to watch gun practice, the firing of the guns was found to derange her standard compass, demonstrating the necessity for means of adjustment of the correcting magnets.

Time and again did Sir William bring before the scientific world, and before nautical men, his improvements. They were praised, they were criticised, they were condemned by the prejudiced, but they were received with deadly apathy by the Tapers and Tadpoles of official circles. At the British

Association of 1876 at Glasgow he read a paper on "Compass-correction in Iron Ships." In April 1877 he addressed the Institution of Engineers in Scotland on "Compass-adjustment on the Clyde." At the British Association the same year he described a Marine Azimuth Mirror for use with his compass, and gave another communication on the mariner's compass, with correctors for iron ships. Lord Walter Kerr, after trying a borrowed compass on H.M.S. *Minotaur*, wrote in September that he had suggested to the Government to purchase it. Gradually the shipowners began to realize the advantages of the new compass as the captains of the vessels on which they had been tried sent in favourable reports. Even foreign Governments began to give them a trial. Still the British Admiralty proved impervious. In 1878 Sir William Thomson addressed to the then First Lord, the Rt. Hon. W. H. Smith, M.P., the following letter:—

*April* 11, 1878.

SIR—I take the liberty of calling your attention to my compass and sounding machine, and asking you kindly to consider the question of introducing them into the Royal Navy for practical use.

The compass has been now amply tested at sea in upwards of 60 large iron steamers and sailing ships.[1] After eighteen months of very varied experience in all seas and all weathers it has, as the accompanying printed reports show, been found to work well, and to possess some considerable advantages over the other forms of compass hitherto in use. After six months' trial on board the German ironclad *Deutschland*, a second has been ordered by the German Imperial Admiralty, who have also recently

[1] I enclose a list.

ordered two more sounding machines after a short trial of a first one. The Russian Admiralty have recently ordered both compass and sounding machine. The compass has also been supplied to the Italian Navy and to the Brazilian ironclad *Independenzia* (now H.M.S. *Neptune*).

From results of trials on board ships of war firing heavy guns I have recently made some improvements in my compass, by which I hope it will be rendered thoroughly available for the navigation of a ship of war in action, and will so have a great advantage over the compasses hitherto in use. My compass on board H.M.S. *Minotaur* is not provided with these improvements, as the experience on which they are founded has been acquired since the time when it was purchased by the Admiralty, but if its action under gun-fire is not found altogether satisfactory it will be easy to add them to it without disturbing its adjustment on board.

I venture to suggest that more extensive trials of my compass in the Navy might now properly be made, and particularly that it should be tried in ships of several different classes both as an azimuth compass for ordinary navigation and as a between-decks steering compass for use in action.

The enclosed printed reports regarding my sounding machine leave no doubt as to its general usefulness for ships of all classes, and from them it may be seen that in ships of war in trying circumstances it might be of vital importance. You will not, therefore, I trust think me too sanguine in expressing the hope that the British Admiralty will early resolve to have it adopted for general use in the Royal Navy.—Your ob^t. serv^t. WILLIAM THOMSON.

The First Lord's reply was non-committal, but a year later Thomson made a proposal to put one of his compasses at the disposal of the Admiralty for trial—at his own expense. This offer produced the following response :—

ADMIRALTY, S.W., 23 *May* 1879.

SIR—With reference to your letter of the 30th ultimo, I am commanded by my Lords Commissioners of the Admiralty to inform you that your offer to supply one of your improved compasses at your own expense for trial in any ship or ships in which their Lordships may give orders to have it tried, and to allow it to remain on trial without charge as long as they may desire, is accepted.

2. I am further to acquaint you that the first suitable opportunity for a trial of this machine will be taken advantage of, and a further communication will be made to you on the subject.—I am, Sir, your obedient servant,

ROBERT HALL.

Sir William Thomson,
The University, Glasgow.

As, however, Sir William in his summer cruises on his yacht made the acquaintance of influential naval officers, there arose gradually a movement within the Navy for the adoption of the compass. Amongst those who warmly advocated the Thomson compass, and indeed became its champion against the official party who opposed its introduction, was Captain Fisher (now Admiral of the Fleet Lord Fisher). When H.M.S. *Wye* grounded in the Red Sea the court-martial declared that this would not have occurred had she been fitted with Thomson's compass and sounding machine. One of the captains, writing of this to Thomson, declared that in advocating his inventions at headquarters they had to approach the subject like burglars approaching a safe.

Twice Sir William brought his compass before the Royal United Service Institution, on February 4,

1878, and again on May 10, 1880. On the second occasion he described certain mechanical improvements for combating unsteadiness due to vibrations caused by the ship's engines; a device for ascertaining the heeling error; and improved adjustments of the correcting magnets in the binnacle. In the discussion which followed Thomson referred to a trial of his instrument on the *Glatton* in 1878, when it had remained perfectly steady during gun-fire. It was now on trial on the *Northampton* by order of the Admiralty.

At the bombardment of Alexandria on July 11, 1882, were two ships, H.M.S. *Inflexible* and H.M.S. *Alexandra*, that had been fitted with Thomson's compass. Captain (afterwards Admiral Sir Charles F.) Hotham, who commanded the latter ship, reported the superiority of Thomson's compass over other patterns, while Captain Fisher (now Admiral of the Fleet Lord Fisher), of the *Inflexible*, stated that during the bombardment this compass was particularly useful for steering a course, as it was the only compass that would stand the tremendous concussion caused by firing the eighty-ton guns. He reported it far superior to any other compass both with regard to its steadiness and general reliability, and had had perfect satisfaction with it also on board H.M.S. *Northampton*.

It was, however, not until November 1889 that the then superintendent of the Compass Department of the Admiralty, Captain E. W. Creak, F.R.S., could write to Sir William that he had received official intimation that his 10-inch compass was to be adopted in future as the standard compass for the Navy, and that it was to be at once placed on twenty ships.

THE KELVIN COMPASS.

The standard form of Thomson's compass, as adopted by the Admiralty in 1889, is depicted in the accompanying figure. Eight small needles of thin steel wire from $3\frac{1}{2}$ to 2 inches long, weighing in all 54 grains, are fixed (like the steps of a rope-ladder), on two parallel silk threads, and slung from a light aluminium circular rim of 10 inches diameter by four silk threads, through eyes in the four ends of the outer pair of needles. The entire weight is about 170 grains. It has double the period and one-seventeenth of the weight of the 10-inch compass previously in common use, and its frictional error is not more than one-quarter of a degree.

The following more recent letter to Admiral of the Fleet Lord Fisher deals with the modern method of adjustment without the labour of "swinging" the ship :—

GLASGOW, *March* 3/92.

DEAR FISHER—In answer to your question I enclose a paper giving some details as to 50 ships in which the compass adjustment has been performed solely by means of the deflector.[1] This was 50 out of 180, the adjustment having been done by sights in the other 130 of that set. The proportion 50 out of 180 is, I should think, pretty nearly our average for deflector. Ships supplied with my compass are generally adjusted under weigh in any convenient place where there is sufficient sea room. In such places it is rarely the case that we have objects on shore of known bearings available for the adjustment. So, unless we get the sun, we always adjust by deflector; and if the sun fails during an adjustment, we finish it by the deflector. The adjustment by deflector is *quite thoroughly as trustworthy* as by sights of the sun or by shore bearings; and a ship never waits for sights after her compasses have been adjusted by deflector. In many cases, however, after the adjustment has been completed by deflector, bearings of the sun have been got, and the ship swung to verify the adjustment on all points. An error of 1° on some of the courses may be found; but for all practical purposes the compass is found correct.

[1] The deflector and the azimuth mirror are described in *Popular Lectures*, iii. p. 322 and p. 329.

Residual magnetism and magnetic sluggishness affect adjustment by sights quite as much as by deflector. All sizes and classes of ships are adjusted by the deflector, including large passenger steamers of the P. & O., of American lines, Cape mail ships, and cargo steamers, and yachts. The adjustment is done by day or night indifferently, in fog and in clear weather, or in a gale of wind or rain or snow, according as circumstances require; and immense savings of money are made for the shipowners and builders by avoidance of otherwise unnecessary detentions. The benefit to our Navy from the use of the deflector would be relatively much greater if the time ever came when it should be engaged in large action of war. However battered a ship may be, or however suddenly required out of port after repairs, an hour and a half would always quite suffice to make quite sure of her compass. A ship with us is never detained on account of weather for the adjustment of her compasses.—Yours very truly,

KELVIN.

*P.S.*—Our adjustments here are generally done by a compass adjuster, who learned from Mr. W Bottomley and myself in the course of a few days. A week's instruction and practice is quite enough to allow any capable adjuster to adjust by deflector in a thoroughly trustworthy manner. No navigating officer in the Navy would be unable to learn thoroughly in a few days; and, once learned, its use would let him always in any part of the world correct his compass with perfect confidence.

From the first, Lord Kelvin's compasses were manufactured by the firm of James White, Optician, of Glasgow, in which business he became the leading partner. As subsequent improvements were made in detail further patents were taken out to protect the invention. This, however, did not prevent imitations more or less close in design from being put on the market by rival firms of compass makers, involving the vexations of legal proceedings to protect the patent rights. In the principal action, *Thomson* v. *Moore*, his fundamental patent was triumphantly vindicated.

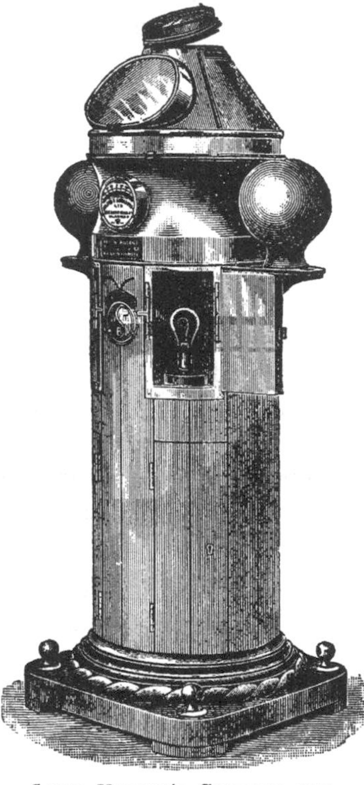

Lord Kelvin's Compass and Binnacle. External View.

Plate XII. depicts Lord Kelvin standing beside his latest compass. In this portrait he is actually taking a bearing by means of the azimuth mirror, and is shading the side light with his hand so as to get a reading of the scale. In the accompanying illustration is shown the form of the compass, with the two soft-iron corrector globes placed one on each side at the level of the card, and with lighting by an electric lamp beneath the compass bowl.

In recent years the Admiralty has shown preference for another sort of compass in which the whole card floats in liquid, and for ships of war it began to supersede Lord Kelvin's form after the end of 1906; the principles of correction remain, however, the same.

## THE SOUNDING MACHINE

To the Society of Telegraph Engineers in April 1874, Sir William Thomson communicated a paper on "Deep-Sea Sounding by Pianofore Wire," in which he described his first success on board the *Lalla Rookh* in June 1872, when he sounded to a depth of 2700 fathoms with a 30-pound sinker hung from a three-mile line made up of lengths of pianoforte steel wire spliced together. He recounted further soundings made on the *Hooper* in 1873 off the east and north coasts of Brazil by the same simple means, and described the device of an auxiliary hauling-in wheel which took off the chief strain from the sounding wheel, which must itself be very light. He pointed out the advantage presented by this plan over the old method of sounding with a hempen rope, its more rapid operation, its compactness, but, above all, the fact that it may be used for flying soundings, since in its use it is not necessary that the ship be stopped.

On November 11, 1875, Sir William delivered a public lecture[1] on "Navigation" in the City Hall at Glasgow. In this lecture he described various nautical instruments, his own deep-sea sounding apparatus, and Napier's pressure-log. He said little about his compass, but spoke of methods of navigation, and of the use of the globes. Referring to a terrestrial globe on the platform, he told his

[1] Published 1876 by W. Collins and Co. under the auspices of the Glasgow Science Lectures Association. Reprinted in *Popular Lectures and Addresses*, vol. iii. p. 1.

audience that he had registered a vow to lose no opportunity of speaking against the neglect of the Use of the Globes. A celestial and a terrestrial

THE SOUNDING MACHINE.

A, Rim holding wire coil. B, Winding drum.
C, Clamp for rim. G, Sinker enclosing gauge.

globe ought to be found in every school of every class. The pressure-log, an ingenious device for indicating the speed of a ship by causing the rush of water against the stern to force up a column of liquid in a tube, occupied attention for several years, and was the subject of lengthy correspondence between Thomson and Wm. Froude down to the end of 1878.

A sounding machine was put on board the Cunarder *Russia* in the summer of 1876, when Sir William was on his way to the Centennial Exhibition at Philadelphia, and he took flying soundings while going at 14 knots, and found bottom in 68 fathoms, quite unexpectedly in a place where 1900 fathoms were marked on the chart. At the British Association meeting the same autumn he expressed his opinion that the old system of sounding by hemp ropes had done its last work on board the *Challenger*.

In connection with this use of steel wire the story is told that Joule, visiting White's shop, found Sir William surrounded by coils of wire which he was inspecting, and inquiring their use was told they were pianoforte wire for sounding. "For sounding what note?" inquired Joule. "The deep C," was Sir William's reply.

Nevertheless, officialdom remained unconvinced and apathetic. In lectures to the United Service Institution in 1880, 1881, and 1884, he still continued to advocate the sounding machine and the compass, and in the end his persistency won the day. A letter which he addressed to Captain Wharton (afterwards Sir William J. L. Wharton, Admiral and Hydrographer) is a contribution to history.

*Nov.* 26/88.

DEAR CAPTAIN WHARTON—I send by book post, along with this, a volume of the *Society of Telegraph Engineers' Journal*, in which you will see on p. 207 details including latitude and longitude of my first pianoforte wire sounding.

On page 218 you will see some evidence of how anxious I was to give freely all the assistance I could to the Admiralty in respect to the deep-sea survey soundings. And on pages 223-228 you will see that I was also anxious to make known to the public the possible use of steel wire for flying soundings in ordinary navigation. My letter to *The Times* in May 1875 about that time, on the loss of the *Schiller* on the Scilly Islands, was another attempt to get the public to accept my navigational soundings gratis. When I saw nothing come of all this, I commenced a voyage to New York and back in the summer of 1876, making a serious effort to realise the thing for practical use myself, and at the end of 1876 I took my first sounding-machine patent. But you see from my beginning in June 1872, of which I wrote a particular account to the Hydrographer (Sir George Richards) in July, I all along for more than four years showed my hand and kept back nothing, in letters and in conversation and publications, both to the Admiralty and general public; and ultimately I was obliged to take a patent and work the thing out myself if the public was to get the benefit of it at all.

The enclosed little article from the *Phil. Mag.* for Nov. 1874 shows that I began similarly in respect to the compass. From that time till the end of the summer of 1875 I tried, on shore and afloat, to realize the desideratum described in the marked passages at the end of the article, and showed everything I was doing, from time to time, to Sir Frederick Evans. I was most anxious that the Admiralty should take it up, and I should have been most pleased to help freely in any way possible, and to work hard to bring about the result. I could not succeed in getting the slightest encouragement, or in raising any interest whatever in the subject; and at last in despair I resolved, about the end of 1875, to work it out myself and take a patent.

Forgive this long yarn, but I thought you might like to know that I have from the very first been always most anxious to do anything I possibly can in perfect unison with the Hydrographic Office.—Believe me, yours very truly,
WILLIAM THOMSON.

An important feature of the sounding machine was the dynamometric brake applied to the paying-out wheel, to regulate the movement; it had first been patented by Thomson in 1858 as an adjunct to machinery for laying submarine cables. The first form of depth-recorder attached to the sinker consisted of a glass tube closed at the top, containing air which was compressed as the water rose in it higher and higher under the increasing pressure; the amount of compression being registered by the use of chemicals, ferroprussiates at first, chromate of silver later, giving indications by change of colour. In the final form of depth-recorder the degree of compression was indicated by a piston which was forced into a cylinder containing air, and a marker pushed along the piston recorded the depth mechanically.

THE DEPTH-RECORDER.
C, Marker.
D, Piston.

Lord Kelvin used often to tell how in August 1877, returning from Madeira in the *Lalla Rookh* in thick weather, he steered by soundings taken every hour, right up to the Needles, having no sight of land except just a glimpse of the high cliffs east of Portland.

Writing in December 1892 to Captain Lecky, Lord Kelvin remarks :—

> What a sad disaster that was, of the *Roumania*. If they had used my sounding machine to get bottom when they came within 100 fathoms, and then kept it going once every half-hour, they never would have come into danger. On all those coasts (Finisterre to Cape St. Vincent) the danger line is outside 50 fathoms, and safety is to be had by keeping the machine going (two men at it) once an hour, or once every half-hour, or more frequently when circumstances require it. It is wonderful how slow ship-owners and ship-insurers, not to say ship-officers, are to see that they must not only have the machine on board, but must use it systematically, and that if they do so, such an accident as stranding on any of our European coasts, whether outside, on the Atlantic, or in the English or Irish Channels, or in the North Sea, could never happen to a non-disabled ship.

When the battleship *Montague* was lost in 1906 on Lundy Island, he was even more emphatic in denouncing the folly of captains who neglect such simple means of safety.

### Lighthouse Lights

The furnishing of lighthouses with distinctive lights that should prevent them from being mistaken one for another was a matter of keen interest to Sir William Thomson during these years. Flashing and revolving lights were indeed known; and some attempt had been made to give them a distinctive character by assigning to them slower or quicker periods of revolution. His principal contributions to the subject are to be found in an article on "Lighthouses of the Future" in *Good Words*, March 1873; a lecture on the Distinction of Lighthouses, delivered on October 24, 1877, to the Shipmasters' Society, reported in *The Times*, October 31; and a paper on "Lighthouse Characteristics," read at the Naval and Marine Exhibition at Glasgow, February 11, 1881, read also and discussed at the Society of Arts, March 2, 1881, and reprinted in Vol. III of his *Popular Lectures and Addresses*. In the first of these he drew attention to Babbage's suggestion of 1851 of an occulting light, and Captain Colomb's system of signalling by flashlights in the Navy, introduced about 1867. He now proposed that lighthouses should flash out, in short and long flashes, that is, in dots and dashes, some distinctive letter of the Morse code. The letter to Prof. Peirce (p. 659 above) gives a history of his efforts. The following letter, addressed to Dr. John Hopkinson, defines his plan:—

*Jany. 20th, '75.*

DEAR HOPKINSON—I have only this moment fallen upon a paper on group-flashing lights which you had kindly sent me, possibly during my absence in summer. I see on page 4 a mistake which I am sure you will gladly know to be a mistake. You say that the scheme proposed by myself has not received any practical recognition. It has been in action on the Holywood Bank Lighthouse since the first of Nov. The signal there adopted is two short eclipses and a long, making the letter U (dot, dot, dash) of the Morse alphabet. It is exceedingly well liked by the pilots and other practical men who have had experience of it.

Since the time of our last correspondence I have modified my plan from short and long flashes to short and long fixed lights.

I have also a very simple and durable mechanism for producing the same kind of distinction by longer and shorter extinction of gas. I have this going for the sake of illustration in my lecture room for several hours every evening at a window whence it is visible to a considerable part of Glasgow. It is probable it will soon be applied to the Cumbrae Light, Claugh Lighthouse, and a new lighthouse to be made on the Clyde on Roseneath patch. I do hope you will endeavour to move the Trinity Board to look into my proposal. If they will do so I feel perfectly confident that they will adopt it. You will see that it is very much simpler than the group-flashing lights which form the subject of your paper.

I am quite aware, however, of the value of the system of horizontal condensation used in the ordinary revolving lights and in the group-flashing lights. So of course I don't want to urge that they should be given up in every case. The great power of the ordinary revolving light will no doubt secure its being always used at some of the eclipses. My present proposal is to apply to every light which is at present a fixed white light, a simple eclipsing apparatus which shall make one, two, or three eclipses, each eclipse to be for a certain duration, about half a second, or three times as long as that for the dash, and the successive eclipses of the signal are to be separated by half a second interval. The light is to burn brightly and undisturbed for five or six seconds, then is to follow the successive eclipses distinguishing it, then the light is to burn brightly for five or six seconds again, and so on periodically.

The Holywood Bank Light was a fixed oil light distinguished by a red shade. The red shade is now removed, and the eclipses are made in an exceedingly simple manner by a revolving ring carrying screens. Nothing can possibly be more simple than the mechanism, or more easy to apprehend than the result. It can never be then for the reason you stated that my plan shall be not applied to every one of the present most important kinds of lighthouses. To promote its usefulness, however, it is absolutely essential that the period be diminished to a small fraction of what it ordinarily is. This has now, I am happy to say, been done for the Toward Point Light, in accordance with my recommendation. The period was as usual about a minute. It is now about ten seconds. Any one who has looked for one of the ordinary revolving lights at sea, will agree with me as to the painfully unsatisfactory character of the fifty seconds (or longer) waiting for a flash and the chance of missing it altogether, when it does come. A revolving light of a minute or of a longer period is in point of fact much less easily "picked up" in all circumstances of difficulty, particularly in bad weather, rainy and stormy, than a fixed light of much smaller intensity.

In his lecture of 1877 Sir William advocated the superiority of distinctions made by intermittent flashes, or by revolving beams of light rather than by colours, which might be deceptive in fog. He instanced his own Craigmoor Light in the Firth of Clyde, which flashed with a twice-repeated long-short signal, or — · — ·, which was, in fact, the telegraphic letter C. Also the Holywood Bank Light at Belfast Lough, which gave · · — (that is, dot, dot, dash) as its signal. In the 1881 lecture he classified the characteristic qualities of all lighthouses under three heads: Flashing Lights, Fixed Lights, and Occulting or Eclipsing Lights. He pointed out that a fixed light may be mistaken for a masthead light of a steamer; that the Admiralty

List for 1875 gives about 100 flashing lights around the British Islands, all alike except in that they vary in duration from one flash in four seconds to one flash in two minutes. A new and important departure had been made in 1875 by the introduction by Dr. John Hopkinson of a group-flashing light on the *Royal Sovereign* shoal, giving successive groups of flashes, with three flashes in each group. But three dots was simply letter S of the Morse code. On this Sir William expressed the hope that the ignorance of the Morse code by sailors[1] would prove no obstacle. "The great thing," he said, "is to find how lights may be most surely and inexpressively rendered distinctive, so that no sailor, educated or uneducated, highly intelligent or only intelligent enough to sail a collier through gales and snowstorms and fogs all winter, between Newcastle and Plymouth, may know each light as soon as he sees it without doubt or hesitation." Then he turned to statistics, and noting that, of the 623 lights in the British list, 490 were fixed, 112 flashing, and 21 occulting or eclipsing, he urged that the superior apparent brilliancy of a flashing light was dearly bought because of the great diminution of usefulness to the sailor in the intervening periods of darkness. What was gained in brilliancy was lost in time of visibility. He, therefore, advocated

[1] "Sailors, we may hope, are happily ignorant of this truth, otherwise the proverbial captain of the collier would be calling out to his chief officer, 'Bill, was that a S, or a I, or a H, or a E?' Bill, if he was well up in dramatic literature, would reply, 'Captain, them is things as no fellow can understand.'"

long periods of illumination broken up by short groups of eclipses or occultations on some characteristic plan. He had himself erected on the college tower of the University of Glasgow a trial eclipsing light giving the signal · — · (dot, dash, dot), or letter R of the Morse code, by eclipses; the group lasting seven seconds, with thirteen seconds of uninterrupted light between. Finally, he advocated the adoption of similar group signals for the sounding of sirens.

During the summer of 1880 he had much correspondence with the Commissioners of Irish Lights, the authorities of Trinity House, and the Clyde Commissioners, in which he urged the adoption of distinctive characteristics on this plan.

## Tides

Tidal phenomena often occupied Lord Kelvin, and he gave much thought to their elucidation. Perhaps the best account of his work on them is that to be found in the lecture which he delivered at Southampton in 1882, reprinted in Vol. III. of his *Popular Lectures*. He carefully pointed out the distinction between waves due to the action of wind, sudden "tidal waves" caused by earthquakes, and true tides due to the attraction of the sun and moon. He also discovered, from the reduction of tidal observations, a great swell of the ocean from one hemisphere to the other and back again, once a year, and explained it as the result of the sun's heat. He attached great importance to systematic measurement of the rise and fall of the sea at different ports; and himself devised a form of Tide-gauge of great simplicity. The theory of the tides had been worked out many years before by Laplace and Airy, who had shown that if the seemingly irregular variations of rise and fall were observed at any port for a considerable time, the observations could be reduced or analysed by very elaborate mathematical calculations into a number of constituent tides; and, if these have been found, it is possible to predict the height of the tide at any future time, for the same place, by computing the separate constituents and adding them together. Down to about 1876 the only means of analysing the various components which go to make up the complicated variations of rise and fall was a very laborious arithmetical process of harmonic analysis. This was investigated and carried out from 1867 onwards by a Tidal Committee of the British Association, formed at Sir William's instigation, and by the Indian Government for various ports in India. But in 1876 Sir William Thomson discovered a mechanical means of performing this, by the ingenious Tidal Harmonic Analyser, in which use was made of James Thomson's disk-globe-cylinder integrator, mentioned on p. 692. The first working model had five of the disk-globe-cylinders, and served for finding the terms of first and second orders only of the harmonic series. A larger instrument is used at the Meteorological Office. Besides these a synthetic instrument for use as a Tide-Predicter (now in use at the

National Physical Laboratory) was constructed, in which the various periodic components of rise and fall, as observed by the Gauge, and computed by the Harmonic Analyser, could, by an ingenious mechanism, be recombined for any future epoch. The entire tides of any port can be predicted by means of the instrument, which in twenty-four minutes draws the curve showing the rise and fall for a whole year to come. In 1876 Sir William showed the first model of this instrument, at the Exhibition of Scientific Apparatus at South Kensington, to Queen Victoria. The best account of this series of instruments is to be found in the *Proceedings of the Institution of Civil Engineers*, vol. lxv., March 1, 1881, together with the discussion which arose on the paper, and Sir William's reply to the pretensions of a claimant to the invention of the Predicter.

On several different occasions Lord Kelvin took an important part in committees to advise the Admiralty.

### Admiralty Committee (1871) upon the Designs of Ships of War

Early in September 1870 the public was shocked by the news that H.M.S. *Captain*, a newly-built turret ship of 6950 tons, with engines of 900 H.P., while on her experimental cruise, was caught in a gale in Vigo Bay, and went down with all hands, there being about 500 persons on board, including her designer, Captain Cowper-Coles. After the disaster the Admiralty appointed a Committee to examine the designs upon which the *Captain* and other recent ships of war had been constructed. The Marquis of Dufferin was chairman, and the majority of the Committee consisted of experienced naval officers. But a notable departure was made in adding to the Committee several leading scientific men, including Sir William Thomson, Professor Rankine, Mr. W. Froude, and Mr. G. P. Bidder. The instructions to the Committee on their appointment in January 1871, stated that the Admiralty sought a professional and scientific opinion upon the designs of ships of various types. The inquiry touched on the faulty points in the design of the *Captain*, the relative stability and efficiency of the *Monarch*, and the improvements possible in ships of other classes. At that date the first-class ships still carried masts and sails, and one of the contentious points of the inquiry was whether the *Monarch*, a two-turret armour-clad ship carrying full sail equipment, could be really efficient as a fighting engine if built also to have real efficiency under sail. The Committee met fortnightly through the spring of 1871, and made its final Report in July. Several definite scientific questions were remitted to a Scientific Sub-Committee under Thomson as chairman. Rankine made a Report on the stability of ships under canvas, demonstrating that the *Monarch* had a limiting safe angle of heeling in a squall of 27½° as compared with the 9° of the *Captain*. Thomson for the Sub-

Committee drew up a Report on the stability of unmasted vessels such as the *Thunderer* and the *Devastation*, based upon an investigation by Rankine and himself in February. He also drew up Reports on the stability and structural strength of ships of the *Cyclops* class. The minutes of evidence show that Thomson also gave evidence himself upon the inclination producible by wind-pressures and upon exceptionally steep waves. The work of this Commission was very heavy, and its conclusions important, as it had to lay down rules to govern the construction of future vessels. Even definitions and expressions which have since become fundamental (*e.g.* "righting-moment") originated in the course of its work. Sir William used to travel up to town every alternate week on Wednesdays, for four days' absence from his University duties, part of which were delegated to James T. Bottomley at this time.

### Admiralty Committee (1904-5) on Types of Warships

In December 1904 the Lords Commissioners of the Admiralty decided to appoint a Committee to review the types of fighting ships which the Board of Admiralty proposed to adopt for the British Navy. This Committee, presided over by Admiral of the Fleet Lord Fisher, comprised seven distinguished naval officers, together with seven civilian members, of whom Lord Kelvin was one. After considering various general questions such as size of guns, uniformity of equipment, speed, number, spacing and position of guns, the Committee agreed to an interim report proposing outline designs for two special types, a 21-knot battleship carrying ten 12-inch guns, and a 25-knot armoured cruiser with eight 12-inch guns. Lord Kelvin took a continuous and energetic interest in all these steps, frequently corresponding with the President explaining his views. Amongst the great changes, steam-turbines were recommended for propulsion. For the details of design sub-committees of the official members were formed, the non-official members being asked to co-operate in the final conclusions. The Committee had before it a large number of preliminary designs, from which it selected a smaller number for detailed consideration by aid of fuller drawings and models. Lord Kelvin put forward certain proposals as to the under-water form of hull, based upon a suggestion which he had first made (p. 751) to the Institution of Mechanical Engineers in 1887, for effecting a possible addition to the displacement or carrying power of the ship without adding to the wave-resistance or reducing the speed. The Hon. C. A. Parsons was called in to counsel as to the proposed adoption of turbines. Lord Kelvin again took a great personal interest in this question, and towards the end of January 1905 further advised the Committee as to certain technical difficulties, as to arrangements for the compasses and their protection, and as to the due provision and proper placing of the instruments for control of fire, range, and communication of orders. He took a keen

LORD KELVIN'S LECTURE-ROOM IN THE UNIVERSITY OF GLASGOW, AS ARRANGED FOR A LECTURE ON GYROSTATS.

LORD KELVIN'S LECTURE-ROOM IN THE UNIVERSITY OF GLASGOW, AS ARRANGED FOR A LECTURE ON ACOUSTICS.

interest in the operations of the Committee, and his scientific knowledge, no less than his experience in seamanship, was of distinct value in its deliberations. He was disposed on the whole towards favouring the swift armoured cruiser, the *Indomitable* type, rather than the heavier battleship, the *Dreadnought* type, but did not dissent from the unanimous report of the Committee for the adoption of both types. As a result the British Navy found itself equipped with the famous *Dreadnought* (1907), and the even more remarkable *Indomitable* (1908), types which are now being fast multiplied in naval construction.

He also advised the Admiralty Committee on Education of Naval Executive Officers in 1885; and in 1873, at the request of the Plimsoll Parliamentary Committee, he drew up a memorandum on the question of insecure navigation.

The tables for facilitating Sumner's method of finding the position of a ship at sea, which Lord Kelvin drew up, prove the interest he took in seamanship. The present highest authority in the British Navy has assured the writer that he considers Lord Kelvin as the man who has done by far the most for the advancement of navigation in our time. "I don't know," said a sailor in the distant seas of the East, "who this Thomson may be, but every sailor ought to pray for him every night." And the gratitude of the whole world follows him for what he did.

# CHAPTER XVIII

## GYROSTATICS AND WAVE MOTION

FAMILIAR as are the phenomena of rotating bodies in such instances as the fly-wheel and the spinning-top, it is surprising how backward has been the teaching of the dynamics of rotation. Very few of the treatises on mechanics in vogue to-day have anything to say about the all-important principle of the persistence of the axis of rotation in steady motion. Yet it is this persistence which enables the spinning-top to stand upright; which gives the rifle-bullet its deadly fixity of direction; which keeps the earth's axis pointing always to the celestial pole. That scientific toy, the gyroscope, a heavy top spinning between bearings carried on a gimbal ring, will teach more of this subject in an hour than pages of disquisition. It became widely known in the 'fifties, after the researches of Foucault; but its principle was known and taught at Cambridge by Airy and by Earnshaw long before. The strange stiffness shown by the spinning disc; the singular way in which its axis sidles off to right or left when a force is applied to turn it upward or downward; the seeming determination with which it points its axis always in the same direction while one walks about or turns oneself when carrying it, are all consequences of this principle of persistence of axis of spin.

When still an undergraduate, as narrated on p. 79, Lord Kelvin and his friend Hugh Blackburn were struck by the problems of rotation, and at the time when they were reading together for the great mathematical test of the wranglership, they spent a week of their seaside reading-party in spinning tops and rounded stones picked up on Cromer beach. Set up a peg-top on its point, or try to stand an egg-shaped stone upon its end—unless it is spinning it at once falls over. But if it is spinning it will stand up, and will resist being pushed over. Nay, if you spin the egg-shaped stone on its side, it will of itself rise up and spin on its end. Why this acquired stiffness? Why does a body when in rapid movement acquire an apparent power of resisting disturbance which when still it does not possess? All through his scientific life Lord Kelvin was pursued by importunate problems that depend on this apparent stiffness acquired by rotating bodies in virtue of their rotation. Even at the age of fifteen he was familiar with the mathematics of rotating bodies, as his Essay on the Figure of the Earth (see p. 10) shows; and Poinsot's theory of couples had been brought to his notice by Nichol. In 1846 he wrote a mathematical note on the axis of rotation of a rigid body (later embodied in Thomson and Tait, §§ 282-284). He was therefore certainly familiar with d'Alembert's theory of precession. The earth's axis, though in general it points in a constant direction, slowly shifts around in the course of the centuries, exactly as the axis of an obliquely spinning top slowly "precesses" in obedience to the downward force of gravity; though in the earth's case it is the action of sun and moon upon its bulging equatorial matter which causes its axis of rotation to "precess." Thomson, with that instinctive faculty that arose from familiarity with the thing itself, thought out afresh the theory of precessional forces, and refashioned the dynamical formulæ that express the couples which determine the movements of the axis of spinning bodies. To aid his thought he devised new forms of spinning-tops and gyrostatic models. It was one of these which he showed in 1863 to Helmholtz in operation —with disastrous results to his hat—see p. 430. The generic name which Thomson gave to these most instructive toys was the *gyrostat*. A gyrostat is simply a heavy, well-balanced, finely pivoted fly-wheel, like the earlier gyroscope, but enclosed in a hollow metal box or case, within which it is spun. When so enclosed, and suitably mounted, a variety of experiments can be shown that are not readily made with the older gimbal-mounted form of instrument. In one of these forms Thomson mounted the gyrostat within a concentric polygonal rim, so that it could stand on edge, or skate about on a sheet of glass. Others were mounted on stilts, or slung on suspensions to afford various degrees of freedom.

Sundry experiments with them were described in *Nature* on February 1, 1877, and are mentioned in a long paragraph added to Vol. I. Part I. of Thomson and Tait's *Treatise*, § 345, when revised for the second edition in 1879. Piazzi Smyth had twenty years before proposed the use of a gyroscope to steady a platform for astronomical observations on board ship; but Thomson now demonstrated that the gyrostat when spinning not only tended to preserve an invariable axis of spin, but that in this tendency it showed a remarkable quasi-elasticity. If a blow were given to its frame, it quivered to and fro like a stiff spring, and returned to its own position. Thenceforward *kinetic stability*, the stability of position due to the reaction of the moving mass, became a prominent quality to be investigated; and *gyrostatic domination*, the domination exerted upon a system by any revolving masses within it, assumed a new importance. A favourite experiment shown to visitors in the laboratory was made with an enclosed gyrostat, with its axis of rotation vertical, pivoted by its case on horizontal trunnion bearings to a square frame of wood, by which it could be carried about with the frame level. If the experimenter holding this in his hands walks in a straight line, or turns round in the same direction as that in which the fly-wheel is spinning, nothing happens; but should he turn round the other way, the gyrostat immediately turns head-over-heels on its pivots, as if bewitched. Many other curious experiments were shown, in which the gyrostats—

visible or concealed—were made to stand balanced in strange positions, or nodding their axes while sweeping round in precession, thus imitating the astronomical phenomenon of nutation. These strange effects, all capable of being brought under mathematical calculation, had a fascination for Thomson, who saturated himself with the study of their perverse ways.

Noting the philosophical bearing which this question had upon the physical condition of the earth, he constructed liquid gyrostats, ellipsoidal metallic shells, both oblate and prolate, filled with water. The spinning of the oblate ellipsoid about its axis of symmetry was found to be stable—that of the prolate unstable. A hard-boiled egg, if spun on its end, would "sleep" like a top. If spun on its side it would rise up and spin on end. But an unboiled egg could by no means be induced to spin on its end. This was a favourite class illustration. On one occasion a mischievous student slipped furtively into the lecture theatre beforehand, and substituted two fresh eggs for the pair, boiled and unboiled, which were provided for the lecture. When it came to the moment the conspirators waited breathlessly the result of their deed. "Neither of them boiled" was Sir William's unhesitating verdict, as he tried to spin them.

Another of Lord Kelvin's tests for an egg is the following:—Spin the egg on its side. Then lay the finger on it for an instant, just to stop it. On lifting the finger, if the egg is boiled it remains still, if unboiled it goes on turning; for though the shell was stopped the liquid inside was still in motion, and moves the shell when the finger is removed.

In 1875 he gave to the Edinburgh Royal Society the theory of the spinning-top to account for its rising[1] from the oblique position to the vertical attitude in which it "sleeps." His explanation, that this action depended on friction at the rounded point of the peg, had been anticipated, however, by Jellett in 1872.

But it was not spinning-tops alone that exhibited kinetic rigidity.

The following story was told by James White, the optician, within a day of its occurrence in 1868:—

Thomson had been to Sturrock's, the fashionable hairdresser's, in Buchanan Street, Glasgow, some two doors from White's shop. At Sturrock's he had seen an endless revolving indiarubber rope or chain used for driving a rotatory hair-brush. He came excitedly into White's shop as soon as he was free. "White, come with me." And, without a word of explanation, he hurried White to Sturrock's and had the endless chain put into motion, and showed White, by striking it, the rigidity of form which it had acquired in virtue of its motion. "There! get one put up in the class-room at College like that, at once!"—and off he went. In

1 "Hurry on the precession, and the body rises in opposition to gravity"; "delay the precession, and the body falls, as gravity would make it do if it were not spinning," were two of Thomson's aphorisms. The effect of friction at the peg is to hurry on the precession.

a few days the chain was put up; and his senior students remembered the delight with which he showed it to them, and shortly after to Tait who had come over from Edinburgh. He would hit the lower end of the loop vigorously with a stick, and thus force it into various almost fantastic shapes or kinks, which would persist for some minutes. Thomson had, of course, long been keen on problems of kinetic stability, so the chain appealed to him. He devised numerous other experiments on analogous cases of the quasi-rigidity of motion. A circular disk of thin paper spun rapidly around its centre becomes so stiff as to resist a blow of the fist, and resounds when struck. Professor Perry, a student of Thomson's in the 'seventies, has described most of these phenomena in his admirable lecture[1] on Spinning-Tops.

It dawned upon Sir William Thomson that in the quasi-rigidity of spinning bodies there was a simulation of that property of matter known as elasticity, and that it might even be possible to explain the phenomena of elasticity itself by supposing the molecules, or some of them, to be in a state of rotation. He imagined a chain of gyrostats all spinning, and connected together by links; and he laid before more than one of the learned societies[2] mathematical discussions of the properties of a stretched string of gyrostats, along which string a wave might be propagated exactly

[1] Published, 1890, by the Society for Promoting Christian Knowledge.

[2] *Lond. Math. Soc. Proc.* vi. p. 190, April 1875; *Proc. Roy. Soc. Edin.* viii. p. 521, April 1875.

as along a stretched elastic cord. Then it became evident that if a revolving fly-wheel could so behave, a whirling vortex filament in a frictionless fluid would also possess a quasi-elasticity, because it is in rotation. Indeed, the rebound which he had observed with smoke-rings showed such to be the case. Part of his researches on vortex-statics was concerned with this idea. Again, the kinetic theory of gases, in which the "spring of the air," discovered two centuries before by Robert Boyle, was explained as the result of innumerable multitudes of free-flying molecules dashing about in space and colliding with one another, had been demonstrated by Joule, and Clausius, and Maxwell to be profoundly true. If it were true, however, the molecules or atoms must themselves be elastic bodies, and not the infinitely hard round objects which before they had been regarded.

In November 1880 Sir William consented to give another discourse at the Royal Institution. When asked by Dr. De la Rue what the subject was to be, he replied:—

MY DEAR SIR—The title of my lecture might be "On Elasticity viewed as a possible Mode of Motion"—unless Tyndall has patented the expression, and refuses to grant me a license!—Believe me, yours very truly,

WM. THOMSON.

When the discourse was given on March 4, 1881, his opening sentence was:—

The mere title of Dr. Tyndall's beautiful *Heat, a Mode of Motion*, is a lesson of truth which has manifested far and wide through the world one of the greatest discoveries of modern philosophy. I have always admired it; I have coveted it for Elasticity; and now by kind permission of its inventor, I have borrowed it for this evening's discourse.

He went on to explain that the hypothesis would not be raised into a certainty until the elasticity of the molecules themselves should have been demonstrated to be due to motion; and all he claimed now was to point out a possibility. He referred to the well-known examples of the elastic-like firmness of spinning-tops, of rolling hoops, of bicycles in motion, and of smoke-rings and gyrostats. He showed how a flexible endless chain of steel links running round a revolving pulley, and then suddenly caused to jump on to a platform, stood up like a hoop and rolled along the platform until its motion was lost by impact and friction. Also a whirling mass of water in a rotating vase was shown to possess a jelly-like stiffness. "May not the elasticity of every ultimate atom of matter be thus explained?" he asked; but checked the speculation by adding: "But this kinetic theory of matter is a dream, and can be nothing else, until it can explain chemical affinity, electricity, magnetism, gravitation, and the inertia of masses (that is crowds) of vortices." There were difficulties; but he ended optimistically: "Belief that no other theory of matter is possible is the only ground for anticipating that there is in store for the world another beautiful book to be called *Elasticity, a Mode of Motion*."

Two years later, in a paper given to the Royal Society of Edinburgh, he showed, by a mathematical investigation of a gyrostatically dominated combination, that any ideal system of material particles, acting on one another mutually through massless connecting springs, may be perfectly imitated in a model consisting of rigid links joined together and having rapidly rotating fly-wheels pivoted on some or on all of the links. And so he produced out of matter possessing rigidity, but absolutely devoid of elasticity, a perfect model of a spring-balance.

Not only this: for he saw also that by gyrostatic models he could imitate the phenomena of magnetism, and could design even a gyrostatic compass which, unlike the magnetic one, should direct itself always toward the true north pole of the rotating globe of the earth. A favourite experiment was to stand a gyrostat obliquely on his finger, place another gyrostat on the top of it, slanting at a different angle, and a third gyrostat on the top of the second, so making a crooked chain of them. Nor did he stop here; for, as we shall see later, he found it possible, by joining up gyrostatic elements into a network or tissue, to imitate the properties of the ether of space in transmitting transversal waves.

Problems connected with the propagation of wave-motion exercised over Thomson's mind a never-ending fascination. They appealed to him not only by reason of his instinct for mathematical analysis and as problems of hydrodynamics, but by virtue also of his passion for seamanship and all that pertained to the sea.

In discussing the tides, we have seen how Sir William Thomson at the outset distinguished between the periodic movements due to the attraction of the sun and moon, and those—generally much more rapid—due to the driving action of wind. He distinguished no less carefully between the waves in which the restoring force was the weight of the water itself, and those lesser crispations in which the restoring force was the surface-tension of the water-surface; for these last he termed *ripples*. The distinction is well brought out in a communication on Ripples and Waves, printed in *Nature*, November 2, 1871; a result of the first-season cruise of the *Lalla Rookh* (see p. 611). The principal passages in it are here reproduced.

I have often had in my mind the question of waves as affected by gravity and cohesion jointly, but have only been led to bring it to an issue by a curious phenomenon which we noticed at the surface of the water round a fishing-line one day, slipping out of Oban (becalmed) at about half a mile an hour through the water.

What we first noticed was an extremely fine and numerous set of short waves preceding the solid [the line], much longer waves following it right in the rear, and oblique waves streaming off in the usual manner at a definite angle on each side, into which the waves in front and the waves in the rear merged so as to form a beautiful and symmetrical pattern, the tactics of which I have not been able thoroughly to follow hitherto. The diameter of the "solid" (that is to say, the fishing-line) being only two or three millimetres, and the longest of the observed waves five or six centimetres, it is clear that the waves, at distances in any directions from the solid exceeding fifteen or twenty centimetres, were sensibly unforced (that is to say, moving each as if it were part of an endless series of uniform parallel waves undisturbed by any solid). Hence the waves seen right in front and right in rear showed (what became immediately an obvious result of theory) two different wave-lengths with the same velocity of propagation. The speed of the vessel falling off, the waves in rear of the fishing-line became shorter and those in advance longer, showing another obvious result of theory. The speed further diminishing, one set of waves shorten and the other lengthen, until they become, as nearly as I can distinguish, of the same lengths, and the oblique lines of waves in the intervening pattern open out to an obtuse angle of nearly two right angles. . . . The speed of the solid which gives the uniform system of parallel waves before and behind it was clearly an absolute minimum wave-velocity, being the limiting velocity to which the common velocity of the larger waves in rear and the shorter waves in front was reduced by shortening the former and lengthening the latter to an equality of wave-length.

Taking 074 of a gramme weight per centimetre of breadth for the cohesive tension of a water-surface, . . . I find, for the minimum velocity of propagation of surface waves, 23 centimetres per second. . . .

About three weeks later, being becalmed in the Sound of Mull, I had an excellent opportunity, with the assistance of Prof. Helmholtz and my brother from Belfast, of determining by observation the minimum wave velocity with some approach to accuracy. . . . [The mean result of four determinations gave 23·22 cms. per second; and 1·7 cm. as the corresponding wave-length.]

I propose, if you approve, to call *ripples*, waves of lengths less than this critical value, and generally to restrict the name *waves* to waves of lengths exceeding it. If this distinction is adopted, ripples will be undulations such that the shorter the length from crest to crest, the greater the velocity of propagation; while for waves the greater the length the greater the velocity of propagation. The motive force of ripples is chiefly cohesion; that of waves chiefly gravity. In ripples of lengths less than half a centimetre the influence of gravity is scarcely sensible; cohesion is nearly paramount. Thus the motive of ripples is the same as that of the trembling of a dew-drop and of the spherical tendency of a drop of rain or spherule of mist. In all waves of lengths exceeding five or six centimetres the effect of cohesion is practically insensible, and the moving force may be regarded as wholly gravity.

Lord Kelvin's scattered papers on Waves have never been reprinted. If collected, along with those on Vortex Motion, they would form an appropriate fourth volume of his *Mathematical and Physical Papers*. In 1886 he gave, at the British Association meeting at Birmingham, four papers on wave subjects: on Stationary Waves in Flowing Water; on the Artificial Production of a Standing Bore; on the Velocity of Advance of a Natural Bore; and a graphic illustration of deep-sea waves. In the same year he discoursed at Edinburgh on waves produced by a ship advancing uniformly into smooth water, and on ring-waves. To the Royal Society, in 1887, he gave a paper on the Waves produced by a Single Impulse. On August 3, 1887, when the Institution of Mechanical Engineers met in Edinburgh, he delivered an evening lecture on Ship-Waves. This is reprinted in Vol. III. of his *Popular Lectures*.

He began by defining waves, generally, as "a progression through matter of a state of motion." After further definitions he remarked that waves of water have this great distinction from waves of light or waves of sound, that they are manifested at the surface or termination of the medium or substance whose motion constitutes the wave.

It is with waves of water that we are concerned to-night; and of all the beautiful forms of water-waves, that of ship-waves is perhaps the most beautiful, if we can compare the beauty of such beautiful things. The subject of ship-waves is certainly one of the most interesting in mathematical science. It possesses a special and intense interest, partly from the difficulty of the problem, and partly from the peculiar complexity of the circumstances concerned in the configuration of waves.

Before one can follow out the complicated pattern of waves which is seen in the wake of a ship travelling through open water, it is necessary to study the simpler case of waves travelling along a canal. Of these a beautiful investigation was made in 1834 by Scott Russell, who found that, while short waves travel slower than long ones, there is a particular or critical speed for "long waves," that is, waves whose length from crest to crest is many times greater than the depth of the canal. For these the law holds good—supposing water to be perfectly free from viscosity—that their speed is equal to that which a body acquires in falling through a height equal to half the depth of the canal. Thus in a canal 8 feet deep the natural velocity of the "long wave" is 16 feet per second. Now if a boat is dragged along a canal at a speed slower than the critical velocity there is always seen a procession of waves behind it. In fact, if water were a perfect fluid, then a boat dragged along at any velocity less than the natural speed of the long wave in the canal would be followed by a train of waves of so much shorter length that their velocity of travelling is the same as the speed of the boat; and the boat will seem to ride behind the crest of the first wave. These waves wash against the banks of the canal. To produce and maintain them consumes power. Much of the power required to drag a boat along a canal is wasted thus in wave-making. Moreover, water is not a perfect fluid: it possesses a certain amount of viscidity, though less than that of treacle or thick oil. The effect of this viscosity is to cause the cessation of the hindmost waves of the procession, which die out by fluid friction. The velocity of the front of a procession of waves is different, as Stokes and Osborne Reynolds have shown, from the velocity of progress of a wave itself. If the depth be greater than about three-quarters of a wave-length, the rate of progression of the rear of the procession of waves will be half the speed of the boat. Now comes a most wonderful result: if the speed of the canal-boat be more rapid than the speed of the "long wave," it cannot leave behind it a procession of waves; it travels along on the top of a sort of hillock or hump of water which travels along with the boat. As it makes no waves, the wave-making resistance is gone, and it is far easier therefore to draw the boat *fast* than to draw it slower with a whole procession of waves following. If a boat is drawn slowly it makes waves; draw it faster, it makes more waves and requires more effort; draw it still faster, so as to reach the critical velocity. Once that speed is reached, away goes the boat merrily, leaving no wave behind it. This was discovered accidentally early in the nineteenth century on the Glasgow and Ardrossan Canal, by a spirited horse which took fright and ran off, dragging the boat with it, and led to the establishment of a system of express "fly-boats" drawn by a pair of galloping horses. Of this singular circumstance Sir William Thomson found a theoretical confirmation in his mathematical investigations of 1886-87. Applying what is known of canal-waves to the movement of ship-waves in the open sea, Sir William Thomson then showed that the beautiful diverging pattern is comprised between two straight lines drawn from the bow of the ship backwards, with an angle of 38° 56′ between them, with a definite echelon of steep waves across the track following the ship. All this and more he described in the lecture at Edinburgh. He concluded with a suggestion that since wave-resistance depends almost entirely on action at the surface of the water, it might be possible for swift ships to get high speeds of 18 or 20 knots by designing their hulls with a form swelled out below—say with breadth of beam five feet more below the water-line than at the water-line—instead of with vertical sides.

Even the veriest trifles about waves or wave-making claimed Sir William Thomson's attention.

He wrote once to Lord Rayleigh this paragraph :—

> We saw a fine mode of generating a regular procession of periodic ring-waves in water. We were watching a set of ducks yesterday on our arrival here [Netherhall], and the beautiful echelons of waves which they made when swimming. They seemed pleased at being noticed, and one after another stopped swimming and called out to us in their own language; each about twenty very regular periods. By looking at the ring-waves we could count the number of quacks we had heard.

In his last years Lord Kelvin communicated several mathematical papers to the Edinburgh Royal Society in continuation of these researches. They were on the Front and Rear of a Free Procession of Waves in Deep Water (1904); on Deep-Water Ship-Waves (1905); and Initiation of Deep-Sea Waves of Three Classes (1906).

# CHAPTER XIX

## IN THE 'EIGHTIES

ARTIFICIAL though the division of time into decades may be, there are instances where it well fits the march of affairs. If in the development of the electrical industry the 'sixties were notable for submarine telegraphy, and the 'seventies for the siphon-recorder and the telephone, the 'eighties were no less so for the outburst of activity in electric lighting. So soon as sound engineering construction had been applied to the magneto-electric generator, and the mechanical production of electric currents had been thus assured, electric lighting was bound to progress. Already the arc lamp was in limited use. The year 1878 had seen the Avenue de l'Opéra in Paris lit with Jablochkoff's now historic "candles"; and a little later Swan and Edison had brought the carbon glow-lamp into precarious existence. The demand was springing up for improved generators, for better lamps, for switches and safety devices, and, above all, for measuring instruments. From this development it was impossible for Lord Kelvin to keep aloof. His advice and counsel were sought by capitalists who saw in electric-lighting schemes a possible development of commercial importance. With his engineering instincts he could not but busy himself with these things, and the years from 1880 to 1893 are marked by a great devotion to the invention of patented devices. His compass and sounding machine were barely completed, and he was still patenting improvements in them at intervals down to 1890. The long series of his patents for electric measuring instruments, gauges, and meters, begins in 1881, and continued till 1896. Electric lamps he touched only as regards improved means of suspension, in 1884. Dynamo-electric generators claimed his attention from 1881 to 1884; but he made no radical departure. A dynamo with a disk-like ironless armature, and another with an armature of copper bars like the bars of a squirrel-cage, which he designed, attained to no success. He also proposed rotating copper collectors to replace the spring brushes of the ordinary dynamo; but they never came into commercial use. The zigzag winding for alternators, associated usually with the name of Ferranti, he invented independently in 1881; and the Ferranti machines were manufactured under agreement with him with a guarantee of a minimum royalty of £500 per annum. His great ingenuity, command of mechanical devices, and wide experience in precise electrical measurement, combined to bring him success with his instruments for measuring electric currents and potentials. These instruments, manufactured by the firm of James White and Co., in which he was now the chief partner, were not, however, developed without immense labour, and passed through innumerable changes before reaching commercial form. He thought in steel and brass, and must continually have the incipient instrument before him when working out the details; alterations were then made and put to the test. The tests suggested some further change or improvement, and thus many visits must be made to the workshops, and many experimental trials in the laboratory, so that the development, if sure, was both slow and costly.

The first instruments suitable for use in connection with electric lighting were his graded potential galvanometer and his graded current galvanometer. The one was a portable volt-meter, the other a portable ampere-meter, both being developed from the tangent galvanometer modified for new service. They were followed by a magneto-static current-meter and by marine voltmeters and ampere-meters, and by ampere-gauges for use on switchboards. Of the latter a whole series had to be devised to measure currents from ¼ ampere up to 6000 amperes. All these were electromagnetic instruments.

Another series for measuring potentials was based on electrostatic principles, and may be regarded as being developed from the quadrant electrometer of 1860. By the multicellular construction instruments were made, some of them reading as low as 40 volts, others as high as 1600 volts. For higher voltages other types were used; and electrostatic balances, based on the absolute attracting electrometer, were made to read up to 50,000 or even 100,000 volts. In these instruments nothing is more remarkable than the confident way in which, defying the traditions of the ordinary instrument-maker, he insisted on introducing kinematic principles—geometric slides, for example, to give the right number of degrees of constraint. The "hole-slot-plane" arrangement of support, in place of the old three-pointed levelling screws standing on a sole-plate, is due to Sir William Thomson. "His inventions," wrote Professor FitzGerald, "are the direct outcome of the most advanced theory. He is a living example of the necessity for theory, in order to advance practice as much as possible."

A third and later series of instruments is that comprising his standard electric balances, based on the electrodynamic principle of weighing the forces between movable and fixed portions of an electric circuit. Finding in all the earlier electromagnetic instruments inherent limitations as to range of use and as to accuracy within that range, Sir William fell back upon the fundamental discovery of Ampère, and set himself to devise apparatus that should require neither springs nor magnets, and in which gravity should be the controlling force. Furthermore, they could be rendered free from all perturbations due to the neighbourhood of any magnetic apparatus, or to local variations of terrestrial magnetism by astatic duplication of parts. Accordingly, in each balance two ring-coils were set at the ends of a balance-arm, each such movable ring-coil being situated between an upper and a lower fixed coil. The tendency of the current to depress one end of the arm and raise the other could then be counterpoised by weights resting on the balance-arm, or moved along an attached steel-yard. Immense ingenuity was shown in the design of details, in particular in the device for shifting the weights along the graduated arm by a light silk cord running through the case of the instrument. Another notable device was the use of suspending ligaments, consisting of a vast number of very thin parallel wires, for supporting the trunnions of the balance-arm, instead of pivots or knife-edges. It was necessary to provide means of carrying into the movable ring-coils the strong electric currents to be measured, and this would have been quite impossible with pivots or knife-edges, or with any previously known means of flexible connection. The designing of recording meters for electric supply extended over a long period. In the earliest of these the integrator of his brother James was applied, but gave place eventually to simpler devices.

Work for pure science was never dropped during these years of devotion to practical inventions. In the spring of 1880, under the impulse of writing the *Encyclopædia* article on Heat, various researches on steam-pressure thermometry were undertaken, and their results communicated to the Royal Society of Edinburgh. He lectured in March on Magnetism, at Largs, and in May on the Compass, to the United Service Institution. Lady Thomson wrote in April to Professor George Darwin, that they proposed to spend June, July, and August in yachting, and added, "We are in great spirits over the elections; I don't know if you are." The general election of that year had brought a sweeping victory for Mr. Gladstone. In the middle of May Sir William and Lady Thomson paid a visit to Charles Darwin at Down. No record has been preserved of the occasion; but the two men had a good deal of talk, and parted with mutual respect and admiration, though their whole interests lay in different directions.

A letter of Professor James Thomson to Professor George Darwin, to which Sir William added the postscripts here printed in italics, tells of doings on the yacht, whither, after a visit to London to receive the Albert medal[1] of the Society of Arts, he had retreated to "get away from all animate and inanimate bores."

ROSHVEN, FORT WILLIAM,
*Friday*, 25*th June* 1880.

MY DEAR MR. DARWIN—I am along with my wife on a yachting cruise with my brother and Lady Thomson along the West Coast of Scotland, going northwards. We have to-day landed at Prof. Blackburn's place, Roshven on Loch Ailort, having passed Ardnamurchan Point

[1] The terms of this award were: "On account of signal service rendered to Arts, Manufactures and Commerce by his electrical researches, especially with reference to the transmission of telegraphic messages over ocean cables."

yesterday afternoon and during the night in fog and calm, which made us have to lie to during most of the night. My brother has his newest sounding machine and depth-gauge on trial for the first time in their latest forms of development *by natural selection*, and he was quite in earnest feeling his way along the bottom yesterday in the fog by taking soundings and comparing his results with soundings marked on the chart, *in from 30 to 105 fathoms.*

On arrival here I received your letter about ripple marks. My daughter had opened it for me in Glasgow, and she sent me a copy of one of the two passages in Lyell which you referred to. It is from Lyell's *Elements* (page 19), and as it tells a good deal of detailed information and reasoning, I suppose it gives most of what you wanted to refer me to. She was not able to find anything about ripple marks in our edition of Lyell's *Principles.*

I write at present chiefly to acknowledge receipt of your letter. I have thought often a good deal about the ripple mark, but have never yet seen how to clear the matter up. I think perhaps I may be able to offer some one or two suggestions tending towards true theory; but I have not time just now. I would like if manageable to talk over the matter with my brother first, and perhaps I may write to you again before we sail from this place—Loch Ailort.

Mrs. Thomson *and Lady Thomson and Sir William Thomson* join me in kind regards, and I am, yours truly,

JAMES THOMSON.

George H. Darwin, Esq.

[P.S.]—*My wife adds that it has been such beautiful quiet sailing all the cruise that it would have suited you perfectly, and she wishes you had been with us. W. T.*

A month later Lady Thomson wrote from Cowes, asking Darwin to join them on the yacht. "Sir William is supremely happy, and appreciating the life afloat more than ever." After another month Sir William wrote:—

YACHT *LALLA ROOKH*,
SCILLY ISLES, *Aug.* 22/80.

DEAR DARWIN—You see how well the pen does. It has already written many letters and postcards, and has even superseded pencil occasionally in the green book with excellent effect. It should have thanked you before now, but that somewhat incessant cruising—Cherbourg, Torquay, Penzance, Scilly—has given it too much of a scramble for almost every postal opportunity since it came, to let it do so properly.

Don't forget to tell me how much fresh grant for lunar gravity and terrestrial palpitations is to be applied for at Swansea. We sail from St. Mary's Roads here to-morrow forenoon, hoping to arrive there by Wednesday morning, if not sooner. We have had some very good sailing, and if you had been with us I am sure you would have enjoyed our anchorage here, after "standing off and on" all night waiting for daylight, and the visit to Tresco Abbey and its plants of all climates. . .

The "awkward infinity" threatens quite a revolution in vortex motion (in fact a *revolution* where nothing of the kind, nothing but the laminar rotational movement, was even suspected before), and has been very bewildering. Only to-day I have begun to see light through it, and a great deal that it has led to. I hope to be able to say something about it in Section A at Swansea.

I am sorry you did not come with us to Cherbourg. The going and lying at anchor there, and coming thence to Torquay, would all have tested and fortified your sea-going qualities. The sounding machine and depth-gauge worked most satisfactorily in as rough circumstances as they are likely to meet with practically in ordinary use at sea.

Swansea was the scene of the British Association meeting in August 1880; and there Sir William read three papers on Vortex Motion, one on his Depth-gauge, one on the Sounding Machine, one on Contact Electricity, and one on the Critical Temperature of a Fluid.

In October he was back in the Clyde adjusting compasses on the Czar's yacht *Livadia*, and teaching the use of the sounding machine to the Russian sailors.

On December 18 he wrote to Dr. J. Hall Gladstone, urging him to stand as parliamentary candidate for Glasgow and Aberdeen Universities. "The tide has turned sufficiently to make a Liberal majority probable—you ought to have a very good chance. . . . I shall be very glad to look forward to your being our member if you will decide to stand, and I think you may do much good in many ways by being in the House."

Christmas that year was spent as usual at Knowsley, followed by family parties at Druid's Cross and Mere Old Hall, Knutsford.

The year 1881 was one of exceptional activity. The revision of "Thomson and Tait" for the new edition had long been on hand; and the editing of the second part of Vol. I. had been entrusted to Professor (now Sir) George Darwin. On January 17 Lady Thomson wrote to Darwin for Sir William, "who is sitting on the sofa deep in a new depth-gauge, simpler than any you have seen," sending him a message about the effect of continents in setting up harmonics of the second order in tidal equations.

Fleeming Jenkin wrote to him about a new heat-engine, which promised great things, and received the following reply :—

I am much interested in all you tell me of your heat-engine. I think something of the kind, whether for solid fuel or gas (did you see the report of Siemens' lecture here, according to which gas is to be the fuel?) is to be *the* thing of the future, the steam-engine a thing of the past. I am to be in Edin. for the second R.S. meeting of this month (we shall be staying at the Crum Browns' from the Sat. till the Tuesday) and I hope you will be able to show me your furnace and experiments then. I am afraid, however, that I could not join the proposed company. I am more than fully occupied with compass and sounding machine, etc., and I have spent so much money on them (including as a *small* item 3 compass patents for England and America, 2 sounding machines, etc. etc., and sounding machine patent for France, and Germany also) that I am not able to find £2000 for another undertaking, however promising, nor time and thoughts to give to an inventing company. I hope, however, that you will find all the assistance and support you want when you have so good a thing in hand, and such promising results to offer.

On February 11 Sir William lectured on Lighthouse Characteristics (see p. 727) at the Naval and Marine Exhibition at Glasgow. He was interesting himself deeply, too, in Mr. Anderson's Patent Law Reform Bill, then before the House of Commons, and wrote upon it in *The Times* of February 24. March 1 saw Sir William in London reading to the Civil Engineers the famous paper on his Tide-gauge, Tidal Harmonic Analyser and Tide Predicter described on pp. 730-31 above. The reading and discussion of this paper lasted three evenings. On March 4 Sir William described to the United Service Institution his navigational sounding machine and depth-recorder, and in the evening of the same day discoursed at the Royal Institution on Elasticity (see p. 743). Two letters to Helmholtz, who was coming to London to deliver the Faraday lecture to the Chemical Society, relate Thomson's next movements:—

12 GROSVENOR GARDENS, LONDON, S.W.,
*March* 13, 1881.

MY DEAR HELMHOLTZ—For the last three months I have been always intending to write and thank you for having sent me such an excellent *student* as Mr. Witkowsky, but time has flown too fast, and now I write urgently to ask you to come and see us in Glasgow on the occasion of your visit to England at the beginning of April. I am coming up from Glasgow to London on purpose to hear you lecture on the 5th of April, and to meet you at the Chemical Society's dinner on the 6th. I would remain in London till Saturday the 9th, when both you and I having heard Tyndall's lecture on the Friday evening, would be ready to leave London. So I do hope you will come to Glasgow with me on the Saturday and give us as long a visit as you can. Lady Thomson (who is with me this time in London, but will not make the journey again in April) begs me to say that she looks forward with much pleasure to making your acquaintance in Glasgow, and joins me in hoping that you will say Yes to my request.

On Monday (being a holiday) we might go to Largs, where we have built a country house, and for the rest of the week the laboratory will be at full work again. On Monday the 18th we might go to Edinburgh and see Tait, etc., and attend the meeting of the Royal Society there on the Monday evening.

On Wednesday I return to Glasgow, having been here for a fortnight on account of four lectures which I have been giving. Let me have a line addressed to me here, and above all things let it be Yes.—Believe me, always truly, WILLIAM THOMSON.

*March* 29/81, THE UNIVERSITY, GLASGOW.

DEAR HELMHOLTZ—I am very glad to have the prospect of seeing you here on the 15th of April. I hope it will turn out that your Cambridge engagement will not prevent you from coming. I now write on the part of Lady Thomson and myself to say that we hope Mrs. Helmholtz will come with you, and that you will stay with us as long as you can. Perhaps Mrs. Helmholtz would stay here while you and I go to Edinburgh for the R.S. meeting on Monday the 18th, and return here on Tuesday.

I am going to London on Thursday night, and to Portsmouth on Friday evening. I shall be glad to hear from you when we may expect to see you. If you write by return of post address me here; if any day later this week address care of Capt. Fisher, R.N., Crescent Lodge, Southsea, Portsmouth.

I am to stay with the Spottiswoodes in London, where I shall be on Monday the 4th of April.

Excuse haste, and believe me, yours very truly,

WILLIAM THOMSON.

Two days later, Lady Thomson wrote to Darwin that Sir William had had a tide-gauge set up in the Clyde, only about fifteen minutes' walk from the University—a constant source of interest. "He is very deep in electric light too just now—also telephone companies. . . . You would meet Prof. Helmholtz here about 15th or 16th if you come."

Again, a note to Helmholtz:—

PORTSMOUTH, *April* 4 [1881].

DEAR HELMHOLTZ—I go to London to-morrow morning to stay with the Spottiswoodes, 41 Grosvenor Place. I arrive there at 11 o'clk., and will go on immediately anywhere I can find you; so please let me have a line which I may find on my arrival at Spottiswoode's to say where I must go to find you. You will no doubt be busy to-morrow in the prospect of your lecture for the evening, but I must just see you, if only to fix a time for Wednesday, when we can go anywhere and do anything together.—Believe me, yours always affectionately, WILLIAM THOMSON.

It was in the spring of 1881 that Sir William Thomson became interested in the Company which had been formed by Sir Joseph W. (then Mr.) Swan and his friends in Newcastle for the manufacture of the Swan glow-lamp. Though Sir William declined to take any official post as Director or Technical Adviser, he accepted the unofficial position of honorary consultant, and from time to time assisted the Company in that capacity.

Returning on May 11 from London after a brief cruise in the *Lalla Rookh*, Sir William Thomson found a new scientific sensation. News had come to England of the production in Paris by Camille Faure of a secondary battery or accumulator for the storage of electric energy. *The Times* of May 16 contained a letter from Major Ricarde Seaver, announcing that he had brought over from Paris a box containing a million foot-pounds of stored energy, which he was taking to Glasgow to Sir William Thomson to be tested. Sir William, who apparently had never before appreciated the beautiful and patient researches of Gaston Planté, extending from 1869 to 1878, on the formation of secondary cells of lead, and their use in electric storage, was immensely seized with the new invention, and fell upon it with more than his wonted energy.

On May 17 he wrote from Netherhall to Dr. Gladstone :—

We are just setting out for a day or two's cruise in the Clyde, to test sounding machine and depth-gauges, but this address will find me. The "box of electricity" is being kept under continued tests and measurements by James Bottomley in my absence. It is splendidly powerful, but I have yet to find whether it does the whole amount of work specified by Mr. Reynier (and Mr. Faure), and how much actual work must be spent on it each time to renew its charge.

He had, unfortunately, just then been seized with an illness, one of the recurring sequelæ of the accident to his leg twenty years before, which obliged him to remain in bed for over three weeks. This did not in the least diminish his ardours. The whole laboratory staff was kept at work early and late experimenting, manufacturing new cells with red lead and sulphuric acid, charging and discharging the batteries, and reporting every detail at the bedside of the patient. A Swan lamp, fed from batteries made in the laboratory, and hung by a safety-pin to his bed-curtain, created great astonishment in the many visitors who called on him. He wrote from the University to Jenkin on June 3 :—

I am about as well as possible, but the doctors said I should have to stay in bed for a fortnight at least. I shall be all right in a few days with absolutely no residual evil. I think very probably the fortnight will be shortened. Meantime I have plenty on hand in the way of proofs and outlying math[l] work. They are as hard at work as possible in the laboratory on the Faure storage cells, which are doing *splendidly*. It is going to be a most valuable practical affair—as valuable as water cisterns to people whether they had or had not systems of water-pipes and water-supply.

He had a long letter in *The Times* of June 9, which also had a long article on the subject. He stated that he had now subjected the battery to a variety of trials in his laboratory for three weeks, and that he was continuing the experiments at the request of the Société Anonyme la Force et la Lumière, which owned Faure's patents. He was interested in seeing in it "a realization of the most ardently and unceasingly felt scientific aspiration of his life—an aspiration which he scarcely dared to expect or to see realized." The problem of storage was "one of the most interesting and important in the whole range of science." Already the battery in his laboratory had been found useful in surgery for cauterizing. The largest useful application awaiting the Faure battery was, however, to act as a cistern does to a water supply, as in the event of a stoppage or break-down of the machinery it would prevent failure of the light. The Faure accumulator, kept supplied from the engine by a supply main, will be always ready at any hour of the day or night to give whatever light is required. Precisely the same advantages in respect of force would be gained by the accumulator when the electric town supply is, as it surely will be before many years pass, regularly used for turning-lathes and other machinery in workshops and sewing-machines in private houses. All this and more his letter set forth.

On June 17 Lady Thomson wrote to Darwin :—

Sir William sends to tell you that he has been absolutely forced off the air problem just when he was so exceedingly interested in it—he hopes to go back to it again before long (but I don't know when!). . . . He has been so taken up with his "potted energy" that he has had no time for anything else. The laboratory is very hard at work making Faure accumulators and measuring their work. James Bottomley has gone to Paris to help them with their scientific work there. Sir William is *tremendously* interested in it. We are going to have electric light in the *Lalla Rookh* with its aid. I am sorry to say we are *here*. We had about a fortnight of yachting, and had just come up here again to prepare to go south, and have been delayed by Sir William getting laid up. It is nothing serious, but he has been three weeks in his bed! Can you imagine his bearing it patiently? but he does, and keeps everybody busy about him, and for himself the day is never half long enough. We shall be here at least ten days longer—and we are due everywhere!—Paris—London—Portsmouth, and the poor *Lalla Rookh* is lying idle at Largs.

*June* 21. . . . Sir William is still a prisoner. . . . I am general secretary just now, so you may imagine I am pretty busy. Rennie [Sir William's amanuensis at that date] is working in the laboratory. They are so short-handed there just now.

THE UNIVERSITY, GLASGOW, *June* 22/81.

DEAR JENKIN—

. . . . . .

I dare say you saw *The Times*' City letter and Major Seaver's letter in reply *re* Force et Lumière. I hope that the proprietors of Faure's patents, the "Société la Force et la Lumière," are going to buckle to the real work of manufacturing and supplying accumulators, and that no formations of monster companies will continue to form any part of their plans. They have a splendid thing in hand, and I am very anxious to see it worked out well.—Yours, W. T.

THE UNIVERSITY, GLASGOW, *June* 26/81.

DEAR JENKIN—Have you any good man to spare who knows something of electricity and engineering, and *French*, supposing I want to send a man to Paris to take charge of testing and inspecting in the manufacture of Faure accumulators?

J. T. Bottomley is there at present, and I intend to go in about a fortnight, but it is possible we shall have to leave some one in charge.

White has been appointed, under my direction, sole agent for the manufacture and sale of Faure accumulators in the United Kingdom.—Yours. W. T.

By July 9 Sir William was well enough to be in London, on his way to join his yacht at Portsmouth, where he was busy with the *Inflexible*, settling her compasses, and seeing to her experimental equipment with "potted energy" in the shape of storage batteries for electric lighting. He was also busy arranging with the Jay Gould cables for the use of mirror galvanometer and siphon recorder. Then back again in London, he stayed for six weeks at the Grand Hotel, busying himself personally to arrange with Mr. Pender, Dr. C. W. Siemens, Mr. E. P. Bouverie, Mr. I. Lowthian Bell, Sir John Lubbock, Mr. C. Seymour Grenfell, and Captain Douglas Galton, to form a company to exploit the Faure battery. The agent of the French patentees, M. Philippart, had a marvellous genius for drafting prospectuses, and evolved scheme after scheme. All schemes had the feature of showing immense future prospects of profit to the investors, and all were arranged to bring to the vendors a sum varying from a quarter to half a million pounds. He induced Sir William, whose labours up to this point had been purely voluntary, to enter into an agreement to act as technical adviser, receiving in return 10 per cent of the capital. Sir William insisted that the fact and extent of his interest should be published. To this course M. Philippart objected. The owners of the Faure Patents proposed the enormous capital of £400,000, of which they were to receive £120,000 in cash and half the shares. The negotiations lasted through August, as one by one withdrew from taking part in so formidable a venture; but it was not till some weeks later that Sir William was brought to realize the very unsatisfactory position; and he too withdrew—a heavy loser by the affair—from all association with the venture, though he wished the invention success. He was keen to light his own house at the University by Swan lamps and Faure cells.

The year 1881 being the Jubilee of the British Association, its meeting in York—the city of its foundation—was one of unusual brilliance. Sir John Lubbock (now Lord Avebury) presided; and Sir William Thomson was chosen once more as president of the Mathematical and Physical Section. He took as the topic of his Address the Sources of Energy in Nature available to Man for the Production of Mechanical Effect. Beginning by remarking how the science of Energy had practically arisen during the fifty years of the existence of the Association, he reverted to his earlier paper of 1852 (see p. 289), in which he had enumerated the natural sources of available power—food for animals, natural heat, potential energy of elevated masses, natural movements of air and water, natural combustibles, and artificial combustibles. Heat radiated from the sun was undoubtedly the chief source, and a second possible one was the tides due to the earth's rotation. Of the natural sources, tides, food, fuel, wind, and rain, only one, the tides, was not derived from sun-heat. Tide-mills existed indeed; but the vast costliness of dock construction was prohibitory of almost every scheme for utilizing tidal energy. Parenthetically he praised the metric system, "to which we are irresistibly drawn, notwithstanding a dense barrier of insular prejudice most detrimental to the islanders." As the subterranean coal-stores of the world were becoming exhausted, surely and not slowly, the price of coal was upward-bound, and it was not chimerical to think that in some form windmills or wind-motors would again assume ascendency. Even now they might be used for lighting. Now that we had dynamos and Faure cells for storage, the chief want was cheap windmills. Rain-power was not economical except as the natural drainage of hill country. For anything of great work by rain-power, the water-wheels must be in the place where the water-supply with natural water was found. "Such places are generally far from great towns, and the time is not yet come when great towns grow by natural selection beside waterfalls, for power." But the splendid suggestion of Siemens,[1] that the power of Niagara might be utilized by transmitting it electrically to great distances, had given a fresh departure. From the time of Joule's experimental electro-magnetic engines, and by the theory of the electro-magnetic transmission of energy, it had been known for thirty years that potential energy could be transmitted by means of an electric current through a wire, with unlimitedly perfect economy. Adopting Siemens's shortened form of the word, he declared in conclusion that "dynamos" give now a ready means of realizing economically on a large scale many practical applications of Joule's thermodynamics of electro-magnetism; and, in particular, they make it possible to transmit electro-magnetically the work of waterfalls through long insulated conducting wires, and use it at distances of fifties or hundreds of miles from the source, with excellent economy.

One passage of the address, not included in the reprint in Vol. II. of the *Popular Lectures*, is worth preserving:—

> High potential, as Siemens, I believe, first pointed out, is the essential for good dynamical economy in the electric transmission of power. But what are we to do with 80,000 volts when we have them at the civilized end of the wire? Imagine a domestic servant going to dust an electric lamp with 80,000 volts on one of its metals! Nothing above 200 volts ought on any account ever to be admitted into a house or ship or other place where safeguards against

[1] Presidential Address of Sir William Siemens to the Iron and Steel Inst.

accident cannot be made absolutely and for ever trustworthy against all possibility of accident. In an electric workshop 80,000 volts is no more dangerous than a circular saw. Till I learned Faure's invention I could but think of step-down dynamos.

In the section meetings Sir William followed up the outlines of his address by reading papers on Some Uses of Faure's Accumulator in Electric Lighting; on the Photometry and the Illuminating Power of Electric Lamps; on the Proper Proportions to be given to the Resistances in the Different Parts of Dynamos; and on the Economy of Metal in Conductors of Electricity. The last two were of importance. In the one he gave rules for the good design of shunt-wound dynamos; in the other he enunciated the now famous law, that for economy in long-distance transmission of power at a given voltage the best economy is obtained when the conductor is of such a thickness that the interest on the capital expenditure on the copper is equal to the annual cost of the energy lost in transmission through the conductor. Though there is modification needed for cases in which the cost of the line (for example, in extra-high voltage transmissions) other than the mere weight of copper is not proportional to the mileage, "Kelvin's law" is still the foundation of practice. He also read a paper on an electro-ergometer, only to find that his proposal was the same as the wattmeter, shown to the section by Ayrton and Perry the same day.

At this meeting there was a very keen discussion on the report of the Committee on Electrical

Standards, which had been reappointed on the motion of Prof. Ayrton the previous year. An International Congress—the first of its kind—was to be held at Paris in the middle of September, at the Electrical Exhibition, and it was keenly desired that other nations might agree to the system of units and standards which the British Association Committees had evolved, chiefly under Thomson's inspiration, since 1862. Joule had shown the original B.A. ohm (p. 419) to be slightly smaller than its intended theoretical value in the absolute centimetre-gramme-second system; and new determinations of the absolute values of resistances, currents, etc., had been made in different countries by Rayleigh, Mascart, Rowland, Roíti, H. F. Weber, and others. There was also a serious proposal to substitute for the practical units of current and voltage other units ten times as great, so as to make them agree with the "absolute" units. The Committee at York was divided on this point, and the proposal to keep the unit of current (then called the *weber*) at one-tenth of the absolute unit was carried by a bare majority, Sir William Thomson urging that the smaller value was more convenient, being about of the magnitude needed for one Swan lamp. "Alas! no West Highlands this year for us," wrote Sir William to Jenkin on September 9. "We are off to Spottiswoode's, Coombe Bank, this moment, and thence to Paris, Monday; Netherhall a fortnight hence, hurrah!"

At the Paris Congress Sir William was a notable

figure amongst the leaders of the electrical world. M. Cochery, Minister of Posts and Telegraphs, presided. The foreign vice-presidents were Sir William Thomson, Signor Govi, and Prof. Helmholtz. Helmholtz, Werner Siemens, and Du Bois Reymond represented Germany, and there was a strong feeling in favour of abandoning the British Association's unit of resistance, the ohm, in favour of Siemens's unit, the column of mercury one metre long. The debate grew warm. One who was present has narrated the unforgettable scene of comedy of Thomson and Helmholtz disputing hotly in French, which each pronounced *more suo*, to the edification of the representatives of other nationalities. A severe disruption was averted by a timely adjournment, during which, by the efforts of M. Mascart and Mr. (now Lord Justice) Moulton, a compromise was arranged to accept the B.A. ohm, but to represent it concretely by a column of mercury of appropriate length, the precise value of which was to be fixed at an adjourned conference the next year, after further researches. The names of the electrical units, *ohm*, *volt*, *farad*, and *coulomb* were agreed to; but that of the *weber* was changed to *ampere*, in honour of the great French savant. Sir William was the life and soul of the whole business of the Congress.

While staying in Paris Sir William gave to the Académie des Sciences a paper on the Thermodynamic Acceleration of the Rotation of the Earth, which he had recently discovered.

In October he learned that he had been awarded the Diplôme d'honneur at the Exhibition for his fine show of compasses, galvanometers, and other instruments and inventions.

The important part which Sir George Darwin took in the revision of the *Treatise* occasioned much correspondence and is illustrated by the following note:—

THE UNIVERSITY, GLASGOW, *Nov.* 8, '81.

DEAR DARWIN—In the italics of § 569 I think c. of i., though never heard of by either Pappus or Guldinus, ought, as you say, to be put in place of c. of g. But in § 572 it is really the *approximate* theorem of c. of g. that is used, and gravity that is the occasion for using it. Therefore leave c. of g. in § 572 and all similar cases.—Yours,

W. T.

[*P.S.*]—I enclose another sheet for your imprimatur; also a piece of slip of the preceding sheet.

Sending more proofs on November 16, he wrote:—

I am tremendously obliged to you for what you are doing. . . . The truth is your help has been a great spurt to me.

Within three days he was back in Glasgow arranging to light his residence and the University laboratory by Swan lamps, a Clerk[1] gas-engine, and Faure cells.

On October 17 Lady Thomson wrote to Darwin that Sir William had gone off to the *Inflexible* at

[1] Sir William took part in a discussion on the theory of the gas-engine, after a paper by Mr. Dugald Clerk, at the Institution of Civil Engineers, April 4, 1882.

last for a trip about the south coast, to end at Plymouth on the 27th, when he would go to Cambridge:—

Before he left on Saturday he bade me write to you to thank you very much, and to tell you what valuable assistance you have given in T and T′; he says in future it will have to be T and D, not T and T′. . . .

*Nov.* 4.—Sir William says I am to tell you that he felt much humiliated by the number of blunders which you caught out. . . . He gave his first lecture on Wed[y] morning, and is already in full swing of his winter's work.

The same day he wrote to Jenkin:—

You must have engine and dynamo to work the Swans, which, as you say, are lovely. The true *domestic* use of the Faure is for continuing a *small* amount of lighting after the main demand ceases when the engine is (thanks to it) to be let stop. . . . Excuse haste and shakiness of hand standing on my legs before I run out to my class.

On December 8th Lady Thomson sent Darwin word that they had had a very pleasant visit from Mr. and Mrs. Hopkinson. "Sir William and he did nothing but talk electric light and dynamos!" Sir William's house was then being wired for electric lamps. He added the following postscript to the letter:—

Thanks for the Adams [an estimate of geologic time], which will do splendidly for T & T′. I am glad it is not a recantation. Why don't you want the earth consolidated as it is? It is *the probable history* of the earth's emergence into the *consistentior status*: not complete in every detail, but about as correct on the whole as any ordinary history of England.

On December 26th Sir William wrote again from Knowsley :—

I have been engrossed with finishing one patent, and making the provisional of another, for two or three weeks, but I have brought all of both T & T′ and my reprint with me, so I hope to get something done in the holidays. Shall I send you sheets for your final imprimatur after I have done all I can on them, or would you rather wait till after the Tripos?

The following extract from a later letter on the rigidity of the globe has an interest of its own :—

*Dec.* 28, 1881.

. . . That a change from ellipticity of $\frac{1}{230}$ to $\frac{1}{300}$ could take place in a body as rigid as the earth is as a whole, without leaving gigantic traces, I have always thought to be so very excessively improbable, that I have never doubted the validity of the assumption that the earth has not experienced it since consolidation, nor of any conclusion founded on this assumption.

The durability of Africa and America have (*sic*) always to my mind been irrefragable proof of the permanent rigidity of the earth under stresses comparable with those due to angular velocity, giving comparable elevations and depressions of a complex level surface relatively to a surface through the same solid particles. It seems to me exceedingly improbable that the angular velocity can have changed by much more than $\frac{1}{70}$ of its own amount since consolidation, because 1000 or 2000 or 3000 feet of elevation of equatoreal solid relatively to a level surface would not redress itself by the earth's yielding to over severe stress; and even considering volcanoes, or whatever may be the cause of the greatest elevations and depressions that have taken place in geologic history, I do not think a 2000 ft. or 3000 ft. elevation of equatoreal land, by failing centrifugal force, could be lost among the other disturbances. I think it just possible that 1000 feet might be overlooked by us among the great changes that have taken place, and therefore I don't think Adams' datum absolutely inconsistent with $10^6$ centuries of geological time. But if Adams' datum is correct it would give some weight in the balance for arguing something less than that. I should think from 20 to 50 million years, all things considered, the most probable. Underground heat, I suppose (not to speak of biology), makes $20 \times 10^6$ more probable than $5 \times 10^6$ or than 10. There are many reasons that make anything more than $100 \times 10^6$ improbable, so far as I can see.

Consulting Dr. Gladstone about the chemistry of accumulators, he wrote:—

THE UNIVERSITY, GLASGOW,
*22nd Dec.* '81.

DEAR GLADSTONE—James Bottomley showed me your letter, and we are much interested in it, as well as in what you wrote to me about the Planté and Faure batteries. My great difficulty all along has been to understand how it is that the electrolytic action of the peroxide in contact with the lead plate upon which it is placed does not go on oxidizing the plate. I cannot see the conducting connection between it and the plate in contact with it, so as to allow the complete electrolytic surface of the coating to be in regular action, and yet not to cause oxidation of the lead plate in contact with it. I am also much puzzled about the function of the sulphuric acid. It is not simply to render the liquid a conductor. I have tried a Faure element without sulphuric acid, but hitherto with only negative results. It has hitherto been a complete failure in respect to capacity; and yet I believe I have sent enough of quantity through it, to produce a large action on the minium. However, the trials are incomplete, and I may get better results yet.

I want to know why it is that the sulphuric acid does not form sulphate of lead, with the protoxide found in the battery. Is it that there is besides the peroxide, another oxide which is not attacked by sulphuric acid?

The liquid in my cells is sometimes quite destitute of acid, and is then replenished with dilute acid in the proportion of ten volumes of water to one of acid.

The loss of the acid sometimes takes place several times in the process of forming a cell. The cells in which it goes on much are found to be bad ones, but I have not been able to find the cause of the badness. The presence of sulphate of lead on the peroxide, as I find it sometimes coating the peroxide over in places like disease spots, is certainly an evil; not, I believe, merely by obstruction, but by somehow tending to let the cell discharge itself. Cells sometimes go bad, and on being recharged for a long time, become again as good as they were. The whole thing is in immense need of investigation, but the investigation is excessively difficult.

I think it is going to be of great practical value notwithstanding all these difficulties. I am on the point of beginning to light my laboratory, lecture-room, and house with Swan and Edison lamps. As an auxiliary for night lights and dark winter mornings, I shall have about 130 cells (round) of the Faure battery, on the original pattern, which I made here last June and July. Many of them are bad, but on the whole I expect they will give the result I require, although not with perfect economy. I have got my house completely wired from attic to cellar, and I mean to have no gas and no candles. The gas-engine I am going to use will be in my laboratory, and I hope to find it running on my return from England, where we are to spend the holidays in Liverpool and its neighbourhood. . . .

All through the year 1882 the revision of T & T′ occupied Sir William's time, and he wrote every few days to Darwin. On January 1 he sent word that he had "shortened the first day of the year considerably by a great despatch of Mathematical and Physical papers" to the printers; and he was much bothered by § 830 of T & T′. The next day another note adds: "I am on my way to Glasgow for an odious affair in the Phil. Soc. to-night about gas-meters (*awards!*)." Many of these letters tell of other matters beside printers' proofs. Sir William's house was the first in Scotland to be fitted with electric lights. He himself devised many details—switches, fuses, suspensions, and such like:

*Feb.* 15.—I have been overwhelmed with electric lighting. It is going to be a great success in the house. It is already a great pleasure, but wants *much much* to complete it for practical convenience.

In May, after the College session was over, there was a fresh outburst of correspondence on the parts of the book relating to tides and other matters. At last, on June 11, he found himself on the *Lalla Rookh* in Lamlash Bay; but, alas! he had left the proofs behind at Netherhall.

It is most unfortunate. Though we sailed much of yesterday, which is all-absorbing (the chief merit of sailing); but this is Sunday and wet all after church time, an opportunity sadly lost. I did look at proofs in the Pullman on Thursday night, but not by electric light, so detected no errors.

*June* 15 (addr. Yacht, *Lalla Rookh*, Largs)—I shall be here again next Thursday as *cras ingens iter abemus aequor*; *i.e.* the gravity and kinetic equipotential on a line from Gareloch to Largs.

A very characteristic letter to Darwin is the following:—

"*LL.R.*" INGENS AEQUOR, GARELOCH,→
LOCH LONG,→ LARGS,→ GREENOCK,
*June* 21/82.

DEAR DARWIN—P. 302 and the Tables of my 1876 report *are* "utterly right" in respect to the epochs of the fortnightly and lunar semi-diurnal tides; and your "$\xi$—$\nu$" could but be met by an unqualified *not guilty*. . . .

I must have been wrong in one of my marginal "oh no's" on one of your returned MSS. And I suppose the "?" after "hinc" is to be answered in the negative.

But why did you go on attacking § 302, and Laplace and Hopkins? Their astres fictifs and mean suns and mean moons are a most valuable contribution to means of expression (alas! an unintended pun) in physical and descriptive astronomy. It is the combination of aphasia, with heedlessness as to logical clearness and accuracy when aphasia is overcome that makes the *whole difficulty* in the ordinary dealings of astronomical writers with such questions as this that we have had in hand, which by aid of Laplace and Hopkins properly used (and not abused), as I have abused them to some degree, by not saying equatoreal mean moon when I meant it, and orbital mean moon when I meant it, and ecliptic mean sun's longitude instead of even in a " " and for a moment adopting "sun's mean longitude." . . .

I don't know when this will be posted, as we are now (5.15) becalmed, rather far up Loch Long. I intended to be up in Glasgow and have a trial of my potential regulator to-night; but I am resigned. My dynamo at all events will not suffer, as this (*LL.R.* I mean) is the only place for scheming such things.—Yours,

W. THOMSON.

[*P.S.*]—Excuse the disjointed and untidy character of the above, as it was only possible to write during showers, so beautiful are the masses all round above the great equipotential; till the weather became too fine and I put the stylograph in requisition on deck.

Early in July the yacht started for a West

Highland cruise, and on July 9 Sir William wrote from Crinan Bay:—

. . . You must give a paper to the B.A. Southampton on Fortnightly Tides and the Rigidity of the Earth. It is *the fruit* for which the B.A. Tidal Committee of 1869-76 was sown, as you may judge from T & T′ § . . .

My burning question for the last sheet but one of reprint has cost me a good deal more time and work (time chiefly spent in trying to read with complete intelligence §§ 700-732 of *Electrostatics and Magnetism*) than I reckoned. I see my way pretty thoroughly now, I think, and I do hope to get it off, and the last of the vol. imprimatured, almost instantly. I have done a good deal, however, on the T & T′ proofs, and am only waiting to get a sentence or two written for § 830, and possibly an addition from the Rede lecture for 1866 (*Phil. Mag.* 1866, 1st half-year) decided on.

*P.S.*—The necessity for a depth-gauge not founded on the spring of air becomes more and more urgent, and has taken a good deal of my time from T & T′ and the reprint. It slumbers till I get to sea, and then comes on with renewed intensity. It has done so at all events these four years, every time.

July 19 found the yacht at Roshven, and thence Sir William sent four pages about his reprint; next day four more about slope of potential; the same evening eight more, highly mathematical; and on 21st from the Sound of Sleat another missive, a leaf (torn from his green-book) of mathematics, which he describes as "a very pleasing little nugget dug out of old 1847-8 neglected diggings since we left Roshven this morning." But all was not plain sailing, for the next letter tells of dire disaster.

ROYAL HOTEL (alas !), PORTREE,
*July* 25/82.

DEAR DARWIN—We had a beautiful sail out of Isle Ornsay and through the Kyles of Skye and Loch Alsh on Saturday morning, and I unfortunately neglected, in consequence, to finish writing out the piece of theory of the day before. It would have been quite satisfactory and sufficient as to the elastic yielding of rock to weight of water, and would have kept me quite busy enough to keep me from a practical experiment which I made about 20 minutes after high water on its yielding (proving it to be very small) to a much greater weight of wood and iron than the load of air or water that you have been thinking of. The result was that by the time of low water (5 P.M.) the wood had yielded very visibly and the rock not perceptibly. We left it there, on Scalpa, about 10 miles from this, and were brought here in a friendly steam yacht of 650 tons (Mr. Stewart's *Amy*) that chanced to pass by when the result of the experiment was becoming apparent. We were taken back next day, and Mr. Stewart set his crew to help mine and we baled and pumped the *Lalla Rookh* dry, and caulked the deck seams with tow and tallow, it having become apparent that it was through them that she had filled after the two previous low waters, and we had to jetison her by taking about ten tons of matter out of her. She yielded half an hour after midnight to a third determined pull, which Mr. Stewart with splendid resolution and courage gave her, and came off. The elastic limits had been largely exceeded, and there is a large contortion of the strata, giving a mountainous and picturesque character to deck and starboard side after the removal of the stress; but she makes even less water than she did before (which was very little, quite as little as it ought to be), and she left this evening in tow of a tug I telegraphed to Greenock for yesterday morning. She will be made as good as ever very soon, in a week or two I hope, but not in time to come round with us to Southampton. So I am sorry to say our cruising, which was an unalloyed pleasure till the last instant of it when we were looking at a seal with our glasses, is over for the season; and we cannot ask you to come and settle the earth's rigidity on board the *Lalla Rookh* (it having been temporarily settled the other way so far as she is concerned). But we shall carry the war into the enemy's country and have our revenge with you on terra miscalled firma in some of the committee rooms at Southampton.

The theory has, however, been developing itself (though not on paper) very prettily. Let me know if what I posted at Ornsay was sufficient, or at all events intelligible as far as it went. I was going to put into proper shape the reduction to calculable results which with $m = \infty$ would have put into workmanlike shape the clumsy thing of Roshven. This part of the affair is simply done by taking in your crushing as unit of force the mutual attraction between two grammes of matter concentrated at two points 1 centimetre asunder (which for copper makes $n = 450 \times 10^6$ "grammes weight" per square centimetre). I had got on, before or after the seal, I forget which, to find the interpretation and an interesting physical illustration of the correction to annul surface tangential force which I had introduced in the Isle Ornsay despatch. Here it is. [*Here follow fourteen octavo pages of mathematics and diagrams, ending up*] . . . 90°, if the material does not break, as was the case of the *L.R.*, where no doubt there was perfect tangency at bounding line of interface between coppers and rock. Nothing more.—Yours,

W. T.

Other letters followed very quickly.

*July* 30, NETHERHALL.—We are much grieved to hear of the death of Frank Balfour. It is indeed a sad loss to the world, which can ill spare such men. We feel very much for you and quite understand Cambridge being different to you without him. . . . In *haste* to catch post, or rather to be not late for church, as letters must be posted before. . . .

*Aug*. 1, NETHERHALL.—I must settle down now

thoroughly to § 830 and subsequent sheets of T & T′ to the end. . . . You asked me why I chose so moderate a slab of water in T & T′ § 818. Answ. because it is something we know about. Not very unlike, for example, the actual effect of the water in the English Channel on a plummet at St. Alban's Head.

I have always admired Cavendish's experiment as perfectly free from error due to yielding of the foundation. His *two* attracting masses are supported on a pivot midway between, so that the pressure on the ground never is changed during the experiment.

*Aug.* 2, NETHERHALL.—I am returning Title-page, Preface, and Contents of *Math. and Phys. Papers*, Vol. I., for press by this post. . . . Reprint of *Electrostatics and Magnetism* is commenced, and Vol. II. of Papers is to begin as soon as T & T′ let me.

SKIPNESS, ARGYLLSHIRE
(Address NETHERHALL),
*Aug.* 14/82.

DEAR DARWIN—Till the moment of leaving Netherhall on Saturday I was so driven, with a multitude of unavoidables, that I could not write a word on any of the points you have been writing to me about. On Friday I got the very last scrap of Vol. I. of Collected Papers out of my hands for press, so *that* is out of the way, and I have got fairly on to things for finishing the T & T′ vol., particularly the infinite homogeneous solid bounded by a plane (I wish you would give a non-self-contradictory designation for this). It comes out enchantingly, I mean the complete problem of any arbitrarily given distribution of surface force (3 components, one normal and two tangential, for every point). I have been at it all my spare time since I came here, and may possibly get it written out for press to-morrow. . . . I shall look carefully to whether anything should be put into T & T′ as to the effect of the weight of the water on the elastic yielding. . . . The plane-bounded infinite solid is *too* captivating to let me touch anything else till I get it off the stocks.—Yours, W. T.

Southampton was the place of meeting of the British Association in August 1882, under the Presidency of Dr. (later Sir) William Siemens. His brother Dr. Werner Siemens was a notable guest; also the veteran Professor Clausius. Thomson read several papers; on some new absolute galvanometers; on the transmission of force though an elastic solid; and on magnetic susceptibility. He also gave an evening discourse on the Tides (see p. 729).

A distinguished geologist present has preserved the following account[1] of this event:—

. . . Sir William Thomson's lecture on the Tides, which was given to a large audience, was good for all who understood it. But Thomson himself was splendid; he danced about the platform in all directions, with a huge pointer in his hand; he shook in every fibre with delightful excitement, and the audience were as delighted as he.

The adjourned Paris Conference on electric units was to be in October. On September 30 Sir William, then at Netherhall, wrote to Lord Rayleigh asking if the date was to be the 16th, as there appeared to be some misunderstanding.

Dr. C. W. Siemens, who is with us just now, tells me that you cannot be in Paris on the 16th, but I hope this is not the case. We could get on but very badly without you, and in fact I suppose your ohm must be declared the one true ohm for our generation.

To Darwin he wrote on October 12th :—

I am going to-morrow to Newcastle to see Swan's light

[1] *A Memoir of William Pengelly*, p. 264.

factory, and thence to Bolton's copper mills, Cheadle, on my way to London.

Thence he wrote to Lord Rayleigh :—

GRAND HOTEL, TRAFALGAR SQUARE,
LONDON, W.C., *Oct.* 13/82.

DEAR LORD RAYLEIGH—I have at last received a positive official notification that the 16th is fixed for the Units Committee, and I am going accordingly to Paris on that date. I don't leave London till Monday forenoon, so if you have anything more to send by me I shall be able to receive it by post. Dr. Siemens had left us when your letter (which crossed mine to you) and paper for him came, but I am to see him this evening and I shall give it to him. I shall bring the other copies with me and more for the members of the Congress if you will send them.

But will you not come to Paris after all, *taking* leave of absence from Cambridge for the necessary time? If you would come on Monday and remain a few days it would tend much to ease the way of the Congress to the settlement of the ohm, not to speak of other questions. Let me have a line here, unless you will cross over along with us by the tidal train for which we leave Charing Cross at 11 on Monday forenoon.—Yours very truly,

WILLIAM THOMSON.

From Paris he wrote to Dr. Hopkinson about the muddle :—

HÔTEL CHATHAM, RUE DAUNOU, PARIS,
*Oct.* 17/82.

DEAR HOPKINSON—On the 20th of September Lord Granville answered the French authorities that the names of proposed English delegates, and the fixture of the 16th as the commencement of the meeting, had been communicated to the Science and Art Department. So the fault seems to have been in South Kensington. I hope your telegram will bring it home to them, and, still more, that you will yourself be here to-morrow.

Our Government seems to have determined to do as little as possible, and I think you did quite right in respect to the telegram you received yesterday morning.

I think I must write a complaint, because I find myself put down as a delegate from England in an official list without having received any intimation that I had been appointed; and I only hear of it here through the French. I shall not write, however, till I hear from or see you and learn the effect of your telegram of yesterday. I hope you will be here to-morrow, as the practical thing to be done is really important. A commission of 15 on the measurement of the ohm, appointed to-day, meets to-morrow. You of course will be on it.—Yours very truly,

WILLIAM THOMSON.

[*P.S.*]—Helmholtz, Wiedemann, Kohlrausch, Lorenz, Werner Siemens are all here, and all except Wiedemann and Kohlrausch in this Hotel.

The next letter to Lord Rayleigh tells of the doings of the conference:—

HÔTEL CHATHAM, RUE DAUNOU, PARIS,
*Oct.* 19/82.

DEAR LORD RAYLEIGH—We are all very sorry that you have been unable to be at the Conference, and I am charged to express "des vifs regrets" that you had been prevented by illness.

The *sous commission* on the fixation of the ohm has held its last meeting to-day, and it is strongly impressed with the conviction that your number $\cdot 9865 \times 10^9$ is within $\frac{1}{1000}$ of the true value of the B.A. unit. (Would you not rather, however, take as the most probable ·9867, being the mean of your result by the old B.A. method, and your two by Lorenz's method, ·9869 and ·9867?) I communicated this statement to them, from your letter, and also the statement that probably the comparison between the B.A. and the Siemens mercury unit, made independently, is trustworthy to $\frac{1}{4000}$. I also communicated your printed slip, giving the results of your and Mrs. Sidgwick's comparison of the 4 tubes with the B.A. unit. This will all appear in the report of the meetings. I believe they would have decided on $(\cdot 9867)^{-1}$ of the B.A. unit (or corresponding numeric of the S. U.) for the ohm, but that Friedrich Weber was there defending some carefully made experiments of his own, which gave him $9550 \times 10^9$ for the Siemens unit, instead of ·9413, as you and Mrs. Sidgwick make it. It seems after all probable that it was the particular one or two standards *called* Siemens units which he had, that caused the discrepance, and that his results may after all be found to agree closely with yours. Helmholtz proposed and the *sous commission* adopted a resolution to move (*via* the French Gov$^{t}$.) for transmission of individual standards from one to another of persons who have made absolute determinations. This will be very valuable, but I shall try to arrange with Fr. Weber for an immediate interchange of standards between you and him. Could you send me one here, which I could receive by Saturday evening, or at the latest Sunday evening, and give to him to take away with him. That would settle the question, no doubt, in a few days. Or with very little delay there might be an interchange between you and him after he gets back to Zurich. He is himself under the impression that you are right, and so are Helmholtz and Wiedemann.

Our last general meeting will probably be on Monday morning, and terminate the whole business of the present session ; but no doubt there will be another this time next year, and I think it is desirable there should be, because I think, judging from the past, good and useful work and conventions will be promoted by at least one other meeting analogous to this and to last year's one.

Lady Thomson and I intend to be at Cambridge from Friday or Sat. of next week to Monday or Tuesday following, but I have not yet heard of the day for the college meeting (the statutory 29th being Sunday), and I do not know where we shall be staying.

I hope you are feeling much better, and quite well again now.—Yours very truly, WILLIAM THOMSON.

On November 27 Lady Thomson wrote to Darwin to urge him to visit them :—

Sir William says you and he can settle about the Preface—and title-page—when you come. Sir William says he has written to ask T′ to come from a Friday to Monday (*8th to 11th, if you are here at that time*) to help to settle the question (he is writing this moment). Sir William has been *very* busy ever since he came home—never a moment to spare; he has dynamos, *and sound from double and multiple sources, such as tuning-forks*, on the brain. [The clauses printed in italics are in Sir W.'s handwriting.]

In December he went to Portsmouth to see about the compasses of H.M.S. *Polyphemus*, a very difficult ship for compass-compensation. And on Christmas day he and Lady Thomson went to Knowsley for the annual visit to Lord Derby.

The revision of the *Natural Philosophy* lasted on into the spring of 1883, and the designing of electric measuring instruments was still a heavy burden. In February Sir William gave a discourse at the Royal Institution on the Size of Atoms.[1] In March he communicated to the Royal Society of Edinburgh three papers: on the Dynamical Theory of the Dispersion of Light, on Gyrostatics, and on Oscillations and Waves in an Adynamic Gyrostatic System. He wrote to Darwin: "I have been very full of gyrostatics lately, partly or chiefly brought on by my Royal Institution lecture on Size of Atoms." To Lord Rayleigh he wrote for information :—

---

[1] See p. 566; and *Popular Lectures*, vol. i. p. 147.

*Ap.* 11, 1883,
THE UNIVERSITY, GLASGOW.

DEAR LORD RAYLEIGH—I am to give a lecture at the Institution of Civil Engineers on Electric Units on the 3rd of May, and I should be much obliged if you will tell me if there is any later work than what I know already of yours, which made the B.A. unit = ·98677 ohm.

I am now regularly taking the Rayleigh (or true) ohm as 1·0134 × B.A. unit, and making resistance coils accordingly. Also, I use now the Rayleigh (or true) volt as greater in the same proportion, *i.e.* = 1·0134 × B.A. volt. I suppose this is very sure to be as near the absolute truth as can appear in anything I am doing; but for the sake of my lecture I should like to know the very last of what you have done; and particularly to know if you change the last figure from what I have.

I am glad to hear from Darwin, who is with us just now, that you are so much better. I hope you will not be troubled by any return of the enemy; but you must not work too hard in the laboratory, or otherwise.—Yours very truly, WILLIAM THOMSON.

The lecture on Electrical Units of Measurement was given on May 3rd. The printed version in Vol. I. p. 73 of his *Popular Lectures* gives but a faint idea of the impression it made on the hearers. The lecture began by emphasizing the necessity in physical science of finding true principles for numerical reckoning and methods for measuring. "I often say that when you can measure what you are speaking about, and express it in numbers, you know something about it; but when you cannot measure it, when you cannot express it in numbers, your knowledge is of a meagre and unsatisfactory kind." He commented on the advance made in electrical measurement. Ten years ago the scientific instrument-maker scarcely knew whether the conductivity of his copper coils was within 60 per cent of that of pure copper; and the professors did not know the resistances of the electromagnets in their laboratories. Now all that was changed, and clerks and junior assistants could measure the resistance of wires more accurately than you could measure the length of ten feet. Few lecturers, not even himself, knew the capacity of the Leyden jars on his lecture-table! As to electromotive force, "we have scarcely emerged one year from those middle ages when a volt and a Daniell's cell were considered practically identical." And this advance was due to commercial requirements.

There cannot be a greater mistake than that of looking superciliously upon practical applications of science. The life and soul of science is its practical application; and just as the great advances in mathematics have been made through the desire of discovering the solution of problems which were of a highly practical kind in mathematical science, so in physical science many of the greatest advances that have been made from the beginning of the world to the present time have been made in the earnest desire to turn the knowledge of the properties of matter to some purpose useful to mankind.

He then referred to the growth of the absolute system of measurement; to the British Association Committee of 1861, and the "ohm" which resulted from its eight-year-long labours; to the Paris Conference of 1882, and the researches of Weber, of Siemens, and of Lord Rayleigh and Mrs. Sidgwick, leading to the latest accepted values. Then with a passing rap at British units he went off into a disquisition as to the importance of absolute units based on universal gravitation and on the semi-period of an infinitesimal satellite! The better to realize an absolute system of measurement, and to detach their ideas from terrestrial limitations, he asked his hearers to imagine a scientific traveller roaming through the universe. "For myself," he said, "what seems the shortest and surest way to reach the philosophy of measurement is to cut off all connection with the earth." And so he supposed his scientific traveller to have lost his watch, his measuring rod, and his tuning-fork, and to be trying to recover the values of the centimetre and the mean solar second; the former, by the aid of a diffraction grating (which he himself must rule on glass) and the spectrum of the sodium flame seen through it; the latter, by making a re-determination of the ohm, or by experiments on electric oscillations with a coil and a condenser! Well do some of those present remember that bewildering excursus. After thus filling an hour and a quarter, he announced that he had only reached the threshold of the subject, and must now commence the consideration of electrical units. He referred to the B.A. reports; recalled how he had himself constructed sets of conductance coils (*mho*-boxes) thirty years before; praised the "magical accuracy" of Joule, whose refined thermal experiments had first raised a doubt on the value of the

B.A. unit. He ended by naming the things still needed to perfect the system of absolute measurement, and recommended the audience to purchase Everett's book on Units,—price three and sixpence!

May was spent partly in London and partly in festivities at Cambridge. Before going north to join his yacht Sir William wrote to Lord Rayleigh :—

I have nothing, at all events as yet new, in the way of dynamos, and I am afraid most of my other practical affairs, integrators, domestic volt-meters, mho-meters, regulators, etc., are also in the future tense or paulo-post-futurum. (Can there be such a tense, or do I misremember my grammar?)

To von Helmholtz, who had the previous year been ennobled by the German Emperor, he wrote :—

YACHT *LALLA ROOKH*, GARELOCH, CLYDE,
*June* 17, 1883.

DEAR HELMHOLTZ—A note I have from Lockyer seems to imply that you have not yet received a copy of the second part of the new edition of Thomson and Tait's *Natural Philosophy*. I directed a copy to be sent to you, immediately on the publication of the book; but I shall write again to the publishers to make sure that they send it, if they have not done so already.

I hope you received in due time the 2nd edition of the first part, and also the first volume of my collected *Mathematical and Physical Papers*. I directed the publishers in each case to send you a copy immediately on publication.

We have now been afloat and at home on board the *Lalla Rookh* for a fortnight, with a weekly visit to the Laboratory, and we shall be oscillating between the two L's for the greater part of the summer. Will you not come and have a cruise with us, either in July in the

Solent or in August in the Clyde? If you could come to us in August or September you would find us with one foot on shore occasionally, at "Netherhall," a house we have built for ourselves at Largs (but now I remember you saw it), but which the superior attractions of the *Lalla Rookh* keep us from living in so much as we would otherwise like. The British Association meets on the 19th of September at Southport (near Liverpool), and if you would come to us in September early enough to have some good sailing first, we might go there together. We hope Mrs. Helmholtz will be able to come with you, and you must tell her not to be deterred by the idea of the yacht if she is not a good sailor, as any time after the middle of August the house on shore will always be available.

In the Laboratory I have been greatly occupied with electric measuring instruments, chiefly for practical purposes connected with electric lighting. One that I hope to have going very soon—a gyrostatic current integrator, or "coulomb meter"—I think would interest you. I have also a new electrometer and some arrangements of galvanometers to give moderately accurate absolute determinations through very wide ranges. I need not tell you about them now, however, as I hope you will come and see them all.

Lady Thomson joins in kind regards to you and Mrs. Helmholtz, and I remain, yours always truly,

WILLIAM THOMSON.

In the summer of 1883 Sir William took his brother James and his wife and two daughters for a cruise.

On coming on board the ladies were amused to find that each of the two learned professors had provided himself with a copy of *Jack Brag* at the railway book-stall, and had brought it with him as light literature to read on the yacht. The two brothers had long interesting talks, interspersed with reading aloud *Jack Brag*; and there were pleasant walks when the yacht was at anchor. The cruise was delightful; there was good sailing weather, though one of the days was so stormy as to make it advisable to seek shelter behind the Otter Spit in Loch Fyne.

A letter from Lady Thomson to Professor Darwin, dated from Otter Bay, Loch Fyne, on June 24th, takes up the tale of their doings.

. . . We have been on board the *Lalla Rookh* ever since we returned from London at the beginning of the month, sailing about the Clyde and taking a weekly run to Glasgow for a few hours at White's and in the Laboratory. We expect to go to London again for a few days early in July, and we shall probably let the yacht sail round to Portsmouth to wait for us. If you are inclined to offer us a visit in the *L.R.* when we are in the Solent, we shall be very glad. We hope to be at Netherhall for about a month, till the 19th Sept., when we go south for the B.A., and then for Sir Wm. to give his address in Birmingham as President of the Birmingham and Midland Institute; and then we go off to Vienna, where he is one of the Government Delegates to the Electrical Exhibition, so our time at Largs will be limited to about a month from 20th August to 19th Sept. I tell you all this that you may know our movements, and may plan to come to us accordingly. . . .

After going to Ireland for the opening of the Portrush electric railway, Sir William and Lady Thomson went to Southport for the British Association meeting on September 19th. He read two papers, one on Gyrostatics, and one on Asymmetry of Crystals.

The Address to the Midland Institute at Birmingham, on October 3rd, was on the Six Gateways of Knowledge. It dealt with the perception of the senses, the sixth sense in his enumeration being the sense of temperature, which he carefully discriminated from the sense of force or tactile sense, with which it is usually confounded.

Dealing with the possibility of a seventh sense—a magnetic sense—which, in spite of the failure of all attempts to detect it by powerful electromagnets, he still considered just possible, he took the opportunity to denounce "that wretched superstition" of spiritualism. The section dealing with sound and music, and with the mathematical view of sound, is of extreme interest. Mathematics he averred to be etherealized common sense, as logic is etherealized grammar.

After Birmingham came the visit to the Electrical Exhibition at Vienna. Sir William made experiments there with M. Abdank on batteries, and admired Ganz's large alternator. Returning *via* Berlin he spent four days with von Helmholtz, and reached England on November 3rd.

This year he presented to his old college, Peterhouse, in commemoration of the 600th anniversary of its foundation, a complete installation of electric light. The plant consisted of a small Lancashire boiler and horizontal engine, driving a Ferranti alternator and a Siemens exciter. It was used for the first time on the Queen's birthday, 1884; and after twenty-five years the same machines are still doing excellent work. Sir William also designed the needful switches and details for pendants and other fittings. The cost to him of this gift cannot have been less than £2000, and is believed to have been more.

On November 21 Sir William wrote to Darwin from Glasgow that he was going to Cambridge to stay a few days at Peterhouse to settle details about the electric lighting.

> The sad news of the death of Sir William Siemens came on us like a thunderbolt. We had travelled home from Vienna together, stopping in Berlin on our way, and I had just three days before his death answered a letter he had written me (now, as it appears, after he had been suffering from the fall), in which he said nothing of being ill, and was full of new plans for meeting at Sherwood, Cambridge, etc. It is a very great distress to us.

And on the 26th Lady Thomson wrote: "We have been at Sir W. Siemens's funeral to-day. His death has been a great grief to Sir William—so much so that I felt I must come up with him and look after him." An obituary notice by Thomson of his friend appeared in *Nature* of November 29th.

In presenting to Sir William Thomson the Copley Medal on November 30, the President of the Royal Society, Professor Huxley, pronounced the following eulogium:—

> The number, the variety, and the importance of Sir William Thomson's contributions to mathematical and experimental physics are matters of common knowledge, and the Fellows of the Society will be more gratified

than surprised to hear that the Council have this year awarded him the Copley Medal, the highest honour which it is in their power to bestow.

Sir William Thomson has taken a foremost place among those to whom the remarkable development of the theory of thermodynamics is due; his share in the experimental treatment of these subjects has been no less considerable; while his constructive ability in applying science to practice is manifested by the number of instruments, bearing his name, which are at present in use in the physical laboratory and in the telegraph office.

Moreover, in propounding his views on the universal dissipation of energy and on vortex motion and molecular vortices, Sir William Thomson has propounded conceptions which belong to the *prima philosophia* of physical science, and will assuredly lead the physicist of the future to attempt once more to grapple with those problems concerning the ultimate construction of the material world, which Descartes and Leibnitz attempted to solve, but which have been ignored by most of their successors.

Returning home, Sir William wrote to Lord Rayleigh, urging him to take up electrostatic measurement. There was further correspondence between them through the winter on electrical units, as another International Conference was fixed in Paris for April 28, 1884. Moreover, Lord Rayleigh had been chosen as President of the British Association for the meeting to be held in Montreal in August 1884, and Sir William Thomson was once again asked to preside over the Mathematical and Physical Section. The following selections from the letters of this time deal with some of these matters:—

THE UNIVERSITY, GLASGOW,
*Jan.* 20/84.

DEAR LORD RAYLEIGH—I send you the "Units" lecture and two others.

Many thanks for the information about Weber and the spinning method. The apparatus used by the B.A. Committee was made for me by White in Glasgow, and put to some rough tests by myself before it was put into the hands of the Committee, whose first reported experiments with it must have been made about January 1863. I don't know if anything of it was published before Aug. 1863, but I think it must have been in 1861 or early in 1862 that I set about having it made. If I find any notice of it prior to Aug. 1863 (or to Weber's date) I shall tell you, but it is not a matter of importance. I shall look for Weber's paper, however, immediately. I had never seen it and only knew of the *Electrodynamische Maasbestimmungen.* I shall be greatly interested to hear more of your silver weighings and further conclusions as to a standard for E.M.F. I suppose from your letter that you find Clark's all trustworthy to less than ·2 or ·3 per cent?

I am working hard at potential galvanometers to serve up to 200 volts, and electrometers for above that; and mho-meters to serve for practical current-meters, also integrators and regulators. As to electrometers, I had made the cylindric form roughly and heavily, which for absolute determinations has, as you say, the advantage of requiring no guard-border. But I thought the small flat disc easier to realize for work in potentials so small as 200 volts.

I hope you have been profiting by your stay in Bath, and that you are now feeling much better.—Yours very truly, WILLIAM THOMSON.

*P.S.*—Have you decided when and how you go to Montreal?

THE UNIVERSITY, GLASGOW,
*21st Feb.* 1884.

DEAR LORD RAYLEIGH—I should be very glad to hear from you how you use the Clark cell to get such constant results in respect of potential. I find that it varies enormously, and shows great polarization when allowed to flow through as great a resistance as 15,000 ohms. I could not use it with any good results for my potential galvanometers, whether by a zero method such as that of Poggendorff or other, unless it were confined to showing on an electrometer its potential, or the equality of its potential to that of some point of one of my circuits.—Believe me, yours very truly,

WILLIAM THOMSON.

*P.S.*—I am getting results that promise to come within $\frac{1}{5}$ per cent by a new form of Daniells (I described it in the *Telegraphic Review* and the *Electrician* about 3 weeks ago). I was sorry not to be able to look for you in the Laboratory when I was in Cambridge one day about a fortnight ago.

THE UNIVERSITY, GLASGOW,
*March* 5/84.

DEAR LORD RAYLEIGH—The Cunard Company have offered Lady Thomson and myself free passage to America and back in one of their ships. So we leave, in the *Aurania* if possible (that is if times suit), about the end of July. I hope this may also suit you and Lady Rayleigh. It will be very pleasant if we can make the voyage together.

As to my new form of standard Daniells, I have been getting excellent results ever since I wrote to you last. Here is a specimen of some got to-day in a new departure in the way of systematic short-circuiting. . . .

My newest long-range potential galvanometer is, when no resistance is added to its circuit, to have, I think, exactly 1000 R[ayleigh] ohms for its resistance. Its constant I shall *probably* determine by 5 of the new cells in series. I am, however, also making new forms of electromagnetic standardizers (on the fundamental principle of Weber's electrodynamometer), which are promising well, and may possibly in practice prove as accurate as the standard cells.

I hope in a few days to be able to send you a specimen of my new standard cell. I have been waiting to do so till I should get the plan fixed. Now I am only waiting till I can get a few cells made on a plan which I am fairly satisfied with as *fixed*.—Yours always truly, W. THOMSON.

UNIVERSITY, GLASGOW,
*Ap.* 25/84.

DEAR LORD RAYLEIGH—We set out to-night for Paris and hope to arrive to-morrow evening. Hôtel Chatham, Rue Daunou, will be our address. I should be very glad to hear from you your latest results, or to know that there is, in the hands of some member of the Congress in Paris, a printed paper, or papers, describing your work and results, also Glazebrook's, as to *ohm* specially, but also *volt* and electrochemical equivalent. A line or two from you to me, telling me what you would now take as the most probable value of the B.A., and the length of the column of mercury that, with $1^{m}/_{m}$ section, at 0° cent., has resistance equal to the true ohm.

I have an electrodynamometer (current standardizer) in progress, to give very good sensibility with current of 5 amperes through it; but fairly sensitive, I hope, with from 2 to 3 am. I hope next week, or at all events within a fortnight from now, it will be ready to be sent to you. I hope these strengths are not too great to be convenient for measurement by your standard (though I remember you said about $\frac{2}{3}$ of an amp.). The resistance of my new inst$^{t.}$ will be only about $8 \times 10^7$ (= ·08 ohm), so it will be very easy to get a strong enough current through *it* to give good sensibility, but I am afraid you may not get so strong a current easily in series with yours, which has far higher resistance.

Excuse this scrawl, written in a Senate meeting at which I am assisting!—Yours truly, WILLIAM THOMSON.

On June 22 he wrote again :—

I have a very promising table of expts. from my laboratory this morning, which seems quite to promise one very simple ampere-gauge with a range of 1 to 100 amperes.

The last week of June was spent with Sir Anthony Hoskins on board H.M.S. *Hercules*, with the reserve squadron, on a cruise to Heligoland and the Orkneys.

NETHERHALL, LARGS, AYRSHIRE,
*Aug.* 4.

DEAR LORD RAYLEIGH—Please don't forget to tell the printers to send me your Address. I leave Glasgow for L'pool on Friday, and L'pool on Saturday in the *Servia*; and I hope to receive it before I leave. (Will you let me have a proof though not finally corrected, if a finally corrected one cannot reach me in time.) We are very sorry you are not coming with us in the *Servia*.

I have made great advance (I think) in standard-current measures, for all currents from 10 milliamperes to 1000 amp^s. (or as many more as are to be measured in any case). I shall tell you about it when we meet. It is founded on the tendency of a ball or cube of soft iron towards places of greater force ($c \times$ rate of variation of $R^2$ in any direction = the component force in that direction, R being the "resultant magnetic force" of the field). This action, I find, is not sensibly influenced by magnetic retentiveness, and is very closely in proportion to the square of the strength of the current, through the whole practicable range.—Yours very truly,

WILLIAM THOMSON.

THE UNIVERSITY, GLASGOW,
*Aug.* 6/84.

DEAR LORD RAYLEIGH—Many thanks for the proof of your Address. I am very sorry, however, that I have put you to the trouble of sending me one with some of your corrections written on it. I thought the printers could have sent me one or I would have waited. I enclose you a copy of my Section A address. I hope the meeting, and generally your visit to America, will prove rather a rest and refreshment to you than a fatigue, but I shall remember what you say, should there seem to be any occasion for guarding against overwork. I hope the press-men (interviewers or others) will not prove a great plague!

I am a good deal troubled about the new milliamperemeter just now, but I hope to-morrow I may see things going better about it in the laboratory.—Believe me, yours very truly, WILLIAM THOMSON.

These letters show how greatly Sir William's time was taken up with his instruments. Von Helmholtz, too, had been struck with this earlier in the spring when paying a round of visits to his English friends, Tyndall, Huxley, Roscoe, and others.

He found Sir William absorbed in regulators and measuring apparatus for electric lighting, and for electric railways. To Frau von Helmholtz he wrote:—

On the whole, however, I have an impression that Sir William might do better than apply his eminent sagacity to industrial undertakings; his instruments appear to me too subtle to be put into the hands of uninstructed workmen and officials, and those invented by Siemens and von Hefner Alteneck seem much better adapted for the purpose. He is simultaneously revolving deep theoretical projects in his mind, but has no leisure to work them out quietly; as far as that goes I am not much better off.

Then immediately he adds:—

I did Thomson an injustice in supposing him to be wholly immersed in technical work; he was full of speculations as to the original properties of bodies, some of which were very difficult to follow; and, as you know, he will not stop for meals or any other consideration.

Sir William had indeed weighty matters on hand. He had agreed, more than a year before, to give in Baltimore, in the autumn of 1884, a course of lectures, as will be related in the next chapter; and he had chosen as the topic molecular physics, in relation to the wave-theory of light. His address for the British Association at Montreal was also in process of evolution, and the ever-haunting idea of a kinetic explanation of the properties of matter was again uppermost.

The title of the Montreal address was "Steps towards a Kinetic Theory of Matter." Beginning with the molecular processes that go on in a gas according to the kinetic theory, and throwing out a suggestion that the condition for equality of temperature between two gaseous masses is equal average amounts of kinetic energy per molecule, he discussed the nature of the repulsive motion which apparently characterises the action of free molecules during the impact at every collision. We must regard each molecule as being either a little elastic solid or else a configuration of motion in a continuous all-pervading liquid. The first step toward a molecular theory of matter had been this very kinetic theory of gases, of which all we had to the present time would be equally true if a gas consisted mechanically of little round, perfectly elastic solid pieces of matter flying about. But the difficulty remained unsolved that if each molecule is a continuous solid the whole translational energy must ultimately become transformed into vibrational energy. Even apart from this difficulty, the elasticity of a gas was, on this theory, only explained by making an assumption much more complex and more difficult to explain—the elasticity of a solid. This led him on to the next possible step, the possibility of explaining kinetically that property of matter which we call elasticity, as being itself a mode of motion. He therefore restated the principal points of his former lecture (see p. 743 above) on elasticity, and again described the gyrostatic model by which he imitated a common spring balance by use of fly-wheels, themselves devoid of elasticity. Another kinetic model could be imagined having irrotational circulation of a perfect liquid through apertures in solids immersed in it. Vortex rings and coreless vortices, under conditions consistent with stability, could also present phenomena like elasticity, the elasticity being due in reality to motion.

To the sectional meetings at Montreal Sir William made two communications: on a gyrostatic working model of the magnetic compass; and on safety-fuses for electric circuits.

The Montreal meeting over, many of the scientific men in attendance travelled south, *via* Niagara Falls, to Philadelphia, where an electrical exhibition was being held, and where also the American Association for the Advancement of

Science was holding its session. Here Sir William read a paper (the title only appears in the Proceedings) on the distribution of potential in conductors experiencing the electromagnetic effects discovered by Hall.

While in Philadelphia Sir William was induced to deliver, in the Academy of Music, on September 29, a popular lecture on "The Wave Theory of Light." This lecture,[1] as all who were present will remember, gave an excellent summary of points in the elementary theory of light, and touched upon the problem of the propagation of waves in the ether; elucidation being afforded by a model of a molecule embedded in jelly; while the mobility of matter through the ether was illustrated by the ready, if slow, passage of leaden bullets by their own weight through shoemakers' wax. The lecture was enlivened by several characteristic passages:—

"You in this country are subjected to the British insularity in weights and measures." "I hope the teaching of the metrical system will not be let slip in the American schools any more than the use of the globes." "I look upon our English system as a wickedly brain-destroying piece of bondage under which we suffer." "The luminiferous ether . . . is the only substance we are confident of in dynamics. One thing we are sure of, and that is the reality and substantiality of the luminiferous ether." "Some people say they cannot understand a million million. Those people cannot understand that twice two makes four. That is the way I put it to people who talk to me about the incomprehensibility of such large numbers. I say *finitude* is incomprehensible, the infinite in the

[1] Reprinted in *Popular Lectures*, vol. i. p. 300. See also p. 1035 *infra*.

universe *is* comprehensible. . . . What would you think of a universe in which you could travel one, ten, or a thousand miles, or even to California, and then find it come to an end? Can you suppose an end of matter or an end of space? The idea is incomprehensible."

From Philadelphia Sir William and Lady Thomson travelled on to Baltimore for the lectures which he was to give at the Johns Hopkins University.

Lady Thomson wrote on October 7th from the Mount Vernon Hotel, Baltimore, to Professor George Darwin:—

I hope you had a good voyage home. We have had nothing but hot weather since you left. We have never been able to keep in cool regions. We were two days in Boston and Cambridge, which we enjoyed *very* much. The only cool days we have had were at Southampton in Long Island, Sag Harbour and Shelter Island, on our way to Boston. We had to come back to Philadelphia on our way here, as Sir William had promised to give a lecture there on the 29th on the "Wave Theory of Light."

We came on here on the 30th, and we have been quite glad to be quiet in one place, but the heat continues very trying at night as well as in the day. Everything is hot and dry and parched. The leaves of the trees are falling instead of colouring, and the perpetual blue sky is quite fatiguing! We are *very* tired of this hot weather—thermometer rarely below 80° Sir Wm. is enjoying his lectures, I think, as much as his audience. They are going on capitally, and he has a most eager, interested audience of about 60 or 70, which is double what they expected, many of them Professors from all parts of the States.

But the Baltimore Lectures constitute so notable an episode that they require a chapter to themselves.

# CHAPTER XX

## THE BALTIMORE LECTURES

A UNIQUE episode in the career of Lord Kelvin was the course of twenty lectures on Molecular Dynamics which he delivered at Baltimore in the late summer of 1884. At a time when most of the American Universities were mere High Schools, with little or no post-graduate instruction, the Johns Hopkins University of Baltimore stood out alone in realizing that the glory of a university lay in advanced post-graduate studies and research. Not only did she fill her chairs with the leaders of thought, but she sought to free them from academic routine and administrative drudgery that they might the better labour for the few ripe students capable of high instruction. And to this she added the function of inviting European professors of distinction to give short courses in the lines of their own individual work. An invitation to deliver such a course was addressed to Sir William Thomson in 1882:—

PRESIDENT D. C. GILMAN to SIR WILLIAM THOMSON.

JOHNS HOPKINS UNIVERSITY, BALTIMORE,
*Aug.* 13*th*, 1882.

DEAR SIR WILLIAM THOMSON—I am very sorry that so long an interval has elapsed since your kind reply to an overture which Professor Sylvester made to you by my request, in respect to your coming to Baltimore and delivering some lectures. I beg you to hold me responsible for this delay, and to exonerate him entirely. There was a strong possibility that I should visit your country during the present summer, and if I had been able to do so, a personal interview would have enabled me to explain to you some circumstances which it would be tedious to dwell on in writing, particularly in regard to the time of your coming. But as my going is postponed, I write to say that the warm weather is a serious obstacle to work at the late period which you named, and the highest advantages of your visit would be promoted by an earlier appointment, if you can possibly make one. Prior to April 15, and after October 1, would be the periods most useful, and we should hope that you would lecture to advanced students in physics on such topics as you would choose yourself. If, during the winter of '82-3 or of '83-4 you could be persuaded to visit Baltimore, I would at once ask the Trustees to make you a formal proposition in place of this unofficial and informal letter. From what has been said by my colleagues, Prof[s]. Sylvester and Rowland, and by Prof. Wolcott Gibbs, I believe that the announcement of a University Course of lectures by you in Baltimore, on topics of your own choice, would give a strong impulse to the study of Physics in this country. I should also hope that your visit would be made pleasant in many ways.

If Lady Thomson has not forgotten an American visitor who took lunch with you in 1875, I beg leave to present my high regards, and I am, dear sir, very truly yours,

D. C. GILMAN,
*President J. H. U.*

Sir W. Thomson.

The letter is endorsed thus in Sir William's hand: "Rec[d]. Sep. 2, Netherhall. Answ[d]. Sep. 5, accepting for Oct. 1883, cond[l]. on permission to defer, by notice not later than July 1/83, to Oct. 1884." The response which follows was handed to Prof. Rowland for delivery to Sir William, whom he expected to meet in Paris :—

JOHNS HOPKINS UNIVERSITY, BALTIMORE,
*Oct.* 12, 1882.

DEAR SIR—I have the pleasure of replying to your note of Sept. 6 by entrusting these lines to the hand of my colleague Prof. Rowland, who hopes to see you soon in Paris, and to talk over with you the subject of your proposed lectures. I beg leave to assure you that your acceptance of our overtures has given great pleasure to all of our company who have heard of our correspondence. We hope it will prove to be possible for you to come to us in the autumn of 1883, but if not, the welcome will be in store for you in 1884.

This is an unofficial note. I hope in a few days (after their meeting, Nov. 6) to send you a formal note from our Trustees. Meanwhile I am, dear sir, yours with high regard, D. C. GILMAN.

JOHNS HOPKINS UNIVERSITY, BALTIMORE,
[*November*, 1882].

DEAR SIR—My colleague, Professor Rowland, has returned from the Electrical Congress, and I am extremely sorry to find that he did not deliver the enclosed letter, nor have any conversation with you in respect to your proposed visit.

I have been formally requested by the authorities of the University to invite you to come here and lecture at any time in the course of the next academic year convenient to you. We are somewhat at a loss as to what pecuniary compensation to offer you for this service. Supposing that you could hardly be persuaded to stay in

Baltimore more than a month, the Committee proposed that I should offer you your travelling expenses to and from Glasgow, and the sum[1] of $1000.

Allow me to repeat the assurance that we will do all in our power to make your visit pleasant, and we believe that your influence in the promotion of science in this country, by such a visit to this University, would be very strong, and would extend far beyond the immediate company of your hearers, and far beyond the period of your visit.—I am, dear sir, yours with high regard,

D. C. GILMAN.

Sir W. Thomson, F.R.S., etc.

Lord Kelvin's own account of the matter is given in his preface to the printed volume of 1904 :—

Having been invited by President Gilman to deliver a course of lectures in the Johns Hopkins University, on a subject in Physical Science to be chosen by myself, I gladly accepted the invitation. I chose as subject the Wave Theory of Light, *with the intention of accentuating its failures* ; rather than of setting forth to junior students the admirable success with which this beautiful theory had explained all that was known of light before the time of Fresnel and Thomas Young, and had produced floods of new knowledge splendidly enriching the whole domain of physical science. My audience was to consist of professional fellow-students in physical science ; and from the beginning I felt that our meetings were to be *conferences of coefficients*, in endeavours to advance science, rather than teachings of my comrades by myself. I spoke with absolute freedom, and had never the slightest fear of undermining their perfect faith in ether and its light-giving waves by anything I could tell them of the imperfection of our mathematics ; of the insufficiency or faultiness of our views regarding the dynamical difficulties of ether ; or of the overwhelmingly great difficulty of finding a field of action for ether among the atoms of

[1] [The sum eventually agreed upon was £400.]

ponderable matter. We all felt that difficulties were to be faced and not to be evaded; were to be taken to heart *with the hope of solving them if possible*; but at all events with the certain assurance that there is an explanation of every difficulty, though we may never succeed in finding it.

The lectures were delivered to an audience of twenty-one regular attenders, including some of the first physicists of the day, as may be seen from a few of the names:—

Lord Rayleigh (England); H. A. Rowland (Baltimore, Md.); Eli W. Blake, jun. (Providence, R.I.); Cleveland Abbe (Washington, D.C.); Albert A. Michelson (Cleveland, Ohio); Fabian Franklin (Baltimore); Arthur S. Hathaway (Baltimore); George Forbes (England); Henry Crew (Wilmington, Ohio); Louis Duncan (Baltimore); A. T. Kimball (Worcester, Mass.); Arthur L. Kimball (Baltimore); T. C. Mendenhall (Columbus, Ohio); Edward W. Morley (Cleveland, Ohio); R. W. Prentiss (Baltimore), etc. There was also a considerable number of casual attenders.

The course began on October 1, and ended on October 17. The lectures were either at 3.30 P.M. or at 5 P.M.; on five days at both times. Generally each lecture was divided into two parts, with a ten minutes' interval, during which the lecturer chatted with his students. From the first Sir William adopted an unconstrained conversational style, treating his hearers rather as fellow-workers in the problems under investigation than as an audience to be addressed *ex cathedra*. They, in their turn, worked for him by hunting up references, searching the University library for books, and tabulating solutions of problems; informally constituting, as he himself jokingly suggested, an arithmetical laboratory. His "twenty-one coefficients" they were called by one of themselves, in pleasant suggestion of the twenty-one coefficients by which in the most general case the elasticity of a body is defined.

Sir William Thomson had written out none of the twenty lectures beforehand; he had not indeed formulated a systematic skeleton of the series. Part of the extreme interest of the course arose indeed from his unpreparedness. Admitted to the very laboratory of his thoughts, his hearers became eye-witnesses of his methods, his amazing intuitive grasp, his headlong leaps, his mathematical agility, his perpetual recurrence to physical interpretations, his vivid use of mechanical analogies, and his incessant resort to models, sometimes actual, sometimes only mentally visualized, by which his meaning could be conveyed. His audience began to see that here was a man who, instead of taking at second-hand what other workers had found or written, thought things out for himself from first principles, making discoveries even while lecturing; and enjoying the surprise of finding that some of the things he was newly discovering for himself had already been discovered and published by others. For instance, on page 282 [1] we find:—

[1] The references are to the pages of the original papyrographed edition of 1884.

I was thinking about this three days ago, and said to myself, "There must be bright lines of reflexion from bodies in which we have those molecules that can produce intense absorption." Speaking about this to Lord Rayleigh at breakfast, he informed me of this paper of Stokes's, and I looked and saw that what I had thought of was there. It was perfectly well known, but the molecule first discovered it to me. I am exceedingly interested about these things, since I am only beginning to find out what everybody else knew, such as anomalous dispersion, and those quasi-colours, and so on.

And again (p. 120):—

I am ashamed to say that I never heard of anomalous dispersion until after I found it lurking in the formulas.

Happily for all concerned, one of the audience, Mr. A. S. Hathaway, who acted as reporter, was not only a trained mathematician, but an expert stenographer. Within three months a verbatim report of the lectures was produced by the papyrograph process in a limited edition which is treasured by its fortunate possessors. A graphic account of the lectures was also published in *Nature*[1] by one of the "coefficients," Professor George Forbes.

The difficulties in the way of accepting the wave theory of light Sir William stated as four in number:—

*First. Dispersion.*—The difficulty is to explain how the period of vibration of light of different colours can affect the velocity of their propagation through a medium, and can cause some colours to be more refracted than others. Of this phenomenon

[1] *Nature*, vol. xxxi. pp. 461-463, 508-510, and 601-603; March 19, April 2, and April 30, 1885. The author has, by the kind permission of Prof. George Forbes, drawn freely on these articles.

two explanations have been offered. The first is that of Cauchy, who ascribed it to heterogeneousness: some of the molecules in the structure must have sizes not infinitely small compared with that of the wave-length. The second is due to Helmholtz, who supposes the molecules to have a compound structure such that they have natural periods of vibration of their own. The space occupied by a molecule must be filled with a substance differing from the ether either in rigidity or in density, or in both respects. Thomson preferred Helmholtz's view, and in his first lecture set himself to devise a new model molecule consisting of a thin rigid shell, to the interior of which masses were fastened by springs. Several varieties of the spring-shell molecule were devised during the course of the lectures, and models of them shown. "It seems to me," he summed up, "that there must be something in this, that this, as a symbol, is certainly not an hypothesis, but a certainty" (p. 12).

*Second. The Ether.*—Here the difficulty is in mentally conceiving a medium so highly rigid that it propagates vibrations with the enormous velocity of 186,000 miles a second, and yet is so perfectly mobile that the earth and the heavenly bodies can apparently sweep through it without being retarded. This difficulty Sir William brushed aside as "not so very insuperable." To him there had long been an explanation, drawn from the behaviour of solid pitch or of Scotch shoemakers' wax. This substance is a brittle solid, of which you might make a rod that can vibrate like a rod of glass. But if a slab of it two inches thick is laid on the top of water in a glass jar twelve inches in diameter, and leaden bullets are placed on the top of the slab, in a few weeks they will be found to have passed right through it. The difference between the ether and the wax was a question of *time*. Let the luminiferous ether be looked upon as a wax which is elastic for excessively rapid vibrations, but capable of yielding to stresses that are of longer duration. "We do not know at this moment whether the earth moves dragging the luminiferous ether with it, or whether it moves more nearly as if it were through a frictionless fluid" (p. 8). If we consider the exceeding smallness of the period of a wave of light, we need not despair of understanding the property of the ether.

It is no greater mystery at all events than the shoemakers' wax. That is a mystery, as all matter is; the luminiferous ether is no greater mystery. We know the luminiferous ether better than we know any other kind of matter in some particulars; we know it in respect to the constancy of the velocity of propagation of light of different periods (p. 9).

Luminiferous ether must be a body of most extreme simplicity. It may perhaps be soft. We might imagine it to be a body whose ultimate property is to be incompressible; to have a definite rigidity for vibrations in times less than a certain limit, and yet to have the absolutely yielding character that we recognize in wax-like bodies when the force is continued for a sufficient time. It seems to me that we must know a great deal more of the luminiferous ether than we do. But instead of beginning with saying that we know nothing about it, I say that we know more about it than we do about air or water, glass or iron—it is far simpler; there is far less to know. That is to say, the natural history of the luminiferous ether is an infinitely simpler subject than the natural history of any other body (p. 10).

Another striking passage from the first lecture is—

In the first place, we must not listen to any suggestion that we must look upon the luminiferous ether as an ideal way of putting the thing. A real matter between us and the remotest stars I believe there is, and that light consists of real motions of that matter—motions just such as are described by Fresnel and Young, motions in the way of transverse vibrations. If I knew what the magnetic theory of light is, I might be able[1] to think of it in relation to the fundamental principles of the wave-theory of light. But it seems to me that it is rather a backward step from an absolutely definite mechanical motion that is put before us by Fresnel and his followers to take up the so-called electromagnetic theory of light (p. 6).

*Third. Refraction and Reflexion.*—The difficulty here is of a different order. Green's equations for the refraction and reflexion of polarized waves of light do indeed approximately yield results for the relative amounts of light that are refracted and reflected at different angles of incidence; but when followed up not qualitatively only, but quantitatively, they differ considerably from the facts. For example, they fail to account for the almost complete extinction observed in the reflexion of polarized light at the polarizing angle.

*Fourth. Double Refraction.*—In this case the

[1] This passage and others of similar import led his audience to conclude that at that date the lecturer had never read Clerk Maxwell's book!

difficulty is that according to the elastic theory, when the medium is displaced during wave-propagation, the forces tending to restitution must depend on the plane of the distortion, whereas if the Huygens wave-surface is true they must depend, not on the plane of the distortion, but on the direction of the vibration. And this is inadmissible, because Stokes has found by minute experiment that the Huygens wave-surface is most accurately obeyed by light.

The difficulties to be faced having been thus stated, the rest of the course was devoted to their discussion. But each difficulty had to be regarded from three points of view: (1) the propagation of a disturbance through an elastic medium regarded as a whole, or from the *molar* standpoint; (2) the character of *molecular*, as distinguished from molar, vibration; (3) the influence of molecules on the propagation of waves. Some of the lectures were definitely divided into two parts, one devoted to the molar, the other to the molecular view.

With the second lecture began a systematic mathematical exposition of the equations of motion in an elastic solid, with their three principal distortions and three principal dilatations, and the twenty-one coefficients by which they are related in the quadratic function expressing the general equation of energy. With the second half of the lecture an abrupt change was made to consider an apparatus consisting of three heavy masses suspended below one another by spiral springs, and which was shown to possess three independent periods of vibration, depending on the stiffnesses of the springs and the values of the attached masses. The dynamical problem presented by the apparatus led to the equations of motion for an elastically embedded molecule. In the third lecture the two divisions of the subject were each further elaborated. The treatment is intensely characteristic of Sir William Thomson's methods. He scorned all formulas (he himself always said formulas, not formulæ), however neat, if they merely served as mathematical exercises, or gained their apparent precision from the adoption of unwarranted assumptions. All his formulas must have a physical meaning, even if he had to load them with the symbols of undetermined quantities,—seeming, indeed, to glory in the possibilities of future discoveries which lay behind the equations. He would write them out on the black-board in full, though they might have been left to be inferred from considerations of symmetry. "The expenditure of chalk," he remarked, "is often a saving of brains." Again and again he referred to the *aphasia* of mathematics, its inarticulateness to express physical ideas. "The old mathematicians used neither diagrams to help people to understand their work, nor words to express their ideas. It was formulas, and formulas alone. Faraday was a great reformer in that respect, with his language of 'lines of force,' etc." Again, after a mathematical argument, turning on the circumstance that the differential coefficient of a certain continued fraction with respect to the period was essentially negative, he illustrated the inference by drawing attention to the effect of applying a push or a pull at the end of the movement of one of his vibrating models, and remarked: "From looking at the thing, and learning to understand it by making the experiment if you do not understand it by brains alone, you will see that everything I am saying is obvious." Yet he quoted approvingly from Green the words, "I have no faith in speculations of this kind unless they can be reduced to regular analysis"; and he praised Stokes that he "speculates, in a way, but is not satisfied without reducing it to regular analysis."

In the fourth lecture he dealt with the difficulty that in the solid elastic theory a displacement must give rise not only to lateral vibrations (propagated by shears or distortions in the medium), but also to a longitudinal or pressural wave (propagated in virtue of compressions and rarefactions in the medium). Yet apparently in the luminiferous ether there is nothing found that corresponds to this compressional wave.

We ignore this condensational wave in the theory of light. We are sure that its energy, at all events, if it is not null, is very small in comparison with the luminiferous vibrations we are dealing with. But to say that it is absolutely null would be an assumption that we have no right to make. When we look through the little Universe that we know, and think of the transmission of electrical force, and of the transmission of magnetic force, and of the transmission of light, we have no right to assume that there may not be something else that our philosophy does not dream of. We have no right to assume that there may not be condensational vibrations in the luminiferous ether. . . . The fact of the case as regards reflexion and refraction is this, that unless the luminiferous ether is absolutely incompressible, the reflexion and refraction of light must generally give rise to waves of condensation. Waves of distortion may exist without waves of condensation, but waves of distortion cannot be reflected at the bounding surface of two mediums without exciting in each medium a wave of condensation . . . and may after all the law of electric force not depend on waves of condensation? (p. 41).

Suppose that we have at any place in air, or in luminiferous ether (I cannot distinguish now between the two ideas) a body that, through some action we need not describe, but which is conceivable, is alternately positively and negatively electrified; may it not be that this will be the cause of condensational waves? Suppose this, that we have two spherical conductors united by a fine wire, and that an alternating electromotive force is produced in that fine wire, for instance with an alternating dynamo-electric machine; and suppose that sort of thing goes on away from the disturbance—at a great distance up in the air, for example. The result of the work of that dynamo-electric machine will be that one conductor will be alternately positively and negatively electrified, and the other conductor negatively and positively electrified. It is perfectly certain, if we turn the machine slowly, that in the neighbourhood of the conductors we will have alternately positively and negatively electrified elements with reversals, perhaps two or three hundred per second of time, with a gradual transition from negative, through zero, to positive, and so on; and the same thing all through space; and we can tell exactly what the potential is at each point. Now, does any one believe that, if that revolution was made fast enough, the electrostatic law would follow? Every one believes that if that process be conducted fast enough, several million times or millions of million times per second, we should be far from fulfilling the electrostatic law in the electrification of the air in the neighbourhood. It is absolutely certain that such an action as that going on would give rise to electrical waves. Now it does seem to me probable that those electrical waves are condensational waves in luminiferous ether ; and probably it would be that the propagation of these waves would be enormously faster than the propagation of ordinary light waves (p. 42).

This notion of a pressural wave haunted the lectures almost to the end. In the seventeenth lecture, dealing with reflexion and refraction at an interface in a case where the forces across the interface are balanced, he continued :—

That leaves a clean simple problem of dynamics, and yet people have been working at it for fifty years and have left it in a very sadly muddled condition, with the exception of the clear, accurate, and very comprehensive papers of Green and Rayleigh. The thing that has introduced the difficulty, and makes this a more complicated difficulty than the other cases, is the pressural wave. The pressural wave, in fact, has been the *bête noir* of this problem. I do not know how Cauchy treats the animal. Somehow, he introduces fallacious terms involving consumption of energy. MacCullagh and Neumann killed the animal with bad treatment. Sam. Haughton yoked it to an Irish car, and it would not go. Green and Rayleigh have treated it according to its merits, and it has escaped whipping at their hands (p. 233).

In the fifth lecture, after a preliminary discussion of the periods of vibration of one of his spring models, Sir William suddenly interrupted himself in a passage which, as corrected, runs :—

No more of this now, however. It is fiddling while Rome is burning to be playing with trivialities of a little dynamical problem when phosphorescence is in view, and when explanation of the refraction of light in crystals is waiting. The difficulty is, not to explain phosphorescence and fluorescence, but to explain why there is so little of sensible fluorescence and phosphorescence. This molecular theory brings everything of light to fluorescence and phosphorescence. The state of things as regards our complex model-molecule would be this: Suppose we have this handle P moved backwards and forwards until everything is in a perfectly periodic state. Then suddenly stop moving P. The system will continue vibrating for ever with a complex vibration which will really partake something of all the modes. That, I believe, is fluorescence (p. 55).

From this there was developed a consideration of the question of the actual constitution of ordinary light. Lord Rayleigh, at Montreal, had emphasised the distinction between the velocity of a group of waves and that of the waves themselves:—

It seems to be quite certain that what he said is true. But here is a difficulty which has only occurred to me since I began speaking to you on the subject; and I hope, before we separate, we shall see our way through it. All light consists in a succession of groups. . . . Take any conceivable supposition as to the origin of light, in a flame, or a wire made incandescent by an electric current, or any other source of light; we shall work our way up from these equations which we have used for sound to the corresponding expression for light from any conceivable source. Now if we conceive a source consisting of a motion kept going on with perfectly uniform periodicity, the light from that source would be plane-polarized, or circularly polarized, or elliptically polarized, and would be absolutely constant. In reality there is a multiplicity of successions of groups of waves, and no constant periodicity. One molecule, of enormous mass in comparison with the luminiferous ether that it displaces, gets a shock, and it performs vibrations until it comes to rest or gets a shock in some other direction; and it is sending forth vibrations with the same want of regularity that is exhibited in a group of sounding bodies consisting of bells, tuning-forks, organ-pipes, or all the instruments of an orchestra played independently, in wildest confusion, every one of which is sending forth its sound, which, at large enough distances from the source, is propagated as if there were no others. We see thus that light is essentially composed of groups of waves (p. 56).

I want to lead you up to the idea of what the simplest element of light is. It must be polarized, and it must consist of a single sequence of vibrations. A body gets a shock so as to vibrate; that body of itself then constitutes the very simplest source of light that we can have; it produces an element of light. An element of light consists essentially in a sequence of vibrations. . . . One of you has asked me if I was going to get rid of the subject of groups of waves. I do not see how we can ever get rid of it in the wave-theory of light. We must try to make the best of it, however (p. 66).

I have tried to represent a sudden start, and a gradual falling off in intensity. Why a sudden start? Because I believe that the light of the natural flame, or of the arc light, or of any other known source of light, must be the result of sudden shocks from a number of vibrators (p. 95).

One exceedingly subtle point was raised, to which no answer appears yet to have been given, viz. whether, at the commencement of the impact of a beam of light for, say, the first thousandth of a second, there will not be an initial state of things different from the subsequent steady state, with a kind of initial fluorescence, and possibly a different refraction or reflexion from that subsequently persisting.

Sellmeier had deduced from Fizeau's experiments that in each train of waves there was no serious falling off for 50,000 vibrations. Is the diminution of amplitude in the course of several million vibrations practically nil? Possibly not: it was a dynamical question. Helmholtz had introduced into the equations certain terms to explain possible diminution by viscosity.

I must still say that I think Helmholtz's modification is rather a retrograde step. It is not so perhaps in the mathematical treatment; but at the same time Helmholtz is perfectly aware of this kind of thing that is meant by viscous consumption of energy. He knows perfectly well that that means conversion of energy into heat; and in introducing it he is throwing up the sponge, as it were, so far as the fight with the dynamical problem is concerned (p. 98).

A characteristic passage indeed, showing the constant habit of appraising a mathematical argument by its physical bearings. But, again, the formulas must be studied from another point of view, as is shown by the following passage relating to the determinants of the equations of connected masses:—

Now, as to the calculations. I do not suppose anybody is going to make these calculations;[1] but I always feel in respect to arithmetic somewhat as Green has expressed in reference to analysis. *I have no satisfaction* in formulas unless I feel their arithmetical magnitude—at all events, when formulas are intended for operations of that kind (p. 72). . . . I should think something like an arithmetical laboratory would be good in connection with class work, in which students might be set at work upon

[1] Some of them were, in fact, made by Prof. E. W. Morley during the course, to Lord Kelvin's great satisfaction.

problems of this kind, both for results and in order to obtain facility in calculation (p. 73).

Throughout the lectures there are frequent references to Lord Rayleigh's investigation of the polarization of light in the sky, and its blue colour, attributable to the presence of minute particles. This has a bearing on the question of density and rigidity of ether :—

> The observed polarization of the sky supports the supposition (which is as much as the incertitude of the experimental data allows us to judge) that the particles, whether they be particles of water, or motes of dust, or whatsoever they may be, act as if they were little portions of the luminiferous ether of greater density than, and not of different rigidity from, the surrounding ether.

By the end of the ninth lecture the spring-shell molecule had undergone many modifications to afford an explanation of anomalous dispersion and other optical properties. With the tenth lecture the propagation of waves, and the energy equations for waves advancing in a medium, came into discussion. In connection with this the audience was recommended to read the memoirs of Poisson and of Cauchy. "The great struggle of 1815 (that is not the same idea as *la grande guerre de* 1815) was, who was to rule the waves, Cauchy or Poisson?" He therefore suggested to the arithmetical laboratory to take up the case of wave-propagation in which the velocity is dependent on the wave-length. The spring-shell molecules were again to be taken and put into the ether, and the question examined, what will be the velocity of propagation under various hypotheses as to the masses of the attached molecules, and how much it will be modified by their presence? Cauchy and Poisson give only symbols, and occasionally numerical results: they do not give any diagrams or graphic representations; it would repay any one going into the subject to work out graphically all varieties of the problem of deep-sea waves.

At this point of the course the advance was made from elastic solids that are isotropic to those which are aeolotropic,[1] or have unequal properties in different directions. This led back again to Green's twenty-one coefficients or moduluses in all their generality; and again there were new models produced to illustrate aeolotropic conditions, to facilitate a dynamical theory of the propagation of light in crystalline media. One of these was a parallelepipedal structure with eight rings at the eight corners, connected by cords running through the rings along the edges of the parallelepiped.

Now Navier's and Poisson's theory of elasticity give as an essential a fixed relation between the compressibility and the rigidity, such that if it were true it would make an incompressible elastic solid impossible. On this the comment is:—

It is curious that they did not notice that jelly is

[1] In a later passage (p. 214) we find: "In a structure as a whole, properties are produced in virtue of the manner of the structure. In fact, all structures of iron-work, ties, and bracings, etc., are such that if we imagine a myriad of them put together—built up, as it were, like bricks—we should have an aeolotropic elastic solid."

practically incompressible. It is a wonder that they did not try it, and see that it did not fulfil Poisson's ratio. Their mistake was due to the vicious habit in those days of not using examples and diagrams. In the *Mécanique céleste* you find no diagrams, nor in Lagrange, nor in Poisson's splendid memoir on waves (p. 129).

Then follows a passage intensely characteristic :—

Although the molecular constitution of solids supposed in these remarks and mechanically illustrated in our model is not to be accepted as true in nature, still the construction of a mechanical model of this kind is undoubtedly very instructive, and we could not be satisfied unless we could see our way to make a model with the eighteen independent moduluses. My object is to show how to make a mechanical model which shall fulfil the conditions required in the physical phenomena that we are considering, whatever they may be. At the time when we are considering the phenomena of elasticity in solids, I want to show a model of that. At another time, when we have vibrations of light to consider, I want to show a model of the action exhibited in that phenomenon. We want to understand the whole about it ; we only understand a part. It seems to me that the test of "Do we or do we not understand a particular point in physics ?" is "Can we make a mechanical model of it ?" I have an immense admiration for Maxwell's mechanical model of electromagnetic induction. He makes a model that does all the wonderful things that electricity does in inducing current, etc. ; and there can be no doubt that a mechanical model of that kind is immensely instructive, and is a step towards a definite mechanical theory of electromagnetism (p. 132).

This use of the model as affording a mental picture of a system endowed with particular properties was now to be directed to elucidating the theory of the propagation of light through crystalline media. The way was not yet quite clear.

But if the war is to be directed to fighting down the difficulties in the undulatory theory of light, it is not of the slightest use towards solving our difficulties for us to have a medium which will kindly permit distortional waves to be propagated through it, even though it be aeolotropic. . . . What we want is a medium which, when light is refracted and reflected, will under all circumstances give rise to distortional waves alone (pp. 141-142).

In short, the difficulty of the compressional wave, which Green's theory did not exclude, was again cropping up. While pointing out that this difficulty could be disposed of by simply assuming incompressibility in the medium, and reiterating that the lack of any corresponding facts of observation proved that any such action, if it exists, must be exceedingly small, the lecturer affirmed his belief, not "as a matter of religious faith, but as a matter of strong scientific probability," that such waves exist, and that the velocity of this unknown condensational wave is the velocity of propagation of electrostatic force.

In discussing the properties of aeolotropic solids Sir William Thomson referred to Rankine's nomenclature by which he sought to elucidate the elasticity of solids :—

I must read to you some of Rankine's fine words. . . . Any one who will learn the meaning of all these words will obtain a large mass of knowledge with respect to an elastic solid. The words "strain" and "stress" are due to Rankine, "potential energy" also. Hear the grand words : "Thlipsinomic, Tasinomic, Platythliptic,

. . . Plagiotatic, Euthythliptic, etc." (p. 161). . . . I explained to you yesterday Rankine's nomenclature of thlipsinomic and tasinomic coefficients. In a certain sense these may be all called moduluses of elasticity. I have defined a modulus as a stress divided by a strain, following the analogy of Young's Modulus. If we adhere to that then the tasinomic coefficients are moduluses, and the thlipsinomic coefficients are reciprocals of moduluses (p. 183). . . . Rankine was splendid in his vigour and the grandeur of his Greek derivatives. Perhaps he overdid it, but I do not like to call it an error. We cannot all use his words, but we learn from them in reading his papers. Instead of his "platytatic" and "platythliptic" coefficients I use the much less grand and more colloquial expressions "sidelong normal" and "sidelong tangential" coefficients (p. 185).

Shortly after the middle of the course a new model was introduced, which was immediately dubbed the "wiggler." A steel wire was hung vertically, and five or six horizontal wooden laths, two feet long and two inches wide, were attached across it one above the other. These laths were loaded at their ends with weights, the weights on each lath being smaller than those on the one above it. The lowest of the laths was connected to a bifilar pendulum by which forced vibrations of various periods could be impressed on the system. It illustrated the action of a compound loaded molecule when subjected to vibrations from an external source. By this model it was shown that the system possessed several critical periods, each corresponding to the natural period of some one of the successive vibrators. At the critical period for any one vibrator all those below it are vibrating in one direction, while the particular vibrator and all above it are moving in the opposite direction. The model is admirably adapted to explain absorption and also anomalous dispersion.

And so the lectures went on from day to day with delightful discursiveness. Here and there a reference to Stokes—"I always consult my great authority, Stokes, whenever I get a chance"—or a hit at the "brain-wasting perversity of the insular inertia which still condemns British engineers to reckonings of miles and yards and feet and inches, and grains and pounds, and ounces and acres"—or an ejaculation that "there are no paradoxes in science"! Questions by the hearers set the lecturer off on new trains of thought. Discussions started at the lecture were continued at the supper table. The whole seemed one animated conference. But the continual discursiveness of the lecturer became ominous. "How long will these lectures continue?" asked President Gilman one day of Lord Rayleigh, while walking away from the lecture-theatre. "I don't know," was the reply; "I suppose they will end some time, but I confess I see no reason why they should."

Lecture sixteen introduced the question of the mass of the ether. "We have not the slightest reason to believe the luminiferous ether to be imponderable." To this in November 1899 Lord Kelvin added an emphatic note, "I now see that we have the strongest possible reason to believe that the ether is imponderable. . . . But is there any gravitational attraction between different portions of ether? Certainly not. . . . We must believe ether to be a substance outside the law of universal gravitation."

In the later lectures there was a discussion of the so-called rotation of the plane of polarization by quartz, discovered by Arago, which was explained by ascribing to the medium a spiral molecular structure; and of the true rotation by magnetism of the plane of polarization, discovered by Faraday. These properties again were illustrated by models —the latter effect being simulated by a new kind of molecule with a gyrostat spinning within the shell. But having thus devised a crude model Sir William set it at once aside:—

> Why do I not go into it, and try to make it a part of our molecular dynamics? I answer, because I cannot bring out the law of inverse proportionality to the square of the wave-length, which observation shows to be somewhat approximately the law of the phenomenon. Until a week ago, I thought that by putting a fly-wheel somehow or other into our molecule I could get a rotary effect, according to which the magnitude would vary according to two terms, one inversely as the wave-length, the other inversely as its cube. . . . But, alas! my results give me another law; not more effect with greater frequency, but less effect with greater frequency, according to the square of the wave-length. I therefore lay it aside for the present, but with perfect faith that the principle of explanation of the thing is there (p. 243).

In an appendix, added November 1, 1884, and incorporated in the stenographic Report, Sir William Thomson described an "improved gyrostatic molecule" having, within a massless spherical shell, two fly-wheels on an axis jointed with a ball-and-socket joint between them, producing a gyrostatic effect when the molecule is accelerated in any direction except along the axis. If one imagines minute gyrostatic molecules of this sort embedded in the ether, the influence of their rotation on the translational motion will, if they are small enough, give the same law as that observed by optical experiment.

Most of the following lectures were subsequently rewritten in later years by their author; but as originally given they well deserve careful study. The twentieth lecture, in particular, is too good to be lost. It began by a reference to Rankine's suggestion that in a crystal its inertia might be different for forces in different directions. This suggestion, for which there is no observed foundation in physics, he connected with Rankine's theory of molecular vortices, and remarked on the suggestiveness of the title :—

Rankine was that kind of genius that his names were of enormous suggestiveness ; but we cannot say that always of the substance. We cannot find a foundation for a great deal of his mathematical writings, and there is no explanation of his kind of matter. I never satisfy myself until I can make a mechanical model of a thing. If I can make a mechanical model I can understand it. As long as I cannot make a mechanical model all the way through I cannot understand ; and that is why I cannot get [this is probably the reporter's Americanism for the word "accept"] the electromagnetic theory. I firmly believe in an electromagnetic theory of light, and that when we understand electricity and magnetism and light we shall see them all together as parts of a whole. But I want to understand light as well as I can, without introducing things that we understand even less of. That is why I take plain dynamics. I can get a model in plain dynamics; I cannot in electromagnetics. But so soon as we have rotators to take the part of magnets, and something imponderable to take the part of magnetism, and realize by experiment Maxwell's beautiful ideas of electric displacements, and so on, then we shall see electricity, magnetism, and light closely united and grounded in the same system (pp. 270-271).

If the nett result of these lectures was to leave the impression that the difficulties in the way of the wave-theory of light—and in particular those connected with reflexion, refraction, and double-refraction—were still unsurmounted, it certainly left on the hearers no sense of despair. The optimism which would leave an undecided problem in perfect faith that the ultimate solution would be forthcoming in due time was as notable as the determination to attack the difficulties all the more strenuously *because* they were difficult. As the record of a piece of living intellectual effort it would be hard indeed to find a parallel to these twenty Baltimore lectures. They left their mark on physical science in two continents, and influenced Lord Kelvin's thought for the rest of his life. His own farewell to his audience was brief:—

I am exceedingly sorry that our twenty-one coefficients are to be scattered; but though scattered far and wide, I hope we will still be coefficients working together for the great cause we are all so much interested in. I would be most happy to look forward to another conference, and the one damper to that happiness is that this is now to end, and we shall be compelled to look forward for a time. I hope only for a time, and that we shall all meet again in some such way. I would say to those whose homes are on this side of the Atlantic, come on the other side, and I will welcome you heartily, and we may have more conferences. Whether we have such a conference again on this side or on the other side of the Atlantic, it will be a thing to look forward to, as this is looked back upon as one of the most precious incidents I can possibly have. I suppose we must say farewell (p. 288).

At a dinner-party the previous evening, given by President Gilman to Sir William Thomson and his band of hearers, the following humorous verses were read by one of their number, whose identity is sufficiently attested by the signature :—

THE LAMENT OF THE TWENTY-ONE COEFFICIENTS IN PARTING FROM EACH OTHER AND FROM THEIR ESTEEMED MOLECULE.

An aeolotropic molecule was looking at the view,
Surrounded by his coefficients, twenty-one or -two,
And wondering whether he could make a sky of azure blue
With platytatic *a b c*, and thlipsinomic *Q*.

They looked like sand upon the shore, with waves upon the sea,
But the waves were all too wilful and determined to be free;
And in spite of *n*'s rigidity they never could agree
In becoming quite subservient to thlipsinomic *P*.

Then web-like coefficients and a loaded molecule,
With a noble wiggler at their head, worked hard as Haughton's mule;
But the waves all laughed and said, "A wiggler, thinking he could rule
A wave, was nothing better than a sidelong normal fool."

So the coefficients sighed, and gave a last tangential skew,
And *a* shook hands with *b* and *c* and *S* and *T* and *U*,
And with a tear they parted; but they said they would be true
To their much-belovèd wiggler, and to thlipsinomic *Q*.

Signed (*g.f.*) A CROSS COEFFICIENT NOW ANNULLED.

Before leaving Baltimore, Sir William gave, by request, a lecture to students in Hopkins Hall on The Rigidity of the Earth.

About a year later the twenty-one coefficients subscribed for one of Professor Rowland's fine diffraction gratings, and sent it as a present to Sir William Thomson. The pleasure this gave him is recorded in the following letter to Prof. T. C. Mendenhall:—

GLASGOW, *5th Dec.* 1885.

DEAR PROF. MENDENHALL—I wrote to Prof. Rowland acknowledging the receipt of the grating, but I ought before now to have thanked all the other Coefficients for their kindness in giving it to me. I should feel greatly obliged if you would transmit to those of the Coefficients who are in America my heartiest thanks for their great kindness, and say to them that it will be a permanent memorial to me of the happy three weeks of 1884 when we were together in Baltimore.

One of the Coefficients, Prof. Davies, is here just now in Glasgow, but I am sorry to say he is very unwell. . . .

After the British Association Meeting at Aberdeen I was delighted to be able to show the grating to some of our English appreciators, including one of the Coefficients, George Forbes, and Lord Rayleigh (whom we may consider as at all events a partial Coefficient), and to Prof. Fitzgerald of Trinity College, Dublin, Oliver Lodge of Liverpool, Glazebrook of Cambridge, and Capt. Creak of the Compass Dept. of our Admiralty, who came to stay with us at Netherhall, our country house, for a few days on their way south. We had no sunlight to work with, but we got the double sodium line in the first and second spectrums from a salted spirit-lamp flame exceedingly well, and we were all delighted with the result. I had never myself seen anything like it before.

Lady Thomson joins me in kind regards.—Yours very truly, WILLIAM THOMSON.

On December 25, 1886, Sir William wrote to Professor Mendenhall, sending to the Coefficients a little Christmas greeting in the shape of copies of a series of articles on Stationary Waves in Flowing Water. To the last he looked back with unalloyed pleasure to the days spent at Baltimore.

# CHAPTER XXI

## GATHERING UP THE THREADS

SIR WILLIAM THOMSON was now sixty years old. He had, as we have seen, declined in 1871, and again in 1879, to occupy the Cavendish Professorship at Cambridge. In the autumn of 1884 it became known that Lord Rayleigh, who had filled it with great distinction since Clerk Maxwell's death, would not longer hold the Chair. For the third time overtures were made to Sir William Thomson. His letter to George Darwin speaks for itself:—

THE UNIVERSITY, GLASGOW,
*Nov.* 20/84.

DEAR DARWIN—I am afraid it cannot be—alas, alas—The wrench would be too great. I began taking root here in 1831, and have been becoming more and more fixedly moored ever since. I have things in train here to allow me to make the most of my capacities for work, and to make a new departure such as the Cavendish La^by^ and Prof^p.^ would be a life's work again. Who ever becomes professor must devote himself to the work—an excellent thing for any one beginning life, or looking for a fixed place in scientific work. To me it could not be otherwise than wholly a diminution of effective work. Similar questions have been before me, on several occasions, since, in fact, the first foundation of the Cambridge Chair, and in some other important directions. Each time I felt forced to the conclusion that Glasgow was the place for me. I have continuously felt ever since that the conclusion was right: and I feel more strongly than ever that my work is cut out for me here, and that any change would be a loss, however tempting in every way the change might be. . . .

I am frightfully under water about work for Baltimore (supplements I want to send off, at the latest, next week) for the papyrograph of my lectures there. This has kept me from answering your letter about Canadian tides. I thoroughly agree with your letter as far as I can judge, but I want to look at charts and tide tables to see if I have anything else to suggest. Next week I hope to do so. I hope meantime you are not inconvenienced by my delay.

My wife joins in kind regards to you and the Plumian Professorin.—Yours always truly,

WILLIAM THOMSON.

On December 7 Sir William again wrote, suggesting that Stokes should be urged to take the chair, and discussing possibilities. He added:—

I am still at high pressure for Baltimore Lectures report; sending a despatch of supplement every mail. I am at the moment in great trouble about thin metallic films. The rotation of the plane of polarization due to magnetism in a thin translucent iron film (discovered by Kundt, see *Phil. Mag.*, Oct. '84) *will not* come out right, to agree with Kerr's result for reflection at a polished pole, and my formula for both. I have been under torture nine days about this.

On 12th Lady Thomson wrote to Darwin that they were coming to Cambridge on 19th, and this time to Peterhouse, because of the electric lighting and the impending Sexcentenary. "Sir William," she said, "is much exercised over the Cavendish

Professorship. . . . My man is Professor Tait. . . . Sir William never goes to Edinburgh and to his laboratory without saying what a splendid experimenter he is: but he could not take it even if there were any chance of his getting it. The emolument would be an insuperable barrier. He has much more where he is."

To Sylvester, Sir William wrote on December 18:—

Till a few days ago I have been desperately hard at work on supplements to my Baltimore Lectures, which I have been sending to Mr. Hathaway to be incorporated with his shorthand report. I got my last despatch away only last Saturday, after which I felt delightfully free; but of course shoals of other things came on which had been delayed, so I have not had much breathing time yet.

Among these other matters were three papers read to the Edinburgh Royal Society, and details of electric lighting inventions, which form the topic of the following letters of December 3 and December 9 respectively, to Professor J. A. Ewing:—

THE UNIVERSITY, GLASGOW,
*3rd Decr.* 1884.

DEAR EWING—I have been thinking over your form of spring for the lamp-holder, and I have come to the conclusion it could not be patented usefully. . . . Curiously enough, in connection with my flat spring holder for lamps, I have, in all my hanging lamps at Peterhouse, a spiral spring on the wire which holds the lamp glass, with exactly the same kind of elasticity acting, as you have chosen for giving the horizontal forces I want in the platinum rings of the lamp. My flat spring is of course the very simplest way of producing the forces required. . . .—Yours truly, WILLIAM THOMSON.

THE UNIVERSITY, GLASGOW,
9 *Decr*. 1884.

DEAR EWING—None of the patents are yet published, I believe, but I shall enquire about them and have them pushed on as soon as possible. I have dropped the electromagnetic safety break, because after I had put up 20 or 30 in Peterhouse I found that the slamming of doors, etc., upset them. They had worked perfectly in my laboratory. I regret this the less, however, because J. T. Bottomley and I have worked out an improved safety fuse which is simpler and cheaper, and in ordinary circumstances surer than the electromagnetic safety break. . . .—Yours very truly, WILLIAM THOMSON.

Peterhouse held high celebration on December 22, 1884, of the Six-hundredth Anniversary of its Foundation. The evening banquet was graced by Prince Edward of Wales, the Marquis of Hartington, and many celebrated men, amongst them the U.S. Ambassador Mr. J. Russell Lowell, Mr. Matthew Arnold, and Sir William Thomson, himself its most distinguished Fellow. The eighth toast,[1] proposed by Thomson, was that of "Other Seats of Learning," to which Lowell responded.

The year was concluded with visits to Lord Derby at Knowsley, to Lady Thomson's brother-in-law, Mr (now Sir) Alexander Hargreaves Brown,

[1] The toast list was long and the banquet lasted late. Near the close, Professor (now Sir James) Dewar proposed the toast of "Pure and Applied Science," calling on Sir Frederick Bramwell to reply. Rising slowly to his full height, and with a twinkle in his eye, the eminent engineer thanked them, and explained that at this late hour of the evening the only example of applied science that occurred to him was the application of the lucifer-match to the domestic candle. Seizing a pencil, Lowell instantly wrote on the back of his menu the following epigram :—

Oh ! brief Sir Frederick, would that we could catch
His happy humour, and could find his match.

at Druid's Cross, near Liverpool, and to Mr. William G. Crum, Knutsford.

*Jan.* 1, 1885,
MERE OLD HALL, KNUTSFORD.

DEAR HELMHOLTZ—I have not yet seen the Baltimore Lectures, but I believe they will come very soon. One of the Johns Hopkins mathematicians, Hathaway, who is also a shorthand writer, undertook to make a report of the lectures and to bring it out in "papyrograph"; and I had a telegram from him a few days ago saying it was nearly ready. I have no doubt that Mr. Hathaway's work will be well done, but I am afraid to see *my part* of the result! I hope to make something better of it in the course of a year or so, and to bring it out in a printed volume. Till this is done it may be advisable *not* to let the labour of translation be undertaken. We enjoyed our visit to America very much (notwithstanding a degree of heat which, we were told, was "quite exceptional"; but the American climate seems exceptionally hot except when it is exceptionally cold). The Baltimore Lectures were a great pleasure to myself, because I had twenty or thirty most agreeable and interesting "coefficients," many of them from distant parts of the States, who came to Baltimore with leave of absence from their Colleges and Universities for the three weeks of the lectures, and I felt myself stimulated to an interest in the subject that I had never felt before, and which forced me to learn something of it of which I had till then been exceptionally ignorant.

I think I have at last (since about midsummer) hit upon a convenient and simple plan for electromagnetic measuring instruments, in which the "constant" of each instrument will be truly constant; or rather as nearly so as is the earth's mass and rotational period. I have been working at the thing incessantly since our return from America, and I hope very soon to have some instruments made which will be convenient for ordinary use.

We are here spending our holidays with my brother-in-law W. G. Crum and his family, but return to Glasgow on Monday. My wife joins in kind regards to you and Mrs. von Helmholtz and the boys, and best wishes to you for the New Year, and I remain yours always truly,

WILLIAM THOMSON.

At Bangor, at the University College of North Wales, Professor Andrew Gray, so long assistant to Sir William Thomson, had, as Professor of Physics, organized new Physical Laboratories, and besought his former chief to come to the formal opening of them. The ceremony took place on February 2, 1885, with an Address[1] by Sir William Thomson. The new laboratories which he now came to open had been built on the space formerly occupied by the stables of the Penrhyn Arms Hotel. A reference to p. 233, *ante*, will show what associations must have been conjured up by his visit.

He spoke of the laboratory of a scientific man being his place of work, and of the history of laboratories; personal reminiscences of those at Glasgow, and of the students who had composed his own volunteer corps of laboratory workers. He laid great emphasis on the functions of laboratories in the true work of a university. "In university work teaching and examining must go side by side, hand in hand, day by day, week by week, together, if the work is to be well done. The object of a university is teaching, not testing . . . and examining should only be part of its work, and that only so far as it promotes teaching. The credit of the university should depend on good teaching." Bangor College

[1] Printed in *Popular Lectures*, vol. pp. 473-501.

was not yet a university; yet he looked forward hopefully to the time when it would be—if not an independent university of itself—a constituent college of the University of Wales.

From Glasgow, on February 8, he wrote to Mrs. King:—

I am just now, and indeed have been from the beginning of the session, hard at work on a new set of measuring instruments for electric potentials and currents, to meet modern practical and scientific requirements. I had been for four or five years trying for something of the kind, and getting something made and brought into use to partially supply the want, but it is only since last May that I have got on the right plan, and now I am just on the point of getting several different instruments to do, much more completely than ever, what is wanted. . . . We were at a beautiful concert last night—a popular concert, crammed with working people enjoying Beethoven, Schubert, Weber, Mendelssohn, to the utmost, on a programme chosen by universal suffrage of themselves.

Though Sir William was busier than ever over measuring instruments, he found time to entertain at his house the Lord Rector of that year, who was no other indeed than his former colleague, Edmund Law Lushington, Emeritus Professor of Greek.

*Nature* of May 14, 1885, contained a review by Helmholtz of Vols. I. and II. of Sir William's *Mathematical and Physical Papers*. It concluded thus:—

Let us hope for an early continuation of this interesting collection. There are still nearly thirty years to be accounted for. When we think of that, we cannot fail to be astonished at the fruitfulness and unweariedness of his intellect.

Then Sir William wrote :—

YACHT *LALLA ROOKH*, LARGS,
*June* 18, 1885.

DEAR HELMHOLTZ—I wanted before now to write and thank you for the very appreciative account of my Reprint of Collected Papers which you gave to *Nature*, and which we read with great interest, but was prevented by our unsettled life, between London, visits at Cambridge (Prof. Stokes) and Lord Rayleigh's, and some other friends in England. We came on board yesterday and slept on board last night, but even yet have not attained to the quiet and settled life of the *Lalla Rookh* which we have been looking forward to, as I must be in Edinburgh and Glasgow this week, and in London to give evidence before a Committee of the House of Commons on the proposed Manchester Ship Canal next week. Except for these disturbances we should live under a triangle of forces, pulling to the laboratory, Netherhall, and the *Lalla Rookh*, and having their turns of preponderance with some regularity. I have a new depth-gauge to test on board, and had some very promising trials at our anchorage here yesterday evening. To-day I hope to find it working well, when sailing with a good breeze in deeper water. I am very busy in the laboratory with electric measuring instruments. I am making standardizers on the principle of Faraday's law, which I worked out mathematically 35 years ago, to the effect that a globe, or any not too elongated piece of soft iron, experiences a force in the electromagnetic field. . . . I have recently noticed theoretically, and verified experimentally, that the force on the soft iron in these circumstances is exceedingly little influenced by magnetic retentiveness. A short bar, or a globe, experiences very nearly the same force, and produces very nearly the same magnetic effect externally to itself, as if its inductive susceptibility were infinitely great. I find slight effects of residual magnetization if R has been very great, and I provide with the instrument a little reversing key. A rapid succession of half a dozen or a dozen reversals of the current (changes of sign of R) brings the force on the iron to a perfectly definite condition, in which I find also (as theory indicated) the force experienced by the soft iron is very rigorously proportional to the square of the strength of the current by which R is produced. I believe that I can thus have a secondary standard for the strength of a current accurate to $\frac{1}{10}$ %. . . .

I have also an instrument on the same principle, but with an inspectional scale from 80 to 170, and practical scale of weights, for different grades, from 5 milligrammes to 80 milligrs.: giving, for instance, 1, 2, 4 for ratios of the currents corresponding to the same mark on the scale. . . . The movable iron is a short solid cylinder, weighing about $1\frac{1}{2}$ grams, and is attracted upwards by the electromagnetic force. The standardizing point of this instrument is 80 of its scale. I have had several of these made, and they promise to be very useful for laboratory and for practical purposes (for instance, to serve, with an added external resistance of 2500 ohms, for an engine-room voltmeter, to measure from 80 to 120 volts, or 170 with less accuracy, on an inspectional scale).

I am also making iron-clad magnetostatic galvanometers to serve from $10^{-9}$ of an ampere to 150 amps., and a hectoamperemeter (on the attraction of iron principle, but across, not along the lines of force) to serve from 30 amps. to 1000 amps. I have made absolute idiostatic electrometers for laboratory and practical purposes to serve for 400 volts to 20,000 volts, which are already working well. My aim is a two-branched chain to measure currents from $10^{-9}$ of an amp. (*i.e.* a mikro-milliampere!!) to a 1000 amps., and from $\frac{1}{10000}$ volt (or as much less as you please) to 80,000 volts, all connected by proper standardizers and comparers, and susceptible (I hope) of an accuracy of $\frac{1}{10}$ % in every case, when the requisite care is given. All these are secondary standards, or magnetostatic instruments adjusted by standardizers themselves secondary standards. For a primary standard of current I have made an absolute sine-galvanometer, which works well. I think I can rely on the measurements we make of the horizontal component of the terrestrial force to $\frac{1}{10}$ %, and I am sure I can keep the error of the sine-galvanometer well within this figure. For primary electrostatic standard I am now going on to make an attracted-disk electrometer to work with about 10,000 or 20,000 volts difference of potential. I have a simple plan by well-insulated condensers (arranged like the old Leyden phials "in cascade") by which I can multiply an hundredfold, from 100 volts (measured by galvanometers and resistance coils, in electromagnetic measure) to 10,000 volts, to be measured by absolute electrometer. Thus I expect to get a satisfactorily accurate evaluation of the number of electrostatic units in the electromagnetic unit, which *may* be more accurate than the measurements made hitherto.

I am afraid I have wearied you with this long letter. I ought really to have waited till I could give it you better told, in print; but I thought you would be interested, and now I am appalled to find how much I have taxed your eyes. Is it possible to persuade you to come to Scotland in summer, and to bring Mrs. von Helmholtz? You know your "swimming home" here if you will come, and Netherhall you know also. Tell your wife that she will not be required to be a moment in the *Lalla Rookh* if she does not like, and if you care to come for a sail with me or a cruise to a little distance the ladies will come or stay on shore, as they please. The British Association is at Aberdeen on the 10th of September. We would, if you please, go to it, but not to be troubled to read papers.—Yours always truly,

William Thomson.

In July he was cruising about the Solent. He wrote from Cowes on July 23 to Lord Rayleigh, inviting him for a cruise before he should sail north on August 6.

Will you come on Saturday the 1st and remain till

Wednesday the 5th (on which day I must go up to London to dine with Cyrus Field at the Star and Garter, Richmond, in commemoration of several fifths of August combined[1] to make the first successes of Atlantic telegraphy). But we shall see two days or 2½ days of the Cowes Regatta, before I must leave. We may ourselves have a good sail on both the Monday and Tuesday, and the preliminary Sunday will allow a settlement of Clark cells and other electric subjects, to make way for hydrodynamics. I write in the middle of disturbances—flags flying, guns at Spithead sounding from the distance, and royalties flying past in steam launches, and visitors on board, excuse incoherence.

Sir William and Lady Thomson were the guests of Lord Aberdeen at Haddo House on the occasion of the British Association meeting at Aberdeen on September 9. At this meeting Sir William took part in a lively discussion on electrolysis, opened by Professor (now Sir) Oliver Lodge. He also read a paper on electric measuring instruments, and another on the method of multiplying potentials, of which he had written to Helmholtz earlier in the year. Lady Thomson wrote to Darwin that it was a very pleasant meeting, much superior to that at Montreal; and invited him to come to Netherhall, where there was to be a party of electrical folk—Captain Creak, Oliver Lodge, George FitzGerald, R. T. Glazebrook, Sir George Stokes, and later Lord and Lady Rayleigh.

On September 30 Sir William wrote to Mrs. King:—

Our party is now all dispersed; the last of them, her brother John and his wife, having left us this morning.

[1] See pages 343, 360, and 364 *supra*.

We go up to Glasgow for two days' work to-morrow and Friday, and we set out from here on Monday morning for Malvern. I hope you are still enjoying your villegiatura, and that you all three will bring back a good durable stock of health when you return home. I am exceedingly busy with new electric measuring instruments (special ones, on which I have been hard at work), and a new depth recorder for my sounding machine, which after 9 years hitherto unavailing attempts now promises success.

In October, for a change before the winter session should open, they went to Malvern, and then to Cambridge.

For some years James T. Bottomley had taken over as Deputy-Professor all the less-advanced teaching, so that Sir William's labours as Professor were much lightened. The programme for the winter of 1885-86 shows how large a proportion of the work was now devolved upon him.

MATHEMATICAL COURSE OF NATURAL PHILOSOPHY

SIR WILLIAM THOMSON

Every Wednesday at 12 noon.

A special subject is chosen in each session and treated mathematically: such as—The Wave Theory of Light, Hydrodynamics, Theory of Magnetism, etc. etc.: for session 1885-86 the subject is Vortex Motion.

J. T. BOTTOMLEY

Tuesday and Thursday at 11 A.M.

Dynamics, so far as can be treated without the aid of the Differential Calculus.

Monday and Friday at 12 noon.

Dynamics with the aid of the Differential and Integral Calculus.

Examinations are conducted weekly and home Exercises given from time to time.

Sir William also gave experimental lectures on Thursdays and Fridays, at 9 A.M., to the whole large class; and Mr. Bottomley gave two others on Tuesdays and Wednesdays at the same hour.

He wrote in November to Lord Rayleigh :—

The fibre suspension for balances is doing splendidly. I think it is going to answer well for ordinary weighings. I am making a delicate balance, suspended on several lines of single silk fibre, which will weigh safely up to two grams, and will not only be very delicate, but exceedingly easily used. I am also making one to weigh anything up to 50 lbs., to the nearest quarter-ounce—not refusing to act according to ozs., stones and pounds—and perhaps you may see it in all grocers' shops, or at all events in all enlightened grocers' shops, by and bye. But the thing I wanted the fibre suspension for has been a complete success. I have got over all my difficulties with the milliamperemeter. Two of them keep together within $\frac{1}{10}$ per cent.

"Capillary Attraction" was the title of a discourse given by Sir William Thomson at the Royal Institution on Friday, January 29, 1886. This lecture, which is printed in *Popular Lectures*, vol. i. pp. 1-55, began with a discussion of the gravitational attraction between two minute portions of matter at very small distances apart. At a distance of one one-hundred-thousandth of an inch (or 250 micromillimetres) their mutual attraction would be insensible. But at one-fifth of that distance (as estimated by Quincke) it would begin to be sensible. All capillary phenomena might be explained without assuming any other law than that of gravitation. He had shown[1] in 1862 that, provided only we may assign a sufficiently great density to the molecules themselves, heterogeneousness of structure will suffice to account for any force of cohesion, however great. By such cohesive forces one could explain the presence of surface tension, acting like a contractile film of infinite thinness over the surface of a liquid, although one must not fall into the paradoxical habit of thinking of the surface film as other than an ideal way of stating the resultant effect of mutual attraction between the different portions of the fluid. One could even calculate from the surface tension the period of vibration of a dewdrop. A sphere of water of radius 1 centimetre would have a period of vibration of $\frac{1}{4}$ second. He then referred to the explanation given by his brother James, in 1855, of the phenomenon, due to surface tension, known as the "tears" of strong wine, and to the graphic solution of a number of problems of the equilibrium of the surfaces of pendent drops and other free liquid surfaces, which problems are too difficult for solution by the methods of the calculus. This graphic solution had been carried out with great perseverance and ability by Professor John Perry in 1874, when he was a member of the laboratory corps. These drawings were now exhibited. A fine experimental illustration was afforded by a large-scale model of a pendent drop, enclosed not in an ideal film, but in a real film of thin sheet

[1] Note on Gravity and Cohesion, *Roy. Soc. Edin. Proc.* vol. iv. p. 604, April 21, 1862.

india-rubber. Suspended from the roof was a stout horizontal metal ring, about 60 centimetres in diameter, covered on its lower face by a tightly-stretched sheet of india-rubber. When water was poured in, the flexible bottom bulged downward, and, as more and more water was poured in, the form changed and elongated, just as a pendent drop might do. The growth of this "drop," its successive changes of form, with their varying stabilities and their vibrations when disturbed, furnished an exciting episode in the lecture,[1] which culminated when finally the elastic film gave way and the drop burst over the lecture table, splashing the nearer members of the fashionably attired audience.

One result of this lecture was a communication to the Edinburgh Royal Society, on March 1, of a paper on the magnitude of the mutual attraction between two pieces of matter at a distance of less than ten micromillimetres. In April he gave another paper on a new form of portable spring balance for the measurement of terrestrial gravity.

[1] This experiment of the dewdrop was a favourite one in Sir William's regular university course from early times, when his assistant Tatlock used to mount a ladder to pour water into it. Sir William (as Mr. Bamber, a former student, has narrated in *Engineering*) had previously drawn upon the piece of sheet-rubber certain lines and curved figures which, as the drop swelled up, changed their forms and proportions. Most enthusiastically he would point out these changes and write down equations of curvature on the board; he himself dodging about, now standing below the drop, now returning to his place behind the table. Then he would go back to the drop, and poking it with the pointer would say: "The trembling of the dewdrop! gentlemen." Meantime the drop itself was increasing in size and thinness in an ominous way. "More water, Tatlock. Now, gentlemen, you see how this line has altered. Go on, Tatlock, more water." The class awaited the *dénouement* breathlessly. At last it came, the "drop" having expanded from about 9 inches in diameter to nearly 2 feet. Fortunately the learned Professor was on the right side of the table, with his head not as much under the "drop" as usual. There was no more lecturing that day after the "drop" burst.

On February 21 Sir William wrote to Darwin :—

I am off to London to-night to elect a professor of physics for Sydney, and return to-morrow night. The new electric measuring instruments are doing well, but not *yet* ready to give out. It has been a truly tremendous piece of work for me these four years.

The following letter to Mr. (now Sir William) Preece tells of the electric lighting :—

*16th March* '86.

DEAR PREECE—In answer to yours of yesterday I began electric lighting in my house in June 1881. By the end of that year I had 106 lights in my house, and soon afterwards I had 52 lights in my class-room, 12 in my Laboratory, and 10 were fitted up in the Senate Room of the University. My lamps are Swan, 85 volt, 16 candles.

I made a Faure battery for myself of 120 cells arranged in three parallels, and used it incessantly for about 18 months, till I found it worked to death.

Last November I got an improved Faure-Sellon-Volckmar battery from the Electric Power Storage Co. By the accompanying copy of a letter I have just written to Mr. Drake, you will see how splendidly satisfactory this has been. I scarcely ever have the engine going, except during the day to charge the battery. Without the engine I get from the battery alone ample current for 40 lights: when I want more than 40 lights I use engine and battery together in the usual manner, and then I have ample current for from 70 to 80 lights.—Yours very truly, W. THOMSON.

The next letter, to the Bursar of Peterhouse, brings new aspects into view :—

*22nd March* '86.

DEAR DODDS—Thanks for cheque. I enclose receipt. The dividend is lamentably small; but what a lesson the state of landed property in our region ought to be to Gladstone and other would-be remedialists of the agricultural distress in Ireland. I wonder it has never occurred to the great mind of Gladstone himself that some moderate encouragement and help towards business habits and arrangements between Landlords and Tenants, coupled with defence of honest dealing on each side by all the power of the empire, is the only remedy possible, and all the remedy that is needed, to make the best of bad times in Ireland as everywhere else.

Thanks for your information about the electric light. I hope it is doing well, and that it will be fairly satisfactory also in respect to economy. Have the Undergraduates got any of the lowering lamps yet? I think with a lamp that can be readily lowered down to the table every one will feel the single light in his room always sufficient and quite pleasant and agreeable for reading.

With kind regards to all my friends at Peterhouse,—I remain, yours truly, WILLIAM THOMSON.

Sir William Thomson viewed with fierce hostility the Home Rule Bill and the Irish Land Bill which Mr. Gladstone had introduced that session into Parliament. As an Ulsterman he strongly disapproved of both schemes, the danger of which to Ireland and to the Empire had been at the very outset urged upon him by his brother James. Hitherto in such part as he had taken in politics he had been ranked as a moderate Liberal. But now he flung himself with intense eagerness into the Unionist cause, and became a leader amongst the Liberal Unionists of Scotland.

He was still trying to pick up the threads of his Baltimore Lectures, as will be seen by his reply to two of the most distinguished of his American colleagues.

CLEVELAND, OHIO, *March* 22, 1886.

DEAR SIR WILLIAM—You will no doubt be interested to know that our work on the effect of the motion of the medium on the velocity of light has been brought to a successful termination. The result fully confirms the work of Fizeau. The factor by which the velocity of the medium must be multiplied to give the acceleration of the light was found to be 0·434 in the case of water, with a possible error of 0·02 or 0·03. This agrees almost exactly with Fresnel's Formula, $\frac{n^2 - 1}{n^2}$.

The experiment was also tried with air with a negative result.

The precautions taken appear to leave little room for any serious constant error, for the result was the same with different lengths of tubes, different velocities of liquid, and different methods of observation. We hope to publish details within a few weeks.—Very respectfully, your obedient servants,
ALBERT A. MICHELSON.
EDWD. W. MORLEY.

*3rd April* 1886.

DEAR PROFS. MICHELSON AND MORLEY—Thanks for your letter of 22nd ult. I am exceedingly interested in what you tell me, and am more eager to see in print your description of your experiments. The result is clearly of the greatest importance in respect to the dynamics of the luminiferous ether and of light.

I am working all I can (but with quite an almost desperate amount of interruption) to have my *Baltimore Lectures* brought out in print. I hope before the volume is completed to have results from you that can be incorporated as an appendix.

My time is very much engaged with electric measuring instruments, which I have been working incessantly at for the last four years. I have now nearly come to a conclusion in respect to forms of instruments, but I have still a great amount to do. However, I do see daylight clearly now through it.

Lady Thomson joins with me in kind regards—and I remain, yours very truly, WILLIAM THOMSON.

THE UNIVERSITY, GLASGOW,
*27th April* '86.

DEAR DARWIN—I scarcely think we shall attend the B.A. this year. I think I must take a holiday when I get these electric measuring instruments, which I have been so hard at work upon for 5 years, off my hands.

I think *colour vision* would be an excellent subject for discussion at the B.A. I think the Electromagnetic and the Elastic-solid theories of light will be a very good subject. There would be a great deal of rubbish talked on that or on any conceivable subject of discussion ; but I think that there is more hope of some good metal flowing out from that than from any other crucible-full that occurs. . . . —Yours truly, W. THOMSON.

*P.S.*—I hope Cambridge is going to do its best to stop the two Bills on second reading. Was such madly mischievous legislation ever proposed in the History of England before?

In May came visits to Lady Siemens at Tunbridge Wells, and to Lord Rayleigh at Terling Place. Thomson wrote to von Helmholtz :—

SHERWOOD, TUNBRIDGE WELLS,
*May* 23/86.

DEAR HELMHOLTZ—. . . I am still very much engrossed with electric measuring instruments. I am afraid I cannot now say that I am nearer than, but I think I can safely say that I am now as near as I thought I was, to satisfactory results when I wrote to you a year ago! This is not much to say, but I do seem happily to see daylight through an affair for which I have been struggling without intermission for 5 years. If you see Dr. Werner Siemens, will you tell him that I have never sent him the notice I promised of the new instruments which I showed him in Paris two years ago, because I found them unsatisfactory, and that I have been working ever since to get something better? I little thought that an electrometer would be the first of the set to be got into practical working shape: but so it is. I enclose a notice of it which I have only received from the printers a few days ago, and I shall write to White to send it (the notice) to Dr. Siemens immediately. My other instruments are electromagnetic. The standardizing current meter of which I wrote to you (founded on the tendency of a piece of soft iron to move from places of weaker to places of stronger force) has done very well, and has proved very satisfactory in my laboratory and on our house circuit, but I am keeping it back, because for most purposes, if not for all, I have got something better—an intrinsic current meter founded on the tendency of an oblate ellipsoid of revolution (about 8 mm. equatoreal and 5 mm. axial diameter) to place the plane of its equator along the lines of force. The little oblate is supported on a stretched platinoid wire, balanced relatively to gravity, so that it will work as well at sea as on land. It has given me very good results, within 1 (one-tenth) per cent of absolute accuracy, and will serve very conveniently for directly standardizing currents of from ·1 of an ampere to 20 or 30 amperes.

All these things have left me very little time for other work, but I begin now to feel a little freer, and I hope soon to get on with the Baltimore Lectures (preparing for the press, etc. etc.), which have been too long deferred. We are staying here for a few days with Lady Siemens. I have told her the news of the little grandson, which interests her much. She sends her kindest greetings, and "particularly to the little grandson." Lady Thomson joins in the same, and particularly to yourself and your wife. We return to Scotland next week for laboratory, *Lalla Rookh*, and Netherhall.—Yours always,

WILLIAM THOMSON.

A postcard to Darwin deals with matters they had discussed two days before at Cambridge.

Chemical energy is infinitely too meagre in proportion to gravitational energy, in respect to matter falling into the Sun, to be almost worth thinking of. Dissociational ideas are wildly nugatory in respect to such considerations. Hot primitive nebula is a wildly improbable hypothesis—so now it seems to me at all events, and it can help us nothing. Langley, I suppose trustworthily, finds a somewhat greater (? 7 to 6) radiation of the Sun than Pouillet, whose result I took.

W. T. Glasgow, *June* 6/86.

The rejection of the Home Rule Bill was followed by a general election, and Sir William Thomson threw himself into the thick of the fray, travelling about the West of Scotland, speaking and presiding at meetings for several weeks. He combined with this the recreation of sailing, and visited several places to take part in political gatherings.

He wrote to Tait, 16th June 1886:—

I am too busy, not only with the election, but with measuring instruments and with my Baltimore Lectures, to be able at present to give a thought to the Boltzmann "sibboleth," or anything else about the kinetic theory of gases.

From Roshven he sent to Lord Rayleigh, on July 18, a number of equations about the propagation of harmonic waves, one of them full of complicated roots. He wrote:—

The following, due to a quartette of circumstances, (1) release from election work, (2) bondage to Baltimore Lectures, (3) incessant surroundings of deep-sea waves, and (4) a wet Sunday, will, I think, interest you. . . . It is a surprisingly curious formula, and what Hopkins used to call *very prickly*. . . .

The very end of my electioneering was at Oban on Tuesday night, where and when Craig Sellar and I held a meeting of Liberal Unionists in support of Col. Malcolm (C.), candidate for Argyllshire. There was much joy on board the *Lalla Rookh*, happily storm-stayed at Tobermory, on Friday night, to hear "Majority 613 Malcolm." What *is* to be done to keep Mr. Gladstone from mischief after all? We hear that it is supposed "his game is to discredit the government and make it appear in the eyes of the country that he is indispensable, and then come back on his own terms." . . . I don't for a moment believe he will succeed, but there is too much reason to believe *that* is his game, and more than enough to show that Unionist organization must be kept up vigorously and Unionists must act well together in Parliament; keeping AS FAR AS POSSIBLE all subjects on which, as conservatives and liberals and radicals, they may differ among themselves, until a time comes when they can be honourably fought out and fairly settled, which certainly they cannot be until the two imps of mischief, Parnell and Gladstone, are finally deprived of all power for evil. What a blessing it would be if we could have Lord Salisbury, Lord Hartington, Chamberlain and Jesse Collings all in one government. . . .

Three days later, being storm-stayed at Roshven, he sent Lord Rayleigh some new proofs of the proposition that the wave-velocity is double of the group-velocity, and added :—

We are delighted to see one Nationalist beaten by a Unionist in Ireland. So if dear old Scotland behaves not like a fool once more, in Orkney and Shetland, we shall have 117! What shall we (they, I mean, Lord Salisbury, Hartington, Bright, Trevelyan, Chamberlain and Co.) do with it? Wisely, I hope.

Again, after three days, he sent a postcard with more about groups of waves, and this note :—

We leave for Glasgow to-day by rail for six days' Laboratory and White's, and Glasgow for Portsmouth Saturday night. Will you join us at Cowes next week to look for hydrodynamic problems and illustrations at the R.Y.C. Regatta?

September brought the British Association meeting at Birmingham. Here Sir William read four papers bearing on wave-problems (see p. 748), and one on an instrument to measure the differences of gravity at different places.

The yachting ended early this autumn, the last cruise being on September 30. He was very busy with calculations on deep-sea waves and canal waves, required to complete certain points promised in the *Baltimore Lectures*. Politics also made demands on his time, and he went over to Belfast with his brother James to take part in Mr. Chamberlain's Anti-Home Rule meeting there on October 11, 1886. The last half of October was spent at Great Malvern, and in a brief visit to George Darwin at Cambridge.

Endless patience on minute detail characterized Sir William Thomson's prolonged efforts to bring to perfection the standard electrical instruments to which he had set his hand nearly five years before. The following letter to Lord Rayleigh enables one to understand a little the nature of such protracted toils:—

THE UNIVERSITY, GLASGOW,
*4th Nov^r^. '86.*

DEAR LORD RAYLEIGH—Many thanks for your volume of Electric Measurements which I was delighted to find on our return here three days ago.

I have at last realized my elastically suspended balance. I have one now with a strong enough elastic suspension to bear 80 kilogrammes, and with 2½ or 3 kilogrammes whole weight hung on it, it is amply sensitive to three milligrammes. The suspension consists of 400 parallel copper wires, each about 0·1 mm. diam. and 4 millimetres free in the vertical bearing part. For a convenient weighing balance to accurately weigh 4 to 5 kilogrammes, I would probably use 100 bearing wires, but of platinoid instead of copper, possibly with 5 to 10 cms. lengths in the vertical bearing parts, and I would hang the weight pans from one end of the balance also by fine wires instead of knife edges. From what I have seen I am perfectly sure I could easily weigh with it 4 kilogrammes to the nearest 4 milligrammes, and very likely to the nearest milligramme. I mean now to make such a balance for ordinary use.

What I now have and have tried for the first time to-day up to 150 amperes is a hecto-amperemeter. The current enters the movable coil by 200 of the bearing wires and leaves it by the other 200. It is intended for 400 amperes, but I am pretty sure it will be quite available up to 500 or 600 amperes. Here are results of the first rough trials made this afternoon :—

| | | New Hecto-Amperemeter. | |
|---|---|---|---|
| | Current in Amperes. | Observed. | Calculated from No. 3. |
| (1) | 50 | 167·4 | 167·33 |
| (2) | 100 | 669·2 | 669·33 |
| (3) | 150 | 1506·0 | 1506·0 |

The unit is 6 mgrms. Thus the weight for 100 amperes is 4·015 grammes.

The currents were fixed at the 50 and 100 amperes by a test comparison with a hecto-amperemeter of the same kind and same gauge for which the coefficient had been determined by electrolysis of copper. The previous one had been compared with a deka-amperemeter for currents of from 12½ to 100 amperes, and had shown agreement within $\frac{1}{5}$ per cent, which, for a first attempt on two instruments of such very different gauges, and through so considerable a range of current, was fairly satisfactory. It, and what I know otherwise, settles that the new hecto-amperemeter is available for currents of from 12 to 600 amperes. The instrument is going to be perfectly portable: no amount of knocking about it can have in travelling can damage the bearing ligaments.—Yours truly, WILLIAM THOMSON.

On November 9 he sends to Lord Rayleigh a mathematical discovery.

The force required to tow a boat uniformly, whether in a canal or in an open expanse of water (supposed inviscid), is zero when the velocity exceeds $\sqrt{gD}$. This anticipates ($49\frac{1}{2}$ years after date!) Scott Russell's discovery as to towage in a canal. I hope in Part III. of a set of articles on Stationary Waves, which I am sending to the *Phil. Mag.*, to give a table of the forces required for towage, at different speeds, up to $\sqrt{gD}$, of a boat whose length is short in comparison with the wave-length corresponding to its speed, and comparable with, or much greater than, the breadth of the canal.

About the end of 1886 there arose a rather angry public dispute about priority in connexion with the invention of the seismometers used for making automatic records of earthquakes. To one of the disputants, who had brought the matter to Sir William Thomson's notice, he wrote:—

I am very sorry if any one concerned has said anything that could be personally disagreeable to any other; and I cannot but think that all these questions of fact, so far as the public can take any interest whatever in them, could and should be discussed without coming within a hundred miles of anything acrimonious. If any one person has said anything acrimonious I will advise himself and every other person concerned to determinedly avoid doing the same again.

On January 3, 1887, Sir William Thomson wrote from Druid's Cross, Wavertree, to Lord Rayleigh, telling him that he was going to give a Friday evening discourse at the Royal Institution on January 21, on the origin of the Sun's Heat. The letter tells of more work on the balances and their suspensions, and of more articles on deep-sea waves. It ends with a note of politics :—

And how do you feel about Randolph? Not distressed, I hope. I hope Mr. Goschen's acceptance will do good. Surely Ireland is already feeling a benefit from something like stability of government, with promise of continuity.

The Royal Institution discourse on the Origin and Age of the Sun's Heat gave an epitome of views which have been stated in Chapter XIII. It appeared in substance in *Good Words*, 1887. It was an attempt to explain the enormous radiation of the sun—equivalent to 78,000 horse-power for every square metre of his surface—on Helmholtz's theory that the energy to maintain this output is furnished by a gravitational contraction of his mass. The lecture is a striking example of Lord Kelvin's method of illustrating recondite problems, such as those of the dynamics of a nebula, by imaginary models in which billiard balls or marbles are supposed to be flying about a room and striking against its walls or floor. He referred to the development of the mathematical theory of the gaseous nebula by

Mr. Homer Lane, and indulged in the customary diatribe against the "awful and unnecessary toil and waste of brain-power" involved in the use of the British system of inches, feet, and yards. At the close he pictured most graphically the primitive formation of a gaseous nebula by the falling together of "twenty-nine million cold solid globes, each of about the same mass as the moon," and then sketched the possible subsequent history of such a nebula as it settled down into a rotating system.

This is just the beginning postulated by Laplace for his nebular theory of the evolution of the solar system, which, founded on the natural history of the elder Herschel, and completed in details by the profound dynamical judgment and imaginative genius of Laplace, seems converted by thermodynamics into a necessary truth. . . . Thus there may in reality be nothing more of mystery in the automatic progress of the solar system from cold matter diffused through space, to its present manifest order and beauty, lighted and warmed by its brilliant sun, than there is in the winding up of a clock and letting it go till it stops. I need scarcely say that the beginning and the maintenance of life on the earth is absolutely and infinitely beyond the range of all sound speculation in dynamical science. The only contribution of dynamics to theoretical biology is absolute negation of automatic commencement or automatic maintenance of life.

In the reprint of this discourse in Vol. I. of the *Popular Lectures*, he added, at the word "clock" in the antepenultimate sentence, the following footnote :—

Even in this, and all the properties of matter which it involves, there is enough, and more than enough, of mystery to our limited understanding. *A watch spring is much further beyond our understanding than is a gaseous nebula.*

Amongst the papers read this season was a communication to the Glasgow Philosophical Society on a double chain of electrical measuring instruments to measure currents from the millionth of a milliampere to a thousand amperes, and to measure potentials up to forty thousand volts. To produce such a series of instruments demonstrates an extraordinary devotion to the attainment of scientific precision.

On August 2nd the Institution of Mechanical Engineers held a Conference in Edinburgh, and here Sir William gave the lecture on Ship-Waves noticed elsewhere (see p. 748).

The following letter is entirely characteristic in matter and manner :—

*Aug.* 6/87,
NETHERHALL, LARGS, AYRSHIRE.

DEAR LORD RAYLEIGH—I have had a very bad time since your last letter came, or I should not have let so long time pass without writing in return. The worst on the whole has been, wave-making far in the rear of an infinite fleet of equal and similar ships moving "in line abreast." This has tortured me chronically since last December and acutely for about three weeks till two days ago. But I have been afflicted by a complication of ailments, among which turbulent motion of water between two planes has given me much suffering.

I asked Stokes three questions yesterday. He answered each unhesitatingly.

(1) Given an inviscid fluid moving in a pipe with the same laminar (steady) motion as it would have if it were viscous and were moving steadily under the influence of

gravity (or a distant piston pressing it through the pipe). Do you think the motion of the inviscid fluid would be stable? *Answ.* No.

(2) Do you think a needle balanced on its point resting on the bottom of a basin of very thick treacle would be stable? *Answ.* No.

(3) Do you think viscosity of the fluid could give stability to the motion specified in Quest. 1?

*Answ.* Yes.

This seems to prove my case. We are tremendously interested in your coloured photo[g.] of spectrum and your theory of the affair.

My wife joins in kind regards to you and Lady Rayleigh. I looked for you, both in the *Jumna* and *Serapis*, at the Review but could not see you. I have no message to give from Stokes as he tells me he has been writing.—Yours truly, W. THOMSON.

*P.S.*—Stokes refuses to bet that a mass of water falling into water will develop finite slip. Yes. Vertically, he says no.[1] "If falling with horizontal component, motion requires further consideration." But he won't take 3 to 1 in half-crowns.

August 31st brought round the meeting of the British Association at Manchester, and here Sir William had four papers to read. Two of these were on his electric balances, one was on the turbulent motion of water flowing between two planes, and one on the vortex theory of the luminiferous ether. This was a development of ideas broached in the Baltimore Lectures, and bore a second title, phrased in the purest Kelvinese, "On the Propagation of Laminar Motion through a Turbulently Moving Inviscid Liquid."

A long letter on stability of wave motion, to

[1] He does not admit this unqualifiedly. He says probably no.

Lord Rayleigh, on September 26, winds up with a reference to the acceptance of the Irish Secretaryship by Mr. Arthur J. Balfour.

I hope the Secretary is being well refreshed by Scotland. Very soon (a week or two more or less) I trust there will be an uneventfulness in Irish History that will sadly baffle the most ingenious and venomous of Gladstonites, and that may cause their great leader to endeavour *not* to remember Mitchelstown.

On July 27, 1887, the Jubilee of the Electric Telegraph—that is, of the first British line, from Euston to Camden Town—was celebrated by a dinner in London presided over by the Postmaster-General, and at which Sir Charles Bright, Sir William Thomson, Mr. J. Brett, Mr. John Pender, Sir James Anderson, and others of the old cable and telegraph services were present. Responding for the toast of telegraphic science, Sir William Thomson referred to his comrades and associates in the pioneering work of 1857 to 1866, most of whom had passed away. He eulogized the names of Edward and Charles Bright, Whitehouse, Werner and William Siemens, Canning, Clifford, Varley, Jenkin, Willoughby Smith, the navigators Moriarty and James Anderson, and the financiers Cyrus Field and John Pender, whose steadfast perseverance had led to success. After pointing out the great improvements of recent date in land telegraphy he concluded with the words :—

I must say there is some little political importance in the fact that Dublin can now communicate its requests, its complaints, and its gratitudes to London at the rate of 500 words per minute. It seems to me an ample demonstration of the utter scientific absurdity of any sentimental need for a separate Parliament in Ireland. I should have failed in my duty in speaking for science if I had omitted to point out this, which seems to me a great contribution of science to the political welfare of the world.

Much time was spent this season by Sir William Thomson at Netherhall, which in previous years had been somewhat neglected in the autumn in favour of the *Lalla Rookh*. An extract from the diary of his niece, Miss Agnes G. King, gives a picture of the family doings :—

*July 21st*, 1887.—Uncle William is going off to-day to Portsmouth to be present at the grand review of the fleet in honour of the Jubilee. . . . I have been out all morning since breakfast watching the two brothers lopping branches from the great tree near the front of the house where the garden chair is placed.

Mrs. King wrote to her daughters from Glasgow :—

*Nov.* 6, 1887.—When I arrived here yesterday Uncle William and Aunt Fanny met me at the door, Uncle William armed with a vessel of soap and glycerine prepared for blowing soap-bubbles, on a tray with a number of mathematical figures made of wire. These he dips into the soap mixture and a film forms or adheres to the wires very beautifully and perfectly regularly. With some scientific end in view he is studying these films. I was at once taken into the study to see them, and while Uncle William was showing them to me and giving me such explanations as I could take in, he was teaching Dr. Redtail to whistle "Merrily danced the quaker's wife"—Dr. Redtail being in the dining-room, and the teacher and pupil carrying on the lesson at the top of their voices. The whole evening was occupied with the soap mixture, the wire figures, and the tobacco pipe, while Aunt Fanny read aloud Mr. Balfour's speech.

On Friday afternoon five big stones were thrown through the study window and four panes of plate glass smashed as well as the fern case. It turned out to be the deed of a maniac who gave as a reason for the mischief he had done that Sir William was a villain who had rained electricity on his head. He has been sent to an asylum.

*Nov.* 13, 1887.—Yesterday Uncle William was at Dundee on business. When Aunt Fanny and I sat down to table, and Dr. Redtail perceived Uncle William's place vacant, he immediately shouted out, "Where is Sir William"?

*Nov.* 15, 1887.—The excitement about the election became very wild, procession after procession of reds (for Lord Rosebery) and blues (for Lord Lytton) headed by a couple of pipers in kilts marched round the courts, and, as they passed the house, halted to hurrah for Sir William.

The next letter is an echo of an unsolved difficulty met with in the *Baltimore Lectures* (p. 819).

NETHERHALL,
*Nov.* 20/87.

DEAR LORD RAYLEIGH—I am afraid I am a month already your debtor for your last letter. I had a bad time (at the beginning of which I hoped I might have seen you in London) after receiving it and tried hard to make out an unexceptionable proof of stability, but without success. I quite see the force of your objection, and admire the mathematical theory of various wonderful Maskelyne and Cooke performances which it suggests. It certainly vitiates my seeming proof of stability; but still I think it might be explained away and my proof completed. This kind of seeming test for stability does not detect (as you show it does not) the instability of a system disturbed from a position of static equilibrium. But it does detect instability (as I maintain and I think you will admit) in the parabolic laminar motion of an inviscid fluid, and it proves that viscosity annuls this kind of instability, but I cannot maintain that it proves stability. There can be no peace or happiness in this position of affairs, so you may think of me as very miserable; but I have become involved in another affair (see *Phil. Mag.* Dec., or enclosed scrap) which George Darwin characterizes as utterly frothy; and having launched it the waves in crystals have come on with very interesting, almost happy result. I find that if a homogeneous, isotropic, elastic solid (suppose for the moment it be incompressible) be drawn out with different forces in different directions; so that with reference to O$x$, O$y$, O$z$ fixed relatively to the solid, and remaining at right angles to one another, the co-ordinates $xyz$ of a point in the solid when unstrained, become $\alpha x$, $\beta y$, $\gamma z$ when strained ($\alpha\beta\gamma = 1$). . . . The velocities are the same for the same directions of vib$^{n.}$, and we have, as shown by Green, Fresnel's wave-surface. But the condition (1) is absolutely natural and simple; being essentially true for infinitesimal values of $\alpha - 1$, $\beta - 1$, $\gamma - 1$, and certainly true approximately, *very approximately, perhaps approximately enough to fit our* knowledge if not also your result for such values as would correspond to Iceland Spar (? is it the most doubly refractive of all crystals): and certainly very approximately (not so far as I can yet see rigorously, rather probably not) for foam pulled unequally in different directions. This is a great contrast to "Green's Second theory," and Stokes's report on it. Stokes felt the assumption needed to bring out Fresnel's wave-surface *so strained that* the whole thing was unacceptable. It involved an independent aeolotropy of the solid, which it modified into fitness for Fresnel, by a precise adaptation seemingly unconnected. I always felt Green's dynamical preliminary investigation unintelligible (indigestible as I see I called it at Baltimore), and it is really quite wrong in its seeming result, which would give laminar wave propagation in a fluid by EQUAL negative (although not said to be negative) in all directions. Still one cannot but feel that Green had really divined the secret, and wanted only to write it out better (by aid of T and T′)! to make it perfect.

What I have made out just now (since coming here yesterday—this is the best place for *produce* I always find) is, I think, really interesting and important: particularly as showing *all* we want, and no possibility of anything else, produced by unequal pull in different directions on an *isotropic solid* (Force will do it all) instead of having 18 moduluses to whittle down to two. Nothing I can yet see, however, diminishes the difficulty of the polarization by reflexion.

Have you been getting good results in weighing, etc. etc.?

I hope you and Lady Rayleigh have been well, and have still a good report to give from Ireland. Mr. Gladstone seems rather frightened that he has gone too far in urging maltreatment of the police, and I think we may fairly hope that the tide has turned. My wife joins in kind regards.—Yours always truly, W. THOMSON.

A communication to the Royal Society of Edinburgh on December 8, on Cauchy's and Green's doctrine of extraneous forces, was the outcome of the ideas revealed in the above letter.

About this time the possible arrangements of molecules in crystalline and other solids began to claim Sir William Thomson's attention, and his "green books" for three or four years abound in pages of notes and calculations on the packing of molecules in different arrangements, and in the partitioning of space by assemblages of primitive particles of different shapes grouped so as to make homogeneous solids. Many physicists had tried their hands at different kinds of grouping. Piles of cannon-balls, groups of cubes, assemblages of six-sided cells like a honeycomb, and many others had been suggested as models of molecular grouping ; and many such are to be found sketched in the note-books with calculations about the elastic and other properties of such structures, including piles of equal bubbles assembled as foam. One form to which Sir William gave great attention was that known as the tetrakaidekahedron—the solid with fourteen faces—which may be looked upon as a transition form between the cube and the octahedron, and which he came to regard as a fundamental form as constituent of a homogeneous grouping. It will be seen later how, by models of various groupings of molecules, he sought to elucidate the problem of the ultimate structure of matter.

The collection of his scattered mathematical and physical papers and their revision for reprinting proved a lengthy process. Vol. I. had been finished in July 1882, and Vol. II. in November 1884 ; but Vol. III. still dragged on. In February 1888, Sir William was reported to be deep in a last new article to finish the volume. This work he interrupted, however, to give to the Geological Society of Glasgow a lecture, on February 16, on Polar Ice-Caps and their influence in changing Sea-Levels. The main point in this paper was the influence of the ice-cap of the Antarctic continent on the sea-level and climate of the globe. Croll had estimated the thickness of the Antarctic ice as 12½ miles. From experiments on the plasticity of ice, and the estimated area of the Antarctic continent, Sir William reckoned it to be three miles thick. If, owing to slow astronomical changes, the average annual snow-fall at the Antarctic were to be increased, the result after a few hundreds of years would be, to increase the cap, lower the level of the sea all over the globe, diminish oceanic circulation, and tend also to cool the Arctic ocean, increasing the amount of glaciation in the northern hemisphere simultaneously. The ocean is the great carrier of heat; the most potent influence for altering the climate in any part of the world.

The following extract from a letter to a correspondent gives a contemporary view on the question of electric lighting, then beginning to be widely adopted :—

*28th Feby.* 1888.

. . . As to the Electric Light; I quite expect that it will altogether supersede gas lighting in cities, although it is impossible to say how soon. If it were at present supplied to houses at twice the price of gas, the light bill would be less in most private houses than at present, because the electric light can be extinguished in an instant in any place where it is not wanted, and relighted in an instant, with the greatest possible ease. The storage of electricity is too costly for very general use, but it is quite unnecessary in any very large-scale electric lighting.

On February 9 he wrote Lord Rayleigh that he was trying hard to find "a convenient and useful hydrokinetic analogue for electromagnetic induction," which was wanted to complete Vol. III. of the Collected Papers, the last sheet of which was waiting for it half finished. He wrote again on March 2:—

I am in great agony to get out Vol. III. of my Papers. It has hung fire 5 or 6 weeks, with the last sheet half printed and waiting for a hydrokinetic analogy for electromagnetic induction—(a very trumpery affair I am afraid you will think it—*not* even including soft iron, but still somewhat useful). I hope to get it done in a few days, but I have been much disturbed with the delay and difficulty of getting time to fix up the affair. Lady Thomson bids me say "we hope that Mr. Balfour is keeping well, because we think the whole nation depends on it"; and I say so most decidedly. We had great excitement and joy over Deptford.

In April he wrote: "I am still in the agony of getting my Vol. III. finished."

On May 3 he lectured on Waves to the Cork Literary and Philosophical Society.

The diary of Miss Agnes G. King tells of doings at Netherhall.

*July* 28, 1888.

Uncle W. and Aunt F. arrived about eight, bringing M. du Bois (a gentleman from Holland) with them. During dinner and in the evening Uncle W. talked with great animation to M. du Bois, drawing from him all possible information about his kind of work in the laboratory in Strasburg, also about public feeling in that part of Germany with regard to the late and present emperors. He also drew out all he could about the social state of Holland, political and private, and its relation to Belgium. At ten he asked me to play the piano, and immediately his whole attention was given to the music, detecting the composer by the difference in style and quite surprised if he made a mistake. The moment a few chords of Der Freischutz were struck he got quite excited, and rising from his chair he came and stood behind me not to lose a single note. "There is nothing," he said, "more beautiful than Der Freischutz; there is stuff in it for at least four operas. Weber is a most original and perfect composer." Bits in which he used to take part long ago with his horn gave him special delight. . . .

It was wet in the afternoon, but in spite of the rain Uncle William, Aunt Fanny, M. du Bois and I went out for a scramble. First we scrambled up to the loft to see the carrier pigeons, and Uncle W. asked the coachman many questions as to why the various pigeons were separated from one another. In the evening I played again to Uncle William.

. . . After breakfast Uncle W. stood for a while with a delighted expression gazing at the reflexion of the window in the polished oak table in the turret window, nodding his head from side to side. "Come all of you quick and look! I see Heidinger's brushes better than I ever saw them before." We all came and Aunt Fanny was able to discover them for the first time, but I do not think any of the rest of us saw them at all. Uncle W. told us how he had given a lecture on them one evening on board a ship going to America, and next morning it was very funny to see all the people nodding their heads from side to side and gazing intently at the clouds.

When sending to Lord Rayleigh, on August 25, a piece of Iceland-spar showing striations due to artificial twinning, he added a political postscript:—

Did you not remember the political prisoners chained to the most atrocious felons in Naples thirty-seven years ago? I should think Mr. Gladstone is the only person of over forty-three in the United Kingdom who does not. But if he had forgotten the state of things in Naples thirty-seven years ago, he should have made himself acquainted with it again, before comparing the régimes of the two B's, as he did to the ignorant potters last week at Hawarden.

September 5th brought the meeting of the British Association at Bath. At the Mathematical and Physical Section, under the presidency of Fitz-Gerald, there were notable discussions. Sir William contributed three papers, and joined in a fourth along with his former pupils, Professors Ayrton and Perry, concerning a new determination of "$v$," the ratio of the electromagnetic and electrostatic units. His papers were: A Simple Hypothesis for Electromagnetic Induction of Incomplete Circuits; On the Transference of Electricity within a Homogeneous Solid Conductor; and On Five Applications of Fourier's Law of Diffusion, illustrated by a diagram of curves with absolute numerical values. This last, reprinted in Vol. III. of the *Mathematical and Physical Papers*, p. 428, was a discussion of the equation which Fourier had found for the "linear motion of heat," as now applied to other physical "qualities"; the velocity of motion as propagated into a viscous fluid; the diffusion of liquids into one another; the diffusion of induced electric currents into a homogeneous conductor (which Heaviside had shown to be analogous to the setting of water into motion by friction on its boundary); and the diffusion of electric potential in the conductor of a submarine cable. This short but useful paper was strongly reminiscent of the article on the "Uniform Motion of Heat," which Thomson had given forty-six years before (see p. 42) when an undergraduate. The other two papers belong to the burning question of 1888, electromagnetic theory. Sir William, as we know, had never accepted Clerk Maxwell's electromagnetic theory of light, nor the notion of "displacement currents," on which that theory is based. His reasons for this attitude are discussed on pp. 1021-1025 below. But in the recent years Maxwell's pupils and disciples, Lord Rayleigh, FitzGerald, J. J. Thomson, Hopkinson, Glazebrook, Poynting, O. Heaviside, Oliver Lodge, and others, had been actively pushing Maxwell's theory and testing its applications. Lodge, in particular, had studied the phenomena of the propagation of electric waves along wires, with many beautiful experimental illustrations of their reflexion, and the establishment of nodal points, when periodic surgings were reflected. For two years, indeed, the application of these considerations to lightning protection had engaged the British Association. But now, on the top of all this, came the splendid work of Hertz, proving experimentally that electric waves, generated by the sudden discharge of charged conductors, as suggested by FitzGerald in 1883, could be propagated across open space without wires, reflected and refracted, just as waves of light can be, and herein obeyed Maxwell's equations.

Sir William's part in this discussion is recounted on p. 1041 *infra*.

To the *Philosophical Magazine* of November 1888, Sir William contributed a striking paper, which has not been reprinted elsewhere, "On the Reflexion and Refraction of Light."

Litigation about the compass (see p. 718) kept Sir William Thomson in London for more than a week in the autumn of 1888, and he stayed with his widowed sister, Mrs. King, then residing in Hamilton Terrace, St. John's Wood. The diary of Miss Agnes G. King records some of the events of the time in the following extracts:—

*Nov.* 20, 1888.—Case of Infringement of Compass came on at last; should have begun on Friday. Uncle William and mother dined at Sir Antony Hoskins's. Sir A. said to mother, "I cannot tell you how all we naval men love your brother." Mother answered, "I think everybody who knows him likes him." "Likes him!" said Sir A. "Like is not the word; we all *love* him."

*Nov.* 21, 1888.—We had great excitement at breakfast dressing up eggs boiled and raw in little net bags, and suspending them by threads, to demonstrate the steadying effect of the oil under the compass box. The arrangement was quite a success, and Uncle W. went off in delight with his new toy, intending to call on Sir A. Hoskins on his way to Court and show it. . . . Uncle W. came home in the evening looking very bright; everything had gone well, and the warm and hearty evidence which his friends among the naval men had given, pleased him exceedingly. Captain Fisher's evidence was specially strong and very amusing from all accounts. He described the absolute steadiness of the compass in the bombardment of Alexandria, when great guns were going off under it and beside it.

On his return to Glasgow he wrote, on November 27, to Miss Agnes G. King:—

Tell your mother that I look back with great pleasure to my time in Hamilton Terrace last week and the week before. It brightened for me a very anxious and irksome duty in London, the latter part of which, however, became

Photograph by Miss Agnes Gardner King    Emery Walker Ph. sc.

[1888] Yours very truly
William Thomson

a pleasure, thanks to Captain Fisher and the other naval and nautical men who gave much pleasing and amusing evidence, and to the happy conclusion of the case.

Sir William's insatiable curiosity to know the latest data of science is illustrated by the following postcard written to Professor Ewing, in the train, on December 21, 1888, on his way to Knowsley, for the usual Christmas visit to Lord Derby :—

What is the *very greatest* magnetic force in air that *you* have got? I suppose no one in the world has got greater. What is your very greatest intensity of magnetization? Wishing you a happy Xmas and "good New Year."

In 1888 the Society of Telegraph Engineers was reconstituted on a wider basis as the Institution of Electrical Engineers, and Sir William Thomson was invited to assume the Presidency of the Institution for the year 1889. On January 10 Lady Thomson wrote to George Darwin :—

Sir William has been very busy over his inaugural address as President of Electrical Engineers, which he gives to-morrow night. He goes to London to-night and returns here Friday, by day or night.

The main part of this address[1] was devoted to the subject of Ether, Electricity, and Ponderable Matter, and the relations between them. It was prefaced, however, by allusions to the addresses of some of the former Presidents of the Society, and with a little note as to personal history. He had spoken of the perplexity caused to early

[1] *Journal of Institution of Electrical Engineers*, xiii., 1890, pp. 4-37; reprinted in *Mathematical and Physical Papers*, vol. iii. pp. 484-515.

workers in submarine telegraphy by the inductive embarrassment met with in cable-signalling, and of Faraday's prediction that the Leyden-jar-like action of the gutta-percha coating would cause retardation, and of Varley's further investigation of 1854.

And then came on the great Atlantic cable question. I always remember how that question came upon me. I see in Professor Stokes's presence with us this evening a reminder of the circumstances. I was hurriedly leaving the meeting of the British Association [Liverpool, 1854], when a son of Sir William Hamilton, of Dublin, was introduced to me with an electrical question. I was obliged to run away to get to a steamer by which I was bound to leave for Glasgow, and I introduced him to Professor Stokes, who took up the question with a power which is inevitable when a scientific question is submitted to him. He wrote to me on the subject soon after that time, and some correspondence between us passed, the result of which was that a little mathematical theory was worked out,[1] which constituted, in fact, the basis of the theory of the working of the submarine cable. In that theory electromagnetic induction was not taken into account at all. The leaving it out of account was justified by the speed of signalling which the circumstances of a cable exceeding 200 or 300 miles in length dictated.

Sir William then went on to say how this further question of electromagnetic induction—the retardation due to the magnetic field created around the conductor by the current itself—had been investigated by him, and found to be absolutely imperceptible at the highest speed suitable for working the cables then proposed. But now that old question was being revived.

[1] By Thomson himself, see p. 331 *ante*.

I had myself laid it aside in some corner of my mind and in some slight corners of my notebooks for forty years. Within the last forty days I have really worked it out to the uttermost, merely for my own satisfaction. But in the meantime it had been worked out in a very complete manner by Mr. Oliver Heaviside, who has pointed out and accentuated this result of his mathematical theory—that electromagnetic induction is a positive benefit: it helps to carry the current. It is the same kind of benefit that mass is to a body shoved along against a viscous resistance.

This is at once followed by a highly characteristic passage, in which Sir William illustrated the result of mathematical theory by imagining a dynamical model, in this case a carriage supposed to be travelling through a viscous fluid :—

Take a boat not floated but partially supported on wheels, so that when loaded more heavily it will not sink deeper in the fluid. . . . We will shove off two boats with a certain velocity—the boats of the same shape; but let one of them be ten times the mass of the other: it will take greater force to give it its impulse, but it will go farther. That is Mr. Heaviside's doctrine about electromagnetic induction. It requires more electric force to produce a certain amount of current, but the current goes farther. . . . Old telegraphists remember that they always used to say three or four good leaks in a cable, if they would but kindly remain constant, and not introduce extra trouble by earth currents, would make the signalling more distinct. . . . Heaviside's way of looking at the submarine cable problem is just one instance of how the highest mathematical power of working and of judging as to physical applications helps on the doctrine, and directs it into a practical channel.

He then added "one little piece of practical information," in the shape of a formula for the ratio which represents the increase in the resistance of a copper wire, when used for rapidly alternating currents, arising from differences in the current-density at the axis and at the periphery of the wire. The formula was in terms of Bessel functions (real and imaginary), and was accompanied by a numerical table of values worked out. He then turned to the hydrokinetic analogies by which he had been endeavouring to elucidate electromagnetic theory. It was, he said, merely a mathematical working analogy, and as such exceedingly useful and instructive, and a very potent one in helping us in guessing out and in thinking out the practical problems of electromagnetic induction. But there was another analogy to which he gave the preference—the elastic solid idea which he had broached in 1846 (see p. 198), only twenty-eight days after he had entered on the work of his professorship. He had then attempted to represent electric forces by the displacement, and magnetic forces by the resultant rotations of an elastic body; and had then declared that a special examination of the states of a solid body, representing various problems in electricity, magnetism, and galvanism, must be reserved for a future paper.

As to this last sentence, I can say now what I said forty-two years ago—"*must be reserved for a future paper*"! I may add that I have been considering the subject for forty-two years—night and day for forty-two years. I do not mean all of every day and all of every night; I do not mean some of each day and some of each night; but the subject has been on my mind all these years. I have been trying many days and many nights to find an explanation, but have not found it.

If we had nothing but electricity and ether to consider, the problems of the electromagnetic analogy could be realized by a model of an elastic solid body having tubular pores filled with a dense viscous fluid. But there was also magnetism to be accounted for; and he asked the audience to consider the case of a solenoid having an electric current circulating around its coils. "Whatever the current of electricity may be, I believe *this* is a reality: *it does pull the ether round* within the solenoid." And yet though the ether could thus be pulled, as if it were an elastic solid, the earth moves through it. This led him to discuss the idea of an incompressible medium which yet possessed elasticity, and to suggest a sort of imaginary model of a web-like structure in which the rigidity was given to the medium by putting a gyrostat into each element of the structure. But now, although such a medium would afford a dynamical explanation of magnetism, it left out unexplained the electrostatic forces. And so the end was not satisfying.

And here, I am afraid, I must end by saying that the difficulties are so great in the way of forming anything like a comprehensive theory, that we cannot even imagine a finger-post pointing a way that can lead us towards the explanation. I only say we cannot now imagine it. But this time next year—this time ten years—this time one hundred years—probably it will be just as easy as we think it is to understand that glass of water, which now seems so plain and simple. I cannot doubt but that these things, which now seem to us so mysterious, will be no mysteries at all; that the scales will fall from our eyes; that we shall learn to look on things in a different way—when that which is now a difficulty will be the only common-sense and intelligible way of looking at the subject.

On several occasions during this session of his Presidency, Sir William Thomson took part in the discussions at the Institution of Electrical Engineers; amongst the papers discussed being those of Dr. (now Sir) Oliver Lodge on Lightning and Lightning Protectors; of Mr. W. M. Mordey on Alternate Current Working; of Mr. G. Hookham on Electric Current Meters.

He lectured in the spring of 1889 at the Royal Institution on Electrostatic Measurement; he also delighted the Belfast Natural History Society with an exhibition of gyrostatic experiments. In May he was busy thinking about systems of mutually repelling particles and groups of molecules. All this work issued in an important paper read to the Royal Society of Edinburgh, on July 1, on the Molecular Constitution of Matter (see p. 1050 *infra*).

In view of the approaching Electrical Congress in Paris, M. Mascart wrote to Sir William Thomson, inviting him to be present. Sir William's reply was negative.

12 GROSVENOR GARDENS, S.W.,
*June* 4, 1889.

DEAR MR. MASCART—Your most kind letter has been forwarded to me here. Alas, alas, I cannot have the pleasure of being in Paris to take part in the Congress of Electricians in August and to see the great Exhibition. I have been kept very much in London by my duties as

President of the Institution of Electrical Engineers, and I shall be detained still till near the end of the present month, with official business and duty, connected with the establishment of an electrical standardizing laboratory, which we hope is to be taken in hand by our Government. After so long an absence, I must return to Scotland and remain there steadily to get through arrears of work which have been of necessity accumulating all this time.

Lady Thomson and I have enjoyed all our visits to Paris so much, and particularly the last one, that we shall certainly go again when we can; and we look forward as a pleasure to come, to see you, and Mr. Bertrand, and your families in your homes, as we saw you all this time last year, and again to attend a meeting of the Academy and see our colleagues there.

But in the meantime will you and Madame Mascart not come to Scotland after your labours in the Congress are over, and take a little holiday, for which by that time you will, I am sure, be quite ready? Scotland is still very good in September, and it would be a great pleasure to us to have a visit from you in our country house, Netherhall, about 30 miles from Glasgow. We can run up by rail with great ease any day to my laboratory, which is always open, all the year round, if you would care to look in to it for an hour and see what we may have in hand.

My wife joins in kindest regards to you and your family—and I remain, yours always truly,

WILLIAM THOMSON.

Mascart, however, wrote again, urging him to come, and he consented.

NETHERHALL, LARGS,
*June* 23, 1889.

DEAR MR. MASCART—I cannot resist your letter of the 18th, kindly insisting on my presence at the Congress of Electricians, and I have therefore arranged to come to Paris in time to attend at the opening meeting on the 24th of August, and to remain for, at all events, several days.

Lady Thomson will be with me, and looks forward

with pleasure to seeing you and Mme. Mascart again in Paris, since you will not come to see us here this year.—Believe me, yours very truly, WILLIAM THOMSON.

The next day he wrote to Lord Rayleigh.

I am going to Paris after all to be present at the opening of the Electrical Congress on the 24th of August, and to remain a few days of it. I had such an appealing letter from Prof. Mascart that I could not resist! I wish you were coming too. Will you not come? It would be much better fun with you. . . . I found on our way from London to Glasgow last Thursday a complete settlement of the Boscovich theory of elastic solid. A homogeneous system of single points in equilateral tetrahedral order, each attracting or repelling its next neighbour, according as the distance between them is greater than, or is less than, $a$ (the edge of the static tetrahedron) gives a stable elastic solid. . . .

On August 5 he wrote again to Lord Rayleigh on the subject of surface-tension, and makes a reference to the capillary curves given in Vol. I. of his *Popular Lectures*.

Before I knew of Neumann's having done anything in the matter I saw his theorem in soup (every time I had it) of a particular kind. The pictures, p. 27 of my "Pop" (by the bye this title has been abused unmercifully by Critics in the newspapers), were drawn from observation.

In August came the Electrical Congress in Paris. Sir William took the opportunity to make three short communications to the Académie des Sciences, one on the molecular tactics involved in the artificial twinning of Iceland-spar; another on the equilibrium of atoms and on the elasticity of solids in the theory of Boscovich; the third being a description of one of his models to visualize the physical properties of the ether. Sir William and Lady Thomson returned from Paris on September 4. On the 11th they were at Newcastle attending the British Association meeting, to which he gave a paper on Boscovich's Theory. After the meeting they stayed on, during which they were the guests of Captain (now Sir Andrew) Noble, for a few days with Lord Armstrong at Cragside. In November came the news that Sir William had been made a Grand Officer of the Legion of Honour.

Mrs. King, staying in her brother's house at the University in November, sent letters to her daughters which afford glimpses of his activities and engagements.

*Nov.* 22, 1889.— . . . Uncle William went to London last night. He dined comfortably at 7.30 on all the courses, including entrées and game and his coffee, and set off in good, quiet style, and the ladies went up to the drawing-room when, lo and behold! there was the "green book" lying on the sofa. The telephone was instantly in requisition to summon a cab with all speed, and Margaret the housemaid was ordered to meet it, and being a capable woman she did so and succeeded in putting the precious book into Uncle William's hands before the train started, and he got it before he had missed it.

*Nov.* 23, 1889.—Uncle William came back this morning quite fresh after his busy day in London. It was to meet with the Lords of the Admiralty that he went. It is marvellous how easily travelling is conducted now. He goes so often by night to London the railway attendant knows exactly how to make his bed, and all the little arrangements he likes, and attends most carefully to his comfort. He drove at once to Admiral Fisher's, where he had his bath before 8.30 breakfast, and then set off about his various business. The meeting with the Admiralty was most satisfactory to him, for it is now ordained that his be the standard compass, and be used throughout the Navy.

Much mean and underhand work has been brought to light. For instance, of 60 letters (I suppose in answer to inquiries) from Captains, one from the Captain of the *Euryalus* spoke of some slight objection to the compass, eight said they had not had sufficient experience, and the remainder spoke of it in terms of unbounded admiration and appreciation. Of these letters the 51 had never been produced, but were hidden away in pigeon-holes in the Hydrographic Office, and the disapproving one was made a great deal of. I believe this has been going on for years and that Admiral Fisher has been instrumental in exposing the abuse. . . . There is much of the Circumlocution Office in the whole affair. Uncle William does not want it talked of. . . .

In December 1889 there was held in the Bute Hall of the University of Glasgow a great bazaar in aid of the funds of the Students' Union. Sir William and Lady Thomson took a very active part in this affair, which resulted in a profit of over £11,000. Lady Thomson's stall alone took over £2000.

In this year Sir William was a member of the Royal Commission upon the University of London, which up to that time had been merely an Examining Body. For some years an agitation had been proceeding to create a teaching University in London. The majority of the Commission declared in favour of reconstructing the existing University. Sir William Thomson, along with Sir George Stokes and the Head Master of Harrow, presented a minority report, favouring the creation of a separate Teaching University.

His year of presidency of the Electrical Engineers having expired, Sir William Thomson, on January 9, 1900, duly installed Dr. John Hopkinson as his successor. Before retiring from the Presidential Chair, he called the attention of members to the fact that electrical engineering originated with telegraphy, and specially with submarine telegraphy. He reminded them that he himself had been a shareholder in the first Atlantic cable of 1858, and a co-director with Sir John Pender, who in the hour of crisis gave a guarantee of a quarter of a million sterling for the funds required for the enterprise.

On February 1 he wrote to Lord Rayleigh:—

THE UNIVERSITY, GLASGOW,
*Feb.* 1/90.

DEAR LORD RAYLEIGH—See Maxwell, vol. ii. § 537. Is the last paragraph of this § very approximately true in practical cases?

*And*; have you *evidence* that on suddenly breaking a primary circuit, the total quantity flowing through a secondary circuit (with galvanometer coil in it) is independent of the suddenness with which the primary circuit is broken; and that it *is* that calculated from the self-induction of the secondary circuit and the ohmic resistance of the secondary circuit supposing the current to be running *full bore* through it?

It seems to me that ordinary practical suddenness of breaking a primary circuit must be such that the ohmic effective resistance in the secondary, even if of wire ·1 cm. diam., or less, must be largely augmented in virtue of the current not running full-bore through it during a sufficient part of the whole time of the discharge to consume a large proportion of the whole energy. I suppose I might find an answer to this by reading through all your papers and all Oliver Heaviside's, but I prefer this method as a shorter cut. I hope you are long ago quite out of the "grippe."—Yours, W. THOMSON.

This inquiry was the basis of a paper in the *Philosophical Magazine* in March 1890.

Three days later he wrote again, on Oscillations of the Atmosphere, suggesting that the large semi-diurnal term in the variations of the barometer may be due to one of the modes of free vibration of the atmosphere being in period not very different from twelve hours. The next letter is to von Helmholtz.

NETHERHALL, LARGS,
*Feb.* 24, 1890.

DEAR HELMHOLTZ—I have just this morning heard from Tait that the Senate of Edinburgh University has invited you to be Gifford Lecturer for the next period of two years, and I lose not a minute in writing to beg you to accept. We have so long wished to see you again in Scotland, that now, with such an occasion as this, we hope very much that you will be induced to come. During the month of the engagement in Edinburgh each year you must live part of the time with us in Glasgow, as the railway journey is quite short ($1\frac{1}{4}$ hour). The actual lecturing would, I am sure, be interesting to yourself if you feel that you can undertake it, which I hope will be the case; and I need not say that it will be greatly appreciated in Scotland, and by a far wider public than those who will hear you, or than Scotland.

Lady Thomson and I are here from Friday to Monday, as we are as often as we can get away from Glasgow at the end of the week, but Glasgow is our home during the winter session.

It chances that Prof. and Mrs. Max Müller are with us and I have told him of the wish of Edinburgh. He allows me to say that he is *most anxious* that you should accept. He will write and tell you about the conditions. We hope that Mrs. von Helmholtz will come with you. My wife joins in urging that you should accept, and in kind regards to you both, believe me, yours always truly, W. THOMSON.

To the Edinburgh Mathematical Society in February he sent a paper on the Moduluses of Elasticity in an Ideal Elastic Solid, constructed according to Boscovich's theory of an assemblage of points; and in March, another on a Mechanism for the Constitution of the Ether, to the Royal Society of Edinburgh.

In April and May he was discussing electric oscillations at the Glasgow Philosophical Society and at the Institution of Electrical Engineers.

The remarkable discovery by Hopkinson of a nickel steel which could assume either a highly-magnetic or a non-magnetic state, according to treatment, excited him greatly, and on receiving a sample he wrote :—

GLASGOW UNIVERSITY,
*March* 27/90.

DEAR HOPKINSON—I shall be glad to do anything I can in respect to your law case should it go on and my assistance be desired.

I have not hitherto been able to make any experiments on the specimen of nickel iron wire which you sent me; but I had just within the last three or four days been arranging to take it in hand. So I hope very soon to see for myself, and to let my laboratory corps see *something* of your wonderful result.—Yours very truly,

WILLIAM THOMSON.

[*P.S.*]—I shall be very glad to hear what you find with the new samples.

Sir William was still busy gathering up the threads to complete his third volume of Collected Papers, and *at last*, in May 1890, he was able to fix on a mechanical representation of magnetic force, which had baffled him ever since November 1846. The account of it constitutes Article xcix. of Vol. III., which he was now able to finish. Its preface is dated from Peterhouse Lodge, June 2, 1890.

A project for utilizing the powers of Niagara was now on foot, and the American promoters turned for advice to the highest authorities in Europe. They first approached Sir William Thomson; and the following letter which he wrote to Mascart shows the stage at which the proposition had arrived:—

PETERHOUSE LODGE, CAMBRIDGE,
*June* 21/90.

DEAR PROF. MASCART—I have to-day given a note of introduction to Mr. Edward D. Adams (President of the New York Cataract Company, controlling the Niagara Falls) which he will probably present you in Paris about ten days hence. He will be accompanied by Dr. Sellers, of Philadelphia, who is consulting engineer to the Company. They have come to Europe for the purpose of seeing and learning about electrical and hydraulic works. The Company is on a very solid foundation, being connected with some of the American bankers of highest character, and supported by all the money they require, so that they are not asking for subscriptions either in this country or in America.

They wish to form a consulting Commission of four,

and have asked me to preside. They wish to have you as one of the four, and they will explain the conditions and terms to you when they see you. Dr. Sellers is to be one of the Commission, and for the fourth they think of asking Mr. Turrettini (I am not sure if I spell his name correctly), who has done great hydraulic work at Geneva.

I have consented to act on the Commission for them, and I hope you will see your way to join it also. The work and plans which will be put before us will be most interesting; and the Company is of such a character that it will be thoroughly satisfactory and creditable to us to be associated with it as advisers.

Lady Thomson joins in kindest regards to you and Madame Mascart, and reminds her of her promise to bring you to see us in Glasgow and Netherhall soon.—Believe me, yours very truly, WILLIAM THOMSON.

The Niagara Commission[1] of experts, as finally constituted, consisted of Sir William Thomson (chairman), M. Mascart, Col. Turrettini, Professor W. C. Unwin, and Dr. Coleman Sellers. Professor George Forbes was appointed as the official engineer. The promoters invited projects from the chief European electrical contractors, as well as from American engineers. Twenty-six different plans were submitted. After many months the Commissioners were in a position to report; and eventually none of the plans proposed was adopted, though some use was made of fourteen of them. The engineer, in conjunction with Dr. Coleman Sellers, drew up the designs for the Cataract Company, and the contracts were given to American engineers to execute. Sir William Thomson was

[1] See *Cassier's Magazine*, June 1892 and July 1895; also *Electrical Review*, Feb. 9, 1894.

throughout and to the last entirely opposed[1] to the use of alternating currents in any shape, but was in a hopeless minority. The machines for the first power-house were two-phase alternators, of about 3500 kilowatts each, and of a special type designed by George Forbes, having an internal fixed armature and an external revolving field-magnet system.

In September 1890 the British Association met at Leeds. Sir William Thomson made five communications to the sectional meetings; on Contact Electricity, on Alternating Currents in Parallel Conductors, on Anti-effective Copper in Parallel Conductors, on a Method of Determining the Magnetic Susceptibility of Diamagnetic and Feebly Magnetic Substances, and on some of his newer electrical instruments.

Sir William had purposed to go in October to Cambridge to attend the statutory meeting of the Fellows of his College, but for once he was absent. Lady Thomson wrote to Darwin, "Sir William has been, and is, so busy over endless things, University Commission, etc. etc., that he has written to the Master if they can do without him this time." He

[1] He wrote, on January 6, 1892, when the first machines were nearly ready for work: "I am much interested in what you tell me about the progress of the Niagara work. I am very glad that contracts are being made for the power, and I hope soon to hear of a commencement of its actual use having begun. I do not believe that alternating currents will be found to be the right solution of the electric transmission to a distance. I have no doubt in my own mind but that the high-pressure direct-current system is greatly to be preferred to alternating currents. The fascinating character of the mathematical problems and experimental illustrations presented by the alternating current system, and the facilities which it presents for the distribution of electric light through sparsely populated districts, have, I think, tended to lead astray even engineers, who ought to be insensible to everything except estimates of economy and utility."

had a prospect of being even more busily occupied; for not only were the meetings of the Niagara Commission in prospect, but it was well known in the Royal Society that Sir George Stokes would resign the Presidency of that body in November, and the general consensus of the Fellows was fixed upon Sir William as the next President.

St. Andrew's day, November 30, is the date fixed by the statutes of the Royal Society for the election of Council and of a new President. Sir William had long been marked out for this, the highest honour which scientific men in Great Britain can offer to the most distinguished of their number. He had had no fewer than ten important memoirs in the *Philosophical Transactions*; he had been a Fellow since 1851; he had received a Royal Medal in 1856, and the Copley Medal in 1883. He was now to step into the Presidential Chair, to occupy it for the five years which custom has ordained as the period of its occupancy.

Sir William Thomson was now much more in London than in previous seasons; and by reason of his position as President he was invited to many functions, such as the annual banquets of the Royal Academy of Arts, the Institution of Civil Engineers, and of other learned societies and other professional bodies. This necessitated many journeys between London and Glasgow, and his ability to sleep soundly while travelling at night stood him in good stead. He had other matters, too, to keep him busy. He had invented a new water-tap to supersede those which in domestic use gave so much trouble by dripping when worn, or when their packing had hardened with age; and he was troubled by threatened litigation over his patent compass. Also his thoughts were running on the old problems (see page 157) of electric and magnetic screening. On this topic he wrote an inquiry to Lord Rayleigh on March 16, 1891, adding the news: "My compass case is settled between the solicitors at complete surrender, plus £500 towards damages and costs." He was to lecture in April at the Royal Institution on Electric Screening, and prepared to send to the Royal Society two papers on the same subject. Respecting the second of these he wrote on March 29 to Lord Rayleigh, saying that it would include the following matters:—

> Let a closed copper shell (it need not be a figure of revolution) rotate about any axis. Let a second, enclosing it, rotate about a second axis exactly perpendicular to the first. Let any magnet be held at rest inside the first shell. It will exert no influence on space outside the second shell if the rotations are sufficiently rapid.
>
> Variational action, if not too excessively rapid, promotes electrostatic transparency. It promotes electromagnetic opacity. A sheet of black paper in ordinary hygrometric condition is a perfect screen against steady electrostatic force. If the force alternates $\pm$ with moderate frequency, say from 10 periods to 10,000,000 per second, the paper is no screen at all. If the frequency be greater than $200 \times 10^{12}$ the paper screen is again opaque.

On April 24 his students presented him with an address congratulating him on his election as President of the Royal Society. In acknowledging the presentation he said: "I have been a student of the University of Glasgow for fifty-five years to-day;[1] and I hope to continue a student in the University of Glasgow as long as I live."

SIR WILLIAM THOMSON REPLYING TO THE STUDENTS' ADDRESS, April 24, 1891.

This season he rented a suite of rooms at No. 127 Ebury Street, London, from the end of the Glasgow session. Thence he wrote on May 1 to his sister Mrs. King: "We are going to-day to the

[1] This impression seems not to correspond accurately with the records; for he matriculated in October 1834 (see page 8). The accompanying illustration, which appeared in the *Daily Graphic* of April 28, 1891, is reproduced by the courtesy of the proprietors of that journal.

private view of the Academy, where we are delighted to hear that Elizabeth [Miss King] has a picture. To-morrow we go to the opening of the Naval Exhibition, and, as P.R.S., I dine in the Royal Academy in the evening."

A recent development in the kinetic theory of gases, known as the Boltzmann-Maxwell doctrine, had given him matter for thought of late. According to this doctrine the kinetic energy in a mass of gas under given conditions of pressure and temperature is distributed in equal proportions between the translatory motions of the molecules as they fly about in space and their rotatory motions within themselves. Sir William Thomson was unable to admit that this law was anything but empirical, and persistently disputed its foundation. Lord Rayleigh, on the other hand, was its stoutest upholder. There was much discussion between them, and many were the notes and postcards which Sir William sent, raising case after case to contest its generality. At the Royal Society on June 11 he stated for discussion a number of test-cases. He had written on June 6 to Lord Rayleigh a problem of this kind, and added:—

> I am confining my "test-case" to placing it before the Royal Society for consideration; and it is not this question, but the Baltimore molecule that is to constitute it. I am sorry you are not to be at Greenwich [at the Annual Visitation of the Royal Observatory] to-day. If you had come we might have got the question settled.

The Board of Trade had recently appointed an Electrical Standards Committee to settle for legal purposes the denominations of Electrical Standards in the United Kingdom, and so give legislative weight to the decisions of the International Congress of 1889. Sir William Thomson was naturally consulted in this matter. At the meetings of the Committee, from January to July 1891, he took an active part in the discussions, and in the examination of scientific witnesses. He was strongly in favour of making the standard ohm of solid metal. He opposed the acceptance of the Clark cell as a standard of electromotive force on the ground that it was not permanent. He also signed the supplementary reports of 1892 and 1894.

In referring to the *ampere*, Sir William Thomson said that the grave accent on the first "e" in the name should be omitted, as the word had now become adopted into the English language. This was agreed; and a reference to the Order in Council issued on August 23, 1894, will show that the law recognizes this to be the spelling.

The season over, Sir William and Lady Thomson sailed for Madeira. Lady Thomson wrote on August 19 to Mrs. King, "William sits out all day under the trees very happy with his green book, and deep in a very difficult problem which he is working hard to solve! It is a grand time for him."

On September 1 Sir William addressed to Lord Rayleigh a letter, dated "In a hammock, Camacha," Madeira, telling him of a manuscript he had sent to the *Philosophical Magazine* about a model with a jointed pendulum and a jumping clock which is "a perfectly decisive test-case against the Boltzmann-Maxwell doctrine"; and adds:—"This (Madeira) is perpetual Senate-house. Book-work and problem papers to be done with no books of reference."

By September 12 they were back in England; but the British Association meeting at Cardiff was over.

In November 1891 he was repeatedly writing to Lord Rayleigh on a mathematical matter—Hill's determinant—on which he had doubts. To one of these notes he added a postscript: "I have been seeing Parker's great electric factory at Wolverhampton (now belonging to the Electric Construction Co.). It is splendid—far ahead of anything I have seen before." To another letter was appended a paragraph on the installation, on November 26, of Mr. A. J. Balfour as Lord Rector at Glasgow, when he delivered an address on Progress. "Prince Arthur has been most charming and potent in philosophy yesterday and politics to-day. His reception was splendid; classes and masses unitedly enthusiastic and appreciative." Mr. Balfour stayed with Sir William in the University. There was afterwards a banquet in the City Hall, at which, replying to the toast of the University, Sir William spoke of himself as "a child of the University," and described the changes in its constitution which himself had witnessed by the action of two successive Royal Commissions. The next day, acting as the spokesman of a number of friends and subscribers, Mr. Balfour presented to the University a portrait by Herkomer, of Sir William, which had been subscribed for in commemoration of his election as President of the Royal Society. In presenting the portrait Mr. Balfour eulogized the unceasing, the astonishing activity of mind which had secured him the gratitude of this generation and the title to fame in the generations to come. Sir William, in acknowledging the presentation, said :—

I have felt that to live in the University of Glasgow, and to remain at the University of Glasgow to the end of my life, is the career that has been most of all conducive to my happiness and appropriate to give me the means, the position, and the surroundings—to provide for me the surroundings—that have allowed me to go on with my scientific work. The facilities that the University has given me for experimental research, the splendid assistance that I have obtained voluntarily, offered voluntarily, and continued voluntarily, and persevered in through a great deal of hard and straining work voluntarily, by students of the University of Glasgow, has done more for me than words can tell, and without which it would have been impossible for me to realize any of the results which have been spoken of in so appreciative a manner to-day. I can only say, in the name of Lady Thomson and myself, that we are deeply grateful.

Sir William's Presidential Address to the Royal Society on St. Andrew's day dealt with the most recent geodetic observations, in which certain changes in the position of the axis of rotation of the earth's globe had been detected. He also spoke of the recent Faraday centenary, and of the relations between the Royal Society and the London County Council in the matter of purity of water-supply. A note of personal reminiscence was struck in the following paragraph :—

The name of Becquerel has been famous in science since the days of Biot, Davy, De La Rive, Faraday, Ampère, and Arago. I well remember going to the Jardin des Plantes, in Paris, in January 1845, with an introduction from Professor James Forbes to Antoine César Becquerel, who, even at that remote time, was a veteran in physical science; and finding him in his laboratory there, assisted in work regarding electrolytically deposited films on polished metallic surfaces and their colours, by his son Edmond, a bright young man who had already commenced following his father's example as an active worker in experimental physics. . . .

There was a little family dinner party that Christmas at the University in Sir William's house, at which his sister, Mrs. King, and Professor James Thomson were present. Mrs. King sent to her daughters an account of it :—

We had a lively dinner yesterday. Uncle James and Uncle William were very bright, and it was most amusing and interesting. The question of the franchise for women, and the action of New Zealand in the matter, arose, and Uncle James referred to Anacreon's opinion of women. Uncle William challenged him to quote, and Uncle James recited the ode in Greek, and then translated for the benefit of the unlearned. It enumerated the gifts of the gods to various creatures, as the lion, the horse, etc. etc.—to Man courage, power, wisdom; to Woman beauty, which is stronger than all and can vanquish all. There was a great deal of the gaiety and wit of the College parties in the olden time that I remember so well, and yet there were only the Florentines [the James Thomsons]. There is certainly a great deal in this house to quicken one's wits. An interesting man, an electrical friend of Uncle William's, took tea here yesterday, and I listened with extreme interest to their talk, and watched Uncle William's keenness of perception and brilliance of inspiration. I do enjoy this sort of thing.

This was the last time that the two brothers and their sister were together.

# CHAPTER XXII

## THE PEERAGE

New Year's Day, 1892, brought the announcement that a peerage of the realm had been conferred by Queen Victoria upon Sir William Thomson. The offer of this honour was made by the Prime Minister, Lord Salisbury, in a private and personal letter written with his own hand. It was couched in very friendly and personal terms, expressing the anticipation that the influence of the eminent man of science in the House of Lords would be a great access of strength to them, especially on certain subjects in which their Lordships were not particularly strong. The announcement was received with general satisfaction in the public press, as a well-earned recognition of the pre-eminent position which Sir William Thomson by common consent held in science.

*The Times* spoke of his "unquestioned distinction." The *Daily Telegraph* hailed him as "universally regarded as the first physicist, and one of the profoundest mathematicians, most suggestive thinkers, and most original inventors of the age." The *Daily News* spoke of his "European reputation as a man of science before he became President of the Royal Society," but hinted that he had other claims upon the present Unionist Government by reason of the very prominent part he had taken on behalf of the Liberal Unionist cause. The *Globe*, after declaring that Science has not generally been hitherto recognized as leading to the House of Peers, and that there were "many reasons why it should not, like the high careers of government, statesmanship, and high command, be made a regular highroad thither," added patronizingly that "it should not be without its representatives, when exceptionally qualified, as is assuredly the case in respect of Sir William Thomson." It quoted the opinion of Faraday that a great man of science could not be honoured by a title any more than a great poet; but added that it was surely better to have men of intellect in the peerage than men of money. The *Spectator* described him as "the first man of science, we think, who ever received the distinction of rank for work in that domain." The *Birmingham Post* observed that the peerage conferred on Sir William Thomson "is awarded for work about which there can be no question as to its extent or character. Not for services connected with the destruction of human life, but for those which elevate its character, enlarge its capacities, and increase its happiness, the new peer has deserved his honour." The *Manchester Examiner* considered that there was a certain appropriateness in the destiny by which Lord Tennyson owes his peerage to Mr. Gladstone, and Sir William Thomson his to Lord Salisbury. "The man of words and the man of deeds—both being men of ideas—are again placed in contrast with each other." The Scottish newspapers, without exception, applauded the elevation to the peerage of one so highly esteemed throughout Scotland. "A peerage has at length been conferred upon a scientific man because he is a scientific man," said the *British Medical Journal.*

Sir William was himself sincerely pleased with the token of official appreciation which thus had been publicly offered to him; and there were great rejoicings at his house in the University. But rejoicings were strangely mingled with sorrow, when the very next day brought from Madeira news of the death of Mrs. Blandy, the mother of Lady Thomson. Mrs. King, arriving on January 3rd to congratulate her brother, found two piles of letters on the table, condolences and congratulations alike testifying to the sympathy and regard of friends and relations. Already the question of selecting a title had been discussed. The name " Kelvin, " taken from the Kelvin River, which winds its way through the Kelvin Grove below the (new) University buildings, had been favoured by those most concerned; and, when his sister independently suggested it, Sir William, turning to his wife, said: " Did you hear that? Elizabeth thought of it too; Elizabeth agrees with us; that decides it; it shall be Lord Kelvin; I will write to Lord Salisbury at once." Mrs. King wrote to her elder daughter: " Uncle William showed us many of the letters he has received; the kind tone of all goes to his heart and fills him with happy wonder."

A fortnight later Lord Sandford, visiting Glasgow as one of the Scottish University Commissioners, called at Sir William's house, and took tea with them. His father had been Professor of Greek in Glasgow, before Lushington, and little William Thomson had attended his classes in 1834-35. As boys they had played together on the banks of the Molendinar Burn, which flowed through the old College Green. Now they renewed their childish reminiscences, and Lord Sandford promised to act as one of the peers to introduce Lord Kelvin to the House of Lords when he should attend in his robes to take the oath of allegiance. His other introducer was to be Lord Rayleigh; but the patent of peerage was not yet issued.

Lady Thomson had written on January 6th to George Darwin:—

GLASGOW UNIVERSITY, *Jan.* 6/92.

MY DEAR MR. DARWIN—You will probably have heard before this that our joys and sorrows have come very near together, and so will forgive my not having written to you sooner to thank you and Maud for your very kind congratulations. The day after we heard of the Peerage we heard by telegram from Madeira of my mother's death—it came upon us as a great blow, as we had been having letters from her up to two days before the telegram, telling us how well she was keeping. Everybody has been so kind and said such kind words to us over this honour to Sir William that we have been quite overpowered. We have no new name yet. We have proposed Kelvin as being connected with Glasgow and the University, and not being too territorial—but it waits the Queen's sanction. So in the meantime we are still Sir Wm. and Lady T.—Yours most sincerely,

F. A. THOMSON.

Sir William had, meantime, written to various friends—Lord Rayleigh, Dr. J. Hall Gladstone, and others.

GLASGOW UNIVERSITY, *Jan.* 3/92.

DEAR LORD RAYLEIGH—I enclose C.'s paper, which, if you approve, may be put in type for the 21st, with slight pencil corrections I have made. It is scarcely worth while sending it back to the author for his approval.

I had a very bad time all the holidays over $\frac{d^2u}{dx^2}+(\alpha+\beta\cos x+\gamma\cos 2x+\ldots)u=0$. It works out beautifully, and with great ease for all values of $\alpha$ and $\beta$ (unless $\alpha$ is *too* very great), where $\gamma=0$, etc. But with anything beyond $\alpha+\beta\cos x$ I don't yet see anything satisfactory on the plan of which I wrote to you.—Yours truly,

$x$ (? $x$ = Kelvin).

[*P.S.*]—With best wishes for the New Year from my wife and myself to you and Lady Rayleigh.

THE UNIVERSITY, GLASGOW,
*Jan.* 9/92.

MY DEAR GLADSTONE—I thank you heartily for your kind congratulations, and for what you say of science in connection with them.

We have proposed "Kelvin" as the title for the sake of association with the University and city of Glasgow: but till the Queen approves nothing can be decided.

I am very glad to hear you are all well, in these trying and dangerous saturnalia of microbes.

Lady Thomson joins me in kind regards to all and best wishes for the New Year.—Believe me, yours very sincerely, WILLIAM THOMSON.

THE UNIVERSITY, GLASGOW,
*February* 13, 1892.

DEAR LORD RAYLEIGH—I suppose the Bakerian Lecture[1] will be taken on Thursday the 10th of March, as that has been fixed for the Council meeting, and I shall be present. If, however, it is more convenient to take it on the 17th, you could do the reading and exposition at least as well as I. It will be fully written out and illustrated with at least two large diagrams. My brother, though he is not able for much work or use of his eyes, hopes to have the paper complete before the end of February.

Will you be inclined to dine at the Royal Society Club on Thursday evening next? I would like to do so, and if you would also we might have some better opportunity than the business at Burlington House allows of talking over non-official questions, and particularly high frequency and high potential.

Was Tesla's light in vacuum tubes more brilliant than that of ordinary vacuum tubes excited by coils capable of giving sparks half a foot long? I can conceive of its being so, because of the greater time-integrals of the + 100,000 volts and of the − 100,000 volts than that of the high potential of one sign in the ordinary induction

[1] On the Grand Currents of Atmospheric Circulation, by Professor James Thomson. See p. 918.

coil illuminating vacuum tubes. It seems to me, in short, that the time-integral of the light will be greater and therefore the appearance brighter than with induction coils: but not greater, that I can at present see, than that of vacuum tubes lighted by an ordinary electrical machine giving 100,000 volts.

The marvel to *The Times* and others is there being no wire going through the glass, but the reporters can scarcely be expected to know of the electric light by the ordinary induction coils with nothing but plates of tin-foil outside the glass and no electrodes going through the glass, though this is, I suppose, one of the earliest things known about the electric light in vacuum.

The most interesting thing to me in what I have seen in the report is the impunity of touching one electrode in Tesla's arrangement. . . .—Yours truly,

WILLIAM THOMSON.

*P.S.*—Tesla's light in ordinary vacuum tubes, and in tubes containing Crookes's pieces of Iceland Spar, toasted shells, etc., will be time-integral of what we see (shadows and all) with the current from an ordinary electric machine, or from induction coils, with current alternately in opposite directions, but long enough in each direction to let us see the peculiar effect. The difference of the effects in the two directions is, of course, integrated out in Tesla's light.

You will see in *Nature*, Jan. 6, 1881, vacuum tubes without electrodes by J. Bottomley; and a second note on the subject in *Nature* of Jan. 13, 1881 I see Glazebrook ("G." in *Nature* of this week) seems to think vacuum tubes without electrodes a novelty. J. Bottomley long ago in his lectures showed a completely closed glass vessel of vacuum glowing when he held it in his hand above the spark between two poles of a Ruhmkorff.

GLASGOW UNIVERSITY, *Feb.* 13/92.

DEAR LORD RAYLEIGH—Have we (R.S.) never given Crookes anything except F.R.S.? If not, I think we ought to give him the best we can (? a Copley). His torrent of particles, toasted shells, and Iceland spar converted to malachite and emeralds and rubies and topaz (Iceland spar), and his radiometer and wheelbarrow, etc. etc., are really the greatest things, by far, in deep-going physics of the nineteenth century.—Yours,

W. THOMSON.

On January 28 Sir William Thomson presided at a meeting of the West of Scotland Liberal Unionist Association.

In his opening remarks the Chairman said the prospect of another struggle was now imminent. They should enter upon that struggle not at all discouraged. Twenty Rossendales could not discourage them. They had that best of all foundations for courage, a good cause. The course of the past six years had proved, every month, every year, that they were right in 1886. He could not believe so ill of his countrymen as to think that a majority of them could now be found to send representatives to Parliament pledged to whatever bill Mr. Gladstone might bring in. It would be an awful grief to every person who valued the prosperity of this country and the happiness of Ireland, and who distinguished between right and wrong, should there be such a majority. It would be intolerable for Ireland not to be fully represented in the Imperial Parliament. Let them give every possible facility for the Irish people to manage their own local affairs, but they must have Irish co-operation in the Imperial Parliament. They had every confidence that right would prevail over wrong in the long-run, and reason over unreason.

At a meeting of electors of the College Division of Glasgow, to promote the candidature of Sir John Stirling Maxwell, the Conservative candidate, on February 4, Sir William Thomson, in moving a vote of confidence, took occasion to defend the policy of Mr. Balfour as Chief Secretary for Ireland.

It was now known that what was wanted to cure the ills of Ireland was good, honest, resolute, kindly government. He ventured to say that to-day Mr. Balfour was the most popular man in Ireland. Notwithstanding the torrents of opprobrious epithets hurled at him by some Irishmen, Mr. Balfour had not turned one hair's-breadth from that path that had put Ireland into a condition of peace and prosperity.

Sir William was in London on February 19 to open the electrical engineering laboratory which had been founded at King's College by Lady Siemens in memory of her late husband. In making the formal presentation on behalf of Lady Siemens he said that he was probably the oldest friend of Sir William Siemens amongst those present; and he referred to his achievements and to those of his brother, Werner von Siemens, and of their brother-in-law, Professor Lewis Gordon, in the scientific application of telegraphy, and in the laying of submarine cables. Electrical engineering had in one respect an advantage over other branches of engineering—it had begun on a thoroughly scientific foundation. Beginners in modern electrical engineering thoroughly knew their science before they began practical work. What had grown in several thousand years in ordinary mechanical engineering, had come into existence in the case of electricity in as many days or weeks. Thus it was that this creation of the nineteenth century could hold its own with all the best mechanical engineering in the world. The development of the "dynamo" (the name given by Sir William

Siemens) was largely owing to the investigations of the brothers Siemens.

The same evening Sir William gave a discourse on Motivity to the Peterhouse Science Club, of which he had been elected president. A large gathering of scientific and mathematical visitors attended in the Combination Room of that ancient house. The lecture consisted mainly of a historical account of the theories of conservation and transformation of energy. "Motivity" was in effect the operation of "available energy," as distinguished from such portions of molecular energy (in the form of heat) as cannot be employed in producing mechanical work.

On February 23 Sir William Thomson was gazetted as Baron Kelvin[1] of Largs, in the county of Ayr. He took his seat in the House of Lords on February 25, being introduced, with the prescribed ceremonies, by Lord Rayleigh and Lord Sandford. Those who have not witnessed the introduction of a new peer can scarcely conjure

[1] Lord Kelvin's change of designation was regretted by some who honoured the name of the man who had achieved so much. M. Taine, the distinguished *littérateur*, asked Mr. J. E. C. Bodley to explain something very curious. "I suppose it is still considered an honour to sit in your House of Peers; but why, as a penalty for doing so, should my friend Sir William Thomson, whose reputation belongs to Europe and not to England alone, bury his illustrious identity in an unknown title? Even if he had followed the example of Tennyson and Macaulay, and called himself Lord Thomson, that would not have been distinctive enough; he was celebrated as William Thomson, and he ought to have called himself Lord William Thomson." "I explained," says Mr. Bodley, "that that style could not be applied to a peer, as it indeed indicated that the person so addressed belonged to a certain small section of commoners." "Well," said Taine, "it is another peculiarity of the British constitution."

About the same time a foreign electrician of distinction wrote to the late Professor Ayrton to ask him who this person Kelvin was, who was claiming to have invented the galvanometer that all the world knows to have been invented by Sir William Thomson.

up the scene when he and his supporters, clad in their gorgeous robes, and preceded by the Gentleman Usher of the Black Rod, the resplendent Garter King of Arms carrying the patent, the Earl Marshal, and the Lord Great Chamberlain, troop into the House, bowing thrice; and the new peer, kneeling, presents his writ to the Lord Chancellor and retires to the table, while the clerk reads the patent and the writ. The Oath is then administered to the new peer, and he signs the Test-roll, when he and his supporters bow first to the Cloth of Estate, a second time to the Bishops, and again to the House as he crosses it to take his seat. Being seated there between his supporters they must then bow thrice to the Lord Chancellor, "taking off their hats at each time of bowing"; and when the Lord Chancellor returns the salutation, the new peer is conducted to the Woolsack to shake hands with its occupant. One wonders what Lord Kelvin's thoughts were during the performance.

The following arms were assigned to the barony of Kelvin of Largs:—

KELVIN.

ARMS.—Arg., a stag's head caboshed, gules, on a chief, azure, a thunderbolt ppr., winged or, between two spur revels of the first. *Crest*—A cubit arm erect, vested, az., cuffed arg., the hand grasping five ears of rye, ppr. *Supporters*—On the dexter side a student of the University of Glasgow, habited, holding in his dexter hand a marine voltmeter, all ppr.: and on the sinister side a sailor, habited, holding in the dexter hand a coil, the rope passing through the sinister, and suspended therefrom a sinker of a sounding machine, also all ppr. *Motto*—Honesty without fear.

A dinner was given by the Glasgow University

Club on the 19th of March to Lord Kelvin of Largs in honour of his elevation to the peerage; when, under the chairmanship of Dr. Joseph Coats, they drank his health and commemorated his forty-six years of service to his University.

In his reply Lord Kelvin spoke of the fostering influences of University life, and it was for more than mere instruction that he must always thank the Universities of Glasgow and Cambridge. In Glasgow there certainly was a very stimulating course. But there were differences in quality and subject in different universities. The thorough teaching of the University of Cambridge, following on the initiatory study in which he had engaged in Glasgow, had been of the greatest possible value. He should be sorry to think that Greek was killed out of Scotch University studies by any preliminary examination. Universities ought not merely to be a means of advancing towards a profession and earning a livelihood. They should lead in this direction no doubt, but they should do more—they should give a possession for life that rust could not corrode, nor moths eat, nor thieves break through and steal. Referring to the Atlantic cable, he remembered the services of his now venerable assistant Donald MacFarlane, and those of members of his volunteer laboratory corps in the old days, Professor Ferguson, and Messrs. Medley and Deacon, who had helped to hook up the lost cable of 1865.

THE UNIVERSITY, GLASGOW,
*March* 22/92.

DEAR RAYLEIGH — I am very glad to hear from Dr. Traill of T.C.D. that there is a prospect of your being at the Dublin Tercentenary, and his guest. My wife and I are to be there and his guests in the College, so we hope you will be with us. The time is from the 5th to the 8th of July. . . . Will Lady Rayleigh not come too, to T.C.D.?—Yours, KELVIN.

*P.S.*—I have instructed Potter, Admiralty Chartseller, to send you a copy of Weir's diagram, that you may see whether, as a most wonderful piece of "applied" mathematics, it is not deserving of recognition by an award from the Gunning Fund of the R.S. It is quite unique as a graphic process for solving a problem of finding the value of a function of three independent variables (lat. of the place, decl$^{n.}$ of the sun or star, hour-angle).

In April Lord Kelvin was nominated a member of the Joint Committee to consider private bills proposing electric railway schemes for London. On the 11th of the month he gave an address on Navigation to the Mercantile Marine Association at Liverpool. Much of the lecture was technical, but the introduction and conclusion are of more than technical interest. He said:—

The remarks of the Mayor of Liverpool had revived his early recollections of Liverpool. His mind went back to 1842, when he visited the docks of Liverpool: his recollections of those days were principally of skysail masts and yards. He had a companion—a young officer in the service of one of the great firms trading to India—who taught him, in his ignorance of seafaring matters, to look with admiration on the tall, tapering masts of the American ships then in Liverpool. There seemed to be something of American pride in those skysail yards, whilst the English ships in dock were contented with royals, never aspiring to the greater elevation. Yet they survived their American competitors, and practically ran them off the sea. There was something much greater than skysail yards now. There were three-masted and four-masted iron sailing ships carrying 3000 tons or more—twice as much as any sailing ship of Liverpool could carry 50 years ago. There had been an enormous advance in the carrying business, owing to the prosperity of England; and every part of the world was represented in Liverpool now as it was then: but England's share was relatively much greater now than then. It is, however, no pleasure to us to think that our friends on the other side of the Atlantic declined so greatly in the work of carriage by sea. It was not due to want of ability and seamanlike qualities, or of appreciation of the management of ships at sea, or of the power to build the best kind of ships for the sea, that the Americans were so much off the sea, and confined their business to three-masted schooners sailing up and down the American coast. It was due to bad politics. Liverpool had always been to the front in sound politics in regard to that great principle of Liberalism—free trade. People seemed to forget that free trade meant two things—freedom and trade, both very good things. He had always thought that that kind of Liberalism which was represented by free trade had been an honour to England, to Lancashire, and to Liverpool. They might well be satisfied that Liverpool had played so great a part in promoting its results. Great changes had taken place since the period he had referred to. The greatest carrying work of the world was now done by steamers, and Liverpool was not behind the rest of the world.

The bulk of the lecture was devoted to "lunars," Sumner's method of finding the position of a ship, the use of the globes in education, Weir's diagram, and the advantage of taking soundings.

Every navigation school, for young or old, should always have a globe at hand, and the work should be done by reference to a globe, to explain the meaning of spherical trigonometry, which, without a globe to help, was as veritable a twisting and screwing of the brain, as severe a mental torture, as could well be conceived. The "Use of the Globes" annulled the mystery of "great circle sailing"; and was truly valuable to the navigator in helping him to judge as to permissible, or eligible, or necessary deviations from the great circle. With respect to dead-reckoning and trusting to it for the safety of the ships, he must say that it was literally true that more ships had been lost by bad logic than by bad logs; and that had largely been promoted by Board of Trade findings in Courts of Inquiry. . . . Now that ships need not stop to take soundings, when they got within the 100-fathom line they should use the lead, and keep sounding every ten minutes or every five minutes, according to judgment, from hour to hour. If this had been done in some recent cases disasters would have been avoided.

About this time, the Board of Trade, in draughting electric-lighting orders, proposed to adopt the term "kelvin" as the name of the unit of supply, in place of the term "kilowatt-hour." Sir Courtenay Boyle communicated the proposal to Lord Kelvin; but he put from him the suggestion in favour of the term "supply-unit," giving, amongst other reasons against the proposal, the ambiguity that might arise if a supply-meter were described as a "kelvin-meter."

On the 8th of May 1892 Professor James Thomson died.[1] He had held the chair of Engineering at Glasgow from 1873 to 1889, when, owing to failing eyesight, he retired. He had been active in science up to the last: his Bakerian lecture on "The Grand Currents of Atmospheric Circulation" having been read (by Lord Kelvin for him) at the Royal Society on March 10, 1892. Lord Kelvin, writing on May 12 to their brother Robert in Australia to tell him of the events, said: "James had been very well, and since last March feeling very happy that he had finished and communicated to the Royal Society a great work of years on the trade winds and other great motions of the atmosphere. He was still occupied with a continuation of work of this kind, in which he took great and never-failing interest, when he died." Lord Kelvin's sorrow was very real and very deep. Throughout their lives the two brothers had been associated together not only in much scientific work, but in closest personal association; not a tinge of jealousy ever clouded their more than brotherly affection. In thermodynamics particularly their work was so closely related that it was not always easy to distinguish the parts due to each. Though Lord Kelvin was himself scrupulously careful to

[1] An appreciative memoir of Professor James Thomson, by his nephew Dr. James T. Bottomley, will be found in the *Proceedings of the Glasgow (Royal) Philosophical Society*, 1892 3, vol. xxiv. pp. 220-236.

mention his brother's name in any matter that originated with him, his own fame is apt to overshadow the important contributions made by the elder brother. It has been quaintly said that William would not have been William without James, and that James would not have been James without William.

Lord Kelvin, who had hurried to Glasgow to the bedside of his dying brother, was due to attend Parliamentary and scientific committees later in the week, but wrote to Lord Rayleigh (then Secretary of the Royal Society) that he must put off the reading of a paper on electrostatic capacities till June. He added:—

I would like before the end of the session to communicate, from my brother, an explanation of the beams of light seen when looking with partially closed eyes at a lamp flame, unless we find that the explanation has been already given. They are usually, but as my brother showed me wrongly, attributed to some effect of the eyelashes. He explains them by the [capillary] curvature of liquid between the eyelids and the cornea. They are seen indifferently in virtue of lower or upper eyelid. Do you know anything [1] of this, or whether the subject has been treated by any one?

[1] [I cannot forbear adding here a note of an incident illustrative of Lord Kelvin's personality, though it concerns myself. At the Conversazione of the Royal Society of June 15, 1892, Lord and Lady Kelvin were receiving their guests at the head of the staircase, and it came to the turn of Mrs. Silvanus Thompson and myself to be received. Lord Kelvin literally seized me, and hurriedly said to me, pointing to an electric glow-lamp hanging a few yards away, "Look at that lamp: now half shut your eyes: tell me what you see." I said, "I see irregular luminous streaks extending in somewhat oblique bundles above and below." "What are they due to?" he asked. "Oh, I have always supposed them to be due to the film of moisture at the edge of the eyelids, acting as an irregular cylindrical lens." "Who told you that? Where did you find that?" he asked excitedly. But just then a hand was laid on his sleeve, and a gentle voice behind us said, "William, there are

I shall come on Wed[y] night if possible: but my sister-in-law and two nieces are both very ill and I do not know if I shall feel able to leave.

Alas, the sorrow was increased by the death of James Thomson's widow and of one daughter, within a week of his own death.

On May 28, introducing the Duke of Devonshire at a Unionist meeting in Glasgow, Lord Kelvin told the audience that personally he spoke for the same reasons as those which, in 1886, impelled him to enter upon political subjects that were very far indeed from his ordinary and natural avocations.

As an Irishman he felt then, as an Irishman he felt now, that a frightful damage was threatened against his native land:—to take away Ireland from its grand position as a constituent equal member of the British Empire,—to make it a naval and military station of the neighbouring island of England and Scotland. . . . He had many friends whom he admired and for whom he had the most sincere respect, whom he regretted to have unhappily followed Mr. Gladstone in that crazy departure from good political principles. . . . He was Irishman enough to like the Irish Parliamentary party; to admire their mode of procedure

people waiting." Later in the evening he resumed the subject, telling me how his brother, while lying in bed ill, had studied these apparent rays and given him this explanation; and he asked me whether I had written anything upon the phenomenon. It was not till months afterwards that I found the following note printed in the *Proceedings of the Royal Society*, vol. lii. p. 74, at the end of the paper of his deceased brother, which he had communicated:—

"*Note by the President, of date June* 16.—I had asked many friends well acquainted with optical subjects whether they knew of this explanation of the luminous beams, and all said 'no' until yesterday evening at the *soirée* of the Royal Society, when Professor Silvanus Thompson immediately answered by giving the explanation himself, and telling me that he had given it to his pupils in his lectures on optics, as an illustration of a concave cylindrical lens. He did not know of the explanation ever having been published otherwise than in his lectures. I have myself looked in many standard books on optics, and could find no trace of intelligence on the subject. It seems quite probable, therefore, that of all the millions of millions of men that have seen the phenomenon, none within our three thousand years of scientific history had ever thought of the true explanation except Professor Silvanus Thompson and my brother."]

in some respects; to consider that according to their lights they were doing quite right. He thought their lights were false, and he thought they were doing wrong. A bill which took the Irish members out of Westminster and sent them to Dublin would be an unmixed evil to the British Constitution, an unmixed evil to Ireland.

The summer was a busy one; for besides holding the Presidency of the Royal Society, and attending meetings of the Scottish University Commission, Lord Kelvin accepted the Presidency of the Institute of Marine Engineers, and on June 7 paid a visit to its premises in the Romford Road. He was also advising the Board of Trade as to Electrical Standards.

He wrote to von Helmholtz:—

*June* 20, 1892,
6 CADOGAN PLACE, LONDON, S.W.

DEAR VON HELMHOLTZ—Your letter of the 19th of May has remained too long unanswered: because I wished to have authority from the Board of Trade in answering it. It had been resolved to issue at the middle of the present month an Order defining, for practical purposes, standards of electrical resistance, current, and "pressure" (potential). I am now authorised by the Board of Trade to say that in deference to your wish, the proposed Order will not be issued until next November.

The advising Committee of the Board of Trade in respect to Electric Standards (of which Rayleigh, Glazebrook, Carey Foster, Ayrton, and myself are members) were unanimous in wishing that you, and one or more of your colleagues in the work of electric measurement, would come to the British Association, which meets in Edinburgh on the 3rd of August, and thus give us the opportunity of conferring on the standards to be adopted, and of coming to an agreement on all questions such as those referred to in your letter of May 19th.

I was asked by my colleagues to do what I could to persuade you to come yourself. You know I would compel you to come if I could! I hope the time will be convenient to you (more so than the later times of meeting which for many years back have been chosen for the B.A. meeting), and I wish very much that you would consider whether a trip to Scotland in August would not for one year at all events be as health-giving as Pontresina or elsewhere in the Alps.

It will be a great pleasure to Lady Kelvin and myself if you and Mrs. von Helmholtz will come to Scotland and give us as long a visit at Netherhall as your time before or after the meeting in Edinburgh, or both before and after, will allow.

With our united kind regards to you and your wife, and to (the now matronly) Ellen, I remain, yours always truly, KELVIN.

*P.S.*—For nearly a year I have been greatly exercised over the equation—

$$\frac{d^2n}{dx^2} + \pi(x)u = 0$$

where $\pi$ is such that $\pi(x+c) = \pi(x)$; which occurs in connection with the question of the Stability of Periodic Motion. I have spent more time than I would like to confess on the case $\pi(x) = \alpha + \beta \cos(x)$; and at last, within the last few days, I see my way to answer all those questions regarding it which I had felt difficult.

The results are (to me) wonderful and exceedingly interesting. I hope soon to send you a rough proof in print. I believe we shall find that all periodic motion is essentially unstable.

Various honours fell to Lord Kelvin about this time. On the occasion of the Galileo Tercentenary the University of Padua gave him the Doctorate of Natural Philosophy. He was also nominated by the Berlin Academy to receive the first Helmholtz Medal.

He wrote again to von Helmholtz :—

*July* 12/92,
6 CADOGAN PLACE, LONDON, S.W.

MY DEAR HELMHOLTZ—We are delighted that you and Mrs. von Helmholtz will come to the British Association, and Netherhall. I have written to Tait telling him that you are coming, and asking him to arrange about your being received in Edinburgh.

I am very much pleased to be one of the first four Helmholtz medallists! I have written to the "Akademie" expressing my thanks, but I feel that I must also thank you for having thought me worthy of so great and so especially interesting an honour.

I shall be glad to do what I can to assist with the other Medallists in proposing to the "Akademie" candidates for new elections.

My wife joins me in kind regards to you and Mrs. von Helmholtz, and I remain, yours always truly,

KELVIN.

*P.S.*—We leave London this week, and hope to be in Netherhall soon after.

In July came the Tercentenary Celebration of Dublin University, the brilliant fêtes of which were attended by Lord and Lady Kelvin, who were afterwards entertained at Birr Castle by the Earl of Rosse.

On the last day of July Lord Kelvin unveiled at Stirling the memorial bust of Murdoch, the inventor of gas-lighting, and gave an address on the work of that pioneer of road-engines.

August 4 brought the meeting of the British Association at Edinburgh under the presidency of Sir Archibald Geikie, whose Inaugural Address reverted to Hutton's Theory of the Age of the

Earth—a topic which had claimed Lord Kelvin's attention (p. 605) when the Association met in Edinburgh in 1871. Sir Archibald now admitted that the age-long loss of heat which demonstrably takes place from both earth and sun, makes it quite certain that the present could not have been the original condition of the system. This steady diminution of temperature was no speculation, but fact. It pointed with unmistakable directness to that beginning of things of which Hutton and his Uniformitarian followers could find no sign.

Lord Kelvin, in proposing a vote of thanks, referred to a conversation he had had with a distinguished geologist, after a former lecture of Sir Archibald's, when the geologist demanded for the age of the earth not one hundred million, or one thousand million, nor even a million million years, but would admit absolutely no limit. Then he continued :—

> I have listened with very "great interest to the conclusions that we have heard this evening, and to the developments within these last thirty years, according to which geologists find it possible to hurry up the action without abandoning any fundamental principle of the Huttonian theory, except the perpetual motion and the want of a beginning; how it is possible to understand a little more the evolution of what we see around from preceding states; and how geologists are now struggling intelligently, and not denuding anything from any part of science. I am sure we all feel that geology has made great progress, dating from the Huttonian period of a hundred years ago, and that it has not been stagnant but has had its evolution, and has made great progress within the last thirty years.

This meeting was notable for the presence of von Helmholtz and other foreign electricians who came to thrash out the disputed details about electrical standards, and who also participated in the proceedings of the Sections. Lord Kelvin himself read three papers: on the stability of periodic motions, on the graphic solution of dynamical problems, and on the reduction of degrees of freedom to the problem of drawing geodetic lines. The proposition to found a National Physical Laboratory was also discussed.

A week later Lord Kelvin wrote to Lord Rayleigh:—

NETHERHALL, LARGS,
*Aug.* 23/92.

DEAR RAYLEIGH—I have been very hard on geodetics and dynamics, and just this morning got to the end of this particular affair *most* satisfactorily.

I am getting near the end of all I can do for "instability," too. It has a bewildering variety of illustrative cases. One of the clearest of all is the consequences of Mr. Gladstone's Home Rule Bill becoming law. His majority would be immediately reduced from + 40 to − 40. The act would be repealed, and so on for ever, if "*e*" is real. If "*e*" is imaginary, one or two dissolutions would send Mr. Gladstone off to infinity.

We had a splendid time in Edinburgh. Helmholtz (including Lindeck and Kahle) was most satisfactory as to units, and he and Michelson (and Glazebrook, who in about 20 words made M's. splendid investigation clear to those who, before he (G.) spoke, had not an idea what it was) and Smithells, by the condemnation of his "oxygen theory" which he elicited from Stokes and Helmholtz and Schuster, etc., etc., made the meeting about the best I have ever known in point of instructiveness. But I must say no more of it; it would make you so unhappy not to have been in it.

Their Excellencies were very good here also. *He* has well matured, and sound views regarding Mr. Gladstone and his doings. We got a great deal out of him, all round.

I wish I had telegraphed to you urging you to come to Edinburgh. I wonder if it would have *drawn*?—Yours truly,
KELVIN.

The autumn was mostly spent at Netherhall with a constant succession of visitors to enjoy the hospitalities so lavishly offered within its walls. An intimate account of a visit by a grand-daughter of Mrs. King, Miss Margaret E. Gladstone (now Mrs. Ramsay MacDonald), affords a number of details which help to make up the picture of the life in Lord Kelvin's country home, as seen through the eyes of a naïve young girl. The extracts kindly supplied begin with a reference to the British Association meeting of August.

At Lord M'Laren's I was taken down to dinner by Copeland, the Scotch Astronomer-Royal, a pleasant old gentleman, who talked about astronomy, etc. He told me, without knowing that I was related [to Lord Kelvin], that he had been dining there a few nights before, and when the Madeira wine was served, its flavour had reminded him of a visit to Madeira just about the time of Uncle William's marriage. This made him look up, and he caught Uncle William and Aunt Fanny raising their glasses to their lips and looking at each other from opposite ends of the table. He said he thought it such a pretty incident, just like one of Hans Andersen's fairy tales.

In one of Uncle William's papers "On the Stability of Periodic Motion," it was very lovely to see him trying to balance the long pointer on the tip of his finger.

Helmholtz and Uncle William were inseparable, and both spoke a good deal in the sections, and Tait sat and smiled serenely at everything.

At Netherhall (about September–October), I had a very nice time. . . . As I was so long at Netherhall, I got to know them much better, and to be very devoted to them; at least I don't like Uncle William when he is haranguing upon politics, as if there were no sense and no goodness on the Gladstonian side, and I should dislike it still more if I were on his side myself: but I admire his earnestness, and it is really that which carries him away.

Generally the house was full of company: 24 different people stayed in it while I was there, and endless others came to dinner and tea and lunch. Aunt Fanny likes company very much; and as for Uncle William it doesn't seem to make much difference to him what happens: he works away at mathematics just the same, and in the intervals holds animated conversation with whomever is near. They were both very good to me; and the time I liked best of all was one day when there were no visitors at all, and we were quite by ourselves for about thirty hours.

Uncle William was busy with some great mathematical question; and a side-issue connected therewith was the making of Mercator charts of objects of various shapes. An anchor-ring (shape of a curtain-ring) can be made into a single oblong sheet by being cut round the circumference and also at right angles. But to mercatorize, or skin, a double anchor-ring, or treble, or multiple, was more complicated, and for several days we all were employed on that problem. It was very amusing: Uncle William was always breaking out in the middle of meals, or driving, etc., "Margaret, I've thought of a better way to skin the double anchor-ring," and then describing it. We had no double anchor-ring, so I suggested getting one made in dough; and I believe they had great fun in getting the cook to make a loaf like this 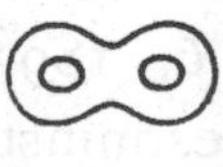 ; but I was out on a steamboat excursion at the time. When I got home I found Uncle William wanted to cover this all in one piece somehow; so I got my paint-box, and we painted the loaf Prussian blue, leaving white lines where it must be cut. That blue loaf was Uncle William's constant companion for some days; and Aunt Fanny said, a fortnight ago, that it was still knocking about the drawing-room at Netherhall. Some of my suggestions were squashed at once because they didn't fit the high mathematics of the subject; the mathematics went on vigorously in the "green book." That "green book" is a great institution. There is a series of "green books"—really notebooks made specially for Uncle William,—which he uses up at the rate of 5 or 6 a year, and which are his inseparable companions. They generally go upstairs, downstairs, out of doors, and indoors, wherever he goes; and he writes in his "green book" under any circumstances. Looking through them it is quite amusing; one entry will be in the train, another in the garden, a third in bed before he gets up; and so they go on, at all hours of the day or night. He always puts the place, and the exact minute of beginning an entry.

. . . . .

Another thing on Uncle William's mind during my visit, especially at the beginning, was his stop-cock patent. He worked at a stop-cock whenever he was within a few hundred yards of one. It was strange; for when I was staying there eight years ago, he and Uncle James were always running about the house with taps in their hands and discussing them. It gives one rather a common-sense way, or scientific way, of looking at things to stay with Uncle William: you begin to feel that everything has a reason, and that that reason may be found out; and things should not be slurred over, or left to chance, when you can direct them by taking a little trouble, and using a little thought.

In September and October Lord Kelvin had two letters in *Nature* on the problems of "Mercatorizing." He wrote to Lord Rayleigh on this, saying that both Jellett and Maxwell were "wildly adrift" as to lines of flexure. He and Lady Kelvin were to pay a visit to Inveraray, "where I expect to find the Duke full of fight, which will be very agreeable and interesting."

Tennyson died on October 6, 1892. At his funeral on October 12, in Westminster Abbey, Lord Kelvin was one of the twelve pall-bearers, along with Lord Salisbury, Lord Dufferin, the Duke of Argyll, Lord Rosebery, and the Master of Trinity.

NETHERHALL, LARGS,
AYRSHIRE, *Oct.* 16/92.

DEAR RAYLEIGH—The appointment 10.30 A.M. Oct. 27, for Board of Trade Committee on Electric

Measurement, holds. Cardew writes me that he will have a reminding circular sent.

The anti-mercatorizing (the reduction to drawing geodetic lines) of the problem of a pair of masses connected by a spring, moving in the line joining them, between two soft planes, is curious. It may possibly help towards the question of distribution of kinetic energy between that of their c. of g., and of their relative motion.

We were greatly delighted with the victory by 3 of yesterday. What do you think of Mr. Morley's beginning? The dullest of apprehension are beginning to learn, but the most shocking part of the thing is that such object lessons are required to teach them. What *does* Mr. Gladstone deserve? The 22 cows with tails cut off, the murder, and the effects of the assault on Inspector Lily are a first instalment of the grand sacrifice to his vanity, entailed by the resolution of electors to give the old man another chance.

I hope Mr. Balfour is getting a good rest, and getting it *fast*, because it seems less likely to be long than it did 6 weeks ago.—Yours, KELVIN.

November 30 brought round the annual meeting of the Royal Society. In his address the President spoke of magnetic storms, terrestrial magnetism, and the elastic yielding of the earth. A week later he read a paper on the velocity of the cathode stream as observed in highly-exhausted vacuum-tubes, and followed this by a letter to Sir William (then Mr.) Crookes:—

*Dec.* 19, 1892.

DEAR CROOKES—I am exceedingly obliged to you for the copy of your paper which you sent me. I have been studying it in the large volume of Transactions, but it is much more convenient and really more useful to have the separate copy.

Have you any measurement of the pressure on the

metal disc produced by the cathode stream? I have great difficulty in seeing how the force of the magnet can suffice to produce the curvature of the trajectories which you have found.

Have you ever found that the curvatures are less where the stream leaves the negative electrode, and greater at the greater distances? as might be the case if the stream experiences resistance such as you suggest. I suppose quite a moderate magnetic force suffices to produce a sensible deflection. My difficulty is to find that any conceivable amount of transverse force exercised by the magnet upon the observed portion of the cathode stream can be large enough to produce the observed curvature, in a stream of which the momentum suffices to produce such pressure as is produced on a vane or other fixed surface on which they impinge.—Yours very truly,

KELVIN.

Prescot, near Liverpool, had recently witnessed a revival of the watchmaking industry, by the establishment of the factory of the Lancashire Watch Company. Here, on January 12, 1893, at a dinner celebrating the extension of the works, Lord Kelvin proposed the toast of success to the undertaking.

Personally, he said, there was nothing in the whole of mechanism in which he took more interest than a watch. He was quite sure the Lancashire Watch factory was going to demonstrate that good Englishmen and good English work wanted no protective duties to keep them to the very front, and to make them successful in the way of finding customers who would pay for the work and take pleasure in its possession, not only in England, but all over the world: "Overcome your enemies with well doing."

He had been all his life engaged more or less in scientific experiments on measurement and on instruments of precision. They were thankful if in electrical instruments of precision they could attain an accuracy of $\frac{1}{10}$ or $\frac{1}{20}$ per cent; but what did watchmaking do? The commonest, cheap, good watch from the Prescot works would keep time to a minute a week, which was something like $\frac{1}{100}$ per cent of accuracy.

He need scarcely say to those present that no workmen could get good wages by saying that they "must have them," irrespective of whether they produced good work and plenty of it. Good wages for all, and the best wages for the best, was a fair and good law; and all the trades unions in the world, and all the legislation by Parliament, would never undo that law of nature.

At the end of January 1893 a slight attack of pleurisy following a chill caught at a concert, kept Lord Kelvin in bed for several days, and prevented attendance at the opening of Parliament. He occupied his enforced leisure by working at hydrodynamics, "making hay while the sun shines," as he wrote, but adding, "I hope this kind of haymaking will only last a day or two more."

In March occurred the trial of the notorious swindle of the Harness Electric Belts. Lord Kelvin, at considerable personal trouble and expense, came forward to give evidence on the worthlessness of the much-advertised article.

At the Royal Academy banquet on April 29 he responded to the toast of Science, which, he said, some people considered austere, but which was full of wonders. Science was not an *entgötterte Natur*, a God-forsaken nature, a soulless nature, with only force and light, chemicals and crystals. Science brought us to the threshold of life, and knew its own incapacity to subject life to the laws of force and electricity. A natural scene without any life in it would lack interest; life would give it interest, dignity, and beauty.

He was busy all that spring with molecular physics and problems involving the principle of least action. His Royal Institution discourse on "Isoperimetrical Problems," and the Robert Boyle Lecture in May at Oxford on the "Molecular Tactics of a Crystal," are reported elsewhere (see pp. 1053, 1054). His crystal models were shown at the soirées of the Royal Society.

In June he took part in the Jubilee dinner of the Cambridge University Musical Society, of which (p. 69) he had been one of the founders; and was present at the Jubilee Concert where Max Bruch, Tchaïkowsky, Saint-Saëns, Boïto, and Professor (now Sir Charles) Villiers Stanford took part in rendering or conducting their own works.

Many banquets and reunions were attended this season, including a Ministerial banquet at Downing Street. July over, they went down to Netherhall to rest till the Home Rule Bill should come up to the House of Lords. A trip to Aix-les-Bains for Lady Kelvin's health was arranged for the autumn —"though Lord Kelvin hates it!" as she wrote to Darwin; adding, "Lord Kelvin is busy with the molecular tactics of cleavage planes and faces of a crystal, and he wishes you were here to help." But the August holiday was saddened by the sudden death of Lord Kelvin's brother-in-law, Alexander Crum.

With September came the debates in the House of Lords on the Home Rule Bill. While Viscount Castletown, a Liberal Unionist, was speaking against the Bill, the House was startled by a most unusual sound. Lord Kelvin, sitting immediately below Lord Castletown, was so carried away by his approval of the speech that, forgetting that the orthodox method of approval consists in shouting "Hear, hear," he vigorously clapped his hands, and almost petrified other peers who sat near him by the unaccustomed demonstration. The Bill was defeated on the second reading, but Lord Kelvin had left for Aix-les-Bains.

In October Lord and Lady Kelvin were in Paris, where he read to the Academy of Science a paper on piezo-electricity, returning early that he might open the Blackpool electric lighting station on October 14. He was going to Cambridge for the statutory Fellows' meeting at Peterhouse, and wrote beforehand to Professor Ewing:—

LONDON, *Oct.* 27/93.

Do you still have magnetic molecules in large numbers, ready placed on points and stands, or easily so placed, just to show to any one looking direct without lantern what they do? I want to help myself about crystals by looking at them. If without too much trouble you could let me see them some time between 9.30 to-night and 12 on Monday, when I have my College meeting, I shall be much obliged.—Yours, KELVIN.

On October 28 he visited the Leys School, for the opening of the new science laboratory, when he delighted the boys by advocating the free use of translations in learning classics, so as to get more time for science. There were more political meetings and speeches in November, and then came the Royal Society anniversary. The Presidential Address on the researches of Hertz and of Crookes is considered elsewhere (p. 1058 *infra*). It fell to the President also to present the Copley medal to Sir George Stokes. He accompanied the presentation by a brief but masterly summary of Stokes's achievements in physics, dwelling particularly on his accurate measuring work, his codification of the optical laws divined by Huygens and Fresnel, and his clue to the dynamics of the undulatory theory of light.

On December 8 Lord Kelvin unveiled the fine statue of Joule in the Town Hall of Manchester, and delivered the striking address on Joule's work which is reprinted in *Popular Lectures*, vol. ii. p. 558. (See also *Nature*, vol. xlix. p. 164.)

Writing a few days later to Lord Rayleigh, to arrange for the reading in January of a paper on the Homogeneous Division of Space, Lord Kelvin briefly narrated how, since the defeat of the Home Rule Bill, he had been to Aix, was busy with three patents and with crystal models; and he concluded with Christmas greetings. Lord Derby, with whom so many Christmas holidays had been spent, had died in April. Henceforth Christmas days were passed at Netherhall.

Many were the speeches made by Lord Kelvin in the year 1894. In January he presented the prizes at the Blackburn Technical School, and spoke on the aims of education and the right attitude to take toward foreign competition in skilled industries: if foreigners do well, let Englishmen do better; but to prevent the people of this country from exercising their good judgment in buying in the best and cheapest market would be grossly unfair. In February he made two speeches to Unionist associations, and on the 23rd of that month presided at a dinner at Birmingham to celebrate Founder's Day of the Mason College. He urged strongly that a case had been made out for a Midland University or a University of Birmingham. "Birmingham should take the lesson of James Watt[1] to heart, and remember that a University would not only be a crown of glory to the city, but would be of continual practical benefit to the workshops of the mechanics." In March he spoke at the Civil Engineers. In April he opened Coatbridge electric light works; spoke at two Unionist meetings, at the Mechanical Engineers, and at the Institute of Journalists. In May he opened the new Engineering Laboratory at Cambridge organized by Professor Ewing; spoke at the Institute of Chemistry on scientific cataloguing and indexing; and at the Royal Society on the electrification of air. He had been busy, too, on the theories of the shrinkage of the globe as it cools, and wrote several letters to George Darwin on the subject. One of these, on March 29, ends thus:—

I have had a very broken time with various engagements, and finishing up our Glasgow Session, and an Oxford lecture (of last May) on Tactics of Crystals, for

[1] James Watt, in 1756, was not permitted by the trade combinations of Glasgow to set up a workshop in that city, but found a place in Glasgow College, in the post of mechanician to the University, which still possesses some of his models. Later he founded in Birmingham the firm of Boulton and Watt.

which I am now in the painful position of meeting a bill overdue, under threats from successive secretaries of the Oxford Junior Scientific Club.

He was still working at crystals, and at the Royal Society's soirée in June had a fresh exhibit of models to illustrate the properties of quartz.

He took little part that session in the work of the House of Lords, but paired in favour of the Bill for Legalizing Marriage with a Deceased Wife's Sister, and against the Tenants' Arbitration Bill.

In June the Ampère gold medal was awarded[1] to Lord Kelvin by the Société d'Encouragement. This medal is awarded to the Frenchman or foreigner whose labours have had the greatest influence, during the six preceding years, on the progress of French industry.

Von Helmholtz, who, in the previous autumn, had had a severe fall, became seriously ill in the summer of 1894, and Lord Kelvin wrote:—

34 HANS PLACE, LONDON, S.W.,
*July* 6/94.

MY DEAR MRS. HELMHOLTZ—We are much distressed to see in to-day's papers that your husband, my old friend, has been taken ill. I hope very much that the illness is not so severe as may have been apprehended. It will be very kind if *you* will write and tell me of him.

You must give him my most affectionate messages and warm sympathy. We shall be very anxious to hear what you can tell us.

[1] The terms of the award refer to the Atlantic cable work and to his union of practical engineering with mathematical power. "He has been the educator of our time. He has renewed theories, transformed scientific methods, opened out unforeseen paths, provoked the adoption of a logical system of units and measures, and contributed in an incalculable sense to the progress of industrial electricity."

Lady Kelvin sends her love, and I remain, dear Mrs. Helmholtz, yours most truly, KELVIN.

*P.S.*—It happens that at this moment I am engaged with a problem that he started 47 years ago, the oscillatory discharge of a Leyden jar.

The British Association meeting of August 1894, at Oxford, was made notable by two events: the announcement of the discovery of argon by Lord Rayleigh and his coadjutor Professor (now Sir William) Ramsay, and the sending of wireless signals between the Clarendon laboratory and the Museum by Professor (now Sir Oliver) Lodge by means of Hertzian waves received by coherer and tapper. Lord Kelvin read a paper on hydrodynamics, and two others on electrical experiments made in his laboratory. There was a discussion also on Maxim's flying machine, which Lord Kelvin (who had himself in July had the opportunity of taking a "flight" in it) described as a "kind of child's perambulator with a sunshade magnified eight times." He did not believe in the aeroplane, and thought the problem of flight might be better solved with a platform having a vertically working propeller at each corner: an aeroplane was like an eagle which had to take a run before it could rise.

After leaving Oxford, Lord and Lady Kelvin went abroad. Lady Kelvin stayed at Aix, while Lord Kelvin took a cruise on the s.s. *Electra*, in the Mediterranean, visiting Syra, Constantinople, Sevastopol, Odessa, Athens, Corinth, and Genoa, back to Aix—returning to Netherhall in the middle of September to spend the autumn in quiet enjoyment of his home.

With the death of von Helmholtz, on September 8th, 1894, Lord Kelvin lost one of his oldest and most-cherished friends.

PARIS, *Oct.* 11, 1894.

MY DEAR MRS. VON HELMHOLTZ—The sad news came to us here yesterday evening, and we sent a telegram to tell you of our sympathy with you in this great grief. Your letter to Lady Kelvin, which she forwarded to me while I was still away on my cruise, made me very anxious, but I still hoped to hear of recovery and to be able to look forward to other happy meetings with my friend. But it was not to be.

I am full of recollections of happy meetings in the past, of which the first was when he came to see me at Creuznach in 1856, and all of which were to me unalloyed pleasure. The loss to myself is so severe that I cannot speak of it, and I feel that I must not intrude on your own most sacred sorrow. But I feel that I must write this now on our way home to Largs, to tell you of my heartfelt grief, and to express for my wife her deepest and warmest sympathy with you.—Believe me, dear Mrs. von Helmholtz, ever your affectionate friend,

KELVIN.

NETHERHALL, LARGS, AYRSHIRE,
*Oct.* 12/94.

DEAR MRS. HELMHOLTZ—I have been most deeply touched by your kind letter which I received a fortnight ago, with all that it tells me of my friend who is gone from us, and of his feeling for me. To know from you that the sympathy and loving friendship which grew in me from my admiration of his work before I had seen him, in those early times in Creuznach, and Bonn, and Arran, when we were together, has been so perfectly reciprocated by him, is to me not only a pride and honour, but a happy possession.

When the grief is so fresh I feel that it is almost cruel to speak of happiness. But even the grief, never to be lost, is brightened, not diminished, by the recollections of his life, and the knowledge of how beautifully and perfectly it worked for good and happiness. And believe me, it will be for all your life a happiness to know how his time in this world has been spent, and how much his memory is valued in his own country, and by all the scientific people of the whole world.

We shall be much interested to hear what you resolve about a house. If one can be found near Ellen I am sure you will like to be near her and your grandchildren. We shall be much pleased to have one of the reproductions from Lembach's portrait if you find that you can have it done satisfactorily.

Lady Kelvin sends her kindest love, and I remain, yours always affectionately, KELVIN.

To Lord Kelvin, as President of the Royal Society, fell the duty, at the annual meeting on November 30, 1894, of recording the deaths of several celebrated Fellows, including Tyndall and Helmholtz. Of both he spoke with sympathy none the less sincere because brief. Then he turned to enumerate the scientific triumphs of the year, chiefest among them the discovery of argon.

. . . . . . .

Rayleigh, persevering in the main object which he had promised in 1882, "a redetermination of the densities of the principal gases," attacked nitrogen resolutely and, stimulated by most disturbing and unexpected difficulties in the way of obtaining concordant results for the density of this gas as obtained from different sources, discovered that the gas left by taking vapour of water, carbonic acid, and oxygen from common air was denser[1] by $\frac{1}{230}$ than nitrogen obtained from chemical processes from nitric oxide, or from nitrous oxide, or from ammonium nitrite,

[1] "On an Anomaly encountered in Determinations of the Density of Nitrogen Gas," *Roy. Soc. Proc.*, April 1894.

thereby rendering it probable that atmospheric air is a mixture of nitrogen and a small proportion of some unknown and heavier gas. Rayleigh, and Ramsay, who happily joined in the work at this stage, have since succeeded in isolating the new gas, both by removing nitrogen from common air by Cavendish's old process of passing electric sparks through it, and taking away the nitrous compounds thus produced by alkaline liquor, and by absorption by metallic magnesium. Thus we have a fresh and most interesting verification of a statement which I took occasion to make in my Presidential Address to the British Association in 1871:[1] "Accurate and minute measurement seems to the non-scientific imagination a less lofty and dignified work than looking for something new. But nearly all the grandest discoveries of science have been but the rewards of accurate measurement and patient, long-continued labour in the minute sifting of numerical results." . . .

After the address came the presentation of medals; and this year the Darwin medal was awarded to Huxley, the youngest of the three who had "kept the bridge" in defence of Darwin's *Origin of Species.* Lord Kelvin, in handing the medal, made the following pronouncement:—

To the world at large, perhaps, Mr. Huxley's share in moulding the thesis of "Natural Selection" was less well known than was his bold, unwearied exposition and defence of it after it had been made public; and, indeed, a speculative trifler, revelling in problems of the "might have been," would find a congenial theme in the inquiry how soon what was now called "Darwinism" would have met with the acceptance with which it had met, and gained the power which it had gained, had it not been for the brilliant advocacy with which in its early days it was expounded to all classes of men. That advocacy had one striking mark: while it made, or strove to make, clear how deep the view went down and how far it reached, it never shrank from striving to make equally clear the limits beyond which it could not go. In these latter days there was fear lest the view, once new, but now familiar, might, through being stretched farther than it would bear, seem to lose some of its real worth. They might well be glad that the advocate of the "Origin of Species by Natural

[1] Republished in Vol. II. of *Popular Lectures and Addresses*, p. 156.

Selection," who once bore down its foes, was still among them, ready, if need were, to "save it from its friends."

The geological controversy (see p. 535) was reopened in January 1895, in *Nature*, by Professor John Perry, who pointed out that Lord Kelvin's estimate of 100 million years for the age of the earth, as deduced from thermal conductivity, must be increased if the rocks at higher temperature in the interior of the globe conduct better than those in the crust. If the conductivity is increased so that the temperature-gradient is 1° centigrade in 45 feet, instead of in 90 feet, the age is lengthened two hundred and ninety times. Lord Kelvin, to whom the matter had been privately submitted, wrote to Perry on December 13, 1894, asking further data about conductivity, and adding :—

The subject is intensely interesting; in fact, I would rather know the date of the *Consistentior Status* than of the Norman Conquest; but it can bring no comfort in respect to demand for time in Palæontological Geology. Helmholtz, Newcomb, and another are inexorable in refusing sunlight for more than a score or a very few scores of million years of past time (see *Popular Lectures and Addresses*, vol. i. p. 397).

So far as underground heat alone is concerned you are quite right that my estimate was 100 millions, and please remark (*P. L. and A.* vol. ii. p. 87) that that is all Geikie wants; but I should be exceedingly frightened to meet him now with only 20 million in my mouth.

And, lastly, don't despise secular diminution of the earth's moment of momentum. The thing is too obvious to every one who understands dynamics.

He wrote to M. Mascart :—

IN TRAIN, LONDON TO GLASGOW,
*Feb.* 2, 1895.

DEAR MASCART—I have been looking back to my old estimate of the time which must have passed since the consolidation of the earth supposed to have been previously melted rock, and trying to find closer limits founded on definite information, if it is to be got, as to the following qualities :—

1. The melting temperature of rock of any time or lava under ordinary atmospheric pressure ;

2. The thermal capacity of rock at different temperatures up to melting ;

3. The thermal conductivity, do., do., do., do.;

4. The change of volume of rock on melting, do.

I have been looking in all the text-books and books of tables in vain for such information.

Has not Deville given really trustworthy measurements of high temperatures, including melting points of glasses and stones ; and as to (4) also ? Moissan too has been working at high temperatures, and probably has results as to several of those four questions. You are sure to know, if any one knows, where information can be had from any part of the world (Italian observers of lava on Vesuvius ? American or others on the lava lake of Hawaii ?), and I should be greatly obliged by a few lines from you to tell me if you have of your own knowledge some of the information I want, or where I might be able to find it.

The mathematical problem of finding the proper *case* of solution of the equation

$$\frac{d}{dx}\left(\frac{\mathrm{K}dv}{dx}\right)=\mathrm{C}\frac{dv}{dt}$$

with K and C given functions of $v$, is, I find, quite *doable* and very interesting.—Yours always truly,

KELVIN.

THE UNIVERSITY, GLASGOW,
*February* 4, 1895.

DEAR MASCART — Since my arrival home I have found, in the following papers, *Philosophical Magazine*, Oct. 1891, p. 353, Messrs W. C. Roberts-Austen and A. W. Rücker, "On the Specific Heat of Basalt"; (2) *Phil. Mag.*, July 1892, March 1893 (p. 1 and p. 296), Carl Barus, "On the Fusion Constants of Igneous Rocks," a good deal of valuable information on the subject on which I wrote to you from the train on Saturday. I write just to save you the trouble of referring me to any of these papers, but I shall be very glad if you can give me other references or information.—Yours very truly, KELVIN.

Two months later Lord Kelvin wrote to *Nature* (vol. li. p. 438), that R. Weber having made measurements on sandstone and slate, saw no reason to think that thermal conductivity increases with temperature; and that his own experiments showed the conductivity to be less, not greater, at high temperature. He also referred to a paper of January 1893, in the *American Journal of Science*, by Clarence King, who, reasoning from thermal data supplied by Carl Barus, concludes: "We have no warrant for extending the earth's age beyond 24 millions of years." On which Lord Kelvin remarked: "I am not led to differ much from his estimate of 24 million years."

The Duke of Argyll came to Glasgow on January 15, 1895, to address the Liberal Unionist Association at the City Hall, and stayed with Lord Kelvin in his University house. Lord Kelvin presided at the meeting, which was densely packed. The Duke, who was far from well, had only spoken for about fifteen minutes when he fainted, and the meeting was dismissed. He was removed to Lord Kelvin's house, and recovered with a few days' rest. Lord Kelvin had to go to Crewe on the 22nd to present the prizes at the Mechanics' Institute in that town. In replying to a vote of thanks he remarked on the new electrical development which was taking place in the railway works at Crewe, where cranes and drilling machines were now being worked by power supplied electrically through wires. He went on to consider the possibility of an electrical train being turned out at Crewe, remarking that although such a thing might not be probable, still neither Mr. Webb nor the North-Western Railway would be behind any one else in the event of the steam locomotive being superseded by electricity. Mr. Webb, who presided, thereupon rejoined that should his company order one of their trains to be run by electricity, he would undertake to complete it in two months.

On the 30th of January Lord Kelvin was at Wolverhampton to open the electric lighting station, where he complimented Mr. Parker on the machine that had been constructed at his works for transforming continuous currents. In the evening he was entertained at a banquet by the Wolverhampton Chamber of Commerce, and spoke on the progress of engineering from the time of the Pyramids to the modern steam-engine of to-day.

There was one science, he said, which was most fascinating, which until recent times was purely scientific, a subject for scientific investigation—the science of electricity. But its application to practical uses had grown up wholly within the present century. A little before the middle of this century Faraday brought out his great discovery, and in his own modest way anticipated some of the great success which followed. He virtually gave us the dynamo. The first practical application of electricity was to electrometallurgy. It was with reference to electricity for electroplating as well as for lighthouses that Faraday had said: "I gave you this machine as an infant; you bring it back as a giant." They had seen it realized in Wolverhampton that day.

The next day Lord Kelvin presided at the special meeting of the Royal Society called to hear a discussion of the discovery of argon by Rayleigh and Ramsay. The day after, he was in Cambridge attending the funeral of his old friend Professor Cayley.

Lord Kelvin had now entered on the fifth year of his Presidency of the Royal Society. At the end of February 1895 he wrote to Professor (now Sir Arthur) Rücker, a Member of the Council:—

Another thing I wanted to speak to you about, if I had had the opportunity at our last Council meeting, is the next election to the Presidency. I shall have had my five years, and it will then be time to choose my successor. It has been a great pleasure to me to be President, but for many reasons I feel that however kindly the Council might feel in respect to re-electing me a sixth time (as indeed has been suggested to me as an idea that might possibly be entertained), it would be in all respects better that another should be elected. I should be glad, therefore, if members of Council would be considering in good time the question of who is to be elected.

Then there were more Unionist meetings to

attend, and papers to be read in Glasgow and London on the Dis-electrification of Air and on the Thermal Conductivity of Rocks.

June and July brought the round of society dinners, and then he fled to Aix-les-Bains for rest and change.

Lord Kelvin's *Popular Lectures* had been completed in 1894 by the belated publication of volume ii. on "Geology and General Physics," and a second edition of volume i. was called for. He sent a copy to Professor Tait, asking him for criticisms, and received a sheet bubbling over with incoherent fun couched in the old fantastic phrases, and furious comments like those so freely bandied between the two friends in their earlier intimacy of the 'sixties, when the great *Treatise* was in the making. Lord Kelvin returned Tait's sheet scored over with lively rejoinders in blue pencil, and Tait posted it after him to Aix with repartees in red. The following letter is couched in less exuberant terms :—

*July* 2, 1895.

DEAR TAIT—I thank you heartily and without reserve of any percentage for all the corrections and ameliorations which you have given me for volume i., my chief motive for doing so being that I want you now, without any delay, to do likewise for volumes ii. and iii. Don't waste your time in trying to mitigate in any way your expression of disapprobation, however extreme. However keenly or even wildly you may express yourself I shall not be angry, not even sorry, except when I feel that the condemnation is merited.

What you say as to *Heat, a Mode of Motion, Elasticity, a Mode of Motion*, the title of Tyndall's beautiful book, and the nature of the Royal Institution Friday evening audience (which always includes such people as Stokes, Rayleigh, Schuster, Dewar, Lord Rosse, and all the best and strongest scientists who chance to be in London at the time with no more amusing evening engagement to keep them away) does not pain me at all. "My withers are unwrung." So they are in respect to "centrifugal force."

I accept "Scottish-built," but it must be kept for 3rd edition. As to page 338, l. 3, you will see that I have yielded. But in doing so I have spoiled my sentence, which was quite correct as it stood. Putting them together, with one part turned through zero angle relatively to the other, is much more ingenious than the choice of any other angle would have been.

We (Gray ii. and I) have been making history here (? more important than that made by Mr. Brodrick and Campbell-Bannerman in another place) during the last 10 days. We have done graphic histories of 22 encounters between a free particle (Boscovich) and a vibrator with from one to four impacts in each encounter. We are nearing to-day a test of the Maxwell-Boltzmann law of distribution of energy.—Yours, K.

From Aix-les-Bains Lord and Lady Kelvin took a short trip to Switzerland, returning in time for the British Association meeting at Ipswich on September 11. Then a full month was spent at Netherhall. He was much occupied with calculations about the new Pacific Cable.

In October 1895 the *Institut* of France celebrated the centenary of its creation. Lord Kelvin, as one of the foreign associates, was invited to take part in the four days' celebration. It began on October 23, with a religious service in the church of Saint Germain des Prés, followed by a greeting of the members to the foreign associates and correspondents, and a formal reception by the Minister of Public Instruction, M. Poincaré. The great day was the 24th, beginning with a meeting in the great hemicycle of the Sorbonne, at which the President of the Republic, M. Felix Faure, was present, and an oration was made by M. Jules Simon, and speeches by M. Ambroise Thomas and M. Poincaré. In the afternoon there was a reception at the Élysée; in the evening a grand banquet at the Hôtel Continental, at which Max-Müller was the spokesman of the foreign associates and correspondents. Later, Lord Kelvin presented an address from the Royal Society, and in reply to an allocution of the President of the Institut, pronounced the following speech:—

Personnellement les mots me font défaut pour dire combien j'apprécie le grand honneur que vous m'avez conféré, d'être Associé de l'Institut de France. Mais je dois à la France une dette encore plus grande. Elle est vraiment l'*Alma mater* de ma jeunesse scientifique, et l'inspiratrice de l'admiration pour la beauté de la science qui m'a enchaîné et guidé pendant toute ma carrière.

Dans la bibliothèque du roi, pendant l'été de 1839, j'ai fait la connaissance d'une petite partie de la *Mécanique céleste* de Laplace, pour un *Essai sur la figure de la Terre*, qu'il m'a fallu écrire pour l'Université de Glascou, comme exercice d'étudiant. Avant que je quittasse l'Université de Glascou, mes professeurs m'y avaient montré la splendeur de Fourier.

Six ans plus tard, le vénérable Biot m'a pris par la main et m'a placé dans la laboratoire du Collège de France sous la direction de Regnault; ainsi j'ai vu ce grand physicien, de jour en jour, travaillant sur les propriétés physiques des gaz. A Regnault et à Liouville je serai éternellement reconnaissant pour la bonté qu'ils m'ont témoignée, et pour les méthodes qu'ils m'ont enseignées sur la physique expérimentale et mathématique dans l'an 1845.

Un an plus tard encore, la *Puissance motrice du feu*, le travail de l'immortel Sadi Carnot, m'a révélé les résultats si pratiques et si profondément ultra-théoriques de son génie pénétrant.

Ainsi j'ai été nourri de la science la plus solide et vous comprendrez, mes chers Confrères, pourquoi je regarde, avec une reconnaissance profonde, la France comme mon *Alma mater* de science.

Monsieur le Président de l'Institut et chers Confrères, je vous remercie de tout mon cœur de votre bonté pour moi, et pour l'honneur que vous m'avez fait en m'appelant à répondre au nom de la Société Royale.

The fêtes concluded with a performance of *Le Cid*, at the Théâtre Français, and a visit to Chantilly, where the aged Duc d'Aumale still resided.

Lord Kelvin hurried back to Glasgow for the duties of his Chair, but was interrupted by professional work. A letter to Mrs. King on November 13, from the Athenæum Club, tells her that he will come to see her about 7 o'clock that evening.

I should have to leave at 9, as I am very busy with a law case [1] (bicycle tyres), for which I shall be all day in

[1] The case, relating to the "Clincher" tyre, lasted several days, and was heard by a learned Judge who was as guiltless as Lord Kelvin of having ever learned to ride a bicycle. After hearing his evidence the Judge turned to him and said :—"Lord Kelvin, I have heard a great deal in this case about the tendency of the tyre to come out of the rim ; now, in this patent the inner tube is canvas-covered and inexpansible, and the pressure on the road tends to press the tyre into the rim." The answer came solemnly and with great deliberation :—"Suppose, my Lord, that I were riding a bicycle" (with the chief accent on the second syllable) "at a high velocity, and suppose that in order

Court, and I shall have a good deal to write and work out for it here in the evening. It is a very interesting case and I shall tell you all about it.

On November 27 Lord Kelvin presided at the Huxley Memorial Committee. On the 28th he was spokesman at a deputation to the Duke of Devonshire on behalf of the movement for a Teaching University for London.

On the 30th was the anniversary meeting of the Royal Society. Lord Kelvin's Address opened with an enumeration of the names of those Fellows who had died within the year—including Cayley,[1] Neumann, Huxley,[2] and Pasteur. He then spoke of the great catalogue of scientific papers undertaken by the Royal Society; of the centenary of the Institut; of the *Challenger* Expedition publications; and of the discovery of argon and helium. Amongst the medallists of the year to whom he handed the medals were his former assistant Professor Ewing, and Professor (now Sir William) Ramsay, at one time a student in his laboratory. The following

to decrease the velocity very quickly, I were suddenly to turn the front wheel so that its plane was substantially at right angles to the direction of motion, I think that—there would be—a—tendency—for the tyre to come out of the rim!"

[1] In his reference to Cayley there is a personal note that cannot be omitted here:—

"In Cayley we have lost one of the makers of mathematics, a poet in the true sense of the word, who made real for the world the ideas which his ever-fertile imagination created for himself. He was the Senior Wrangler of my freshman's year at Cambridge, and I well remember to this day the admiration and awe with which, before the end of my first term, just fifty-four years ago, I had learned to regard his mathematical powers. When a little later I attained to the honour of knowing him personally, the awe was evaporated by the sunshine of his genial kindness; the admiration has remained unabated to this day, and his friendship has been one of the valued possessions of my life."

[2] See p. 1088 *infra* for a quotation from the appreciation of Huxley.

are his closing words on vacating the presidential Chair :—

Five years have now passed since you elected me to be your President. Living at a distance of 400 miles from London, I felt that it could not be possible for me to accept the honour when the possibility of its being offered to me was first suggested. I accepted, with much misgiving as to my ability to perform the duty which would fall upon me; and now, after having been re-elected four times, I feel that if the interests of the Society have not suffered under my presidency, it is chiefly because they have been so faithfully and unintermittently cared for and worked for by the other officers, the Treasurer and the Secretaries, who have left nothing undone that could be done to promote the welfare of the Royal Society. For their unfailing kindness to myself I can only offer my heartfelt thanks. I soon found that what I looked forward to with apprehension—the Council meetings, and as many of the ordinary meetings as I could attend, during my University session in Glasgow—were the reverse of fatiguing; and I am only sorry that I have been so many times obliged to forgo the pleasure of performing that part of my presidential duty. I look back otherwise with unmixed pleasure to all the meetings at which I have presided, and my sole regret now is—I cannot disguise it, and it is a very keen regret—that these five years are passed, and that to-day I cease to be your President. I thank you all, my colleagues of the Royal Society, for electing me five times to be your President, for forgiving me all my shortcomings, and for the inestimable benefit which you have conferred on me by giving me your friendship.

The following letter to Lord Rayleigh brings us back to mundane affairs :—

*Dec.* 14/95,
THE UNIVERSITY, GLASGOW.

DEAR RAYLEIGH—My secretary is busy drawing a new (old) form of water-tap, for which I am immediately going to take out a patent (to get over the possibility of failure in hard water such as we have in the Athenæum, and in some other places in the south of England, in the existing pattern which has been bringing discredit on it, and which, since I ceased to be P.R.S., I saw to be due to geometrical and mechanical ill-conditionedness, producing no bad result with Loch Katrine water), and this is my apology for less clear writing than you should otherwise have had respecting encounters.

. . . . . .

I think the post of scientific adviser to the Trinity House would be particularly interesting to you, and if you think the conditions as to duty expected and remuneration are satisfactory, you would, I think, be quite right to accept it.—Yours, KELVIN.

He wrote on December 24 to Professor Hugh Blackburn :—

You were no doubt amused by letters of arm-chair enthusiasts for the beauties of Nature, about the Falls of Foyers, and (I hope) pleased by the final quietus to the fever given by the County Council. The turbines, I hope, will be going, and the aluminium coming, by February or March.

Did you receive a *Comptes Rendus* containing my "allocution" at the centenary banquet in Paris? If not, it will be sent. You will see that I did not forget Glasgow. I am not sure if 31 Rue Monsieur le Prince has not been taken down, to give space for new (Ecole de Médecine) buildings.

The year closes with a letter to his niece, Miss May Crum :—

THE UNIVERSITY, GLASGOW,
*Christmas*, 1895.

MY DEAR MAY—Your father brought me your letter, and we have been greatly enjoying his visit. It is very kind of you to console me on my sad imprisonment. Really I feel perfectly well, and strongly disposed to disagree with Dr. Tennent, but your Aunt Fanny won't allow me to disobey. It will be *too* sad if we lose the delightful time we had been looking forward to at Thornliebank, but I hope when the 3rd comes, or before it, we may find Dr. Tennent more reasonable. . . . I have been reading Lucretius, much helped by Munro's

(*not* the author of THE doctrine) translation, and trying hard on my own account to make something of the clash of atoms, but with little success.—Your affectionate unk(! !)le,
KELVIN.

The year 1896, which was to witness the jubilee of Lord Kelvin's tenure of the Chair of Natural Philosophy, opened inauspiciously enough, for he was far from well that winter. All through December 1895 he complained at intervals of a tired leg, sore and painful at times, but just before Christmas, when he was purposing to leave Glasgow for Netherhall, an œdematous swelling, such as had troubled him much after the accident of 1861, caused his medical adviser to order him to keep his bed for several weeks. At the end of January there supervened a sharp attack of pleurisy. Though suffering much he bore his sufferings with great patience, and was greatly helped by his ability to sleep soundly. Except when the pleurisy was acute he worked at his notebook, dictated letters to his secretary, and took the keenest interest in the events of the hour, both scientific and political, as the following letters show. It had been settled that he was to give another Friday evening discourse at the Royal Institution in April, but he now wrote to the Honorary Secretary, on the 10th of January, that he must abandon the project:—

I have been almost wholly confined to bed since the 16th of December on account of a swelling in my bad leg, which was broken Christmas eve 35 years ago. I am now much better, but the doctor's orders are most peremptory that I am still to keep, very much at least, to the sofa, and as much as possible avoid standing. I need not tell you that I am exceedingly sorry to fail you for the Friday evening lecture to which I had been looking forward with much pleasure.

Eventually M. Lippmann took Lord Kelvin's place, and lectured on Colour Photography.

The papers had been full of the wonders of Röntgen's rays, about which Lord Kelvin was intensely sceptical until Röntgen himself sent him a copy of his Memoir,[1] whereupon he wrote :—

*January* 17, 1896.

DEAR PROF. RÖNTGEN—When I wrote to you thanking you for your kindness in sending me your paper and the photographs which accompanied it, I had only seen the photographs and had not had time to read the paper. I need not tell you that when I read the paper I was very much astonished and delighted. I can say no more now than to congratulate you warmly on the great discovery you have made, and to renew my thanks to you for your kindness in so early sending me your paper and the photographs.—Believe me, yours very truly,

KELVIN.

He wrote to Sir Joseph (now Lord) Lister on January 27:—

Ask Rayleigh if he is not in a state of great excitement about Röntgen's X-rays.

To his sister, Mrs. King, now in her seventy-seventh year, he wrote :—

---

[1] W. C. Röntgen, Ueber eine neue Art von Strahlen, *Sitzungsber. der Würzburger Physik-medic. Gesellschaft*, December 1895.

THE UNIVERSITY, GLASGOW,
*Sunday, Jan.* 26/96.

MY DEAR ELIZABETH—I am very sorry to have so long been prevented from writing to thank you for the lovely new year's gift (I can't look at it and see it as only a new year's card) which Fanny brought to me from you when we were in London. Since our return I have had my secretary at work with me from instantly after breakfast (sometimes even before it was finished) till 7.45 or 8 in the evening, every day except yesterday, when I managed to get him away by 2.30 for a Saturday afternoon walk which he told me he was to have with a friend (this sort of thing will become illegal, I am afraid, when the 8 hours bill becomes law). So you will understand why it is that I have so long been kept from writing to you.

Fanny and I both think the picture quite lovely and very interesting, with the old tower overgrown with hanging creepers, and the ruined village, and the dome of St. Peter's in the distance. We are going to have it framed and hung up where we shall see it.

I am still ordered to keep the bad leg up and give it all the rest possible, and told emphatically that the more thoroughly I obey this order the sooner I shall be quite well and again ready for everything. I believe I am (I mean the leg is) really somewhat better, but the progress is disappointingly slow. I am feeling perfectly well, and am getting on with my work (with secretary's aid), as well as staying in bed till lunch-time, and as nearly as possible in the same position till night downstairs allows, but it is tiresome and disappointing to be so long kept from walking about, and going to the laboratory for my work there which is always going on, etc. etc.

I am afraid this is rather queer writing, as it is done on a table in bed! Newspaper reading has been very exacting of late. What inconceivable *folly* (not to use a harsher word) Dr. Jamieson's bringing those 700 or 800 young men to what they came to seems to have been. It is inconceivable also how officers of the British Army joined in the mad affair. What a *trial* and investigation there must be! One thing must follow, that our Government must never again shirk their duty of governing and hand it over to irresponsible, unqualified people or companies.—Your aff. brother, KELVIN.

*P.S.*—I hope you and Elizabeth, Junr., are both much better than when Fanny saw you, and that Agnes is keeping well.

He sent Stokes on February 1 a long mathematical letter, ending thus:—

In respect of the Röntgen X-rays, are you a longitudinalist, or an ultra-violetist, or a tertium-quidist? . . . I had a slight attack of pleurisy, which came on suddenly on Monday. I am feeling much better now, and the doctor is well satisfied with progress, but he won't let me out of his hands on any account yet, and I don't know what he will say when he hears that I have been dictating this, also a short article for *Nature* on "Maxwell's Infinite Velocity for the Propagation of Electrostatic Force," for which look out next week.

A letter on Röntgen's rays to Oliver Lodge, of February 4, is printed in another connexion on p. 1062.

On February 10 Lady Kelvin sent word to Mrs. King that Lord Kelvin was going on as well as possible, but that she feared the leg was going to tie him down for some time:—

The doctors all insist on the necessity of absolute rest. Even Sir Joseph Lister, who wrote to Dr. Hector Cameron to ask him to see Dr. Tennent, writes begging him, however well he feels, to keep quiet and rest. The remains of pleurisy is keeping him still in bed at present, and the leg is still swelled a good deal. Dr. Tennent always ends up by saying he will be quite well for his jubilee. It is fortunate that James [Bottomley] is so well, for he is obliged to take all William's lectures. William frets very much at not being able to take his lectures. He has had the higher mathematicians twice in the dining-room. . . . He is allowed to do some work, fortunately.

Again he wrote to Stokes, and to Admiral Wharton, and to Professor (now Sir) Joseph J. Thomson :—

*February* 12, 1896.

DEAR STOKES—I am very much obliged to you for your letter of yesterday. Rayleigh gives me a reference to § 301 of his book on *Sound*, volume ii., where there is, particularly in equation (12), something for condensational-rarefactional waves, analogous to what I wrote to you regarding transverse vibrations in an elastic solid in my letter of Feb. 1.

As to the true Kathodenstrom, it is wonderful how the Germans have called it Kathodenstrahl, having allowed themselves to be misled by Hittorf and Goldstein. They, every one of them that I know of, *except Helmholtz*, have insisted with something like strong partisan spirit upon a ray of undulatory light from the cathode, and have blindly refused to accept Varley's conclusion of a torrent of molecules, corroborated as it is by Crookes. I enclose a marked copy of my address of 1893, which may show you, or remind you of some dates and references on the subject.

As to the Röntgen rays, you will see in *Nature* for Jan. 23, page 275, second column, that Röntgen himself made the experiment which you tell me J. J. Thomson thinks of making, and that he found, as nearly as he could test it, an intensity varying inversely as the square of the distance. This, of course, if confirmed, is decisive against the hypothesis of "push" in an incompressible medium.

I feel strongly disposed to Röntgen's own supposition

of condensational - rarefactional waves, but still I see tremendous difficulties.

Thanks for your inquiries. I am now practically well except that the doctor still rigorously orders the leg to be kept horizontal, and so for the present I am kept in bed. The pleurisy is nearly quite gone away.—Yours always, KELVIN.

*February* 12, 1896.

DEAR WHARTON—I am very glad to have the paper which you send me with your letter of yesterday. I shall read it with great interest. I have no doubt that you are right that the prime motor of surface currents is the wind. If you care to look at page 145 of Volume III. of my *Popular Lectures and Addresses*, you will see something on the subject which I hope you will find correct. There certainly is a tremendous shovelling, as it were, of water to leeward in breaking waves under strong wind, and I suppose every gale of wind, lasting for two or three days, leaves a strong surface current for several days after it.

There certainly are also deep - seated permanent currents, due to differences in temperature in the water itself, by which great changes of water, especially between polar and equatorial regions, are produced. There was a great deal that was right in Dr. Carpenter's views on this subject, though I believe he did not sufficiently take also into account the great and rapid surface currents produced by wind, and the widespread distant return surface currents which they entail, as in Clayden's model which you describe.

I suppose I am right in believing that nowhere, either in Antarctic regions or in North Polar regions, is there any great area of ice-bound sea which is not landlocked? I should be greatly obliged by a single line in answer to this.—Yours truly, KELVIN.

*February* 18, 1896.

DEAR THOMSON—I am exceedingly interested in three things you tell me in your letter of the 16th, and much obliged to you for telling me them.

1. The selective absorption of *some* of the Röntgen light which you find with plates of the same metal, but of different thicknesses, seems almost certainly to be precisely analogous to the absorption of ordinary light by coloured glass, and therefore to prove that you have got different colours of Röntgen light. If this is true, you will of course find the different kinds of the light that you get by your sifting through plates of different metals, will, all of them, follow the law of inverse square of the distance, except in so far as some of them may be sensibly absorbed by a metre or two of air. It is most interesting to think that now in metals you have quasi-coloured glasses for Röntgen light, and that dielectrics are nearly white, beside being, as Röntgen found them, wonderfully transparent. One of the next things to try is the relative photographic powers of the different colours of the Röntgen light, and to test different photographic substances so as to find which is most sensitive for one colour and which for another;[1] also similar experiments with different phosphorescent substances. It remains to be seen, too, whether the different colours of the X-light have not different refractive indices, if any one of them has really a refractive index sensibly different from unity. I enclose a paper of Lenard's, as it may save you the trouble of looking it up in Wiedemann, if you have not seen it already. I have not had time yet to read it through. You may perhaps find in it something of the coloured-glass principle proved in numerical results, though not referred to in the text.

2. It is most interesting to find the X-light causing the air to be temporarily electrolytic. I suppose this has already been done for the ultra-violet light. We have done it for flame here, as you will see from one of the enclosed papers. Fumes from a spirit-lamp flame, passing between plates of polished zinc and polished copper (the flame being about a foot below the plates), produce a difference of potentials equal to ·78 of a volt between copper wires connected to the two plates. This is just

[1] It seems quite improbable that the best photographic substance for sensibility for X-light has been found by any one hitherto.

about the same as is produced by a bridge of pure water between the polished plates.

I enclose also another paper (by Maclean and Goto) related to this subject, which has only been published in the *Proceedings* of the Glasgow Philosophical Society.

3. Quite wonderful[1] and most interesting! Was the saturated steel needle the test; or was something else used, more adapted for receiving a succession of signals?—Yours very truly, KELVIN.

The next letter gives Lord Kelvin's reasons for the preference he had long held for the use of continuous currents in electric transmission of power:—

*February* 25, 1896.

DEAR MR. CAMPBELL SWINTON—Notwithstanding that I am confined to bed (and to be so according to doctor's orders for a fortnight), I should have been happy to see you, and I should certainly have been greatly pleased and interested to talk over with you the subject of your letter which I have received this morning.

For electric transmission of power over very long distances, *very* high pressure is necessary for economy. I object to the use of alternate currents for this practical purpose, primarily because 41 per cent higher pressure can be transmitted to a great distance by direct current than by alternate current, with the same conductors and insulation in the two cases. For distances exceeding 50 miles it would probably be advisable to use two wires, one of them at +20,000, and the other at −20,000 volts' difference of potentials from the earth. I believe it would not be possible to obtain, for such distances, as good economy by alternate current as by direct current. The prime cost of direct-current dynamos for such purposes has never been gone into practically.

Let me know if there is any other question that I can

[1] [This refers to Rutherford's "magnetic detector," with which he had been sending wireless signals by means of Hertz waves to a distance of half a mile across the town of Cambridge.]

answer in writing. Should you think it worth while to come here and talk the matter over with me, I should be delighted to see you and to discuss all the pros and cons.

I thank you for your kind inquiries. I am much better, and I am really feeling perfectly well, but medical opinion is quite decided that absolute rest for some weeks yet is necessary for my leg. They say that if I am thoroughly obedient to them in this respect I will be perfectly well. This is satisfactory, but I am disappointed to be kept laid up so long.—Believe me, yours very truly,

KELVIN.

Lady Kelvin was able by the end of February to write to Mrs. King:—"William is really very well in his bed, and very busy. The days are never long enough for all he wishes to do!" Many were the letters he wrote to Stokes and to Lord Rayleigh, discussing keenly the possible explanation of Röntgen's rays. Acknowledging one from Stokes on phosphorescence, he wrote on March 27, 1896:—

I have instructed the Pitt Press to send you advance corrected proofs (which have been in print eight years) of the first 112 pages of my *Baltimore Lectures.* You will find the formula I gave you a few weeks ago on page 106, also a good deal of fluorescence and phosphorescence jangling about in one-half or other of most of the lectures. I am afraid you will be very much shocked to see such a rigmarole as having been given in University Lectures. I hope it will not plague you to have them in your hands and to keep them till you get the rest of the volume from the Pitt Press, which I hope may be within a year from now.

In April and May Lord Kelvin was in London, and gave evidence before the Select Committee on

Petroleum. The Act of 1871 had required that paraffin oil should fulfil a flash-test of 100°, but in 1879 the Home Office had lowered the test to 73°, thereby admitting the free sale of dangerous low flash-point oils. Lord Kelvin told the Committee :

> The principle of safety is that oil should never in a lamp reach the temperature of the close-test flash-point. I advise the Committee to fix a flash-point which shall be higher than oil is likely to reach under ordinary conditions of ordinary use.
>
> I am clearly of opinion that in order to avoid accidents the flash-point must be raised, and that no construction of lamp will meet this difficulty.
>
> I call it terrible that 25 per cent of all the deaths by fire in London during the year were due to paraffin-lamp accidents.
>
> It seems to me that the logical outcome of Sir Frederick Abel's work ought to have been to declare that the 100° test in force in the 1871 Act must be fulfilled by a proper close test.
>
> There was no good reason for reducing the test from 100° to 73°. It seems to me that it was a mistake.
>
> I think that the accidents which have been reported are amply sufficient to justify prohibitive legislation—amply sufficient.

In spite of this, and similar views expressed by such chemists as Sir Henry Roscoe and Sir William Ramsay, nothing has been done.

Lord Kelvin felt very strongly on this question. Formerly no paraffin oil was allowed to be sold unless it passed the flash-test at 110° F., that is to say, unless it was of such a quality that even when warmed to 110° it did not give off explosive vapours. The American oil-dealers in 1865-7 agitated to get this reduced to 100°, although in British Government offices no oil is accepted of lower flash-point than 105°, and many of the American States insist on a test equally stringent. Yet in 1868 the test was lowered to 100°, and an open test-cup was legalized, which in practice proved to be erroneous to an average extent of 27 degrees. In other words, oil which was actually giving off explosive vapour at 73° (Fahr.) did not flash in this open cup until it reached 100°. The number of fires due to paraffin lamps increased owing to the introduction of cheap low-flash oils. In spite of this, in 1879, when a new and more efficient test was adopted, the flash-point was by a scandalous manœuvre reduced to 73°. In the ten years from 1881 to 1891 the fires in London due to paraffin oil went up to 50 per cent. It was chiefly due to Lord Kelvin's evidence that the Select Committee rather reluctantly recommended the raising of the flash-point (Abel test) to 100°. A Flash-point Bill, introduced in 1899, was defeated on second reading by 244 votes to 159, a result mainly brought about by the promise (unfulfilled) of Mr. Jesse Collings (the Under-Secretary for the Home Department) to bring in a Government Bill to deal with the whole matter. A new Government came in in 1900, and another in 1906, but nothing has been done. The scandal of the free sale of dangerous low-flash oil continues.

# CHAPTER XXIII

## THE JUBILEE: RETIREMENT

CELEBRATIONS of birthdays and jubilees of distinguished men are much less frequent in Great Britain than in most Continental states. But on rare occasions even the stolid Briton is stirred into a public demonstration of feelings that he is apt ordinarily to conceal. The Jubilee of Lord Kelvin, which was celebrated at Glasgow University on June 15, 16, and 17, 1896, was one of these rare occasions. His colleagues in the University, his students past and present, his fellow-workers in science and in University life, all united to make the event a memorable one. From the official record published by the University the following account is taken, with but few added details.

It was because of the unique character of the work done and of the personality of the worker that the Jubilee of Lord Kelvin became one of the most strangely impressive functions. It had in it something of the nature of a spontaneous outburst of enthusiasm, as well as of the studied and respectful homage shown by representatives of all the world to a great thinker and actor. At the gatherings held in the University and in the City of Glasgow in 1896 to signalize the fiftieth year of Lord Kelvin's tenure of the chair of Natural Philosophy in the University of Glasgow, no element in our academic or natural life was left unrepresented. Delegates from every seat of learning and from nearly every scientific body in Great Britain and Ireland were assembled, and with them were the men who have made or are making their mark in Glasgow and in the West of Scotland. Representatives there were too from the Colonies, and many brilliant and distinguished foreigners came to do honour to their great scientific fellow-worker.

The students at the University had invited delegates from the Universities of Great Britain and Ireland, and from many foreign Universities as well, and throughout the celebrations it was evident that the undergraduates were as eager to honour their senior professor as were his oldest friends. It was interesting also to notice that among the great number of congratulatory messages received by Lord Kelvin during this week, there were many from men who had formerly been undergraduates in his class, and who were now occupying posts in various parts of the world. As an instance of this may be mentioned the addresses sent from former Japanese students of Lord Kelvin's class, now at Tokyo.

The first gathering, a conversazione in the halls of the University, took place on the evening of Monday, June 15, 1896.

It was a brilliant sight, and very different from the daily routine of academic life. The sombre halls and cloisters and staircases of the College were lit up by electric light, decorated with flowers, and filled with a moving mass of colour; and the summer evening was so beautiful, that in the eastern quadrangle many of the guests strolled up and down listening to the pipes of the Gordon Highlanders. The innumerable differences in the robes worn by the guests represented well the cosmopolitan character of the gathering.

About two thousand five hundred ladies and gentlemen had been invited, embracing the representatives of Universities, Societies, and Institutions, and other distinguished visitors; the members of the University Court and Senate, about four hundred and fifty members of the General Council, and fully two hundred Students; the Lord Provost, Magistrates, and members of the Town Council, and many prominent citizens of Glasgow and residents in the West of Scotland and other parts of the country. The Bute Hall, the Hunterian Museum, and the upper hall of the Library were thrown open, and in the latter there was an exhibition of mechanical, electrical, and scientific apparatus and contrivances designed by Lord Kelvin; and of the diplomas and certificates of membership, as well as medals, presented to him by Universities, Colleges, and Institutions.

In the upper hall of the Library the Eastern, the Anglo-American, and the Commercial Cable

Companies had fitted up siphon-recorders in connection with their cables, and a large number of congratulatory telegrams[1] from all parts of the world were received in the course of the evening, and suitable replies transmitted.

After the Conversazione the Students held a Gaudeamus and a reception of delegates from other Universities in the large hall of the Union.

On Tuesday morning, June 16, 1896, an impressive function took place in the Bute Hall of the University, when the many distinguished men assembled in Glasgow presented to Lord Kelvin addresses from the Universities and Societies whom they represented. A letter was first read from the Prince of Wales :—

MARLBOROUGH HOUSE,
*10th June* 1896.

DEAR LORD KELVIN—The Prince of Wales desires me to offer you his warmest congratulations upon your having attained the fiftieth year of the tenure of your professorship in the University of Glasgow.

[1] Amongst the telegrams received during the evening was one from the Glasgow Jubilee Committee which they sent from Glasgow by the Anglo-American Atlantic Cable to Newfoundland, thence *via* New York, Chicago, San Francisco, New Orleans, Washington and New York back through the cable to Lord Kelvin. It ran :—

"By the Atlantic cable, which represents your unrivalled combination of scientific genius and practical skill, the Glasgow Jubilee Committee send you their warmest congratulations."

This message occupied seven and a half minutes in traversing the circuit of about 20,000 miles; and to it Lord Kelvin replied in a message which took but four minutes to compass the same route :—

"The cable companies have beaten Ariel by half a minute. Warmest thanks to the Glasgow University Jubilee Committee."

The Viceroy of India telegraphed from Simla, the President of the Orange Free State from Bloemfontein, and Earl Grey from Bulawayo.

Lord Glasgow telegraphed from New Zealand, and Sir James Sivewright from Cape Town. Congratulations were received from several prominent Americans, including Professor Elihu Thomson, Mr. Westinghouse, and Mr. C F. Brush.

His Royal Highness is in most cordial sympathy with the eminent representatives of universities, learned societies, and other public bodies in different parts of this empire and in foreign states, who, to do you honour, have assembled in the University which has for a long series of years—eventful through the rapid advance of science and its applications—enjoyed the high prestige derived from your close association with its work, and from the invaluable and brilliant contributions to science resulting from the researches carried on by you during the last half-century within its walls.

The Prince of Wales remembers with much satisfaction that he had the gratification, seventeen years ago, to present you with the medal instituted by the Society of Arts as a memorial of the Prince Consort, and awarded to men who have rendered pre-eminent service in promoting arts, manufactures, and science.

The work which you had at that time accomplished was but an earnest of the important researches to which you have since then devoted yourself so indefatigably, and he cherishes the sincere hope that you may long continue to enjoy the happiness derived from the most gratifying evidence that the high value of the service rendered by you through science to mankind is universally recognised and appreciated.—I remain, dear Lord Kelvin, yours truly, FRANCIS KNOLLYS.

*P.S.*—His Royal Highness desires me to repeat what he has already stated to the University authorities, how greatly he regrets that long-formed engagements in the south prevent him from having the pleasure of being present on the occasion of this interesting celebration. FRANCIS KNOLLYS.

Thereafter the following Congratulatory Addresses were presented to Lord Kelvin by the following representatives :—

### *FROM UNIVERSITIES*

*Aberdeen*, Professor Finlay, M.D., Professor Niven, F.R.S., and Professor Pirie; *Ann Arbor*, Professor R. M. Wenley, M.A., D.Sc., D.Phil.; *Baltimore* (*Johns Hopkins University*), James A. Thomas, M.D.; *Bombay*, Mr. Justice Jardine, Vice-Chancellor, and G. N. Nadkarin, LL.B.; *Cambridge*, Professor A. R. Forsyth, Sc.D., F.R.S., Professor Sir George G. Stokes,

LL.D., D.C.L., F.R.S., and Professor J. J. Thomson, M.A., F.R.S.; *Edinburgh*, Professor Crum Brown, M.D., F.R.S., and Professor Sir William Turner, LL.D., D.C.L., F.R.S.; *Glasgow* (*Senatus Academicus*), Professor Stewart, D.D.; *Glasgow University* (*General Council*), John G. Kerr, M.A., and Archibald Craig, LL.B.; *Heidelberg*, Professor Quincke; *Kasan*; *Lille*, Professors Pinloche and Angellier; *London*, Sir Henry E. Roscoe, F.R.S., Vice-Chancellor, and Professor Carey Foster, F.R.S.; *Montreal* (*M'Gill University*), Sir D. A. Smith, G.C.M.G., LL.D., Chancellor, and W. Peterson, LL.D., Principal; *New Haven* (*Yale University*); *New York* (*Columbia University*), Professor Van Amringe; *Oxford*, Professor Clifton, F.R.S., the Provost of Oriel, and Professor Burdon Sanderson, F.R.S.; *Paris*, Professor Bonet-Maury; *Philadelphia*, Professor G. F. Barker, M.D.; *Princeton*, Professor Woodrow Wilson; *Rome*, General Annibale Ferrero; *St. Andrews*, Professor Pettigrew, M.D., LL.D., F.R.S., and Professor Scott Lang; *Sydney*, Professor Liversidge, M.A., F.R.S.; *Tokyo* (*Imperial University of Japan*); *Upsala*, Professor P. T. Cleve; *Victoria* (*Manchester, Liverpool, and Leeds*), Principal Ward, Vice-Chancellor, and Professors Lodge, Osborne Reynolds, M'Cunn, and Stroud; *Wales* (*University of*), Principal Viriamu Jones and Professor Andrew Gray, LL.D., F.R.S.; *Washington* (*Columbian University*), Professor Cleveland Abbe.

## *FROM COLLEGES*

*Aberystwith* (*University College*), R. D. Roberts, M.A., D.Sc.; *Bangor* (*University College of North Wales*), Professor Andrew Gray, LL.D., F.R.S.; *Belfast* (*Queen's College*), Rev. Thomas Hamilton, D.D., LL.D., President, and Professor Purser, LL.D.; *Cork* (*Queen's College*), Professor Bergin, M.A.; *Dublin* (*Royal College of Science for Ireland*); *Galway* (*Queen's College*), Sir Thomas Moffett, LL.D.; *London* (*City and Guilds Technical College, Finsbury*), Professor Silvanus P. Thompson, D.Sc., F.R.S.; *London* (*Royal College of Science*), Professor W. A. Tilden, D.Sc., F.R.S.; *London* (*University College*), Professor Ramsay, F.R.S.; *Newcastle-on-Tyne* (*Durham College of Science*), Professor Philipson, M.D.; *Oxford* (*Balliol College*); *Paris* (*École Normale Supérieure*), Professor Violle.

## *FROM SOCIETIES AND INSTITUTIONS*

*Amsterdam* (*Royal Academy of Science*); *Baltimore* (*Members of Sir William Thomson's Class of* 1884, *Johns Hopkins*

*University*), Professor Cleveland Abbe; *Berlin* (*Royal Prussian Academy of Sciences*); *Cambridge* (*Bachelors and Undergraduates of University of*), F. W. Lawrence, B.A., and Philip W. Wilson; *Christiania* (*Students of the University of*), Cato Aall; *Copenhagen* (*Royal Danish Society of Sciences*), Professor Christiansen; *Cracow* (*Academy of Letters*); *Dublin* (*Science and Art Department*); *Edinburgh* (*Educational Institute of Scotland*), John Dunlop, F.E.I.S.; *Edinburgh* (*Scottish Geographical Society*), Sir Renny Watson; *Erlangen* (*Physikalisch-medicinische Societät zu Erlangen*); *Glasgow* (*Faculty of Physicians and Surgeons*), Bruce Goff, M.D., President; *Glasgow* (*Geological Society*), Sir Archibald Geikie, F.R.S., and J. Barclay Murdoch, Esq.; *Glasgow* (*School Board of*), Sir John N. Cuthbertson, LL.D., and Rev. William Boyd, LL.D.; *Glasgow* (*Students of University of*), John S. Thomson, President of the Students' Representative Council; *Göttingen* (*Royal Society of Science*), Professor Woldemar Voigt; *Helsingfors, Finland* (*Society of Sciences*); *Liège* (*L'Association des Ingénieurs Electriciens Sortis de l'Institut Montefiore*); *Lille* (*Students of University of*); *London* (*Royal Society*), Sir Joseph Lister, M.B., P.R.S; *London* (*British Association for Advancement of Science*), Professor A. W. Rücker, F.R.S.; *London* (*Royal Institution of Great Britain*), Professor Dewar, F.R.S.; *London* (*Mathematical Society*), Major P. A. MacMahon, R.A., F.R.S.; *London* (*Royal Astronomical Society*), A. Ainslie Common, LL.D., F.R.S.; *London* (*Physical Society*), Captain W. de W. Abney, F.R.S.; *London* (*Chemical Society*), Professor John M. Thomson; *London* (*Institute of Chemistry of Great Britain and Ireland*), Professor Ramsay, F.R.S.; *London* (*Institution of Electrical Engineers*), John Hopkinson, F.R.S.; *London* (*Institution of Civil Engineers*), Sir Benjamin Baker, K.C.M.G., F.R.S.; *London* (*Society of Engineers*), Henry O'Connor, Esq.; *London* (*Society for Encouragement of Arts, Manufactures, and Commerce*); *London* (*Glasgow University Club*); *Manchester* (*Literary and Philosophical Society*), Professor Schuster, F.R.S.; *Milan* (*Reale Instituto Lombardo di Scienze e Lettere*), General Annibale Ferrero; *Modena* (*Royal Academy of Science, Letters, and Arts*), General Annibale Ferrero; *Montreal* (*Canadian Society of Civil Engineers*), James Ross, Esq.; *Moscow* (*Imperial Society of Naturalists*), Professor Oumov; *Munich* (*Der Königliche Bayerischen Akademie der Wissenschaften*); *Newcastle-on-Tyne* (*Students of Durham College of Science*); *New York* (*National Electric Light Association of America*), Thomas C. Martin, Esq.; *Paris* (*Conservatoire National des Arts et Métiers*), Professor Violle; *Philadelphia* (*American Philosophical Society*), Dr. J. Cheston Morris; *Rome* (*Italian Society of Science*), General Annibale

Ferrero; *Rome* (*R. Accademia dei Lincei*), General Annibale Ferrero; *Rotterdam* (*Batavian Society of Experimental Philosophy*), Dr. Elie van Rijckevorsel; *Scottish Amicable Life Assurance Society*, Colin Dunlop, Chairman; *Scottish Universities* (*Students of the four*), J. R. Hunter, Edinburgh University; *Tokyo* (*former Students from Japan in Lord Kelvin's Class, now at Tokyo*); *Vienna* (*Imperial Academy of Sciences*); *Washington* (*National Academy of Sciences*), Professor Simon Newcomb; *Washington* (*Philosophical Society*).

It is impossible to give at length the text of all the addresses mentioned in this list, but the following representative ones are of special interest:—

*From the Royal Society*

DEAR LORD KELVIN—The President, Council and Fellows of the Royal Society desire on the happy occasion of the Jubilee of your Professoriate in the University of Glasgow not only to be represented, as they are, by their highest Officers the President and Treasurer, but also to assure you, by some direct words, of the warm sympathy of the whole Society.

There is no need to dwell on the many ways in which you have contributed to that improvement of natural knowledge to secure which the Society was founded, or on the many valuable communications with which you have enriched the Society's records. Since you first joined the Society, and the Jubilee of that event is not far off, the Society has always known how much your belonging to it has added to its strength: but it has been especially during the recent five years, which went too swiftly by, while you filled in so admirable a manner the chair of President, that the Society has felt how close are the ties which bind it to you and you to it.

We ask you to receive our heartiest congratulations on the present glad event, and our warmest wishes for your welfare in the years yet to come. JOSEPH LISTER, *President*.

*From the Institution of Electrical Engineers*

We, the President, Council, and Members of the Institution of Electrical Engineers, desire hereby to offer to your Lordship our sincere and hearty congratulations on the occasion of the Jubilee of your Professorship of Natural Science in the University of Glasgow. It will ever be a source of pride and satisfaction to this Institution, that one who occupies so pre-eminent a position in the scientific world should have been its First President in 1889, besides having been an original member and President in 1874 of the same Association when it existed under the name of the Society of Telegraph Engineers. Not only have you contributed more than any other living man to our knowledge of the laws of nature, but you have found time to perfect practical applications of Science, wherefrom every branch of the Electrical Engineering Profession has derived special benefit. We desire, in conclusion, to express our fervent wish that you may continue for many years to enjoy the blessing of good health, and that Science may still further benefit from your labours.

J. HOPKINSON, *President.*
F. H. WEBB, *Secretary.*

*From the British Association for the Advancement of Science*

MY LORD—The Council of the British Association for the Advancement of Science desire to offer to you their sincere congratulations on the completion of the fiftieth year of your tenure of the Professorship of Natural Philosophy in the University of Glasgow.

It is unnecessary to recount the triumphs you have won during the last half-century in mastering the difficulties which beset the advance of scientific theory and experiment, and in applying scientific principles to the practical service of man. The record of your achievements is fresh in the minds of those who address you, and can never be effaced from the history of the development of mathematical and experimental physics, of engineering, and of navigation. We would rather therefore recall the long and close connection which has existed between the British Association and yourself.

A regular attendant at our meetings, you have not only enriched our reports with many important papers, but have encouraged the efforts of younger men by never-failing sympathy and interest in their work.

You have been President of the Mathematical and Physical Section of the Association no less than five times. You were President of the Association at Edinburgh in 1871, and have since then been a Life Member of our Council.

As colleagues, then, we wish to tell you of the pride with which we, in common with all your fellow-countrymen, regard your distinguished career; and of the feelings of personal attachment with which we express the hope that you may long be spared to enjoy in health and strength the honours you have so nobly won.

Signed on behalf of the Council,

DOUGLAS GALTON.

The address from the University of Edinburgh contained the following passage :—

We know not whether most to admire in you the acute Mathematician, the unwearied Investigator of Physical Problems, the skilled Electrician, or the resourceful Engineer; to you in all of these capacities is due the success of Long-line Submarine Telegraphy, with the innumerable benefits resulting from the power of practically instantaneous communication between all parts of the globe.

We are grateful for the lustre which your brilliant discoveries have shed upon our Scottish Universities, and we are proud to number you among our Colleagues.

*From the Master and Fellows of Peterhouse, Cambridge*

TO THE RIGHT HONOURABLE LORD KELVIN—We, the Master and Fellows of Peterhouse, on the occasion of the Jubilee of your Professorship of Natural Philosophy in the University of Glasgow, desire to express our profound admiration of the splendid discoveries in physical science, and of the valuable scientific inventions, which have characterised the tenure of your professorship and have conferred signal benefits upon the whole civilised world; as well as our gratification and pride that your name, so indissolubly connected with the progress of science in the nineteenth century, should have been for a period of fifty-five years closely connected with this ancient college as Student, Scholar, Fellow, again Fellow *honoris causa* and Benefactor.

We recall with pleasure your noble enthusiasm as an undergraduate in the pursuit of your mathematical studies, and your important contributions to scientific journals, which led our late Master, Dr. Cookson, and your private Tutor, Mr. W. Hopkins, also a distinguished member of the College, at that early period of your career to predict your future eminence in science, and your keen interest in manly sports shown by your success as an oarsman in winning the Colquhoun Sculls, and in rowing in the College boat, which then occupied the second place on the river.

Many of us, students in your Natural Philosophy Class in

the University of Glasgow, have enjoyed the privilege of listening to your inspiring lectures; all of us, as your colleagues in the governing body of Peterhouse, have warmly appreciated your unfailing courtesy, wise counsels and generous sympathy with all that concerns the welfare of the College. We bear in grateful remembrance your munificence on the occasion of the celebration of the Six Hundredth Anniversary of the foundation of our most ancient House.

We fervently pray that your connection with the College may long continue.

In testimony whereof, we have attached our common seal this thirteenth day of June, in the year of our Lord, One thousand eight hundred and ninety-six.

*From the Students of the University of Glasgow*

My Lord—In the name of the Students of the University of Glasgow we desire to offer you our sincere and hearty congratulations on the occasion of your Jubilee as Professor of Natural Philosophy in our University. While we feel it needless to dwell upon your pre-eminence in the world of science, and would not presume to speak of that genius which has enriched humanity by so many brilliant discoveries, we ask, simply as your students, to be allowed to take our part in the universal congratulation at this time.

We rejoice to have an opportunity of expressing in your presence a feeling of affectionate regard no less strong than the admiration to which others besides ourselves are to-day giving voice.

Above all we desire to refer to your long unbroken connection with our University, a connection which must be endeared to you by many precious memories, and has been to successive generations of students a source of grateful pride. We are proud to think that, year after year, and decade after decade, our University has shared in your ever-increasing fame, and that for so long a period it has been her happiness to retain in her midst one whom all nations have delighted to honour.

The last address that was presented was from

*The Senatus Academicus of the University of Glasgow*

My Lord—The rejoicings which have been arranged to celebrate the close of your fiftieth session betoken the admiration and affection with which you are regarded by your colleagues in the Senate, but it is none the less fitting that on this auspicious occasion these feelings should find articulate expression in an address of congratulation.

The fifty years during which you have occupied the chair of Natural Philosophy in this University have to an extent unparalleled in the history of the world been marked by brilliant discoveries in every department of Physical Science, and by the prompt adaptation of many of these discoveries to meet the practical needs of mankind. We recognise with admiration that in both these respects you have been a leader of the age in which we live. . . . But only your colleagues in University work are in a position to appreciate the versatility of faculty, the exhaustless energy, and the tenacity of purpose which have enabled you to grapple successfully with problems the most varied, and to reveal to us on every side the reign of order and law. In the midst of all, you have endeared yourself to us by the graces of your personal character, notably by that simplicity which, unmarred by honours or success, remains the permanent possession of transcendent genius, and by that humility of spirit which, the clearer the vision of truth becomes, bows with the lowlier reverence before the mystery of the universe.

My Lord, the contemplation of a past so rich in achievements and honours encourages your colleagues to look forward to the future in the hope that you may have health and strength to win new triumphs in years to come, and long to remain among us the ornament and the glory of our ancient University.

Then followed the giving of degrees. In the absence of Principal Caird, through illness, Sir William Gairdner, K.C.B., Professor of the Practice of Medicine, occupied the chair in the earlier part of the meeting during the presentation of Addresses, and also conferred the degree of LL.D. upon Lord Kelvin, who then took the chair and conferred the degree of LL.D. on the following :—

Professor Cleveland Abbe, Washington; Professor Christian Christiansen, Copenhagen; Professor Per Theodor Cleve, Upsala; General Annibale Ferrero, Ambassador from H.M. the King of Italy; Professor Izidor Fröhlich, Buda-Pest; Professor Gabriel Lippmann, La Sorbonne, Paris; Professor Archibald Liversidge, Sydney; Professor Éleuthère Mascart, Collège de France; Professor Henri Moissan, Paris; Professor Simon

Newcomb, Baltimore; Professor Nicolas Oumov, Moscow; Professor Emile Picard, Paris: Professor Georg Quincke, Heidelberg; Professor Woldemar Voigt, Göttingen.

Lord Kelvin then said:—

The University of Glasgow is honoured by the presence to-day of many distinguished visitors from distant countries, from America, from India, from Australia, and from all parts of the United Kingdom. Names of men renowned for their scientific work in foreign lands have been added to our list of honorary graduates. That I have had the honour of conferring these degrees in the name of the University is a subject of keenest regret to all here present, because it is due to the absence of Principal Caird, on account of illness. We hope that the beginning of next session will see him at home in the University with thoroughly recovered health. In his absence the duty of conferring degrees has fallen, according to University law, on me as senior Professor present.

I am also one of the recipients of the degrees, and, in the name of all who have to-day been created Doctors of Laws of the University of Glasgow, I thank the Senate for the honour which we have thus received on the occasion of the Jubilee of my professorship. For myself, I can find no words to express my feelings on this occasion. My fifty happy years of life and work as Professor of Natural Philosophy here, among my students and my colleagues of the University, and my many kind friends in the great city of Glasgow, call for gratitude; I cannot think of them without heartfelt gratitude. But now you heap coals of fire on my head. You reward me for having enjoyed for fifty years the privilege of spending my time on the work most congenial to me and in the happiest of surroundings.

You could not do more for me if I had spent my life in hardships and dangers, fighting for my country, or struggling to do good among the masses of our population, or working for the benefit of the people in public duty voluntarily accepted. I have had the honour to receive here to-day a gracious message from His Royal Highness the Prince of Wales, and addresses from sister universities in all parts of the world; from learned societies, academies, associations, and institutions for the advancement of pure and applied science; from municipal corporations and other public bodies; from submarine telegraph companies, and from their officers, my old comrades in their work; from students, professors, and scientific workers of England, Scotland, and Ireland, and other countries, including my revered and loved St. Peter's College, Cambridge.

I have had an address also from my twenty Baltimore Coefficients of 1884. The term "coefficients" is abused by mathematicians. They use it for one of the two factors of the result. To me the professor and his class of students are coefficients, fellow-workers, each contributing to whatever can possibly be done by their daily meetings together. I dislike the term *lecture* applied here. I prefer the French expression "conference." I feel that every meeting of a professor with his students should be rather a conference than a pumping-in of doctrine from the professor, perhaps ill understood and not well received by his students. The Scottish Universities have enabled us to carry out this French idea of conference. I think in every one of his classes the professor is accustomed to speak to his students, sometimes in the form of *viva voce* examination, and oftener, I hope, in the manner of interchange of thoughts, the professor discovering whether or not the student is following his lecture, and the student, by showing what he knows or does not know, helping the professor through his treatment of the subject.

I have had interesting and kindly addresses from my old Japanese students of Glasgow University, now professors in the University of Tokyo, or occupying posts in the Civil Service and Engineering Service of Japan. I wish particularly also to thank my Baltimore Coefficients for their address. They have been useful to myself in my own keen endeavour—unsuccessful, I must say, nevertheless keen—to find out something definite and clear about light and ether and crystals.

The addresses which I have received to-day contain liberal and friendly appreciation of all my mathematical and physical papers, beginning in 1840 and ending—not yet I hope. The small proportion of that long series of writings which has led to some definite advancement of science is amply credited for its results. A larger part, for which so much cannot be said, is treated with unfailing and sympathetic kindness as a record of persevering endeavour to see below the surface of matter. It has been carried on in the faith that the time is to come when much that is now dark in physical science shall be seen bright and clear, if not by ourselves, by our successors in the work.

I am much gratified by the generous manner in which these addresses have referred to the practical applications of science in my work for submarine telegraphy; my contributions to the advancement of theoretical and practical knowledge of the tides; my improvement in the oldest and next oldest of scientific aids to navigation—the sounding plummet and the mariner's compass; and my electric measuring instruments for scientific laboratories, for the observation of atmospheric electricity, and for electric engineering.

I now ask the distinguished men who have honoured me by presenting to me these addresses, to accept for themselves personally, and for the societies represented by them, my warmest thanks for the great treasure which I have thus received—goodwill, kindness, friendship, sympathy, encouragement for more work—a treasure of which no words can adequately describe the value.

I cordially thank the French Academy of Sciences for their great kindness in sending me by the hands of my loved and highly esteemed colleague, Mascart, the Arago Medal of the Institute of France.

I thank all present in this great assembly for their kindness, which touches me deeply; and I thank the City and University of Glasgow for the crowning honour of my life which they have conferred on me by holding a commemoration of the Jubilee of my professorship.

*Professor Mascart's Address*

Professor Mascart, who was the delegate of the Collège de France and of the Académie des Sciences, Paris, at the Jubilee, had intended to make the following speech when presenting the Arago Medal. He afterwards gave his manuscript to Lady Kelvin, and said that he had been too much touched by the ceremony to be able to deliver his address :—

MILORD ET CHER CONFRÈRE—L'Académie des Sciences de Paris, dans laquelle vous êtes aujourd'hui le doyen des associés étrangers, a voulu se joindre aux savants de tous les pays du monde, à vos admirateurs, à vos amis, pour vous apporter des félicitations chaleureuses à l'occasion du cinquantenaire de votre arrivée comme professeur à l'Université de Glascou que vous avez tant illustrée.

Il y a quelques mois, l'Institut de France célébrait le centième anniversaire de sa fondation, ou plutôt de la reconstitution des anciennes Académies sur des bases plus larges. Nous ne pouvons oublier l'élévation de langage avec laquelle le Président de la Société Royale de Londres vint alors traduire les sentiments de cordialité de cette grande et célèbre Institution.

Dans une autre réunion, où vous parliez en votre nom personnel, vous nous avez causé une profonde émotion en déclarant que vous aviez une dette de reconnaissance envers notre pays, que nos grands esprits tels que Fourier, Laplace et Sadi Carnot avaient été vos inspirateurs et que vous considériez la France comme l'*alma mater* de votre jeunesse scientifique.

Si la dette existe, vous l'avez payée avec usure. Dans la longue série de travaux et de découvertes qui galonnent votre admirable carrière, une des plus nobles que l'on puisse rêver, vous avez abordé toutes les questions de cette science à laquelle la littérature anglaise conserve le beau nom de "philosophie naturelle," soit pour contribuer aux progrès des conceptions théoriques, soit pour en déduire des applications utiles au développement de l'industrie et au bien de l'humanité.

Quoi que l'avenir réserve au génie inventif de l'esprit humain, votre nom restera comme ayant été le guide le plus sûr dans une époque féconde et le véritable éducateur de la génération actuelle dans le domaine de l'électricité.

Je suis particulièrement heureux que l'Académie des Sciences m'ait confié le soin de vous remettre une médaille d'or à l'effigie d'Arago, médaille qu'elle réserve pour rendre hommage aux services exceptionnels rendus à la science et qui porte cette devise : "Laudes damus posteri gloriam."

Vos confrères de l'Institut de France espèrent que vous voudrez considérer ce souvenir comme un témoignage de haute estime et de leurs sentiments les plus affectueux.

Dans une circonstance à laquelle je faisais tout à l'heure allusion, vous avez rappelé aussi qu'au début de votre carrière vous aviez fréquenté les laboratoires du Collège de France, où les professeurs de cette époque, Biot, Liouville et Victor Regnault accueillirent avec empressement le jeune homme dont les premières publications faisaient déjà prévoir le brillant avenir.

L'Assemblée de Professeurs du Collège a bien voulu, par une délibération spéciale, me confier la mission de vous témoigner le prix qu'elle attache à ce souvenir en vous apportant le tribut de ses cordiales félicitations.

J'ai encore comme Président actuel de la Société d'Encouragement pour l'Industrie Nationale, à vous traduire les hommages de cette association, fondée à l'origine du siècle et qui a pour but de faciliter l'application des découvertes scientifiques aux progrès industriels.

Il y a deux ans, la Société avait l'honneur de vous décerner l'une de ses plus hautes récompenses, par l'attribution d'une médaille de platine à l'effigie d'Ampère. Vous estimerez sans doute que les figures d'Arago et d'Ampère, placées côte à côte dans la collection de vos écrins, ne s'y trouveront pas en mauvaise compagnie, de même que les deux savants ont été associés de si près dans leurs immortelles découvertes.

Enfin vous avez eu à diverses reprises l'occasion de témoigner une bienveillance particulière à la Société Internationale des Électriciens en assistant à quelques-unes de ses séances et en honorant de votre concours le Congrès de 1889.

La Société m'a prié d'être son interprète dans la circonstance actuelle, pour vous exprimer ses sentiments de reconnaissance, son admiration pour vos travaux, et pour vous offrir des respectueux hommages.

On the evening of Tuesday, June 16, Lord Kelvin was entertained by the Corporation and University of Glasgow at a Banquet in St. Andrew's Hall. After dinner, Sir James Bell, Bart., the Lord Provost, who presided, read a message from the Queen as follows:—

The Queen commands me to beg that you will kindly express to Lord Kelvin Her Majesty's sincere congratulations on the occasion of the Jubilee of his professorship in the Glasgow University. Her Majesty trusts that many years of health and prosperity may be in store for him and Lady Kelvin. The Queen is particularly gratified at the presence of so many eminent representatives from all countries of the world, who have come to do honour to your distinguished guest.

ARTHUR BIGGE, on behalf of
HER MAJESTY.

The Lord Provost, on rising to propose the toast of the evening, said—

We have received one or two cable messages which I have been desired to read. They are addressed to Lord Kelvin. The first is from Toronto, and reads as follows: "The Councils of the University of Toronto and of University College offer you their heartiest congratulations on your attainment of your fiftieth year of your professorship, and they earnestly wish that you may be long spared to serve science, whose advancement you have so signally promoted.—J. Loudon, President." Then from Quebec there is one: "I send most cordial congratulations on this occasion. Your illustrious fifty years' services have been of great profit to science.—La Flamme." Another telegram, just received from Moscow, is addressed: "To the celebrated Lord Kelvin, famous, learned, we send our congratulations.—The Moscow University students." In addition to these telegrams, I have been requested to say that numerous letters have been received expressing, on behalf of the writers, regret at inability to be present at these celebrations. I have one from Lord Salisbury expressing his great regret. Mr. Campbell has intimated one that he has had from Sir John Gorst, and there are many others. I wish to read one from Principal Caird. Need I say how greatly we all regret the cause by which we are unavoidably deprived of his presence, and of the matchless eloquence with which he would have presented the toast which in his absence falls to me to-night.

I know how far more deeply than can be expressed in words our beloved Principal grieves over his inability to be with us to-night, but we confidently hope that he may soon again be restored to health.

Ladies, my Lords, and Gentlemen,—The toast that I have now the honour to propose for your acceptance is that of "Lord Kelvin, and hearty congratulations on the attainment of his professorial Jubilee." These congratulations are manifold and great—they come from all European countries; they come from India and our Colonies, from across the Atlantic, from the great scientific societies and from the leading scientists of to-day; and in this City we are doing what is for us unique—the University and City authorities are joining hand in hand to show, in the strongest manner possible, our intense admiration and appreciation of Lord Kelvin and his work. A great social and commercial revolution dates from August 1858, when the message was signalled under the ocean, "Europe and America are united by telegraphic communication. Glory to God in the highest, on earth peace and goodwill towards men."

To almost every branch of scientific research Lord Kelvin has given contributions of inestimable value. In regard to the laws of heat, he has been one of the greatest discoverers; while, through the perfecting of the compass and improving the means of sounding, the risk of loss of life or vessel has been so minimized that I do not think I am overstating the case when I say that these two discoveries have saved thousands of lives, and millions of pounds worth of property.

Lord Kelvin's discoveries and appliances have world-wide use, from the most complicated and delicate instrument to improvements on the simplest form of mechanism. His industry is unwearied, and he seems to take rest by turning from one difficulty to another—difficulties that would appal most men, and be taken as enjoyment by no one else. But what has been the result of these great gifts of genius, coupled with this industry? They have resulted in a lifetime of discoveries fraught with good; year by year something has been accomplished; paper after paper that are standards on their subject-matter have been written until we are lost in amazement at what has been done. While concurrently with this active productivity his lordship's university classes have been carefully carried on. How many students in these fifty years of Jubilee have been fired with their teacher's enthusiasm we can never know, but from this class many have gone who have attained great distinction and who look back with pride and pleasure to the days passed in the Natural Philosophy class-room of Glasgow University.

My Lords and Gentlemen, it is given to few men to labour in one place for fifty years; it is given to almost none to do so with the distinction achieved by our guest, a distinction now so great that he may justly be called the greatest living scientist. Lord Kelvin has given in America the Baltimore lectures—lectures not given to students but to professors. He was a member of the Niagara Commission. He has been awarded honours innumerable and of every kind from learned societies in the Old World and the New, and, as you know, he has just demitted the office of President of the Royal Society, after a term of office marked by the greatest brilliancy. This life of unwearied industry, of universal honour, has left Lord Kelvin with a lovable nature that charms all with whom he comes in contact.

Lord Kelvin, indeed, inspires love and reverence in all. His home life is love and melody. His helpmate is worthy of him, and greater cannot be said. Those who have the great privilege of their friendship, with fervent prayer, will in their hearts add to the toast the wish that Lord and Lady Kelvin may long be spared to one another.

Lord Kelvin, who on rising to reply was greeted with prolonged applause, said:—

First of all, I desire to express the deep and heartfelt gratitude with which I have heard the most kind and gracious message from Her Majesty the Queen, which has been read to us by the Lord Provost. But I cannot find words for thanks. I can only, on the part of Lady Kelvin and myself, tender an expression of our loving loyalty to the Queen. My Lord Provost, my Lords, and Gentlemen, I thank you with my whole heart for your kindness to me this evening. You have come here to commemorate the Jubilee of my University professorship, and I am deeply sensible of the warm sympathy with which you have received the kind expressions of the Lord Provost regarding myself in his review of my fifty years' service, and his most friendly appreciation of practical results which have come from my scientific work.

I might perhaps rightly feel pride in knowing that the University and City of Glasgow have joined in conferring on me the great honour of holding this Jubilee, and that so many friends and so many distinguished men—friends and comrades, day-labourers in science—have come from near and far to assist in its celebrations, and that congratulations and good wishes have poured in on me by letter and telegram from all parts of the world. I do feel profoundly grateful. But when I think how infinitely little is all that I have done I cannot feel pride; I only see the great kindness of my scientific comrades, and of all my friends in crediting me for so much.

One word characterises the most strenuous of the efforts for the advancement of science that I have made perseveringly during fifty-five years; that word is FAILURE. I know no more of electric and magnetic force, or of the relation between ether, electricity, and ponderable matter, or of chemical affinity, than I knew and tried to teach to my students of natural philosophy fifty years ago in my first session as Professor. Something of sadness must come of failure; but in the pursuit of science, inborn necessity to make the effort brings with it much of the *certaminis gaudia*, and saves the naturalist from being wholly miserable, perhaps even allows him to be fairly happy in his daily work.

And what splendid compensation for philosophical failures we have had in the admirable discoveries by observation and experiment on the properties of matter, and in the exquisitely beneficent applications of science to the use of mankind with which these fifty years have so abounded! You, my Lord Provost, have remarked that I have had the good fortune to remain for fifty years in one post. I cordially reply that for me they have been happy years. I cannot forget that the happiness of Glasgow University, both for students and professors, is largely due to the friendly and genial City of Glasgow, in the midst of which it lives. To live among friends is the primary essential of happiness; and that, my memory tells me, we inhabitants of the University have enjoyed since first I came to live in it (1832) sixty-four years ago. And when friendly neighbours confer material benefits, such as the citizens of Glasgow have conferred on their University in so largely helping to give it its present beautiful site and buildings, the debt of happiness due to them is notably increased.

I do not forget the charms of the old college in the High Street and Vennel, not very far from the "comforts of the Salt-market." Indeed, I remember well when, in 1839, the old Natural Philosophy class-room and apparatus-room (no physical laboratory then) was almost an earthly paradise to my youthful mind; and the old College Green, with the ideal memories of Osbaldistone and Rashleigh and their duel, created for it by Sir Walter Scott, was attractive and refreshing to the end. But density of smoke and of crowded population in the adjoining lanes increased, and the pleasantness, healthiness, and convenience of the old college, both for students and professors, diminished year by year. If, my Lord Provost, your predecessors of the Town Council, and the citizens of Glasgow, and well-wishers to the city and its University all over the world, and the Government, and the great railway company that has taken the old college, had left us undisturbed on our ancient site, I don't believe that attractions elsewhere would have taken me away from the old college; but I do say that twenty-five of the fifty years of professorship which I have enjoyed might have been less bright and happy, and I believe also less effective in respect to scientific work, than they have been with the great advantages with which the University of Glasgow has been endowed since its migration from the High Street.

My Lord Provost, I ask you to communicate to your colleagues of the Town Council my warmest thanks for their great kindness to me in joining to celebrate this Jubilee. Your Excellency, my Lords and Gentlemen, I thank you all for the kind manner in which you have received the toast of my health proposed by the Lord Provost, and for your presence this evening to express your good wishes for myself.

Professor Sir W. T. Gairdner, in proposing the toast of the Representatives present from other Universities and learned bodies, spoke of Lord Kelvin's personal character as it had appeared to his colleagues, and said :—

I feel very strongly that all that has been said of the scientific eminence of Lord Kelvin, and of his innumerable and most remarkable discoveries in science, leaves still without emphasis one point about him which only those who have been in close association with him can appreciate, and that is his childlike humility of character—his very remarkable power of inspiring affection as well as esteem, his interest and sympathy with every one who is related to him in any way whatever.

His Excellency General Annibale Ferrero, Lord Lister, and Professor Simon Newcomb of Washington responded to this toast; and among the later speakers were Professor Story, the Earl of Rosse, Sheriff Berry, and Sir Henry Roscoe.

The Italian Ambassador's speech was as follows:—

Messieurs—Lorsque l'on m'a fait l'honneur de me charger de prendre la parole au nom de tant d'hommes illustres, j'ai pensé que ce serait une grande témérité de ma part que d'accepter un rôle qui aurait convenu à des illustrations dont le nom a déjà une place dans l'histoire de la science. Cependant j'ai pensé que j'avais l'honneur de représenter la patrie des grands prédécesseurs de Lord Kelvin, tels que Galilée, Volta, et Galvani. En venant représenter les institutions scientifiques de nos pays respectifs, nous avons avant tout voulu rendre hommage à l'homme de génie dont on célèbre le jubilé et à l'Université de Glascou qui a la gloire de le posséder. Mais notre présence est aussi pour prouver que le monde scientifique tout entier veut prendre sa part de l'honneur que Lord Kelvin a rendu à la race humaine La lumière qui resplendit aujourd'hui sur l'Université de Glascou est comme celle du soleil. Elle n'appartient pas à un seul pays, mais s'étend sur toute l'humanité. Nous devons des remercîments spéciaux à l'illustre professeur Gairdner pour la santé qu'il vient de porter aux représentants des institutions scientifiques. Nous devons payer un tribut de vive reconnaissance à la ville et à l'Université de Glascou pour l'accueil si cordial et honorifique qui nous a été accordé. C'est avec le plus grand sentiment d'admiration que nous avons assisté à ce jubilé vraiment grandiose, digne de l'Université qui l'a organisé, et du grand homme qui en a été l'objet. Le spectacle sublime auquel nous avons assisté ce matin a autant élevé notre esprit que touché notre cœur. Je ne puis mieux exprimer notre pensée comme qu'en disant que nous avons assisté à l'apothéose de la science. Je veux finir en exprimant le vœu le plus cher à l'Université de Glascou et à nous-mêmes, c'est-à-dire que la Providence, laquelle lui a confié dans la personne de Lord Kelvin un trésor incalculable, lui conserve pour des longues années ce trésor précieux à l'humanité toute entière.

The University Senate had arranged that the celebrations should terminate on Wednesday with a sail through some of the more picturesque parts of the Clyde. A company of about two hundred and fifty, including Lord and Lady Kelvin and a large number of the delegates and visitors, were conveyed by special trains to Greenock, where they embarked on board the steamer *Glen Sannox.* The route followed was by Largs and Millport, and then northwards along the western coast of the Island of Bute, through the Kyles, and thence homeward.

Perhaps the most striking thing about the Jubilee of Lord Kelvin was its spontaneity; indeed the absence of officials from the ceremonies was the cause of comment at the time. A writer in the *Saturday Review*, drawing a comparison between the careers of Lord Kelvin and of von Helmholtz, each of whom had lived to celebrate his Jubilee, pointed out the difference between England and Germany in the public estimation in which science is held and honoured:—

Though it is the fact that Lord Kelvin and von Helmholtz were each honoured by a title of nobility, the difference in the recognition is truly striking. It was not until Sir William Thomson began to dabble in politics that the great and wise and eminent in official circles discovered those transcendent claims to recognition which had long been patent in the world of science. Whereas the German Emperor, in conferring a patent of nobility upon von Helmholtz, specially commented on his abstention from intermeddling with political questions.

At the Helmholtz Jubilee in 1891, the ceremonials in Berlin were marked by the presence of two Secretaries of State, the Minister of the Interior and the Minister of Education. Thus did Germany set her official approbation upon the honours acclaimed by the assembled representatives of science. At the Kelvin Jubilee official recognition seems to have been studiously withheld. Neither the Lord President of Council nor the Secretary of State for Scotland put in an appearance. The Chancellor of the University of Glasgow was conspicuous by his absence. The Prince of Wales, who, in the absence of political officials, might have represented the Sovereign, was detained "by long-standing engagements . . . in the south." . . . The King of Italy could send an Ambassador. But Italy is one of those countries where science is honoured for its own sake.

The following contemporary narrative from the pen of his grand-niece Miss Margaret E. Gladstone adds a few graphic details :—

The Graduation ceremony next morning was the best of all the functions. It began by the presentation of over eighty addresses from all parts of the world—(at least before that, the students whiled away the time by singing, "Here a poor buffer lies low," and when Uncle William came in "Jolly Good Fellow"), —and then we had a Latin prayer and a letter from the Prince of Wales. The gowns and uniforms of the delegates were most varied and gorgeous : the foreigners came first, and then the representatives of the various British Universities and learned societies, and of the students of different Universities ; each man with his roll or book tucked away under his arm. Each one shook hands with Uncle William and some of them made little speeches. Mascart, who brought several Addresses from France, and the Arago gold medal, had prepared a nice little speech, but was too *touché* to say anything when it came to the time. He and Moissan and Max-Müller wore the French Academicians' dress—green palm-leaves embroidered on a black ground—and looked very fetching. The Lille professors had orange gowns ; but the most gorgeous of all was the Italian Ambassador, with rows of orders across his uniform.

After the addresses, Professor Gairdner gave Uncle William the LL.D. of Glasgow (how pleased old Macpherson, the janitor, must have been to put on his hood !), which was quite illegal, for it ought to have been conferred by the Chancellor, Lord Stair, or the Vice-Chancellor, Principal Caird, both of whom were away ill, or, after them, by the Senior Professor, who is Uncle William himself. However, he took the chair after that, and did the capping of the foreigners who got degrees. The students made a little banging each time the velvet cap descended, and shouted out a few remarks ; but on the whole they were remarkably quiet and suppressed. Sir Joseph Hooker told us he missed the pea-shooting which went on so much at the time he was a fellow-student with James and William Thomson in my great-grandfather's class. After the capping Uncle William made his reply. Some of us were quite afraid he was going to break down, he got so pale, and the effort to find words seemed too much for him, but he drank a little water with whisky in it, and then got on all right. I thought his speech was beautiful. So did every one.

. . . . . .

In the evening the word "Failure" in which he characterized the results of his best efforts seemed to ring through the hall with half-sad, half-yearning emphasis. Some of the people tried to laugh incredulously, but he was too much in earnest for that.

. . . . . . .

Yet at the same time it was not pessimistic, for it was so evident what keen joy he had had in his work, and still has, and how warmly he feels the help and affection of his fellow-workers. I should think on the whole he has had as happy a life as any one can, and happy because of his utter devotion to his work, and his thoroughness in throwing himself into whatever is before him, and constant readiness to learn.

As for the students, I am afraid they laughed, with good cause, when he spoke of the ideal lecture as a conference, because I always hear that he goes up in the heights when he is lecturing to them, and pours forth speculations with great enthusiasm far above their heads. They are, however, very fond of him: they serenaded him at three o'clock in the morning of Tuesday, after their rowdy *Gaudeamus* at the Union; only, fortunately, he slept through it. On Wednesday, when our steamer passed the one on which they were all, on an excursion, and Uncle William got up on our paddle-box to wave to them, the cheering and waving were tremendous; and they came up in the evening after that to his house and he went out and gave a little speech, telling them to go back to their work, for that was the best thing for every one—whereat they groaned and howled; and that it had been good for them to have a jollification—whereat they cheered lustily.

In thinking over Uncle William's speeches, the tone in which he gave them, and his quiet, serious, deferential look when praise was heaped upon him, dwell in my memory. There was something pathetic about it all—a sort of wonder that people should be so kind to him, and a wish that he had done more to deserve it all.

. . . . . . .

I forgot to say that at the close of the banquet we all sang "Auld Lang Syne" at Uncle William's request.

. . . . . . .

Many of the scientific men went off that night or next day, but a good number were left to go on the steamboat excursion next day on the *Glen Sannox*. It pelted all the morning, but we were very happy chattering under an awning or down in the saloon; and we could just see Netherhall looming through the mist as we past it and fired off two guns. Then, while we were down at lunch, it cleared up and we came up to find ourselves close to Arran, with the most beautiful sunshine and cloud effects.

We came back by the Kyles. . . . In the middle of the afternoon some of them got up the Lancers on deck, and it was great fun to see Aunt Fanny leading off with Professor Ferguson, till a sudden squall of rain came on, and they had to fly. . . .

. . . . . .

Uncle William is simply marvellous: he is seventy-three, and has been ill all winter and spring (I think, though, that being kept partly a prisoner by his bad leg has been better for him than rushing round society functions, and endless meetings), and yet after all he had gone through, both of physical and mental (and emotional) exertion in the past few days, and an Address from, and a speech to, the Liberal Unionist Association that morning, he was quietly alert, taking in all the Addresses, and all that was going on in the bustle; and then when we came in to tea talking away with my father about chemical theories, and about the terrible shipwreck there has been, and everything that turned up.

The next day we came up with them in the same train. At the last minute his green-book was left behind, so he had to content himself with answering his telegrams (they had written twenty-two telegrams before we lunched with them at Preston) and writing letters to Quincke, Stokes, and James Bottomley on Röntgen rays, etc. Then by 9 P.M. he was taking the chair at the Royal Institution, and, as Aunt Fanny said, seemed better than before he went down to Glasgow. He appreciated the beauty of the country too on the way up, and thought I was using my time well by looking out of the window a good deal—it certainly was a perfect day, with blue sky and sunshine, and clouds to give rich shadows, and then the grass and young corn such a bright green, and the haymakers in the fields, and the hedges lovely with wild roses.

A week afterwards Lord Kelvin addressed to the various persons who had sent letters or presented felicitations, a lithographed autographic letter, as follows:—

THE UNIVERSITY, GLASGOW,
*June* 23, 1896.

Lord Kelvin is much gratified by welcome letters containing congratulations on the occasion of the Jubilee of his Professorship, and highly valued good wishes, and kind words, which have come to him from many friends.

He is very sorry that it is impossible for him to write in reply to each. He desires now to express his heartfelt thanks to all for the great kindness and sympathy which he has received.

Lord and Lady Kelvin were invited by Her Majesty Queen Victoria to dinner at Windsor Castle on Monday, July 13, 1896, and to remain until the following day. The Marquis and Marchioness of Dufferin and the Duke and Duchess of Argyll were amongst the guests. After dinner they all sat in the corridor, and one after another of the guests was brought up to speak to Her Majesty, at her request. When Lady Kelvin came to her she said to her: "It is a long time since we have met, not since we were at Inveraray Castle." That was in 1875 (see p. 663). She then inquired how Lord Kelvin had been. To Lord Kelvin she talked of his Jubilee. They could not help noticing that the Queen was very lame and infirm, and of failing eyesight.

On Wednesday, July 15, Her Majesty held an investiture at Buckingham Palace, when she conferred on Lord Kelvin the Grand Cross of the Royal Victorian Order. As he approached the presence to receive the decoration, H.R.H. the Duke of Connaught whispered to the Queen that it was Lord Kelvin. At once she said: "Lord Kelvin must not kneel, it might hurt his leg."

Later in the same afternoon Lord Kelvin attended at the Victoria Institute to hear the Address by Sir George Stokes.

In commemoration of the centenary of Robert Burns a statue of Mary Campbell ("Highland Mary") was unveiled by Lord Kelvin at Dunoon, on August 1, 1896.

On August 12, returning to Largs with Lady Kelvin, he was met at the railway station by the Provost and Magistrates of that burgh, and presented with an address of congratulation on his recent Jubilee.

A month later, Lord and Lady Kelvin were staying at Knowsley during the meeting of the British Association at Liverpool, under the presidency of Sir Joseph (afterwards Lord) Lister. Lord Kelvin read three papers: one, an Attempt to explain Chemical Affinities by Molecular Dynamics; another (in association with two of his pupils), on the Communication of Electricity from Steam to Air; a third (in collaboration with J. T. Bottomley and M. Maclean), on the Measurement of Electric Currents through Air.

All this autumn Lord Kelvin suffered at intervals from acute facial neuralgia, seated in the "fifth nerve," which he used to refer to as "the demon," or "number five." The attacks used to come on very suddenly, and again suddenly departed. In October he was kept in bed four days by a bad attack. He was unable to be present at the annual meeting of the Royal Society on November 30.

On December 2 Lord Kelvin was admitted as an Honorary Member of the Philosophical Society of Glasgow, which he had joined as an ordinary member on December 2, 1846. An Address was presented to him, and at the same time a bronze bust of him was presented by a body of subscribers to the Society; a duplicate bust being presented to Lady Kelvin. The same day there were unveiled, in the staircase of the University, two memorial tablets to Lord Sandford and to Professor Veitch. Lord Kelvin, who presided at this ceremony, gave recollections of his childhood, when Frank Sandford and he used to play in the old College Green, and make ships and sail them in the Molendinar Burn. On December 21, Lord Kelvin was in Edinburgh, communicating to the Royal Society of Edinburgh a paper on the Electrification of Air by Röntgen Rays. December 22 saw him in London, attending the opening, by the Prince of Wales, of the Davy-Faraday laboratories at the Royal Institution, endowed by Dr. Ludwig Mond.

After his Jubilee Lord Kelvin had resumed his University duties, and continued to experiment with unabated zeal. On January 13, 1897, he gave an evening public lecture at Glasgow University on the Molecular Dynamics of a Crystal; and within a month read two papers dealing with crystalline forms to the Royal Society of Edinburgh. He gave several other papers to the same body on various subjects, including osmotic pressures, a subject in which he strongly opposed the views of Ostwald, writing to Lord Rayleigh several diatribes thereon. He was active, too, in other directions, taking part in the Jubilee of Queen's College,

Belfast. He entertained Nansen when he visited Glasgow in February to lecture on his Polar travels, and presided at the lecture.

*Nature* of March 25, 1897, contained a long and interesting illustrated article on Lord Kelvin's laboratory at the University, and on White's Instrument Factory, where Lord Kelvin's instruments and compasses were made. James White, the original owner of the optical business, had died in 1884. His skill had been from the early days of the mirror galvanometer of great assistance to Lord Kelvin, who, when the manufacturing part of his works had increased, had found the capital needed for the business. The factory in Cambridge Street, Glasgow, built in 1884, employed about 200 hands, besides a staff of trained electrical engineers in the testing department. At this date the managing partner was Mr. David Reid, while Mr. James Ferguson looked after the electrical instruments; but Lord Kelvin was daily about the works giving detailed instructions to workmen or foremen, and testing the adjustments of instruments and the fitting of the parts. His usual programme, when at home, was, after giving his secretary instructions about his correspondence, and, on certain mornings, lecturing to his class from 9 to 10, to walk or drive into the town to White's, often remaining there until time for a midday lecture, or for lunch.

At this date appeared in *Vanity Fair* a rather cruel portrait of Lord Kelvin, accompanied by a terse but cordial appreciation.

"He is so good a man," wrote the narrator, "that four years ago he was most deservedly ennobled as Baron Kelvin of Largs; yet he is still full of wisdom, for his Peerage has not spoiled him. He is a very great, honest, and humble Scientist who has written much and done more. Yet, with all his greatness, he remains a very modest man of very charming manner."

Lord Kelvin dined at the Royal Academy banquet on May 1st. Replying for Science, he said he thought that if Raphael or Michael Angelo had seen a photograph they would have been both delighted and astonished. So much may be taken for granted, but whether they would, as he opined, have declared it to be "a masterpiece of art" is quite open to question. He aroused a ripple of laughter amongst the artists by telling them that they owed a debt of gratitude to the men of science who had proved to them that yellow and blue do not make green, but that green is unquestionably a primary sensation.

On May 8 the Master of Peterhouse, Dr. Porter, who had been a student at Glasgow in 1846, under Lord Kelvin, was presented with a testimonial for his services to the University and town of Cambridge; and his portrait, painted by Ouless, was at the same time presented to the College. Lord Kelvin, as Senior Fellow of Peterhouse, received the portrait on behalf of the College, and spoke warmly of Dr. Porter, and of the efficiency that had marked his administration.

It was at this time that a very determined, but unsuccessful attempt was made to induce the

University of Cambridge to open its degrees to women. Girton and Newnham Colleges had now been in successful operation for twenty years; and the promoters of the movement claimed that the University would now be justified in granting the titles of degrees to those women who were successful in the Tripos examinations, which had long been open to them. The bulk of the resident members of the Senate were certainly opposed to this step. Lord Kelvin sided with the opponents, and became chairman of a London Committee of Cambridge graduates to resist the proposal.

On May 21 Lord Kelvin gave a Friday evening discourse at the Royal Institution, on the Contact Electricity of Metals. He had worked actively at this subject forty years before, and had employed his electrometers to show the effect, discovered by Volta, of bringing copper and zinc into contact and then separating them, when they were found to be oppositely electrified. For many years a great controversy had raged on the theory of the voltaic cell, some attributing its operation to contact, others (notably Faraday) to chemical action. Volta's experiment had then been discredited until Lord Kelvin's electrometer experiments justified its truth. More recently electricians, while accepting the fact, had also sought to explain it as a chemical action. In the 'eighties there was considerable controversy as to this, Oliver Lodge maintaining that the so-called contact force, of about one volt, was due to the difference between the respective chemical affinities of the oxygen of the air for zinc and for copper. Lord Kelvin would never accept this idea, but attributed it to that affinity of zinc and copper for one another which is shown in their union to form brass. With that curious impenetrability to other men's ways of regarding things that was characteristic of Lord Kelvin's mind, he could never be brought to admit the cogency of Lodge's main proposition,[1] based as it was on the work which Lord Kelvin himself had done in collaboration with Joule in the 'fifties (see p. 401). He remained to the last unconvinced. Part of his Royal Institution lecture was directed to this point, and part to the effect of the Becquerel rays emitted by uranium in altering the apparent electrification due to contact.

In the House of Lords, on May 28, 1897, Lord Kelvin spoke against the imposition of death duties upon works of art and collections of scientific instruments or of objects of natural history, the property of private owners; also private observatories, which had been of the greatest benefit to astronomy. He urged that in such cases the duty charged should be merely nominal.

The Age of the Earth as an Abode fitted for Life was the subject of an address by Lord Kelvin to the Victoria Institute on June 2, 1897. In the Chapter on the Geological Controversy of

[1] The matter was further complicated by Lord Kelvin's method of defining the potential of a conductor as the potential of a point in the air infinitely close to its surface. This, as the present writer once pointed out in one of the discussions, leads to a fallacious conclusion; for then the difference of potentials between copper and zinc in contact, if so defined, is quite different from the true difference of potentials between the metals themselves.

the years 1862-1871 (see pages 535-551 *supra*), an account is given of Lord Kelvin's earlier arguments; and on p. 941 is a brief notice of the revival of the question in 1895. He now reiterated his position in a reasoned address, which will be found in the *Transactions* of the Victoria Institute for 1897, and is here summarized :—

He first recapitulated the old arguments against the Uniformitarian postulate of unlimited time; adding Sollas's recent estimate of 17,000,000 years as the outside limit of time elapsed since the beginning of the Cambrian epoch. He then referred to newer knowledge as to fusion and conductivity of rocks, which had narrowed his own earlier estimates down to something between 20 and 40 millions of years, and to Clarence King's independent estimate of 24 millions. Then, with a passing fling at British weights and measures, he discussed the bearing on the date of consolidation of the globe of recent measurements on the fusion and solidification of diabase, granite, and basalt. First, granite would crystallize out in a crust; through the cracks in this an upper crust of basalt would extrude. When this oozing through cracks ceases, we have reached Leibnitz's *consistentior status* with surface cool and solid, and an internal temperature increasing to 1150° C. at 25, or 50, or 100 metres down. The main features of great continents would become fixed by the granitic formations in the lava ocean before its consolidation. After solidification was reached at the surface, cooling by radiation would continue. Condensation of vapours in the atmosphere would produce lakes, seas, and rivers. But to account for free oxygen in the air one must suppose vegetation to have begun; and only a few thousands, or perhaps hundreds of thousands of years later would there be oxygen enough to support animal life as we now know it, unless, indeed, the earth's primitive atmosphere contained free oxygen. Certainly, if sunlight were ready, the earth was ready both for vegetable and animal life within a few hundred centuries after the rocky consolidation of its surface. But was the sun ready? According to the dynamical theory worked out by Helmholtz, Newcomb, and himself, the sun was probably ready only 20 to 25 millions of years ago. But to account for the commencement of life on the earth mathematics and dynamics fail us. We must pause, face to face with the mystery and miracle of the creation of living creatures.

This was the last pronouncement of Lord Kelvin as to the age of the earth. But it is not the latest pronouncement of science. The discovery by Curie of the spontaneous evolution of heat from radium, and the detection by others of small quantities of radio-active materials in the crust of the earth, have added the knowledge of another internal source of heat. Sir George Darwin has put[1] the matter thus:—

> The researches of Mr. Strutt on the radio-activity of rocks prove that we cannot regard the earth simply as a cooling globe; and therefore Lord Kelvin's argument as to the age of the earth, as derived from the observed gradient of temperature, must be illusory. Indeed, even without regard to the initial temperature of the earth acquired by secular contraction, it is hard to understand why the earth is not hotter inside than it is. . . . The evidence, taken at its lowest, points to a period many times as great as was admitted by Lord Kelvin for the whole history of the solar system.

And again:[2]—

> If we were still compelled to assent to the justice of Lord Kelvin's views as to the period of time which has elapsed since the earth solidified, and as to the age of the solar system, we should also have to admit that the theory of evolution under tidal influence is inapplicable to its full extent. . . . Lord Kelvin contended that the actual distribution of land and sea proves that the planet solidified at a time when the day had nearly its present length. . . . The calculations contained in paper 9, the plasticity of even the most refractory forms of matter under great stresses, and the contortions of geological strata, appear, to me at least, conclusive against Lord Kelvin's view.

Sir George Darwin, whose researches on the dynamics of the globe contain frequent references to Lord Kelvin's researches, sums up the position by saying: "If I dissent from some of his views, I

[1] *Scientific Papers*, vol. ii., Preface, p. ix. [2] *Ibid.* p. vii.

none the less regard him as amongst the greatest of those who have tried to guess the riddle of the history of the universe."

A memorial subscription in honour of Sir John Pender resulted in a gift of £5000 toward the endowment of the laboratory and chair of electrical engineering in University College, London. The presentation to the college was made on July 2, 1897. Lord Kelvin, as a member of the Memorial Committee, alluded feelingly to the work done by Pender—his lifelong friend—for the success of the Atlantic cables of 1858, 1865, and 1866, and in afterwards founding the Eastern and the Eastern Extension Telegraph Companies.

The sixtieth year of Queen Victoria's reign was celebrated on June 22, 1897, by a great procession through London to St. Paul's Cathedral. In that splendid and historic procession Lord Kelvin was assigned no place: science, art, literature, and learning were studiously ignored by the powers that be, in the organization of that vast demonstration of England's progress under her great Queen. He was present on board the White Star liner *Teutonic* to witness the Naval Review on June 26th; and as a peer was invited to attend the Military Review at Aldershot on July 1st.

On June 28th he opened the Shoreditch municipal electric lighting station, and praised the scheme for utilizing refuse in the dust destructor. On July 2nd he wrote to *The Times* warning the public against the dangers of the so-called "dry shampoo." On the 15th he formed one of a deputation from the Royal Society to present a congratulatory address to Queen Victoria at Windsor; and on the 29th he visited the training ship *H.M.S. Worcester*, to which he presented one of his standard compasses.

On August 4th he visited Greenock to open a carbon factory in connexion with the aluminium works at the Falls of Foyers. Some remarks which he made on this occasion evoked much comment:—

There was one splendid application of the electric furnace in Scotland at the Falls of Foyers. That magnificent piece of work of the Aluminium Company was the beginning of something that would yet transform the whole social economy of countries such as the Highlands, where water abounded. He looked forward to the time when the Highlands would be re-peopled to some degree with cultivators of the soil, but re-peopled also with industrious artizans doing the work which that utilization of the water would provide for them. The British Aluminium works were very popular in the locality. It was only at a distance that the sentimental question, "What is to become of the beautiful Falls of Foyers?" was asked. They were going on now in all their beauty, and might go on so for many a year. He did not himself utter the aspiration that they might so go on. He thought when the time came that every drop of water that now fell over the Falls of Foyers was used for the benefit of mankind, no wise man, no man who thought of the good of the people, would regret that the power in the waterfall was developed for the benefit of mankind.

Toronto was the meeting place of the British Association in 1897. Lord and Lady Kelvin sailed on August 7th on the *Campania* for New York, reaching Toronto on the 17th. At this meeting he read one paper, on the Fuel and Air Supply of the Earth. This discourse has never been printed in full. Its main theme was an estimate of the total probable amount of fuel on this planet.

Taking as a standard fuel one requiring three times its weight of oxygen to consume it, Lord Kelvin arrived at the conclusion that the total weight of fuel in the earth is not more than 340 millions of millions of tons; the uncombined oxygen over the globe being 1020 millions of millions of tons. The coal-supply of Great Britain was about one two-thousandth part of the total fuel supply of the world. Great Britain's coal was more than could possibly be burned by all the oxygen in the air immediately over the British Isles. In regard to the effect of sunlight in storing energy and fuel, the present rate of sunshine was equivalent to the production of two tons of vegetation per square metre, per thousand years; an estimate agreeing very closely with the growth of German forests and of English hay-fields. We might, as coal-fields become exhausted, have to think of growing hay for fuel, as more economical than raising coal. But if we burn up our fuel supplies so fast the oxygen of the air may become exhausted, and that exhaustion might come about in four or five centuries.

*The Times* correspondent at Toronto wrote: "Lord Kelvin is above all others the popular favourite here; he is received with acclamation wherever he appears."

On his way to Toronto Lord Kelvin had visited Niagara Falls to see the huge industrial development there arising, which he greatly praised in a statement which he gave to the press on his visit:—

The originators of the work so far carried out and now in progress, hold a concession for the development of 450,000 horse-power from the Niagara waterfall. I do not myself believe that any such limit will be found to the use of this great natural source. I look forward to the time when the whole water from Lake Erie will find its way to the lower level of Lake Ontario through machinery, doing more good for the world than even that great benefit which we now possess in contemplation of the splendid scene which we have before us in the waterfall of Niagara. I wish I could live to see this grand development. I do not hope that our children's children will ever see the Niagara cataract. I look forward to a revival of life and prosperity in the Highlands of Scotland, and to the present crofters being succeeded by a happy industrial population occupied largely in manufactories rendered possible by the utilization of all the water power of the

LORD KELVIN VISITING THE KANANASKIS FALLS, N.W. CANADA, 1897.

country. It seems to me a happy thought that the poor people of the country will be industrious artizans, rather than mere guides to tourists.

In Toronto Lord Kelvin made several speeches —on his university experiences, when he received an honorary degree from Toronto University; on the benefits of the British Association, at the banquet given by the Governor-General of Canada; and on other matters at several minor entertainments.

After the meeting there was an excursion, lasting over a fortnight, across the Canadian continent, by the Canadian Pacific Railway, to Victoria, British Columbia. The party reached Winnipeg on August 28th, and crossed the prairies to Banff Hot Springs, through the Rockies to Laggan, Field, Glacier House, Revelstoke, and Vancouver. Lord Kelvin telegraphed home from Glacier House, "Have had splendid time. Most interesting and varied journey. Are enjoying ourselves immensely." A photograph, taken by Prof. E. M. Crookshank, representing Lord Kelvin standing beside the Kananaskis Fall, which is near the railway between Regina and Banff, is here reproduced as Plate XIV. Wherever he went, the popularity of his reception was universal and striking. Canadians gladly showed their appreciation not only of his position in the world of science, but of the service he had rendered Canada in connexion with the Atlantic telegraph. Returning by the Northern Pacific, they spent four days in the Yellowstone Park, and reached Chicago on September 15th. Thence by Pittsburg to New York, Long Island, and back to New York, whence Lord Kelvin paid a visit to Schenectady to see the General Electric Co.'s works. Passing through Boston he went to Halifax, and on to Montreal to discuss water-power schemes, after which he spent a few days at Lennox with George Westinghouse. On October 11 he was in Philadelphia, visiting the University with Professor Barker; running over to Princeton University; joining in a discussion on "Matter" at the American Philosophical Society; and enjoying social entertainments.[1] On the 16th he sailed in the *Campania*; and by the 29th he was in Cambridge for the statutory Meeting of Fellows at Peterhouse.

The Rt. Hon. Joseph Chamberlain, as Lord Rector of the University, visited Glasgow on November 3rd, to deliver his Rectorial Address[2] on "Patriotism," and stayed with Lord Kelvin.

At the Watt anniversary, on January 23, 1898, Lord Kelvin spoke of the connection of James Watt with the University of Glasgow, and of the improvements which he introduced into the steam-engine. Watt was a leader in science. He recalled how his

[1] An odd occurrence at a dinner party at the house of a gentleman whom for the purpose of biography we will call Mr. Frank Johnson, caused Lord and Lady Kelvin much amusement. A large company had been invited to dine at 7.30. Through some mishap their distinguished visitors were late, and the assembled guests and their hosts began to experience some uneasiness. Suddenly the draperies before the doorway were thrust aside, and in the opening beamed the negro butler, who forthwith announced: "Mistah Johnson, de Lawd am come!"

[2] The reporters record that whilst Mr. Chamberlain kept his eyeglass in his right eye all evening, Lord Kelvin, who wore his monocle in his left eye on entering, shifted it to his right eye during the latter part of the time. He had a favourite theory that the eyes should be rested alternately!

own great master in physical science, Regnault, had "la loi de Watt," the law of Watt as to the latent heat of steam, continually upon his lips. Even telegraphy owed something to Watt, for without the steam-engine and steam navigation they could never have had the ocean telegraphs.

Lord Kelvin returned to thermodynamic studies in two papers read in February and March to the Royal Society of Edinburgh. The first related to the energy theory of volta-contact electricity, being an extension of the ideas put forward in his Royal Institution discourse of the previous year (see p. 996), with experiments conducted in a cycle of operations. The other was on thermodynamics founded on motivity, and in it he called attention to his paper of 1855 (see p. 294), in which he had laid down the doctrine of available energy. He also lectured to the Glasgow Philosophical Society on Mutual Actions between Plastic Solids and Liquids. About the same date he presented to the library of that Society a valuable collection of scientific books.

At the graduation ceremony of April 12, 1898, owing to the illness of Principal Caird, the duty of presiding and presenting medals and prizes fell to Lord Kelvin as Senior Professor. After the ceremony he gave a brief address :—

Graduates and undergraduates, students, you have been busy for many months, or years, with your university studies. You have been creating property. You have not been making money, nor adding field to field, nor building houses. But you have been creating a property more precious than gold or silver, or broad acres, or houses that may be burned or ruined. The property you have created is your very own for ever, indestructible, imperishable, inalienable. The splendid university organization, with its material resources, and the living influence of its teachers and students, has helped you. But every one of you has, by himself and for himself, by the power of God working in him, made the property which he brings away with him. May it to every one of you be a joy and a blessing for ever.

Principal Caird died in July, and there was a rumour in the press that Lord Kelvin would be made Principal of the University. But it was known pretty generally that he would not accept the position; and it appears that no actual offer of it to him was made, the Rev. R. H. Story being appointed to the post.

Before that date Lord and Lady Kelvin had paid a visit to Hallam, Lord Tennyson, at Farringford, Freshwater. Thence Lord Kelvin wrote on June 3 to Darwin that he was still busy with revising the *Baltimore Lectures*. While staying at Farringford he went to see Signor Marconi's installation of wireless telegraphy between Alum Bay and Bournemouth, and insisted on paying a shilling to send "through the ether" a message to Stokes. A week later he agreed to act as consulting engineer to the "Wireless Telegraph and Signal Company" about to be formed. Concerning this he wrote on June 12:—

In accepting to be consulting engineer I am making a condition that no more money be asked from the public, for the present at all events; as it seems to me that the present syndicate has as much capital as is needed for the work in prospect, and as could get a proper return out of earnings on applications which we can at present foresee. I am by no means confident that this condition will be acceptable to the promoters. But without it I cannot act.

At intervals through this spring Lord Kelvin was giving sittings to Mr. (now Sir) W. Q. Orchardson for the portrait presented by subscription to the Royal Society. His passion to fill up available moments is illustrated by the following note to Lord Rayleigh :—

PETERHOUSE LODGE, CAMB., *May* 22 [1898].

DEAR RAYLEIGH—Stokes has been telling me of your investigation proving very extreme nullity of reflection of light at the polarizing angle (? very exactly Fresnel's ?) of water with the roughly pure surface. Will you give me the reference? Will you come to Orchardson's on either Wed. or Friday next, or both, 11 to 1 o'cl., so that we may have a good talk over many important affairs?—Yours, K.

*The Times* of July 15, 1898, contained a short but cogent letter from Lord Kelvin on the Report of the Petroleum Committee (see p. 962). Mr. Jesse Collings, who had always opposed the much-needed reform of raising the flash-point test from the low value of 73° to 100°, had rather foolishly written that if the oil flows out through the lamp breaking, "it matters little whether the oil is 73° or 100°." Lord Kelvin, who was scandalized by this officious defence of the use of cheap low flash-point oils, wrote, "I believe many lives will be saved by the adoption of the Committee's recommendation to prevent, as far as legislation can prevent, the use of oil for an illuminant with flash-point (Abel close test) below 100°."

At the Bristol meeting of the British Association in September 1898, Lord Kelvin read three papers, two of them relating to the dynamics of the undulatory theory of light, and one on the graphic representation of waves.

Lord Kelvin took the chair on October 18, in St. Andrew's Hall, Glasgow, at a great Liberal Unionist meeting addressed by the Duke of Devonshire. Speaking as chairman, he described the Unionist alliance of 1886 as the greatest political event of the second half of the nineteenth century. He prophesied that this alliance would last on into the twentieth century, though the details of government would be different, since history did not repeat itself. "I have," he said, "my ideas as to opposition and supporters of the Government: this is not the time to speak of them. But I do believe in a new mode of action in politics, in which everything shall be discussed according to its merits, and not according to party ins and outs."

In the ancient city of Colchester the annual Oyster Feast is usually made the occasion for inviting distinguished guests of honour. At the Oyster Feast[1] of October 31, 1898, the chief guests were H.R.H. the Duke of Cumberland, Lord

[1] Conviviality at this quaint civic festival never seems to be damped by the circumstance that the menu consists of unlimited oysters and brown bread and butter, all of primest quality; and on this occasion the usual hilarity prevailed. In proposing the toast of Science and Engineering, Lord Claud Hamilton called upon Lord Kelvin, whose name was associated with the toast, to reply to the question: "Have you ever seen an oyster walk upstairs?" Lord Kelvin said: No, he never had; and he did not believe an oyster could walk upstairs. "But if," he said, "we adopt a theory at present very fashionable in modern science, the day will come when the oyster's great, great, great—repeat it five hundred or five million times—great grandson—say a million million years hence—will not only be able to walk upstairs, but also to attend the Oyster Feast, and when he arrives at the top will have to make a speech!"

Kelvin, Lord Rayleigh, the Lord Mayor of London, and Lord Claud Hamilton.

Lord Kelvin attended the Annual Meeting and Banquet of the Royal Society on November 30th, but took no active part in the proceedings. He had been on 28th re-elected for the fifteenth time (with certain intervals between) President of the Royal Society of Edinburgh, an office which he held until his death.

On January 13, 1899, Lord Kelvin paid a visit to Nottingham to see the electric lighting works. In February he was elected an honorary member of the Institution of Electrical Engineers, the first under the new rules of that body.

He was still picking up the threads of his long-delayed publications, and wrote various notes about them to Lord Rayleigh, including the following :—

THE UNIVERSITY, GLASGOW,
*March* 11/99.

The Pitt Press have been profiting by my cold-prevention *régime*. They have been having a despatch of *Baltimore* every day, and must be arranging to engage fresh hands.

I now, since yesterday evening, after a not sleepless night, believe undoubtingly the Green-Cauchy stress-theory of Double Refraction. K.

Easter 1899 saw Lord and Lady Kelvin in Rome. Lord Kelvin was present on April 23 at the sitting of the R. Accademia dei Lincei, where he was welcomed by the President Sig. Beltrami in the following words: "I have the pleasure of announcing to the Academy that this day's sitting is honoured by the presence of its venerated Foreign Associate Lord Kelvin, incomparably the most distinguished representative of physical science—taken in its widest sense—of our time in the whole civilized world." He then recounted Lord Kelvin's achievements, and invited the members of the Academy "to give a salute of honour to our venerated colleague," which they did by rising to their feet. On May 1 a complimentary banquet was given to Lord Kelvin by the Associazione Elettrotecnica Italiana and the Ministry of Posts and Telegraphs. Sig. Mengarini and three other speakers proposed the toast of his health in the warmest terms; and Lord Kelvin in replying spoke with admiration of the extent to which electric engineering had been carried out in Rome. He had been deeply struck by the beauties of ancient Rome and the skill of her builders. Yet, he added, the old Romans, if they could come back to life, and see the transmission of 2000 horse-power by four wires no thicker than his little finger stretching across the Campagna to supply Rome with electricity from the cascade of Tivoli, along nearly the same route as their magnificent aqueducts, would think a feat had been accomplished even greater than those of old times.

Returning northwards Lord and Lady Kelvin were entertained by the Batavian Society of Rotterdam on June 17.

On June 26 Lord Kelvin spoke in the House of Lords against the insertion in mail-steamer contracts of clauses inflicting penalties for delay in arrival,

LORD KELVIN'S LAST LECTURE, 1899.

irrespective of cause of delay. Such clauses were dangerous, as promoting high speed in times of fog. Later in the session he spoke against the Seats for Shop Girls Bill.

He had now reached his seventy-fifth birthday. For some time he had determined on retiring from the duties of his professorship, and had been quietly preparing to withdraw from active service. At a meeting of the University Council on July 11 he presented a petition for leave to retire. The Council granted leave, and accepted his retirement with deep regret, instructing the Principal to prepare a minute[1] to be signed by the members of the Court expressing their sense of the great loss the University was thus to sustain. His resignation took effect on October 1, 1899. In the autumn his former student and assistant, Andrew Gray, was chosen as his successor.

Lord Kelvin had held the Chair of Natural Philosophy since 1846. "It was to him," said Principal Story, "an undisguised pain to sever the tie that bound him to the University." Yet the tie was not completely severed, for he insisted on inscribing his name on the University roll as a "research student."

[1] For the text of the minute and of Lord Kelvin's reply, see *The Electrician*, xliii. p. 689, September 8, 1899.

# CHAPTER XXIV

## THE GREAT COMPREHENSIVE THEORY

Utinam caetera Naturae phaenomena ex principiis Mechanicis eodem argumentandi genere derivare liceret. Nam multa me movent ut nonnihil suspicer ea omnia ex viribus quibusdam pendere posse, quibus corporum particulae per causas nondum cognitas vel in se mutuo impelluntur et secundum figuras regulares cohaerent, vel ab invicem fugantur et recedunt: quibus viribus ignotis, Philosophi hactenus Naturam frustra tentarunt. Spero autem quod huic Philosophandi modo, vel veriori alicui, Principia hic posita lucem aliquam praebebunt. — NEWTON, *Philosophiae Naturalis Principia Mathematica* (Praefatio ad Lectorem).

*SCIENCE is bound, by the everlasting law of honour, to face fearlessly every problem that can fairly be presented to it.* These were Lord Kelvin's words in 1871, in the Address (see p. 599) wherein he held out a prospect of the early completion of the molecular theory of matter—"a great chart, in which all physical science will be represented with every property of matter shown in dynamical relation to the whole." For another quarter of a century he continued—and nowhere more conspicuously than in the *Baltimore Lectures*—to face fearlessly the outstanding problems of physics, the relation of ether to ponderable matter, the inner mechanism of the molecule, and the properties of electricity and magnetism which matter can manifest.

If at his Jubilee his hearers were startled by the frank confession of "failure," it was no note of despair that he sounded. He himself told what he meant in a letter to M. Wilfrid de Fonvielle:—

*July* 25, 1896.

DEAR MR. DE FONVIELLE— . . . I thank you also for the cutting from the *Petit Journal* which came to me from you a few days after your letter. I see in it a slight misunderstanding of what I said in Glasgow regarding "failure of my most strenuous efforts." It was not anything that I had been in the habit of *teaching*, either in my lectures or published papers, to which I referred. I am as firmly convinced as ever of the absolute truth of the kinetic theory of gases. What I feel that I have failed in has been my persevering efforts during 50 years to understand something more of the luminiferous ether and of the manner in which it is concerned in electric and magnetic forces; and it was of this that I said I know no more now than I knew 55 years ago when I became convinced that ether was essentially concerned in all these actions. Trusting you will kindly excuse my making this correction, I remain, yours very truly, KELVIN.

He had, in fact, set before himself very early in his career an immensely high ideal, a noble ambition of so tremendous an import that it would seem as if all his life he had shrunk from exhibiting it in full panoply. Yet there had assuredly haunted him day by day the suggestion of an all-embracing, comprehensive theory of matter. In the Preface to Newton's *Principia*, the great philosopher after stating his claim to have deduced the motions of the Planets, the Comets, the Moon, and the Sea, by mathematical arguments, from the forces of gravity, uttered the aspiration: *Utinam caetera Naturae phaenomena ex principiis Mechanicis eodem argumentandi genere derivare liceret.* That pregnant sentence might well be the symbol of Lord Kelvin's intellectual career.

That he worked often and strenuously at lesser aspects of the subject, that on occasions he deliberately chose to explore some smaller region of the great "chart," is quite compatible with his holding the greater ambition as an ultimate aim. In the years of his boyhood the interconnexion of the physical sciences had been bit by bit revealed. Already Davy had discovered the chemical actions of the electric current, and Oersted and Arago its magnetic properties. Faraday had generated electric currents dynamically by the moving of a magnet; and by a triumphant *tour de force* had proved magnetism to be capable of acting on light. Grove had drawn attention to the correlation of physical forces, a correlation not then amenable to calculation, and requiring the discovery of the great energy-principle to give the clue to the transformations so correlated. But the properties of matter, its ultimate structure, its elasticity and its compressibility, its optical, electric, and magnetic qualities, these were as yet totally unexplained. They *must* be capable of explanation; they must in some way depend upon the arrangement and mutual actions of the molecules of its structure, or upon the structure and properties of the molecules themselves, or upon their relation to the all-pervading ether of space. To be the Newton of the molecular theory which should afford a dynamical explanation of all these properties was a noble and worthy ambition.

Such an idea seems to have come to William Thomson in a partial aspect during the spring of 1846, while he was still at Peterhouse (see p. 159), when he was "often trying to connect the theory of propagation of electricity and magnetism with the solid transmission of force." But in the middle of composing the Inaugural Lecture of his Glasgow course in October 1846, the notion returned to him with compelling vividness, that he could somehow represent by the straining of an elastic solid the phenomena of electricity and magnetism. As his mathematical notebook shows (see p. 197), it was on 28th November 1846 that he at last succeeded in working out the "mechanico-cinematical" representation of electric, magnetic, and galvanic forces formulated in terms of the equations of an incompressible elastic solid.

The paper, published in the *Cambridge and Dublin Mathematical Journal* for 1847, sets forth indeed a mechanical representation;[1] but the theories it propounds are a mathematical skeleton

[1] It was this paper which started Maxwell's investigations. "The distinct conception of the possibility of the mathematical expressions arose in my mind from the perusal of Prof. W. Thomson's papers, 'On a Mechanical Representation of Electric, Magnetic, and Galvanic Forces,' *Cambridge and Dublin Mathematical Journal*, January 1847, etc."; Maxwell, *Camb. Phil. Trans*. Dec. 10, 1855, p. 67.

only, with no physical treatment such as with fuller powers he put forth in later life. He found, in fact, three particular solutions of the equations of equilibrium of an elastic solid. One of these, by mathematical analogy, expressed electric attractions as the result of an elastic displacement; another expressed magnetic forces, statically, as the result of an angular displacement (not a sustained rotational movement); the third, similarly, gave an elastic analogue for the forces in the neighbourhood of a wire carrying a current. He wound up by saying that a special examination of the various states of a solid body "must be reserved for a future paper."

That the main proposition of this paper constituted a mathematical analogy rather than a dynamical explanation Thomson was fully conscious: for, writing to Faraday about it in June 1847 (see p. 203), he himself says that he did not venture even to hint at the possibility of making it the foundation of a physical theory of the propagation of electric and magnetic forces. He added, most significantly, that "*if such a theory* could *be discovered*, it would also, when taken in connexion with the undulatory theory of light, in all probability explain the effect of magnetism on polarized light." Obviously he had been trying to frame such a theory.

Again and again, throughout his lectures and addresses, we meet with the phrase "the properties of matter." The Properties of Matter was the title assigned to a chapter—alas! never written—of the great unfinished *Treatise on Natural Philosophy*. When, at an early date in his career, he was asked what was the object of a physical laboratory, he replied that it was "to investigate the properties of matter." Years afterwards, in the address called "Steps towards a Kinetic Theory of Matter," given in 1884 at Montreal, he said:—

> . . . All the properties of matter are so connected that we can scarcely imagine one *thoroughly explained* without our seeing its relation to all the others, without in fact having the explanation of all, and till we have this we cannot tell what we mean by "explaining a property," or "explaining the properties" of matter. But though this consummation may never be reached by man, the progress of science may be, I believe will be, step by step towards it, on many different roads converging towards it from all sides. The kinetic theory of gases is, as I have said, a true step on one of the roads.—*Popular Lectures*, vol. i. p. 233.

His ideas as to the nature of matter were profoundly influenced by the discoveries of Joule. From 1847 to 1855, when Thomson was working out his mathematical theory of magnetism (p. 211 above), he had (as he stated in 1872 when compiling his reprint of *Papers on Electrostatics and Magnetism*) no belief in the reality of Ampère's theory, according to which magnetism of steel or iron consists of electric currents circulating round the molecules of the magnetized substance. In the note which he appended to p. 419 of that work, he confesses, "but I did not then know that motion is the very essence of what has hitherto been called matter. At the 1847 meeting of the British Association in Oxford, I learned from Joule the dynamical theory of heat, and was forced to abandon at once many, and gradually from year to year all other, statical preconceptions regarding the ultimate causes of apparently statical phenomena."

Already in 1855 Thomson was speculating on the mechanical antecedents of motion, heat, and light, and had suggested that all of them originated in gravitation. And in his Royal Institution discourse of 1856 he suggested that all heat was electric, and all light also, although he gave no arguments in support of the suggestion. But in the same year he laid before the Royal Society certain dynamical illustrations of the rotatory effects of transparent bodies on polarized light. He had been considering Faraday's great discovery that magnetism acting on a piece of heavy-glass through which light is passing, can exert a turning effect on the plane of polarization of a beam of polarized light that is passing through it, and had compared that phenomenon with Arago's earlier discovery that in passing through a plate of quartz there is also a twisting of the plane of polarization. He now emphasized the essential difference between the two cases. If in the case of quartz the light is caused by reflexion to retraverse the quartz in the opposite direction, the rotation of the plane of polarization is annulled; whereas in the case of the magnetized heavy-glass, on reflecting the light back through it the rotation is doubled. He therefore ascribed the action in the case of quartz to a spiral or helicoidal arrangement of the molecular structure; and, in the case of the heavy-glass, to a real rotation impressed on the substance by the influence of the magnet. Not otherwise could one explain the circumstance that the rotatory effect in the magnetic case is independent of the "sense" of the propagation of the light through the substance. To make the matter plainer, he conceived certain dynamical models to show how the vibrations of material systems can be changed in azimuth by the influences of constraints and forces impressed upon them.

From these illustrations it is easy to see in an infinite variety of ways how to make structures, homogeneous when considered on a large enough scale, which, with certain rotatory motions of component parts having, in portions large enough to be sensibly homogeneous, resultant axes of momenta arrayed like lines of magnetic force, *shall have the dynamical property by which the optical phenomena of transparent bodies in the magnetic field are explained.*—*Roy. Soc. Proc.* viii. p. 155, June 1856.

He had a word, too (*ibid.* p. 152), on the probable relations between matter and ether:—

The introduction of the principle of moments of momenta into the mechanical treatment of Mr. Rankine's hypothesis of "molecular vortices" suggests the resultant moment of momenta of these motions as the definite measure of the "magnetic moment." The explanation of all phenomena of electromagnetic attraction or repulsion, and of electromagnetic induction, is to be looked for simply in the inertia and pressure of the matter of which the motions constitute heat. Whether this matter is or is not electricity, whether it is a continuous fluid interpermeating the spaces between molecular nuclei, or is itself molecularly grouped; or whether all matter is continuous, and molecular heterogeneousness consists in finite vortical or other relative motions of contiguous parts of a body, it is impossible to decide, and perhaps in vain to speculate, in the present state of science.—*Ib.* p. 152.

Occasionally he allowed himself to step forward beyond the bounds of the accepted philosophy, and to rush to intuitive conclusions for which afterwards he had to seek for logical demonstration. Witness the following passage[1] from the peroration to his Royal Institution discourse of May 1860 on "Atmospheric Electricity":—

We now look on space as full. We know that light is propagated like sound through pressure and motion. . . . If electric force depends on a residual *surface action*, a resultant of an inner tension experienced by the insulating medium, we can conceive that electricity itself is to be understood as not an accident but an essence of matter. Whatever electricity is, it seems quite certain that electricity in motion IS *heat*; and that a certain alignment of axes of revolution in this motion IS *magnetism*. Faraday's magneto-optic experiment makes this not a hypothesis but a demonstrated conclusion. Thus a rifle-bullet keeps its point foremost; Foucault's gyroscope finds the earth's axis of palpable rotation; and the magnetic needle shows that more subtle rotatory movement in matter of the earth, which we call terrestrial magnetism: all by one and the same dynamical action.

Two points in the above are worthy of note: the suggestion that electricity is an essential quality

[1] *Electrostatics and Magnetism*, pp. 224, 225.

of matter, and that confident assertion that Faraday's experiment proves magnetism to be a dynamical phenomenon akin to the kinetic rigidity of rotation.

It was at this stage that the writing of the Thomson and Tait *Treatise* came in to focus the conception that all the outlying branches of physics were essentially based on dynamics. And though in that great work the correlating notion avowedly put forward is that of the Conservation of Energy, yet it is dynamics, the science of the particular manifestation of energy when it is expended on moving masses, that dominates the treatment. What the oft-promised chapter on the "Properties of Matter" might have brought out in the way of ultimate molecular theory cannot even be conjectured. But here and there some casual phrase reveals a thought, as, for example, in § 340 (p. 275 of Vol. I.), where the authors speak of "the ultimate molecular motions constituting heat, light, and magnetism."

In October 1864 Clerk Maxwell published his famous Electromagnetic Theory of Light, in a paper read to the Royal Society of London.[1] Whether Thomson was present, or whether he made any comment at the time, is not known. What Maxwell propounded, if stated in the briefest possible terms, is this:—that the propagation of light-waves may be explained by supposing that in a plane-polarized wave the minute transverse displacements of which

[1] *Phil. Trans.* clv. p. 459, "A Dynamical Theory of the Electromagnetic Field."

the wave consists are not mechanical displacements but electrical ones, and that these are always accompanied by corresponding minute magnetic displacements, also transverse, but at right angles to the electric displacements, half the energy in each wave being electrical and half magnetic; the medium (the ether) being regarded as having a definite rigidity or electric elasticity (the reciprocal of its dielectric capacity), and as having also a definite inertia (an electromagnetic inertia due to self-induction). Precisely how Maxwell came to this view is not known. Probably it dawned upon him when considering the meaning of the ratio "$v$" (see p. 524), which Weber had shown to be the interconnecting relation between the electric and the magnetic units when reduced to an absolute basis. This velocity had been found by Weber to be $3 \cdot 1074 \times 10^{10}$ cm. per second, being almost identical with that of light; the best determinations of which had given a velocity of $2 \cdot 9992 \times 10^{10}$ cm. per second. Sir William Thomson, in March 1868, had redetermined "$v$," and found the slightly lower value of $2 \cdot 825 \times 10^{10}$. According to Clerk Maxwell, "$v$" bore the physical meaning of that velocity at which any sudden electromagnetic disturbance would be propagated in transverse waves through free space. He himself had been studying such waves, though their actual existence was as yet unknown to experimenters; and he laid down their equations on the theory of electric "displacement" in his memoir of 1864. If waves of light were observed to travel at the same particular speed at which electromagnetic waves were believed, from pure theory, to travel, might not light-waves be electromagnetic in their essential nature? For closer testing of this brilliant speculation Clerk Maxwell himself, in 1869, made a redetermination of the ratio "*v*," and found it $2 \cdot 8798 \times 10^{10}$ cm. per second. But coincidence between the value of "*v*" and that of the velocity of light was not the only argument to support the theory. The opacity of all good electric conductors (except electrolytes) and the transparency of all dielectrics afforded an additional reason in favour of the theory. Moreover, if it were true, then the refractive index of any transparent medium should be proportional to the square root of its dielectric capacity. These things appeared to be broadly true, and Maxwell's pupils and disciples were for some years busy in examining the outstanding discrepancies.

But Thomson, as mentioned on p. 879, would not bring himself to accept Maxwell's views. In the first place, he had in 1860 explicitly denied[1] that electricity as such had any definite velocity of travelling. He had shown in his early telegraphic investigations how the apparent velocity of propagation of an electric impulse along a conducting line or cable is modified by the "embarrassment" due to

[1] Article on "Velocity of Electricity," *Nichol's Cyclopædia* (second edition), 1860; "the supposed 'velocity' of transmission of electric signals is not a definite constant like the velocity of light, even when one definite substance, copper, is the transmitting medium."

It is, however, interesting to note that when reprinting this article in vol. ii. of his *Math. and Phys. Papers*, p. 137, in the year 1883, he added a reference to the chapter on the "Electromagnetic Theory of Light" in Maxwell's *Treatise*.

the charging of the surrounding dielectric (whether gutta-percha, air, or ether); and he was well acquainted with the inductive embarrassment met with in coiled circuits—the retardation due to self-induction. It is true that Wheatstone, in 1834, had experimentally found the speed of propagation of a spark-discharge along copper wires to be 288,000 miles per second, a speed distinctly greater than that of light, and that Fizeau and Gounelle, in 1850, had found the lower value of 112,000 miles per second. But Thomson's pronouncement of 1860 shows that, up to that date at least, he had no confidence that electricity had any such intrinsic velocity. Moreover, he himself regarded Maxwell's theory as a backward step, hindering, not helping the attempt to find a true dynamical explanation of the propagation of light-waves; or, as on a subsequent occasion he put it,[1] an attempt to explain *ignotum per ignotius*. That he did not greet Clerk Maxwell's magnificent hypothesis with the enthusiasm with which it was acclaimed by the younger school of British physicists, is a fact not to be ignored or explained away. Nothing could be further from truth than the unworthy suggestion that he was jealous—he who never throughout his whole career showed the least trace of any shadow of *amour-propre* in matters of scientific discovery—of the fame of his younger compeer. Strict scrutiny of all the circumstances will show that the true explanation is quite other. He was a strong man, intellectually

[1] *Baltimore Lectures*, p. 271, in the passage quoted on p. 836 *supra*.

strong, as we have seen in the matter of Carnot's theory, concerning which he had for three years held out against Joule's view, until he had been able to think out for himself the true scientific theory which brought both views together. So in the matter of Maxwell's theory he held tenaciously to his own way of regarding the fundamental relations between matter and motion, light and electricity; and while not yet prepared to enunciate any comprehensive theory of his own, held back from accepting what seemed to him a partial if not retrograde hypothesis. There is probably also another reason. He once said (p. 827) that he was not satisfied with a formula unless he could *feel* its arithmetical magnitude; at another time that he needed to interpret physically every line of a mathematical argument; and yet again (p. 835), that no theory would satisfy him until he could imagine a model of it. He found himself unable to translate into a dynamical model the abstract equations of Maxwell's theory. It cannot even be said that he actually rejected Maxwell's theory: his own words sufficiently indicate his attitude of mind. In his Presidential Address of 1871, in apparently his earliest comment on the subject, he spoke thus:—

Weber extended the practice of absolute measurement to electric currents. . . . He showed the relation between electrostatic and electromagnetic units for absolute measurement, and made the beautiful discovery that resistance, in absolute electromagnetic measure, and the reciprocal of resistance (or, as we call it, "conducting power") in electrostatic measure, are each of them a velocity. He made an elaborate and difficult series of experiments to measure the velocity which is equal to the conducting power in electrostatic measure, and at the same time equal to the resistance in electromagnetic measure, in one and the same conductor. Maxwell, in making the first advance along a road of which Faraday was the pioneer,[1] discovered that this velocity is physically related to the velocity of light, and that, on a certain hypothesis regarding the elastic medium concerned, it may be exactly equal to the velocity of light. Weber's measurement verifies approximately this equality, and stands in science *monumentum aere perennius*, celebrated as having suggested this most grand theory, and as having afforded the first quantitative test of the recondite properties of matter on which the relations of electricity and light depend. A remeasurement of Weber's critical velocity on a new plan by Maxwell himself, and the important correction of the velocity of light by Foucault's laboratory experiments, verified by astronomical observation, seem to show a still closer agreement. The most accurate possible determination of Weber's critical velocity is just now a primary object of the Association's Committee on Electric Measurement; and it is at present premature to speculate as to the closeness of the agreement between that velocity and the velocity of light.

It will be remembered that in the years 1868 to 1873 Thomson's students—King, Dickson, and M'Kichan—had been set to work in his laboratory (p. 524) on a new determination of "*v*,"

[1] This may refer to Faraday's discovery of the magnetic rotation ot the plane of polarization of light, or it may possibly be a reference to Faraday's remarkable speculation of 1846, entitled "Thoughts on Ray-vibrations," to which Maxwell in 1864 referred, saying, "The electromagnetic theory of light, as proposed by him (Faraday), is the same in substance as that which I have begun to develop in this paper, except that in 1846 there were no data to calculate the velocity of propagation" (*Phil. Trans.* clv. p. 466).

and the result gave a value of $2 \cdot 93 \times 10^{10}$ cm. per second.

But in the meantime Thomson had fallen upon Helmholtz's investigation of vortex motion, and had conceived the idea that in the vortex-ring, with its quasi-elasticity and its indestructibility (if formed in a frictionless fluid), we have the very type of the atom of matter: and he had himself (p. 517 *supra*) worked heart and soul at the vortex-atom theory. He had been searching for a reasonable basis for the kinetic theory of gases, and for exact data of the size, weight, and numbers of atoms, in the hope that the form and motion of the parts of each atom, and the distances separating them, might be calculated. He applauded the notion that the motions by which they produce heat, electricity, and light might be some day illustrated by exact geometric diagrams, and that the fundamental properties of the intervening and possibly constituent medium might be arrived at. He suggested vortices of various forms, re-entrant, columnar, and tubular, to test their physical properties. Anything would be more satisfactory than the incredible old idea of small spheres of infinite hardness and strength: they must be regarded as pieces of matter "of measurable dimensions, with shape, motion, and laws of action." A theory of light which left out the atoms, and sought to explain the light-waves by calling them electromagnetic, seemed to him indeed inadequate, however grand. And indeed Maxwell's own statement of the theory was not without difficulty. It rested on a hypothesis concerning the elastic medium, of which he himself, in his great book on *Electricity and Magnetism* of 1873, was not able to give a very satisfying account; and students of that work were greatly puzzled to attach any consistent physical meaning[1] to the conception of electrical "displacement" which lay at the root of the matter. The fact is, that Maxwell's ideas also were in a state of flux, of which the abrupt transitions and gaps in his exposition remain to testify. Even now it is not possible to give a clear dynamical statement of that which he called displacement, except by reading into it the discoveries of recent years. Thomson's reserve was in entire accord with his temperamental bias toward freeing physical conceptions from untenable hypotheses. He was straining his faculties for a comprehensive molecular theory such as had never "been even imagined" before the nineteenth century; and there could—as he himself explicitly declared in 1871—be no permanent satisfaction to his mind in explaining heat, light, elasticity, diffusion, electricity, and magnetism, when the properties of the atom itself are simply assumed. The vortex theory was to him "a finger-post,[2] pointing a

[1] Maxwell's own statement in his paper in the *Phil. Trans.*, 1864, p. 462, runs thus: "In a dielectric under the action of electromotive force we may conceive that the electricity in each molecule is so displaced that one side is rendered positively and the other negatively electrified, but that the electricity remains entirely connected with the molecule, and does not pass from one molecule to another. The effect of this action on the whole dielectric mass is to produce a general displacement of electricity in a certain direction." The modern conception of the electron enables one to form a clearer view.

[2] On several occasions he used the simile of the *finger-post* to denote hints or possible hypotheses; see pp. 885, 1031, and 1034.

way which may possibly lead to a full understanding of the properties of atoms."

In the same year, 1871, he discoursed to the Royal Society of Edinburgh on the ultramundane corpuscles by which Le Sage had proposed to explain the existence of gravitation, and he connected with that theory certain propositions about the motion of rigid solids in a liquid circulating irrotationally through perforations in them, from which he deduced the gravitational law of inverse squares. After quoting Le Sage's propositions, Thomson remarks that this deduction would be "a perfectly obvious consequence of the assumptions were the gravific corpuscles sufficiently small." Then he continues (*Proc. Roy. Soc. Edin.* vii. p. 588):—

All that is necessary to complete Le Sage's theory of gravity, in accordance with modern science, is to assume that the ratio of the whole energy of the corpuscles to the translational part of their energy is greater, on the average, after collisions with mundane matter than after inter-collisions of any ultramundane corpuscles. This suggestion is neither more nor less questionable than that of Clausius for gases, which is now admitted as one of the generally recognized truths of science. The corpuscular theory of gravity is no more difficult in allowance of its fundamental assumptions than the kinetic theory of gases as at present received; and it is more complete, inasmuch as, from fundamental assumptions of an extremely simple character, it explains all the known phenomena of its subject, which cannot be said of the kinetic theory of gases so far as it has hitherto advanced.

In 1872 he returned to a favourite hydrokinetic

analogy[1] of 1847, in which he had shown that the velocity of fluid along its stream lines may represent from point to point the magnetic forces produced by a system of electric currents distributed on a surface.

In 1875, while under the spell of his gyrostatic studies, he investigated the effect of rotating masses on the properties of any mechanical system in which they might form a part, studying, for example, the propagation of waves along a stretched uniform chain of gyrostats; doubtless bearing in mind all the while the possible explanation it might afford of elasticity in general, or of the propagation of magnetic waves through a medium containing magnetic molecules.

His next pronouncement on Maxwell's theory was in his South Kensington address (see p. 668) on "Electrical Measurement":—

> . . . Professor Clerk Maxwell gave a theory leading towards a dynamical theory of magnetism, part of which suggested to him that the velocity for which the one measure is equal to the other, in the manner I have explained, should be the velocity of light. This brilliant suggestion has attracted great attention, and has become an object of intense interest, not merely for the sake of accurate electromagnetic and electrostatic measuring—the measuring with great accuracy the relation between electrostatic and electromagnetic units—but also in connection with physical theory. It seems, up to the present time, that the more accurate such an experiment becomes, the more nearly does the result approach to being equal to the velocity of light, but still we must hold opinion in reserve before we can say that. The result has to be

[1] See p. 206, his paper on the Electric Currents by which the Phenomena of Terrestrial Magnetism may be produced.

much closer than has been shown by the experiments already made before the suggestion can be accepted.—*Popular Lectures*, vol. i. pp. 442-443.

His Royal Institution discourse of March 1881 (see p. 743), when he exhibited his experiments on the quasi-elasticity of rotating systems, gave him occasion to remark how distant seemed the prospect of any such great comprehensive theory as that he sought.

. . . May not the elasticity of every ultimate atom of matter be thus explained? But this kinetic theory of matter is a dream, and can be nothing else, until it can explain chemical affinity, electricity, magnetism, gravitation, and the inertia of masses (that is, crowds) of vortices.

Le Sage's theory might give an explanation of gravity and of its relation *to inertia of masses*, on the vortex theory, were it not for the essential aeolotropy of crystals, and the seemingly perfect isotropy of gravity. No finger-post pointing towards a way that can possibly lead to a surmounting of this difficulty, or a turning of its flank, has been discovered, or imagined as discoverable. Belief that no other theory of matter is possible is the only ground for anticipating that there is in store for the world another beautiful book to be called *Elasticity, a Mode of Motion*.—*Popular Lectures*, vol. i. p. 145.

When in 1883 Thomson was lecturing on the size of atoms at the Royal Institution he touched upon the argument afforded by the dispersion of light in refracting media, such as glass or water, that the molecular structure cannot be infinitely small compared with the wave-lengths of light, an argument which Cauchy first enunciated as a dynamical theory of the prismatic colours. This theory he had examined and found inadequate to explain the large amount of dispersion actually observed in carbon bisulphide and in dense flint glass. Early in the very morning of his lecture he conceived another explanation—the same which he afterwards amplified in the *Baltimore Lectures* (see p. 817):—

> We must then find another explanation of dispersion; and I believe there is another explanation. I believe that, while giving up Cauchy's unmodified theory of dispersion, we shall find that the same general principle is applicable, and that by imagining each molecule to be loaded in a certain definite way by elastic connection with heavier matter—each molecule of the ether to have, in palpable transparent matter, a small fringe so to speak of particles, larger and larger in their successive order, elastically connected with it—we shall have a rude mechanical explanation, realisable by the notably easy addition of the proper appliances to the dynamical models before you, to account for refractive dispersion in an infinitely fine-grained structure. It is not seventeen hours since I saw the possibility of this explanation. I think I now see it perfectly, but you will excuse my not going into the theory more fully under the circumstances. —*Popular Lectures*, vol. i. pp. 194-195.

He gave a mathematical account of this dynamical theory[1] of dispersion a month later to the Royal Society of Edinburgh.

"Steps toward a Kinetic Theory of Matter" was the title of Sir William Thomson's Montreal address in 1884. This was largely an account of the kinetic theory of gases, as developed up to that date, and in the course of it he touched on the

[1] It is substantially the same as that outlined by Sellmeier (*Pogg. Ann.* cxliii. 1872), and extended by Helmholtz (*ibid.* cliv. 1875).

difficulty arising as soon as one began to consider what happened during the collision or impact between two molecules of a gas. In any attempt to realise a conception of the kinetic theory, one cannot evade a question as to the nature of the force during an impact.

. . . And in fact, unless we are satisfied to imagine the atoms of a gas as mathematical points endowed with inertia, and as, according to Boscovich, endowed with forces of mutual positive and negative attraction, varying according to some definite function of the distance, we cannot avoid the question of impacts, and of vibrations and rotations of the molecules resulting from impacts, and we must look distinctly on each molecule as being either a little elastic solid, or a configuration of motion in a continuous all-pervading liquid. I do not myself see how we can ever permanently rest anywhere short of this last view; but it would be a very pleasant temporary resting-place on the way to it, if we could, as it were, make a mechanical model of a gas out of little pieces of round, perfectly elastic solid matter, flying about through the space occupied by the gas, and colliding with one another, and against the sides of the containing vessel. This is, in fact, all we have of kinetic theory of gases up to the present time, and this has done for us, in the hands of Clausius and Maxwell, the great things which constitute our first step towards a molecular theory of matter. Of course from it we should have to go on to find an explanation of the elasticity and all the other properties of the molecules themselves, a subject vastly more complex and difficult than the gaseous properties for the explanation of which we assume the elastic molecule; but without any explanation of the properties of the molecule itself.—*Popular Lectures*, vol. i. pp. 228-229

But a more serious difficulty, which he had to leave unsolved, was the certainty—he thought it "rigorously demonstrable"—that the whole of the translational energy of the flying molecules, if each is a continuous elastic solid, must ultimately be frittered down into vibrational energy, and therefore become lost by radiation. So he turned to the question of elasticity.

> If we could make out of matter devoid of elasticity a combined system of relatively moving parts which, in virtue of motion, has the essential characteristics of an elastic body, this would surely be, if not positively a step in the kinetic theory of matter, *at least a finger-post pointing a way* which we may hope will lead to a kinetic theory of matter.—*Ibid.* i. p. 235.

This, he continued, we can do in several ways. And he here referred to his own published papers on vortex atoms, and on elasticity as possibly a mode of motion (see p. 743), and to his model spring-balance containing gyrostats (p. 745), and to his model medium imitating the luminiferous ether, built up with portions possessing kinetic stability, due either to gyrostats or to inviscid fluids with "irrotational circulation" through the pores of solids. He imagined in this way a model vortex gas, and another hydrokinetic model composed of vortices in a pure liquid. In the latter case the difficulty about impacts disappeared; and, moreover, so far as he was then able to ascertain concerning the vibration of vortices, there no longer seemed any danger of the translational or impulsive energies of the individual vortices becoming lost or frittered away in energy of smaller and smaller vibrations.

The lecture on the Wave Theory of Light, which Sir William Thomson delivered in Philadelphia on his way to Baltimore, abounds in hints and references to the dynamical explanation of the properties of matter. A few isolated extracts must here suffice. He was concerned with the relation between the luminiferous ether and the molecules embedded in it.

. . . You can imagine particles of something, the thing whose motion constitutes light. This thing we call the luminiferous ether. That is the only substance we are confident of in dynamics. One thing we are sure of, and that is the reality and substantiality of the luminiferous ether. . . . *Popular Lectures*, vol. i. p. 310

. . . So when we explain the nature of electricity, we explain it by a motion of the luminiferous ether. We cannot say that it is electricity. What can this luminiferous ether be? It is something that the planets move through with the greatest ease. . . .—*Ibid.* p. 327.

. . . The fundamental question as to whether or not luminiferous ether has gravity has not been answered. We have no knowledge that the luminiferous ether is attracted by gravity; it is sometimes called imponderable, because some people vainly imagine that it has no weight. I call it matter with the same kind of rigidity that this elastic jelly has.—*Ibid.* p. 329.

In Chapter XX., on the *Baltimore Lectures*, p. 835 above, the passage has been cited in which Lord Kelvin gave as a reason why he could not accept the electromagnetic theory of light, that he could not make a model of it. But a probably more important reason was that Maxwell's theory was essentially molar, not molecular,[1] and did not touch

[1] Maxwell, in his *Treatise*, § 260, vol. i. p. 312, once indeed uses the phrase, "one molecule of electricity," but immediately excuses himself for this lapse

molecular questions, which Lord Kelvin deemed absolutely fundamental. Maxwell had assumed that the mechanical and molecular properties of a material system are independent of one another, and capable of being separately regarded; this Lord Kelvin implicitly denied. And so he went his own way, striving to find other explanations, other modes of handling the problem of the propagation of light which should be essentially founded[1] on molecular dynamics. But it can scarcely be denied that these lectures give the reader the impression of a powerful mind struggling with its own convictions, and hampered by not finding the solution which he was convinced would be ultimately triumphant. No sooner were the lectures delivered than he began to work at supplements to them, and to revise them for a definitive publication.

The first fifteen lectures were not materially altered in the revision, though portions were transferred to them from the later lectures; but from Lecture XVI. to the end they were almost entirely rewritten in the years 1901 to 1903, and their paragraphs numbered consecutively from this point onwards. In Lecture XVII. he quoted from his 1870 article on the size of atoms (p. 566), a difficulty then felt in the kinetic theory of gases, as to the

by adding: "This phrase, gross as it is, *and out of harmony with the rest of this treatise*, will enable us at least to state clearly what is known about electrolysis." The words here italicized amply justify—were justification needed—the reserve which Lord Kelvin felt and expressed.

[1] "It is absolutely certain," he said in Lecture XII., that there is a definite dynamical theory for waves of light, to be enriched, not abolished, by electromagnetic theory."—*Baltimore Lectures* (1904), p. 159.

hard and fast demarcation between infinite force between the atoms when in contact, and zero force when not in contact, arising from the assumption of rigid, inelastic atoms. This hypothesis had seemed to require mitigation. He now wrote that Boscovich's theory clearly supplies the needed mitigation, and proceeded to explain the "Boscovich atoms," which exercise a mutual repulsion at small distances between centres, and an attraction when distances between centres exceed a definite limit. Lecture XVIII., as revised, was devoted wholly to the reflexion of light, and to the difficulty of accounting, by the theories of Fresnel and of Green, for the almost perfect extinction of polarized light when reflected at the critical angle. He referred to his own research of 1888 (p. 872), in which he found that a medium resembling a homogeneous airless foam, held from collapse by adhesion to a containing vessel, would possess a definite rigidity and elasticity of form, and a definite velocity of distortional wave, while fulfilling the condition that had puzzled him so much in 1884 of having a zero velocity for any compressional wave. The difficulty of accounting for metallic and adamantine reflexion, and the mathematical artifice for meeting that difficulty, are considered at length. In Lecture XIX. the reconciliation of the theory of Fresnel with that of Green is effected, by accepting, in order to explain double-refraction, the hypothesis that the inertia of a crystal may be aeolotropic, that is, have different values in different directions. Lecture

XX. was entirely reconstructed by the introduction of his own newer conceptions as to atoms and ether occupying the same space at the same time, and as to electric forces within the molecule, as will presently be seen.

The difficulty which Lord Kelvin felt respecting Maxwell's doctrine of dielectric "displacement" was frequently mentioned by him. It is referred to in the following letter to FitzGerald, who in March 1885 had suggested to the Physical Society a dynamical model of an "ether," in which it was proposed to represent an electrical displacement by a change of structure, and not by a mere shifting of an element of the medium.

*17th April*, 1885.

DEAR FITZGERALD—. . . Any one reading your paper as it stands would imagine that I had ignored a definite velocity of propagation for electric impulses. How wrong this inference would be is illustrated by p. 134 line 9 from foot. I think in any case you ought to add to what you have already quoted from my paper, the last sentence of page 134 and the first of page 135 containing my reference to Kirchhoff.

I have never yet felt any satisfaction in Maxwell's §§ 783, 784, 790, 791, 792, 645, 646, 794, 797, 798, 824 . . . 829. I have never yet met any one who understood a definite dynamical foundation for § 783.

Is there any chance of your being in London or Cambridge during May? We are to be in London for about the first half of the month, and in Cambridge staying with the Stokes's probably during Whitsun week. I shall be very glad if we have an opportunity of meeting so as to fight or agree relatively to the waves of light. —Believe me, yours very truly,

WILLIAM THOMSON.

In 1887 Sir William Thomson himself enunciated a Vortex Theory of Luminiferous Ether,[1] in which he gave pictures of assemblages of vortex-rings to show how they would resist distortion, and so serve as a medium for the propagation of waves. His aim was "to construct by given vortex-motion of an incompressible inviscid liquid, a medium which shall transmit waves of laminar motion as the luminiferous ether transmits waves of light."

In the summer of 1888, while revising the *Baltimore Lectures*, Sir William was still perplexed and haunted by the difficulty of finding an explanation for the propagation of electrostatic forces. His difficulty appears in a post-card which he addressed to Lord Rayleigh on August 12, 1888, from Glasgow, where he had gone on the occasion of a visit of Queen Victoria.

G.C. FOR QUEEN,
*Aug.* 22.

Given two conducting globes oppositely charged (equal quant^y) in blue sky. Let the blue suddenly become conductive; say as conductive as slate or as marble. The discharge will be along the lines of the previous electrostatic force, and $\epsilon^{-tv^2/\kappa}$ will be the time law of subsidence, it being slow enough for no quasi-inertia. Where are the "displacement" currents and where the possibility of expressing the result (with or without quasi-inertia) by any even pseudo circuits, or imagined analogues of incomplete circuits? Alas—alas, the whole thing breaks down the first time it is really put on trial. H. Lamb did find it to fail for such an elementary thing as elect^y flowing to equil^m, but did not notice that this

[1] *Brit. Assoc. Report*, 1887; *Philos. Magazine*, Oct. 1887, . . .

failure shows that the formulas he quoted in the beginning are not "*the* equa$^{ns}$ of electromag$^{c}$ induction." He also (p. 523, ll. 5 and 6 from foot) found $2+2 \neq 4$, and "this peculiarity of Maxwell's has been pointed out by C. Niven." W. T.

It is, however, possible to frame an explanation of the case in accordance with Maxwell's theory, since the subsidence of the electric field will not involve any quasi-inertia, except perhaps inter-molecular, of which Maxwell's theory takes no cognisance.

A fortnight later came the British Association meeting at Bath, where, in the Physical Section under Professor FitzGerald, there was a great discussion upon electromagnetic matters. Within the preceding twelve months Hertz had published the classical researches on electric waves by which he had experimentally realized that which was essentially true in Maxwell's doctrine; and had shown that electric waves can be reflected and refracted exactly as light waves can, and that they travel at the same speed. Rowland of Baltimore was present, and Oliver Lodge (now Sir Oliver). Sir William Thomson was announced to give two papers, one of which at least was directed to electromagnetic questions. He also had, in conjunction with Professors Ayrton and Perry, prepared a new determination[1] of the ratio "*v*."

An excellent account of this discussion was written by Lodge for *The Electrician* (vol. xxi.

[1] Compare pp. 524 and 1022. The result now obtained from measurements of the capacity of a condenser was $2 \cdot 93 \times 10^{10}$ cm. per second. Rowland by a kindred method had found $2 \cdot 995 \times 10^{10}$.

p. 622, Sept. 21, 1888), from which it appears that, without abandoning the solid-elastic theory of light (which in some form or other seems a necessity), Sir William Thomson was assimilating, while still criticizing, Maxwell's theory, and rejecting certain of its excrescences. The main point under discussion related to the mode and rate of propagation of electrostatic potential as compared with electromagnetic potential. Lodge's interesting report contains the following passage :—

. . . On the last day of the meeting Rowland and FitzGerald seemed to come to an understanding, but whether Sir William will coincide with it, or will upset the whole thing once more, remains to be seen. As they put it the matter is now like this : The propagation of electrostatic potential does not go on by end-thrust at all ; it is not really analogous to a pulse of longitudinal compression, though it is apparently and superficially so ; and accordingly its rate of propagation depends not at all on the compressibility or incompressibility of the ether, a question on which it has nothing to say one way or the other. An electrostatic field is not developed *sui generis*, but is always the consequence of a previously existing electromagnetic one, which, on subsiding, leaves it as its permanent record.

Plainly an electrostatic field cannot arise without the motion of some electricity, either with or through a conductor. Now, whenever electricity moves it at once has magnetic properties—its motion generates a magnetic field. When the motion ceases the field at once subsides, and in subsiding it may produce a succession of diffusing and dying away induction currents in neighbouring conductors ; or it may, if the circuit be an incomplete one, leave a permanent vestige of itself in the dielectric as a field of strained ether—this state of strain being what we call electrostatic potential, and the field being familiar to us as an electrostatic field.

Generating it[1] in this way, all distinction between rate of propagation of electrostatic and electromagnetic potential vanishes, they both travel together with the velocity of light; or rather, the thing which travels is the magnetic potential, and its permanent effect *in situ* is the electrostatic potential.

Thus, once more, the difficulty of a longitudinal or pressural wave disappears from the electrical theory of light, into which it had seemed to intrude itself, and $\psi$ is left to enjoy "a long and useful career," though it is not permitted an infinite or any other rate of propagation in its own proper nature. If any one asks how soon will the pull of a suddenly electrified body be felt at a distance? one may answer, "As soon as the charging spark is seen." But if it be asked at what rate electrostatic potential travels, the answer is that it does not travel, but is generated *in situ* by the subsidence of a magnetic potential which travels with the velocity of light.

In November 1888, Sir William communicated to the *Philosophical Magazine* a paper "On the Reflexion and Refraction of Light," in which once more he returned to Green's solid-elastic theory, endeavouring to argue backwards from the facts of the propagation of polarized light to the physical properties which a medium must possess in order to account for the known facts. This important paper has never been reprinted.

As told on p. 881, Sir William undertook the duties of President of the Institution of Electrical

[1] *I.e.* an electrostatic field. Sir Oliver Lodge informs the author that he would now (1909) restate the matter somewhat in the following way: Of the three vectors at right angles to one another—electric force, magnetic force, and motion—the superposition of any two necessarily involves the third. Given, therefore, electric and magnetic induction, motion at once follows, and the wave spreads out with the speed of light. A wave is generated by every electric acceleration, and therefore by a separation of electric charges. Whether any permanent result is left in the region over which the wave has travelled, depends upon whether the electric separation is permanent or not.

Engineers for the year 1889, and wrote to the Secretary of that body as follows :—

*4th Jany.* 1889.

DEAR MR. WEBB—I am sorry it will be impossible for me to have anything of my presidential address in writing. The scientific subject upon which I propose to speak is "Electricity, Ether, and Ponderable Matter," and I hope to show one or at the most two very simple experiments with my gyrostats to illustrate rotating molecules in connection with dynamical theory for magnetism. But I shall have to state very strongly that the difficulties in the way of proving a comprehensive dynamical theory of electricity, magnetism, and light are quite stupendous; and that in the present state of science the imagination is absolutely baffled in attempts to give a mechanical foundation for explaining great laws of nature on which the work of the electrical engineer depends.

I shall call about 11 o'clock on Thursday to arrange about the experiments, for which very little of apparatus, beyond the gyrostats which I shall bring with me, will be wanted. Perhaps one of your young men might be able to assist me in preparing the experiments during the course of the lecture. Whatever record of the address is wanted will depend solely on the short-hand writer.

I should be much obliged by your letting me have in the course of a day or two any statement which it is desirable I should read with reference to the Society under its late title, and with reference to its future existence under the more appropriately comprehensive name now chosen for it.

Reciprocating heartily your kind wishes for a Happy New Year.—I remain, yours truly, WILLIAM THOMSON.

There are several salient features in this address.[1] In one of these he touched on the velocity "$v$," in the following words :—

---

[1] *Journal of Institution of Electrical Engineers*, vol. xviii. p. 4, 1889; partly reprinted in *Math. and Phys. Papers*, vol. iii. p. 484.

But its relationship to the velocity of light was brought out in a manner by Maxwell to make it part of a theory, which it never was before. Maxwell pointed out its application to the possible or probable explanation of electric effects by the influence of a medium, and showed that that medium—the medium whose motions constitute light — must be ether. Maxwell's "electromagnetic theory of light" marks a stage of enormous importance in electromagnetic doctrine, and I cannot doubt but that in electromagnetic practice we shall derive great benefit from a pursuing of the theoretical ideas suggested by such considerations.—*Math. and Phys. Papers*, vol. iii. 490.

Then he returned to the solid-elastic theory that had occupied his thoughts in 1846 (p. 197), and to the difficulty he had found in the forty-two years that had since gone by, in finding any further explanation. He touched on his recent hydrokinetic model of the porous elastic solid of small density having its pores filled with a dense viscous fluid circulating through it, which would realize dynamically the problems of electromagnetic induction, provided that nothing but electricity and ether had to be accounted for. But ponderable matter must be considered also, and there was the mystery that solid bodies could move freely through the ether. So he proceeded to describe an imaginary model ether or medium which should be absolutely mobile to translational movements, and yet possess rotational rigidity. To this end he imagined a sort of network, across the meshes of which were set minute gyrostats spinning each about its own axis, to give the structure the necessary immobility.

Thus we have a skeleton model of a special elastic

solid, with a structure essentially involving a gyrostatic contribution to rigidity. Now do not imagine that a structure of this kind, gross as it is, is necessarily uninstructive. Look at the structures of living things; think of all we have to explain in electricity and magnetism; and allow, at least, that there must be some kind of structure in the ultimate molecules of conductors, nonconductors, magnetic bodies, and non-magnetic bodies, by which their wonderful properties now known to us, but not explained, are to be explained. We cannot suppose all dead matter to be without form or void, and without any structure; its molecules must have some shape; they must have some relation to one another.

So that I do not admit that it is merely playing at theory, but it is helping our minds to think of possibilities, if by a model, however rough and impracticable, we show that a structure can be produced which is an incompressible frictionless liquid when no gyrostatic arrangement is in it, and which acquires a peculiar rotational elasticity or rigidity as the effect of introducing the gyrostats into these squares (*ibid.* p. 508).

In brief, he had abandoned the solid-elastic theory, and substituted a new one, in which he figured to himself a mathematical ether not corresponding to any known physical material, which derived its properties from concealed internal rotations, and which, while absolutely labile, in the sense that it was incapable of propagating a compressional wave, possessed a torsional rigidity. This in fact corresponded exactly[1] to the optical ether suggested in

[1] That MacCullagh's equations for the reflexion and refraction of light may be interpreted on the supposition that what is resisted is not deformation but rotation, was shown by FitzGerald in 1880, and more fully developed by Larmor in 1893 in his great paper, "A Dynamical Theory of the Electric and Luminiferous Medium," in the *Phil. Trans.* vol. clxxxv. (Series A), p. 723; wherein also it is shown that at any point in the medium electrical displacement in the electromagnetic theory corresponds to absolute rotation on MacCullagh's theory, and magnetic force corresponds similarly to velocity.

1829 by MacCullagh, and advocated in 1850 by Rankine, but rejected at the time precisely because it did *not* behave as an elastic solid. This new hypothesis he expanded in 1889 in papers read to the Académie des Sciences and to the Royal Society of Edinburgh, "On a Gyrostatic Adynamic Constitution for Ether."

But even this brilliant suggestion did not satisfy Sir William Thomson. He had turned the optical difficulty and had reconciled Green and Fresnel with truth; but the existence of electrostatic force was still unexplained, likewise the mutual attraction between the iron of an electromagnet and its keeper; nor had he solved the puzzle that the ether must permit solid bodies to move with perfect freedom through it. So when he ended his address to the Electrical Engineers in 1889 he had still to confess that "the difficulties are so great in the way of forming anything like a great comprehensive theory, that we cannot even imagine a finger-post pointing a way that can lead towards the explanation."

But now another turn was coming in the evolution of the theory, by the slowly-forming decision[1] that the vortex theory of atoms, at least in the form in which it had been enunciated, must be abandoned.

[1] In the late autumn of 1883 when Sir William Thomson lectured in Newcastle on atoms and vortex-rings, he and Lady Thomson were the guests of Dr. J. Theodore Merz, from whom he received a copy of the then newly-published Adams Prize Essay of Professor (now Sir Joseph J.) Thomson on Vortex Atom Rings. He travelled back to Glasgow with Dr. Harvey Goodwin, Bishop of Carlisle, with whom he keenly discussed the work. It was some three years later, when revisiting Newcastle to see the early form of Parsons's steam-turbine, that he told Dr. Merz that the vortex-atom theory did not realize his expectations, inasmuch as it did not explain inertia or gravitation.

Lord Kelvin has himself told us of this conclusion in the following words:[1]—

It now seems to me certain that if any motion be given within a finite portion of an infinite incompressible liquid originally at rest, its fate is necessarily dissipation to infinite distances with infinitely small velocities everywhere; while the total kinetic energy remains constant. *After many years of failure to prove that the motion in the ordinary Helmholtz circular ring is stable, I came to the conclusion that it is essentially unstable, and that its fate must be to become dissipated as now described.* I came to this conclusion by extensions not hitherto published of the considerations described in a short paper entitled: "On the Stability of Steady and Periodic Fluid Motion," in the *Phil. Mag.* for May 1887.

He had, however, a gleam of light. Still brooding over his fateful paper of 1847, and the possible mechanical representations which he had then "reserved for a future paper," he began over again to consider an ideal ether which, though incompressible, should have no distortional rigidity, but only an inherent quasi-elastic (gyrostatic) resistance to rotation. He found it to furnish a perfect substitute for the ordinary incompressible elastic solid as to equilibrium and motion throughout the interior, but having a vital difference in respect of the action at any interface between two portions

[1] *Proc. Roy. Soc. Edin.* vol. xxv. p. 565, footnote, June 20, 1904. The words now italicized in the quotation must be regarded as Lord Kelvin's renunciation of his hypothesis of the vortex-atom. Writing in 1898 to Professor Silas W. Holman, Lord Kelvin had said: "I am afraid that it is not possible to explain all the properties of matter by the vortex-atom theory alone, that is to say, merely by motion of an incompressible fluid; and I have not found it helpful in respect to crystalline configurations, or electrical, chemical, or gravitational forces. . . . We may expect the time will come when we shall understand the nature of an atom. With great regret I abandon the idea that a mere configuration of motion suffices."

having different rigidities of the kind considered. And now he was able, by aid of the idea of a circuital displacement of a portion forming in itself a closed curve, to represent magnetic force by the tangential drag on the surrounding medium. If such displacements were periodic in time, the kinetic result would be a propagation of magnetic waves. "We have thus simply *the undulatory theory of light*, as an inevitable consequence of believing that the displacement of elastic solid, by which in my old paper I gave merely a '*representation*' of the electric currents and the corresponding magnetic forces, is a reality." But even so, the theory was incomplete. He had won a step forward by supposing all space filled with a *plenum* endowed with mechanical properties, it is true, but not with those of an ordinary elastic solid or an ordinary incompressible fluid; an ideal ether, satisfactory for optics at least, but unsatisfactory as a basis for an all-embracing theory. It did not represent electrostatic forces; it did not explain why currents heat their conductors; it did not prove that the velocity of light in ether *is* "*v*." How then did Thomson at this date, 1890, propose to escape from the intellectual *impasse*? Let his own words tell:—

All this essentially involves the consideration of ponderable matter permeated by, or embedded in ether, and a *tertium quid*, which we may call electricity, a fluid go-between, serving to transmit force between ponderable matter and ether, and to cause by its flow the molecular motions of ponderable matter which we call heat. I see no way of suggesting properties of matter, of electricity or of ether, by which all this, or any more than a very slight approach to it, can be done, and I think we must feel at present that the triple alliance, ether, electricity, and ponderable matter is rather a result of our want of knowledge and of capacity to imagine beyond the limited present horizon of physical science, than a reality of nature.—*Math. and Phys. Papers*, vol. iii. p. 465.

Note, in this remarkable pronouncement, the implication of the failure of the vortex-atom theory to explain—apart from a something "which we may call electricity"—the properties of matter. He knew—none better than he—that electricity whirling around generated magnetic force along the axis of its spin. If his supposed ether were not the same thing as free electricity, its whirls would not act magnetically; and if it were not the same thing, how explain magnetism? His suggestion now is: *bring in electricity* as "a fluid go-between, serving to transmit force" between matter and ether.

"Looking back on this, however," observes Sir Joseph Larmor,[1] "one can see that the electron was not far off." Nevertheless the time was not come; and so with this suggestive forecast the great comprehensive theory goes out of sight for ten years.

But while the search for a dynamical mechanism which should explain light, elasticity, gravity, electricity, and magnetism was thus in abeyance, Thomson broke out into new activities. In connection with his suggestion of a foam-like structure for ether, in the winter of 1887-88, he had considered how space

[1] Obituary notice of Lord Kelvin, *Proc. Roy. Soc.* vol. lxxxi. (Series A), p. 68.

might be so divided out into a system of adjacent cells that the total partitional area between cells of a given volume should be a minimum; and, guided by experiments[1] on soap bubbles, he discovered, apparently to his surprise, that the form which fulfilled this condition was neither the cube, the octahedron, nor yet the rhombic dodekahedron, but the tetrakaidekahedron, with eight of its faces hexagons and six of them squares, resembling a cube with its six corners truncated, and with all the edges slightly curved. His "green books" for many months from this time are full of sketches and calculations respecting the geometrical and physical properties of assemblages of cells, or units, of this and other shapes. Deeper and deeper he dived into the matter, trying to assimilate all that was known of the systems of crystallography, of planes of cleavage, of twinning, of the integrating molecules of Haüy, of the slipping-planes of Reusch, of the nets of points imagined by Bravais, in fact all that pertained to the physical properties of matter in the crystalline state.

Then, in July 1889, he read to the Royal Society of Edinburgh a paper,[2] destined, surely, to become a classic in physics, entitled "The Molecular Constitution of Matter." He began by a definition of crystalline structure, declaring that any homogeneous isotropic solid, such as glass, was but an isotropically macled crystal. Then he considered a

[1] See the narrative, p. 870, *supra*.
[2] Reprinted in *Math. and Phys. Papers*, vol. iii. p. 395, 1890.

homogeneous assemblage of points, each surrounded by a cell, all space being partitioned out into assemblages of cells, each cell a polyhedron of interfaces. Turning next to Father Boscovich's theory, that the ultimate atoms of matter are points endowed each with inertia, and with mutual repulsions or attractions, dependent only on mutual distances, he discussed the question of the equilibrium and stability of groups of two, three, four, or more atoms. This gave a basis for a Boscovichian kinetic theory of crystals, of liquids, and of gases. In a mere irregular random crowd of molecules it would be difficult to imagine equilibrium, either static or kinetic. Such a crowd might be a liquid — but scarcely a solid. The molecular tactics of crystals depended on certain geometric principles of arrangements which Bravais had laid down on the doctrine of homogeneous assemblages, in which planes of symmetrically distributed points were regarded as *réseaux* or networks. Problems of closest packing in assemblages of globes, like piled shot, or of ellipsoids, were readily treated geometrically. The artificial twinning of Iceland-spar, and the allied phenomena of changes of crystalline structure by pressure, could be accounted for by supposing slip to occur along planes of assemblage. Mathematical investigations of the conditions of equilibrium in such assemblages were reserved; but models were described and shown, to realize elastic solids, both incompressible and compressible, by bows of bent steel wire linked together by rings, and supported by tie-struts.

At the British Association Meeting at Newcastle in September a further paper on the same subject was given by Sir William. Of this a contemporary account by Lodge appeared in *The Electrician* (vol. xxiii. p. 544).

The other communication I have referred to, that on Boscovich's theory, might have been called the Mechanics of Crystallography, or Molecular Statics. It was a powerful and acute piece of pure mechanical reasoning, showing how a great number of the extraordinary behaviours of crystals, including some only recently discovered, and some which at first hearing seem quite incredible, can be accounted for by imagining the molecules built up according to a certain law of force between the parts, a force attractive at some distances, repulsive at others, and not specially contemplating or troubling about how the forces arise, nor of what the substance of the molecules is composed — treating them, in fact, as abstract centres of force (which they certainly are, however much else they also be) and ignoring their inertia, very much as Boscovich in his century-old memoir did. Crystals of various orders, hemihedral crystals, crystals with a periodic structure, and consequent curious selective reflexion for light, crystals that can be sheared, and molecules that are liable to fly to pieces and detonate, were all referred to, and models built up of spiral springs were used to illustrate them.

Further considerations with a Boscovichian explanation of the modulus of elasticity were given in February 1890 to the Royal Society of Edinburgh. Later, he wrote to Lord Rayleigh of matters in his mind:—

THE UNIVERSITY, GLASGOW, *Nov.* 18/90.

DEAR LORD RAYLEIGH. . . .—Did you ever remark that if the density of ether changes as it moves across the interface between glass and vacuum, or between any two transparent bodies, in its actual vibrations of light, and if it is otherwise uninfluenced by ponderable matter, and if the rigidity is equal on the two sides of the interface, we have exactly Fresnel's laws of polarization by refraction and reflection?

And if besides we have, for ether within the space of a ponderable body, a resistance to its motion relative to the ponderable matter, equal to the velocity multiplied by a coefficient which for different metallic bodies is proportional to the density of the ether in them, and for transparent non-conductors is zero, we have Maxwell's equations of electromagnetic induction. This makes the electric conductivity of a metal proportional to the density of the ether in it!—Yours truly,

WILLIAM THOMSON.

Three years passed, and then came in May 1893 the Royal Institution discourse on Isoperimetrical Problems; illustrated by the problem of Queen Dido of Carthage, how to enclose the largest area of territory by an ox-hide cut into an exceedingly long strip, and by the problem of Horatius Cocles, whose patriotism was rewarded with a grant of as much land as he could plough round in a day. In Lord Kelvin's hands these two problems of classical antiquity furnished the text for a discourse on Lagrange's Calculus of Variations, and Hamilton's principle of Least Action, leading to the geodetic exercise how to draw the shortest possible line between two given points on a curved surface. Pappus, who praised the honey-bee for its knowledge of the geometrical truth that a hexagon can enclose more honey than a square or a triangle with equal quantities of building material in the walls, had originated the term *isoperimetrical*, which now serves to denote that large province of mathematical and engineering science in which different figures having equal perimeters, or different paths between two given points, are compared in connection with definite questions of greatest efficiency or least cost, extending even to such questions as the proper laying-out of a railway line in a hilly country, or the dynamical relations between curvature and centrifugal force in the motion of a particle.[1] A few days later this was followed by the delivery, to the Oxford University Junior Scientific Club, of the Robert Boyle Lecture[2] for the year 1893, on the Molecular Tactics of a Crystal. Under this title the geometry of crystalline structure was portrayed in great detail, with discussion of the boundaries of the partitioning cells[3] drawn around the integrating molecules. The most general form of cell was not the parallelepiped but the tetrakaidekahedron. The closest packing of spheres or ellipsoids in a heap left but three degrees of freedom of strain, instead of the six presented by a natural solid; and any distortion of the mass broke some of the contacts and brought about an expansion of it. This was in effect the dilatancy observed by Osborne Reynolds in sacks filled with corn, sand, or small shot.

[1] "I hope you found my lecture interesting," he said afterwards to Lord Alverstone, who was one of the audience. "I am sure we should, my dear Lord Kelvin," was the reply, "if we had understood it."

[2] Reprinted in *Baltimore Lectures* (1904), Appendix H, p. 602.

[3] He amplified this part of the subject in a Royal Society paper in January 1894, "On the Homogeneous Division of Space"; wherein the reader will find discussions of the transformation of the parallelopiped into the tetrakaidekahedron, and of the relation of both to the tetrahedron. Incidentally it contains also the theory of the designing of wall-paper patterns.

Passing on to the subject of twinning, attention was drawn to the beautiful phenomena of internal coloured iridescent reflecting planes observed in some crystals of chlorate of potash, and due to a periodically-twinned structure. A reference to the "chirality" of quartz crystals—according to which some of them are right-handed and others left-handed in their growth—and a suggested explanation of the same, completed the discourse. A mathematical discussion of the elasticity of a crystal, according to Boscovich, presented to the Royal Society in June and July 1893, carried the subject a little further; and at the British Association meeting in the following September a further extension to elucidate from geometry of structure the pyro-electric and piezo-electric properties of quartz and tourmaline was made.

In 1893 was published a translation by Professor D. E. Jones of Hertz's collected researches—*Untersuchungen über die Ausbreitung der Elektrischen Kraft*, to which, on the suggestion of Lord Kelvin, was prefixed the English title of *Electric Waves.* For this volume[1] Lord Kelvin himself wrote as Preface a striking appreciation. He asked the readers to carry their minds back to the time when Newton's doctrine of universal gravitation led to the general belief (not shared by Newton himself) that gravitation and other forces will act

[1] *Electric Waves, being Researches on the Propagation of Electric Action with Finite Velocity through Space*, by Dr. Heinrich Hertz. Authorised English translation by D. E. Jones, B.Sc., with a Preface by Lord Kelvin, LL.D., D.C.L., etc. London: Macmillan & Co., 1893.

between bodies at a distance without the requirement of any intervening medium. Newton's well-known repudiation (in his letter of 1692 to Bentley) of this as an absurdity, was then contrasted with the "infinitely improbable theory" of Father Boscovich, in which all the properties of matter (except heat) were explained solely by action at a distance, by mutual repulsions and attractions between mathematical points. Though Boscovich's theory had been unqualifiedly accepted as a reality before the end of the eighteenth century, a reaction set in in the middle of the nineteenth century, after Faraday's discovery of specific inductive capacity; and before Faraday's death, in 1867, the notion that electric force is propagated by a medium called ether was generally accepted by the rising generation of scientific men. Lord Kelvin then continued :—

Absolutely nothing has hitherto been done for gravity either by experiment or observation towards deciding between Newton and Bernoulli, as to the question of its propagation through a medium, and up to the present time we have no light, even so much as to point a way for investigation in that direction. But for electricity and magnetism Faraday's anticipations and Clerk Maxwell's splendidly-developed theory have been established on the sure basis of experiment by Hertz's work, of which his own most interesting account is now presented to the English reader. . . . It was by sheer perseverance in philosophical experimenting that Hertz was led to discover a finite velocity of propagation of electromagnetic action, and then to pass on to electromagnetic waves in air and their reflexion. . . .

Readers of the present volume will, I am sure, be pleased if I call their attention to two papers by Prof.

G. F. FitzGerald, which I heard myself at the meeting of the British Association at Southport in 1883. One of them is entitled, "On a Method of producing Electromagnetic Disturbances of comparatively Short Wave-lengths." The paper itself is not long, and I quote it here in full, as it appeared in the Report of the British Association, 1883: "This is by utilising the alternating currents produced when an accumulator is discharged through a small resistance. It is possible to produce waves of as little as two metres' wave-length, or even less." This was a brilliant and useful suggestion. Hertz, not knowing of it, used the method; and making as little as possible of the "accumulator," got waves of as little as *twenty-four centimetres'* wave-length in many of his fundamental experiments. The title alone of the other paper, "On the Energy lost by Radiation from Alternating Currents," is in itself a valuable lesson in the electromagnetic theory of light, or the undulatory theory of magnetic disturbance. The reader of the present volume will be interested in comparing it with the title of Hertz's Eleventh Paper; but I cannot refer to this paper without expressing the admiration and delight with which I see the words "rectilinear propagation," "polarization," "reflection," "refraction," appearing in it as subtitles.

During the fifty-six years which have passed since Faraday first offended physical mathematicians with his curved lines of force, many workers and many thinkers have helped to build up the nineteenth century school of *plenum*, one ether for light, heat, electricity, magnetism; and the German and English volumes containing Hertz's electrical papers, given to the world in the last decade of the century, will be a permanent monument of the splendid consummation now realised.

Lord Kelvin's omission to refer to himself as having had any part in the building up of that nineteenth century school of thought is as noteworthy as his recognition of Maxwell's "splendidly-developed theory."

In Lord Kelvin's Presidential Address of November 1893 to the Royal Society, he had much to say that bore upon the foundations of physics. In the first place, he announced as "not the least important of the scientific events of the year," the publication of Hertz's collection of papers on Electric Waves, and he repeated, almost verbatim, the appreciation which he had just written as the Preface to that work. Then, as if to check his own enthusiasm, he continued :—

> But, splendid as this consummation is, we must not fold our hands and think or say there are no more worlds to conquer for electrical science. We do know something now of magnetic waves. We know that they exist in nature and that they are in perfect accord with Maxwell's beautiful theory. But this theory teaches us nothing of the actual motions of matter constituting a magnetic wave. Some definite motion of matter perpendicular to the lines of alternating magnetic force in the waves and to the direction of propagation of the action through space, there must be; and it seems almost satisfactory as a hypothesis to suppose that it is chiefly a motion of ether with a comparatively small but not inconsiderable loading by fringes of ponderable molecules carried with it. This makes Maxwell's "electric displacement" simply a to-and-fro motion of ether across the line of propagation, that is to say, precisely the vibrations in the undulatory theory of light according to Fresnel. But we have as yet absolutely no guidance towards any understanding or imagining of the relation between this simple and definite alternating motion, or any other motion or displacement of the ether, and the earliest-known phenomena of electricity and magnetism — the electrification of matter and the

attractions and repulsions of electrified bodies; the permanent magnetism of lodestone and steel, and the attractions and repulsions due to it: and certainly we are quite as far from the clue to explaining, by ether or otherwise, the enormously greater forces of attraction and repulsion now so well known after the modern discovery of electromagnetism.

Having thus restated his sense of the need of a more comprehensive dynamical theory, he struck a new vein of thought:—

Fifty years ago it became strongly impressed on my mind that the difference of quality between vitreous and resinous electricity, conventionally called positive and negative, essentially ignored as it is in the mathematical theories of electricity and magnetism with which I was then much occupied (and in the whole science of magnetic waves as we have it now), must be studied if we are to learn anything of the nature of electricity and its place among the properties of matter.

He then went on to enumerate the effects in which the vitreous and resinous electricity appear to be fundamentally distinct from one another: frictional electricity; electro-chemistry; pyro-electricity; and piezo-electricity of crystals; electric glow, brush and spark discharges; also in the vast difference of behaviour of the positive and negative electrodes of the electric arc lamp. He referred to Faraday's investigations of discharge in vacuo, to his discovery of the "dark space" in the discharge, and to his remark: "The results connected with the different conditions of positive and negative discharge will have a far greater influence on the philosophy of electric science than we at present imagine." He reviewed the researches upon the electric discharge in vacuo of a number of recent workers, including Gassiot, Plücker, Cromwell Varley, Crookes, Schuster, Joseph J. Thomson, and Fleming. Varley's discovery in 1871 of the molecular torrent from the negative pole, and Crookes's long continued train of brilliant discoveries came in for special mention. The radiometer; the kathode-discharge in a high vacuum; the molecular torrent projected from the negative pole, with its mechanical, thermal, magnetic, and phosphorescent properties; the electrical evaporation of negatively electrified liquids and solids; the convergence of the kathode beam from the concave kathode of a focus-tube; the mutual repulsion between two parallel kathode beams, all were enthusiastically referred to. He concluded thus:—

In the whole train of Crookes' investigations on the radiometer, the viscosity of gases at high exhaustions, and the electric phenomena of high vacuums, ether seems to have nothing to do except the humble function of showing to our eyes something of what the atoms and molecules are doing. The same confession of ignorance must be made with reference to the subject dealt with in the important researches of Schuster and J. J. Thomson on the passage of electricity through gases. Even in Thomson's beautiful experiments showing currents produced by circuital electromagnetic induction in complete poleless circuits, the presence of molecules of residual gas or vapour seems to be *the essential.* It seems certainly true that without the molecules there could be no current, and that without the molecules electricity has no meaning. But in obedience to logic I must withdraw one expression I have used. We must not imagine that "presence of molecules is *the* essential." It is certainly *an* essential. Ether also is certainly *an* essential, and certainly has more to do than merely to telegraph to our eyes to tell us of what the molecules and atoms are about. If a first step towards understanding the relations between ether and ponderable matter is to be made, it seems to me that the most hopeful foundation for it is knowledge derived from experiment on electricity in high vacuum; and if, as I believe is true, there is good reason for hoping to see this step made, we owe a debt of gratitude to the able and persevering workers of the last forty years who have given us the knowledge we have; and we may hope for more and more from some of themselves and from others encouraged by the fruitfulness of their labours to persevere in the work.

Two years more went by, and though others had been working on the recondite problems of the ultimate theory of matter and ether, Lord Kelvin had made no further utterance. At the close of the year 1895 came Röntgen's startling discovery of the new rays generated in a high vacuum by the impact of the kathode torrent upon some solid target; rays that will penetrate even opaque solid bodies. Lord Kelvin's interest in this new revelation was intense, and he discussed keenly with Stokes and others Röntgen's conjecture that these rays might consist of condensational waves in the ether. He himself favoured the suggestion, but saw tremendous difficulties in the way. Several letters written in the spring of the year 1896 show the zest with which he contemplated the bearing of the new discovery upon the ultimate problem of physics. They were addressed respectively to Professor Oliver Lodge (then in Liverpool), Sir

George Stokes, Professor Leahy, and Professor FitzGerald.

*February* 4, 1896.

DEAR LODGE—I thank you for your letter of yesterday and accompanying prints. I had read them in the *Electrician* and thought of calling your attention to "between Hittorf and Crookes." It is curious how generally it has been overlooked, how neatly and thoroughly Varley hit it off in respect to the molecular torrent from the Cathode, and left nothing to hit off about it except what has now been done by Perrin (whose paper I got Lockyer to put into last week's *Nature*): not to speak of the brilliant and splendid and instructive demonstrations by Crookes of his own independent discovery of it.

Do you remark how the nonsense about the "Cathode (undulatory) ray" and perverse rejection of Varley and Crookes (started I suppose by Hittorf and Goldstein but adopted by every one except Helmholtz, that I know of in Germany, including strange to say, Hertz and Lenard), has utterly lost to Lenard the discovery of the Röntgen New Light? I am glad to see that you go for Hertz in this matter, but I think, in all probability he would have come right in it if he had lived and had fallen in with either Varley's or Crookes' papers; but I suppose his pupil Lenard, although very strong and persevering, has more of the German narrowness than Hertz. However I don't judge because of course we have already great things from Lenard (from Hertz's initiation) and may expect more and more from himself. I tried hard, but in vain, to get him to come to Ipswich and told him I wanted to fight him on the Kathoden*strom* but that I was ready to call it Kathoden*strahl* if he could give convincing reason for his view.—Yours truly,

KELVIN.

*February* 24, 1896.

DEAR STOKES— . . . I have been trying almost incessantly since some time before Nov. 28, 1846 (Art.

XXVII. Vol. I. of my *Mathematical and Physical Papers*) for a mechanical-*dynamical* representation of electrostatic force, but hitherto quite in vain. Many others, including Maxwell, have tried more or less for the same, but with no approach to a glimpse of what might become a success. It seems highly probable that there may be longitudinal waves in ether, but hitherto this idea does not help us to explain electrostatic force.

As to the Röntgen light, what you say of high frequency seems to me of very great importance, but you need not confine it to "transversal vibrations." It is equally applicable to waves of longitudinal vibrations if there are any in ether. Excessively high frequency gives, according to the molecular theory which I learned from you before 1852, the same propagational velocity in all transparent mediums as in vacuum.

. . . . . .

*March* 6, 1896.

DEAR PROF. LEAHY—I am sorry to have been prevented day after day from answering your letter of the 27th. The explanation of my sentence which you quote is simply this:—The attraction of rubbed amber or lodestone is exhibited in virtue of their freedom to move, which they could not have if they were embedded in an elastic solid.

Pulsating spheres, or other vibrating bodies on a fluid are free to move: and thus they show the attractions and repulsions due to vibration described by Faraday, Schellbach and Guthrie and others; and can be conceived as capable of showing the attractions and repulsions investigated by various writers on theoretical hydrokinetics. Perforated solids with irrotational circulation of a liquid through the perforations are free to move, and ideally could show the forces described in §§ 733–738 of my *Electrostatics and Magnetism*; and a long pivoted body is free to show the directional tendency described in §§ 739, 740. None of those cases of fluid motion however help us in the slightest degree to a physical explanation of magnetic forces.

You say in your letter of the 27th, "I am aware that at present a hydrodynamical theory holds the field." I cannot agree with this. No one has come within a million miles of explaining any one phenomenon of electrostatics or magnetism by hydrodynamical theory. No mechanical or physical theory of any kind holds the field, though we have really a wonderfully good knowledge of the matter of fact laws of electrostatics and magnetism through a vast variety of phenomena.—Believe me, yours very truly, KELVIN.

*March* 7, 1896.

DEAR FITZGERALD— . . . Meantime as to the Röntgen light, I am more and more disposed to think it is extreme ultra-violet light of transverse vibrations, and I believe Stokes, with whom I have had a good deal of correspondence within these last three weeks, is, I think, also much inclined to the same view.

I was much interested in your last letter about electric action in the neighbourhood of a discharged air condenser. One thing at all events of my rude pencilling I see you have quite convinced yourself of:—"the subject is almost infinitely difficult." And I think you and I may almost agree to delete the "almost." As to what takes place in the axis in each of my two diagrams, it is either longitudinal vibrations or what Stokes calls "push," which is equivalent to instantaneous transmission of pressure. A rigid globe suddenly set to vibrate to and fro in a straight line in an incompressible elastic solid is a mechanical illustration, but does not come within a million miles of being a *dynamical* representation or realisation of either electrostatic or electromagnetic action of any kind. . . .—Yours very truly, KELVIN.

*April* 9, 1896.

DEAR FITZGERALD—I have been prevented each day since I received it from answering your letter of the 4th by unpostponable affairs, of which trying to find out something more about ether and electrostatic stress and magnetism has been one, and (a closely allied subject), a report on india-rubber tyres another.

I certainly intend to add some annotations or appendices to my *Baltimore Lectures* before republication. But there will be nothing to qualify or unsay in respect to page 9 (except perhaps lines[1] 13-19). And I am beginning now to be more hopeful in respect to the last completed paragraph of that page. I think the "ether" of Article XCIX of my collected *Mathematical and Physical Papers*, with perhaps a density (see Art. CIV of *M.P.P.*) greater than $10^{-12}$, and a virtual rigidity greater than $9 \times 10^8$, may really help, and may explain, not explain away, the ideas for ether suggested in B.L. page 10. But whether my to-day's hopes or my despair of the day before yesterday may be to be fulfilled, the paragraph beginning at the foot of page 8 and the succeeding paragraph of page 9 are to be maintained uncompromisingly. It is mere nihilism, having no part or lot in Natural Philosophy, to be contented with two formulas for energy, electromagnetic and electrostatic, and to be happy with a vector and delighted with a page of symmetrical formulas. "*Giebt nur ein Wort* (Mephistopheles) *Ein würdiges Pergament*! (Wagner)."

I have not had a moment's peace or happiness in respect to electromagnetic theory since Nov. 28, 1846 (see vol. i. p. 80 of *M.P.P.*). All this time I have been liable to fits of ether dipsomania, kept away at intervals only by rigorous abstention from thought on the subject. I have been very nearly free from attacks for nearly a year, thanks to Maxwell-Boltzmann and Electrification of Air, till Röntgen's discovery brought on a fearful paroxysm which has lasted without intermission since the beginning of January. I have some hopes that this may be the last attack. This morning I have seen that detached portions of "ether" within steel could not but give it the magnetic retentiveness which it has. Poisson's "coercive force" is

[1] To make what I meant clearer: in line 16, after "electric" insert "waves with," and in line 17 delete "as" (see heading, p. 45). And in line 17, for "simple vibrations" substitute "waves."

simply resistance against sliding between "ether" and ponderable matter.

The greatest of my difficulties to get something towards a physical theory out of the "mechanical representation" (*M.P.P.* Art. XXVII) up to now has been the steel magnet. Next greatest perhaps has been electrostatic stress: next greatest perhaps magnetic induction of currents. The greatest of all *was* the mobility of magnets and electrified bodies, showing the ponderomotive forces experienced by them in virtue of the rigidity of the ether in which they are embedded. But this difficulty (?) is annulled by the ether of Article XCIX, which acts as an incompressible liquid except in so far as its virtual rigidity is called into play by frictionality between it and ponderable matter. The generation of heat by a current through a conductor only adds to electromagnetic theory all the difficulties of the kinetic theory of heat which seem to render nobody unhappy. But, alas! even if we consider this and all those other difficulties as now annulled, another arises which may be greater than all the rest put together:—How is it that the molecules of air or other gas tearing through ether at velocities of 500 metres per sec. or thereabout stiffen it to resist electrostatic stress, and have much less of stiffening effect or perhaps a weakening instead of a stiffening effect when their average velocities are about a 1000 metres per second? (E. Becquerel, "On the Electrical Conductivity of Gases at high Temperatures," *Phil. Mag.* Dec. 1853.) Apropos of this, remark that no experiment hitherto published gives us the slightest evidence as to whether there is or is not a measurable current of electricity between two metals kept at a difference of potentials of 1 volt, 100 volts, 1000 volts or 20,000 volts in the nearest approach to vacuum hitherto produced artificially. If I am wrong give me reference. Meantime Bottomley is arranging to answer by experiment in vacuums down to the 150th of a millionth of an atmosphere of gas not collapsible in the M'Leod gauge.

Returning now to your letter:—With reference to

page 45 of proof of *B.L.* :—True displacement of the ether there must be according to the undulatory theory of light perpendicular to the direction of propagation and perpendicular to the axis of rotational motion and of distortion of the ether. We all agree that the axis of rotation and distortion is really and not merely nihilistically a line of magnetic force. The line of displacement in transverse waves is somehow closely connected with electrostatic force. Witness the production of a spark between the ends of a wire bent into a ring round a solenoid when the ends of the ring touch or nearly touch one another, and when a current through the solenoid is suddenly stopped. But there is also certainly a displacement of matter in the line joining the centres of the spheres in the case which I recently suggested in *Nature*, etc. for consideration.

You will see on page 45, I am very cautious and merely speak of "probable" in respect to condensational waves in ether. There is certainly some displacement of matter in the direction of electrostatic force, but whether it is displacement of ether or displacement of electricity relatively to ether, or displacement of ether and electricity together, we cannot say. Certain it is that the energy of electrostatic force is not simply energy of a stressed homogeneous elastic solid. Your two queries which you "hope" I "will not answer." (1) Referring to spherical waves of transverse vibrations and to page 64 of print of *B.L.* :—Consider a shell between two concentric spherical surfaces very distant from the source, and take the portion of this shell a quadrilateral of nodal cones. Let the distance between two spherical surfaces be a small fraction of the wave-length, and let every diameter of the quadrilateral be very large in comparison with the wave length. The phase of the motion is the same throughout the quadrilateral shell. The diminution of the amplitude of the vibration from maximum in the middle of the shell to zero at every part of the boundary implies infinitesimal thickenings and thinnings in different parts of the shell to keep its volume constant. The case is precisely analogous to longitudinal waves or vibrations in a straight stretched wire. Nobody (not even Jaumann, I suppose,) would puzzle himself into speaking of transverse vibrations in this case, when he thinks of the thickening of the wire in one part and the thinning in another, due to shortenings and stretchings; provided the wave-length is very great in comparison with the diameter of the wire. But people *have* puzzled themselves with the case of wave-length moderate in comparison with the diameter, and have not quite seen what to make of it, though it is really obvious enough. The mathematical problem, involving columnar harmonics as I have called them, "Fourier-Bessel Functions" for wave-lengths neither very great nor very small in comparison with the diameter of a wire or rod of circular section, is really interesting. It gives a velocity of propagation intermediate between the Young's modulus velocity and the velocity of the condensational-rarefactional wave. The extreme case of wave-lengths very small in comparison with the diameter is simply the well-known condensational-rarefactional wave in an infinite solid.

Your query (2) referring to page 106 of print of *B.L.*—The case of T very great puzzled myself about a month ago. It is not very easy to interpret the formula by itself for this case. But I saw easily enough by going back to the process which led to the formula that it gives for $T = \infty$ simply the diminished velocity due to augmented density, as if the masses of all the shells were distributed equally through it without altering its rigidity.

There is another interesting case included in the same work. Let the central nucleus be absolutely fixed: the velocity of propagation is infinitely great for $T = \infty$.

Take T infinitely small and you find the wave velocity $= \sqrt{\frac{l}{\beta}}$, the same as in pure ether. This explains the exceedingly small refractivities of aluminium and vulcanite found by Röntgen, and of other substances found by other experimenters for the X-rays. But the explanation would be better if they were small negative instead of small positive, as found (not confidently) by Röntgen. Stokes and I have had a good deal of correspondence over all this, and about six weeks ago we both concluded that in all probability the Röntgen X-rays are of transverse vibrations of very short period.—Yours very truly,

KELVIN.

*Apri.* 29, 1896.

DEAR FITZGERALD—Your letter of the 17th followed me to London and I have been very busy ever since or I should have written sooner in reply. I am sure you will never find comfort in crystallization or anything analogous to it to explain waves of light in ether, or in anything else than the fundamental doctrine of the undulatory theory of light—true transverse vibrations of moving matter subject to the law of inertia. Hence my "must" to which you object.

Maxwell's expression, "electric displacement" is, I believe, absolutely true so far as it indicates a true displacement of matter, as in the undulatory theory of light, but my difficulty is in respect to the electric quality concerned in this displacement.

Electric force (X, Y, Z) cannot be a mere displacement, because mere displacement does not in an elastic solid or in any conceivable "ether" give rise to energy equal to $R^2/8\pi$ per unit volume of field. I could not in Nov. 1846, nor have I, ever since that time, been able to regard "displacement" as anything better than a mere "mechanical representation of *electric* force." But I have always from that time till now felt, and I now still feel, that somehow or other we shall find rotation of a medium to be the reality of *magnetic* force. We may ideally make a dynamical model with an incompressible inviscid fluid to represent electricity percolating among interstices between ether and ponderable matter of a metallic conductor, and generating heat by the vibrations in ether which such percolation could not but produce, provided all the substances concerned are perfectly free from viscosity. It is not so easy to include in the model electrostatic force and charged condensers. Thus a displacement of electricity relatively to ether, and the strain produced by the fluid trying to get through from vesicle to vesicle in the ether, but in an insulator not succeeding, would be the "electric displacement" and the electrostatic stress.

All this is very crude, but perhaps not absolutely unmanageable for an ideal mechanical model. In transverse vibrations of light and magnetism the ether and the fluid in the vesicles would move together; and the stress of the ether dragging molecules denser than itself in vesicles containing it so as to give it very nearly always the same motion as its own, would be an extreme case of electrostatic stress.

But there can be no comfort whatever in any attempt at a physical theory of electricity and magnetism unless it provides us with a medium capable of giving the prodigiously great forces which we have in electromagnetism, and the smaller, but still very palpable forces of electrostatics, and also at the same time permitting the free mobility of bodies through it, in virtue of which we feel those forces. When I wrote to you my letter of April 9 I thought I saw how it was possible to find this last essential quality in my "ether" of *M.P.P.* Art. XCIX, but a day later I fell back into utter despair, in which I still remain.

You speak of "vectors and symmetrical equations"[1] for investigating "the rate at which an alternating electric current penetrates into a conductor," and "the equations which lead to an exactly similar propagation of electric currents into non-conductors." The former

[1] Symmetrical equations are good in their place, but "vector" is a useless survival, or offshoot, from quaternions, and has never been of the slightest use to any creature. Hertz wisely shunted it, but unwisely he adopted temporarily Heaviside's nihilism. He even tended to nihilism in dynamics, as I warned you soon after his death. He would have grown out of all this, I believe, if he had lived. He certainly was the opposite pole of nature to a nihilist in his experimental work, and in his Doctorate Thesis on the impact of elastic bodies.

are in reality vitally different from the latter; and the latter are merely the equations of motion of an incompressible elastic solid. One of the chief things to be discontented with is the refractoriness of all attempts to bring the two classes of action into dynamical relation with one another, on any hitherto imagined constitution for ether, electricity, and matter. It is not the equations I object to. It is the being satisfied with them, and with the pseudo-symmetry (pseudo, I mean, in respect to the physical subject) between electrostatics and magnetism. I also object to the damagingly misleading way in which the word "flux" is often used, as if it were a physical reality for electric and magnetic force, instead of merely an analogue in an utterly different physical subject for which the same equations apply, see *Electrostatics and Magnetism*, §§ 4, 5, and 6 (first published Feb. 1842). But enough of this carping about words! If we could but get the slightest inkling of how a fragment of paper jumps to rubbed sealing-wax, or a fragment of iron to a lodestone, I could be supremely happy, and would be temporarily content not to ask more of ether, not even gravity.

In your letter of the 17th you hit off exactly the right thing in respect to a sphere or two spheres in incompressible jelly. But you omit to remark that the push along the axis giving a displacement inversely as the square or some higher power of the distance represents the alleged infinitely rapid propagation of longitudinal displacement. What I have called the distant terms don't include longitudinal displacement in the axis. Think first of a rigid globe in an incompressible liquid. Start it instantaneously in motion in the direction of any diameter. Every particle of the liquid will take its proper motion instantaneously. Give now the liquid some rigidity; the *instantaneous* motion all around will be the same as it was with no rigidity: but a distortional wave motion propagated at a finite velocity will follow. This is exactly the state of things represented by Maxwell's equations and worked out in an example by Hertz.

Thanks for page 45, $r^2$ instead of $x^2$. The papyrograph was sometimes not clear, and my eye imperfect in detecting errors in the print.

Your spherule of water moving through ice which melts in front and freezes behind, is virtually the same in result as my despised shoemaker's wax. I stuck to my theory of a turbulent liquid as long as I thought I could get anything out of it, and no longer.

I now abandon everything I have ever thought of or written in respect to constitution of ether . . .—Yours ever truly. KELVIN.

That the veteran of seventy-two should possess the mobility of thought revealed in these hitherto unpublished letters, is wonderful: but it is not more wonderful than the tenacity with which he held to that which seemed to him a real depositum of scientific truth. Though he had abandoned the vortex-atom theory, and was now ready to throw away all preconceptions as to the constitution of the ether, he had never lost faith in dynamics or in the ether itself, or in the possibility of that ultimate solution which had been before him in 1846.

Then came his Jubilee.

Those who have followed the course of his persistent endeavours to see below the surface of matter, as here narrated, will be able to appreciate the true inwardness of his confession: "*One word characterizes the most strenuous of the efforts for the advancement of science that I have made perseveringly during fifty-five years; that word is failure. I know no more of electric and magnetic force, or of the relation between ether, electricity, and ponderable matter than I knew and tried to teach to my students of natural philosophy fifty years ago in my first session as Professor.*"

What was "the most strenuous" of his efforts for the advancement of science? Assuredly not achievement of ocean telegraphy, nor the improvement of the compass; neither the limitation of geological time, nor the establishment of thermodynamic doctrine. Great and strenuous as were the efforts which each of these demanded, none of them can be compared with the fifty-years-long quest for the theory of matter. And here it is worthy of remark that the same integrity of judgment, the same refusal to find intellectual peace at the cost of intellectual sincerity, which in his earlier life had restrained him for three years from accepting Joule's ideas as to the mechanical equivalence of heat and work, had equally, and for thirty years, restrained him from finding satisfaction in Maxwell's theory of light. In the former case his suspense of judgment had been rewarded by the discovery not only of the Dissipation of Energy, but of the still more widely significant doctrine of Available Energy. The everlasting law of honour by which science was bound to face fearlessly every problem that can fairly be presented to it, bound him too. The great dynamical problem of matter was yet unsolved; but it was not insoluble, at least not in his regard. There was not, there could not be, any permanent satisfaction for his mind while the great quest failed of its goal. Was the work hard? Then the harder the more was it worth doing. Did the intricacies of it cause trouble? Then let it be remembered that "tribulation, not undisturbed progress, gives life and soul, and leads to success, when success can be reached, in the struggle for natural knowledge." Not even then, at the age of seventy-two, could Lord Kelvin rest with the great problem still unsolved.

Meantime other workers and disciples had brought new material to bear. Following out the ideas of Faraday and Maxwell as to electricity itself having a natural unit, justifying the expression [1] "one molecule of electricity," von Helmholtz had, in his Faraday lecture of 1881, carried the idea a stage further in emphasizing the signification of the laws of electrolysis. Following on the investigations of Thomson and Loschmidt as to the size of atoms, Johnstone Stoney, also in 1881, had estimated the amount of charge of electricity which is associated with one atom of hydrogen, or other monad element, as being about one one-hundred trillionth part ($10^{-20}$) of a coulomb, and to this natural unit of electricity he, in 1891, gave the name of *electron*,[2] which term is, however, now generally restricted to the unit of resinous or negative electricity.

Refined measurements by Sir Joseph J. Thomson and others showed that the kathode streams in Crookes's tubes were flights of free electrons, disembodied from the atoms of matter with which they are usually associated. Sir Joseph had, in 1881, investigated the magnetic field due to a moving

[1] Maxwell, *Electricity and Magnetism* (1873), vol. i. p. 312.

[2] Lord Kelvin preferred to use the term *electrion*. Sir Joseph J. Thomson prefers the name *corpuscle*. Both mean the same thing as *electron*.

electric charge, and had remarked that the convection of this field involves an addition to its effective mass. More important still in the present connection, Sir Joseph Larmor,[1] in the years 1893 to 1897, developed a new dynamical theory of the electric and luminiferous medium. Adopting the same general line as Lord Kelvin, he proposed a gyrostatic adynamic ether, which would have a rotational rigidity, but allow free movement of solid bodies through it, as the primordial medium. Then he definitely suggested that if the atom be regarded as containing electrons, the phenomenon of magnetism should be explained, not as a rotation of the ether itself, but as a rotation of the electrons. The electron itself was regarded as a strain centre in the homogeneous medium. Radiation is due solely to acceleration or retardation of the electrons, a point which Heaviside had previously elaborated. To this was added the speculation that the mass of each sub-atom is proportional to the number of electrons that it carries, and that each atom is constituted of an orbital system of electrons (as imagined by Sir Joseph J. Thomson), with interatomic forces that are entirely or mainly electric. A system of electrons ranged along a circle, and moving around it with such a speed as to give steadiness constitutes in fact a vortex-ring in the surrounding ether. Lord Kelvin had never postulated the vortex-ring as including an electric charge

[1] *Roy. Soc. Proc.* liv., Dec. 7, 1893; *Phil. Trans.* A. 1894, p. 764; *ib.* A. 1895, pp. 695-743; *ib.* A. 1897, pp. 205-309; also his book, *Aether and Matter*, Cambridge, 1900.

as part of its constitution; but in 1890 he had arrived definitely at the view (see p. 1049) that a triple alliance with electricity as a constituent member was a necessity. Larmor's new theory made the ultimate element of matter not a vortex-ring, but an electric charge or nucleus of permanent ethereal strain. A whirling assemblage of such nuclei might act as a vortex-ring. Here was then a definite idea of a connection between matter and ether, a foundation for a definite and universal theory. It is but fair to add that a very similar view had been independently put forward at almost the same time by Professor H. A. Lorentz of Leyden.

In the later part of this development we are to some extent anticipating, for before it was fully launched Lord Kelvin had himself resumed the quest. At the British Association Meeting of 1896, at Liverpool, he read a paper on the molecular dynamics of hydrogen, oxygen, ozone, and other gases, and of ice, water, and quartz crystal. The object of this communication was to find how much of the known properties of these typical substances can be explained without making any further assumptions than the conferring of inertia on a Boscovich atom.

In 1897 he made several communications to the Royal Society of Edinburgh, amongst them one on some models illustrating the dynamical theory of hemihedral crystals.

Again, in 1898, at the British Association Meeting at Bristol, he spoke of the dynamical theory of refraction and dispersion; of continuity in the undulatory theory between condensational waves, distortional waves, and electric waves. In 1899 he rediscussed magnetism and molecular rotation before the Royal Society of Edinburgh. He showed that an electrified body is set into rotation by the generation of a magnetic field around it.

In the year 1900 Lord Kelvin brought before the Royal Society of Edinburgh, and again at the *Congrès de Physique* in Paris, a new hypothesis of a startling kind, in a paper,[1] on the Motion Produced in an Infinite Elastic Solid by the Motion through the Space occupied by it of a Body acting on it only[2] by Attraction and Repulsion. The very title contradicts the old scholastic axiom that two different portions of matter cannot occupy the same space at the same time. He had been reconsidering the old difficulty of the undulatory theory of light — the motion of ponderable bodies through space occupied by an elastic solid, for so he still regarded the ether. To emphasize the point he appended this trenchant note:—

The so-called "electro-magnetic theory of light" does not cut away this foundation [elastic solid] from the old undulatory theory of light. It adds to that primary theory an enormous province of transcendent interest and importance; it demands of us not merely an explanation of all the phenomena of light and radiant heat by transverse vibrations of an elastic solid called ether, but also the inclusion of electric currents, of the permanent

[1] Reprinted as Appendix A, p. 468, of the *Baltimore Lectures*.

[2] And so Father Boscovich, judged obsolete in 1884, and his theory, pronounced "infinitely improbable" in 1893, was in 1900 "reinstated as guide."

magnetism of steel and lodestone, of magnetic force, and of electric force, in a comprehensive ethereal dynamics.

He conceived the atom as a spherical aggregation of ether of varying density, of the same average value as the density of the ether outside, the internal ether being arranged in concentric shells of varying density, according to some prescribed law of force. It would have no resultant attraction on the ether outside, and could move freely through it at any speed that was small compared with the velocity of light. To reconcile the hypothesis with the experiment in which Michelson and Morley found the ether in the earth's atmosphere to be at rest relatively to the earth, he was prepared to accept the suggestion of FitzGerald and Lorentz that the motion of ether through matter may slightly alter its linear dimensions. There was nothing electrical in this hypothesis.

The year 1901 brought a marked step forward. Under the quaint title of "Aepinus Atomized," Lord Kelvin sought to reconstruct the old one-fluid theory of electricity on a new atomic hypothesis. The doctrine of Aepinus[1] was that positive and negative electrifications consist in excess above, and deficiency below, a natural quantum of a fluid called the electric fluid permeating among the atoms of ordinary matter. Adopting the modern notion of electrions (see p. 1074 *supra*), Lord Kelvin now suggested that

[1] Aepinus, *Tentamen Theoriæ Electricitatis et Magnetismi*, St. Petersburg, 1759.

while electrions permeate freely through all space, whether occupied by ether or occupied also by the volumes of finite spheres constituting the atoms of ponderable matter, each electrion in the interior of an atom of matter experiences electric force towards the centre of the atom, just as if the atom contained within it, fixed relatively to itself, a uniform distribution of ideal electric matter. The electrions themselves were to be exceedingly minute atoms of resinous electricity, while the atom of matter was vitreously electrified. An atom might require one, two, or more electrions to neutralize it. The mathematical laws of the mutual actions of two such atomic structures were deduced, and an explanation found for the attractions and repulsions—the oldest known electrical property—produced by friction. He then calculated the conditions of stable equilibrium for a number of possible configurations of electrions within the atom; two, on a diameter; three, at the corners of a triangle; four, at the corners of a square; four, at the corners of a tetrahedron; six, at those of an octahedron; eight, at those of a cube; and so on up to as many as twenty-one electrions. He found that hypothetical atoms so constituted would realize Faraday's explanation of the phenomenon of dielectric polarization. High temperature would set the electrions into wildly irregular vibrations, causing some of them to be occasionally shot out of their atoms, and either falling back or passing into other atoms, thus explaining the conductivity of solids such as glass or Nernst filaments, when heated. Considerations of the pyro-electric properties of crystals concluded the paper.

The molecular dynamics of crystals was resumed in 1902 in a further paper[1] to the Royal Society of Edinburgh, with elaborate calculations on Boscovichian principles of the stability or otherwise of networks of homogeneous assemblages of atoms.

The completion of the *Baltimore Lectures* was now taken in hand. Chapter XIX was rewritten, with a reconciliation between the formulæ of Green and Fresnel on the basis of the new hypothesis of 1900, by which the character of the action of atoms of matter on ether was reduced to simple attraction or repulsion; and these attractions were rendered effectual by assuming that the ether within the space occupied by the atom could be condensed or rarefied by positive or negative pressures due to repulsion or attraction exerted upon it by the atom and its neutralizing quantum of electrions. This assumption of compressibility removed the last difficulty in reconciling with dynamical theory, provided one might regard the inertia of crystals to be no longer isotropic, an idea which he had formerly rejected as incompatible with the conservation of energy. Lecture XX as thus rewritten, dealt with the chirality of quartz and of optically active liquids, on the new assumption of aeolotropic inertia. He desired to extend the explanation to the magneto-optic rotation, but contented himself by saying: "When we have a true physical theory

[1] Reprinted as Appendix J, p. 662, of the *Baltimore Lectures*.

of the disturbance produced by a magnet in pure ether, and in ether in the space occupied by ponderable matter, fluid or solid, there will probably be no difficulty in giving as thoroughly satisfactory explanation of the magneto-optic rotation as we now have of the chiro-optic." He then returned to the dynamics of ordinary and anomalous dispersion, for which in 1884 he had invented his spring-shell model molecules, to consider what modification must be made to suit the hypothesis of the electrionic structure of the atom. In the new theory each electrion or each group of electrions within the atom could act as a vibrator,[1] which in a source of light takes energy by collision with other atoms and radiates out the energy in waves in the ether. He had been looking all along for a system of vibrators whose motions would explain the dynamics of spectrum analysis, the absorption of the dark lines of the spectrum, and the anomalies of dispersion. He had found them in the electrions. Assuming the electrion to be massless, or rather as possessing virtual inertia only on account of the kinetic energy of its motion through the ether, he solved the equations of motion and deduced that the greater the wave-length of the outgoing waves, the smaller was the proportionate loss of energy per period. This led finally to a new design of model

[1] It is a singular fact that in none of this later work does Lord Kelvin refer to the effect discovered in 1897 by Zeeman, being conclusive as to the electrions being the real vibrators, not the atoms themselves or the molecules. His sole reference to the Zeeman phenomenon is in his 1899 paper on Magnetism and Molecular Rotation (p. 1077, *supra*), where he simply accepts Lorentz's theory.

molecule. In the old model there was as vibrator a central free mass connected by springs to a rigid sheath, the lining of a spherical cavity in ether. In the new electro-ethereal design, the force of the springs was replaced by the electric attraction of the atom on its electrion when the latter was displaced from its position of equilibrium; and the electrion acts directly on the ether in simple proportion to acceleration of relative motion. He had found, then, the *tertium quid* pronounced in 1890 (see p. 1048) to be wanting. Every one of the formulas was found to be applicable, "notwithstanding the vast difference between the artificial and unreal details of the mechanism thought of and illustrated by models in 1884, and the probably real details of ether, electricity, and ponderable matter suggested in 1900-1903."

How far, then, had his aspirations been fulfilled? Let his own words say, as he wrote them in January 1904 in the Preface to the *Baltimore Lectures*.

It is in some measure satisfactory to me, and I hope it will be satisfactory to all my Baltimore coefficients still alive in our world of science, when this volume reaches their hands; to find in it dynamical explanations of every one of the difficulties with which we were concerned from the first to the last of our lectures of 1884. . . .

It seems to me that the next real advances to be looked for in the dynamics of ether are:—(i) An explanation of its condition in the neighbourhood of a steel magnet, etc. (ii) An investigation of the mutual force between two moving electrions, modified from a purely Boscovichian repulsion; as it must be by the composition, with that force, of a force due to the inertia of the ether set in motion by the motion of each of the electrions. It seems to me that, of these, (ii) may be at present fairly within our reach; but that (i) needs a property of ether not included in the mere elastic-solid-theory worked out in the present volume. My object in undertaking the Baltimore Lectures was to find out how much of the phenomena of light can be explained without going beyond the elastic-solid-theory. We have now our answer: *everything non-magnetic; nothing magnetic.* The so-called "electromagnetic theory of light" has not helped us hitherto: but the grand object is fully before us of finding a comprehensive dynamics of ether, electricity, and ponderable matter, which shall include electrostatic force, magnetostatic force, electromagnetism, electrochemistry, and the wave theory of light.

Thus after twenty years the Lectures were completed.[1] With the discovery of radium and the persistent disintegration of its atoms by the emission of electrons and of vitreously electrified atoms of helium (as now known), the old problems had assumed a new aspect. Lord Kelvin sought to explain[2] these phenomena also by aid of his electro-etherial hypothesis.

[1] "The thanks of the scientific world," wrote Larmor, when reviewing the volume in *Nature*, "will surely go out to the veteran author, now by a happy choice Chancellor of the University which he has so long adorned, for this splendid gift, which stimulates and educates even where it fails to convince, and bears on every page evidence of profound and unwearying thought."

[2] Of these latest contributions to science the titles are: "Plan of a Combination of Atoms having the Properties of Polonium or Radium" (*Phil. Mag.* viii. p. 528, Oct. 1904); "Models of Radium Atoms to give out $\alpha$- and $\beta$-Rays respectively" (*Nature*, lxx. p. 514, Sept. 22, 1904); "On the Kinetic and Statistical Equilibrium of Ether in Ponderable Matter at any Temperature" (*Phil. Mag.* x. p. 285, Sept. 1905); "Plan of an Atom to be capable of storing an Electrion with enormous Energy for Radio-activity" (*Phil. Mag.* x. p. 695, Dec. 1905); "The recent Radium Controversy" (*Nature*, lxxiv. p. 539, Sept. 27, 1906); "An Attempt to Explain the Radio-activity of Radium" (*Phil. Mag.* xiii. p. 313, March 1907); "On the Motions of Ether produced by Collisions of Atoms or Molecules containing or not containing Electrions" (*Phil. Mag.* xiv. p. 317, Sept. 1907); "On the Formation of Concrete Matter from Atomic Origins" [posthumously published] (*Phil. Mag.* xv. p. 397, April 1908).

To sum up that which Lord Kelvin effected in this life-long labour were no easy task. To bring all the properties of matter within the range of dynamics involves an implication not readily conceded. Dynamics is the science of matter and motion, or rather of matter and energy, since energy, not motion, is subject to a law of conservation. Is it then possible to reduce all physical phenomena within the duality of matter and energy? Grant merely the duality, and we may explain not only the mechanics of moving bodies, but also sound and heat (except radiation), and, under certain assumptions, elasticity; but not gravitation, nor light. Light demands an ether, as well as energy and matter: it can no-how be explained by postulating a duality, unless matter itself is expressible in terms of ether and energy. And to explain electric phenomena (to say nothing of magnetic) demands four fundamental entities, electrons (or at least something called electricity), matter, energy, and ether. The trend of modern ultra-physics with respect to the constitution of matter is towards the following five categories: (1) the *ether*, that is, the *plenum* filling space; (2) the *electron*, conceived as a plexus in the ether, probably of two species; (3) the *atom*, a complex of electrons in the ether; (4) the *molecule*, a specific group of atoms (or in some cases one atom); (5) the *mass*, an assemblage of molecules. Energy is involved in the construction of any of these out of any other. Now our mechanical ideas and language are all derived from masses and their movement; and our chemical ideas and language mostly from molecules and their constituent atoms. There arises consequently a difficulty, inherent in the terms we use, and the physical implications due to their origin in our experience, when we try to explain or even to describe electrons and ether; for to state the descriptions intelligibly we must frame them in terms of ordinary matter, masses, or molecules. It would seem logical to restate the more complex in terms of the more simple, rather than the reverse, were the simple familiar to us. Lord Kelvin's effort seems to have been to find a theory to reduce the necessary concepts to the smallest number—matter and energy, or, by means of the vortex theory, to ether and energy. In the end he found it necessary to bring in electricity as well. But who shall call this failure? The late Professor FitzGerald voiced the verdict of his fellow-workers in phrases of no uncertain import:—

Though he himself has described these efforts as resulting in failure, his contemporaries and disciples see a succession of brilliant successes, which have not, indeed, fully conquered the citadel of ignorance against which they were directed, but have, nevertheless, conquered many and fair districts, and advanced the armies of knowledge in their reconnaissance of this citadel to an extent that was only possible for a great general, an indefatigable and enthusiastic genius.

# CHAPTER XXV

## VIEWS AND OPINIONS

FROM many passages in the previous chapters some insight is afforded into the views and opinions of Lord Kelvin upon religion, politics, education, and other questions. It were an unseemly thing, and altogether beyond the province of the present work, or the competence of its compiler, to essay to sit in the seat of judgment. Not in our generation will it be possible to exercise dispassionate vision or to disentangle the ultimate from the obvious. Whenever possible, the attempt has been made to cite Lord Kelvin's own words. But the presentation of his character afforded by these scattered utterances would be very incomplete were not some effort made to gather together the stray threads. We shall not look upon his like again; and the things in him which some took for defects must one day be evaluated afresh in the light of that essential nobility of character which shines out the brighter, the closer one comes to his own words and work.

Though in the matter of religious beliefs Lord Kelvin never made any parade of his views, he was a man of earnest convictions, quietly but tenaciously sustained, without bigotry or intolerance. Although brought up in the Established (Presbyterian) Church of Scotland by his father, who had himself at one time studied for the ministry, he conformed while at Cambridge to the Church of England, and subscribed the Thirty-nine Articles both on his entry as an undergraduate and on his admission to the Fellowship in 1845. It is equally clear that in 1846, on his election to the Glasgow Chair, he subscribed to the Westminster Confession. Also, during the lifetime of his first wife, and during his widowerhood, whenever he was at Largs he attended there the services of the Free Church, of which his wife's brother-in-law, the Rev. Charles Watson, was minister. When in Glasgow he regularly attended the services (Established, Presbyterian) of the University Chapel. Later in life he had sittings in the Scottish Episcopal Church in Glasgow, and also attended the Episcopal Church at Largs. He was, in fact, a regular and reverent communicant in the Church[1] of England. His brother James, when quite a young man, had, after great mental struggles, broken from the old Calvinistic faith, and definitely became a Unitarian. A letter of Dr. King's, of date 1850, shows that William reasoned very

[1] Of sacerdotalism and ritualism in all its phases and forms he had an unconcealed detestation. He even went once so far as to write that the only sense in which he could regard the "High" Church as high, was the same as that in which game is said to be "high"—when it is decomposing.

At a meeting of the Ladies' Protestant League, held on July 16, 1902, Lord Kelvin said :—"All well-wishers of England, and of religion in England, must lament that there has been so much of perversion allowed to pass unchecked within the Church of England, with only too feeble remonstrance on the part of the Bishops, to whom they had the right to look for the maintenance of law and order in the Church."

strongly with his brother in favour of the evidence for revelation. Though James remained a staunch, if not at any time an extreme, adherent of the Unitarian faith, William never permitted this frank divergence of beliefs to lessen his affection or his freedom of intercourse—and it was an intercourse of unusual warmth—with his brother. One who from his boyhood knew Lord Kelvin well says of him: "I am quite sure that he was sincerely religious: I would say he was a sincere Christian (meaning by Christianity the religion taught by Christ rather than the religion taught by the churches). I believe he looked deep into essentials, and that he regarded differences of sects as mere matters of form, and looked on the distinctions between Episcopalians, Presbyterians, Quakers, and Unitarians with supreme indifference." To his nephew and nieces who held Unitarian beliefs he never expressed any disapproval; neither did they feel that there existed any religious barrier between them and him; nor did he scruple to suggest that an unbaptized person of sincere convictions should take the communion in church.

In 1895, as President of the Royal Society, it fell to him, in his annual address, to speak of the death of Professor Huxley, to whose views he then referred in the following terms:—

> Even those purely scientific papers contain ample evidence that Huxley's mind did not rest with the mere recording of results discovered by observation and experiment: in them, and in the nine volumes of collected essays which he has left us, we find everywhere traces of

acute and profound philosophic thought. When he introduced the word agnostic to describe his own feeling with reference to the origin and continuance of life, he confessed himself to be in the presence of mysteries on which science had not been strong enough to enlighten us; and he chose the word wisely and well. It is a word which, even though negative in character, may be helpful to all philosophers and theologians. If religion means strenuousness in doing right and trying to do right, who has earned the title of a religious man better than Huxley?

Following the old custom in Glasgow College, he always, as narrated on p. 444, began his morning lecture by reciting, with quiet solemnity, the Third Collect, for Grace, from the close of the Service of the Church of England at morning prayer. As a young man he had thought things out in his own way, and had come to a faith which, not having been received second-hand, but being of personal conviction, was never afterwards shaken. His faith was always of a very simple and child-like nature, undogmatic, and unblighted by sectarian bitterness. It pained him to hear crudely atheistic views expressed by young men who had never known the deeper side of existence.

Though Lord Kelvin did not like to miss attendance at church on Sunday mornings, he was by no means a rigid Sabbatarian, and had no objection to lighter occupations, or to occupying himself with scientific calculations on Sundays. When sailing in his yacht he liked to put in on Saturday at some port, that his captain and crew might have Sunday ashore.

In 1879 he declined to become a Vice-President of the Sunday Society, explaining that if its object had been simply to obtain the opening of museums, galleries, libraries, and gardens on Sundays, he would have been happy to join it; and in 1872 he declined the Vice-Presidency of the Sunday Lecture Society, though he wished well to its efforts. In 1889 he presided at the annual meeting of the Christian Evidence Society; and from 1903 to his death was President of the Largs and Fairlie Auxiliary of the National Bible Society of Scotland.

In 1888, when staying for a week with his sister Mrs. King in St. John's Wood, he himself, during a wet Sunday, suggested a Bible reading, which his nieces remember well, and of which they made notes at the time. He thought that in all Bibles the dates at the tops of the pages should be printed with a query-mark, except in the cases of ascertained historical events. The same evening Mrs. King read from Darwin's works the passage in which he expresses his disbelief in Divine revelation, and in any evidence of Design. Sir William pronounced such views utterly unscientific, and vehemently maintained that our power of discussing and speculating about atheism and materialism was enough to disprove them. Evolution, he declared, would not in the least degree explain the great mystery of Nature and Creation. If all things originated in a single germ, then that germ contained in it all the marvels of creation—physical, intellectual, and spiritual—to be afterwards developed. It was impossible that atoms of dead matter should come together so as to make life.

He was strongly opposed to the "secular solution" of the religious difficulty in primary education. He was equally strongly opposed to denominationalism in schools, particularly for Ireland.

Again and again, in his public career, from his inaugural lecture of 1846 to the end his life, Lord Kelvin declared his belief in Creative Power, and in an overruling Providence. In two points at least his scientific studies brought him to what he considered a direct demonstration of a definite creation: namely, the Fourier equations for the flow of heat, with the mathematical inference, pointed out on p. 111, that there must have been a beginning; and the vortex-atom conception (p. 517), according to which the permanence of the atom proves that no known animate or inanimate physical agency could have originated them. Yet he was never prone to drag in a teleological view as an excuse for shirking a deep-reaching investigation. "If," he said in 1871, in discussing the origin of life on the earth, "a probable solution, consistent with the ordinary course of nature, can be found, we must not invoke an abnormal act of Creative Power." With Paley's *Natural Theology* he had been familiar from his youth; and though he deplored the frivolities of teleology, he considered as solid and irrefragable the main argument of "that excellent old book." He seems in some way to have linked the argument for creative design with his belief in a comprehensive theory of ultimate dynamics. This appears in the closing sentence of his Royal Institution discourse[1] of 1860 (see p. 408), where he had been putting the question whether we are to fall back on facts and phenomena, and renounce all idea of penetrating that mystery which hangs round the ultimate nature of matter. His comment was: "But it does seem that the marvellous train of discovery, unparalleled in the history of experimental science, which the last years of the world have seen to emanate from experiment within these walls, must lead to a stage of knowledge in which laws of inorganic nature will be understood in this sense—that one will be known as essentially connected with all, and in which unity of plan, through an inexhaustibly varied execution, will be recognized as a universally manifested result of creative wisdom."

The limitation of the above statement to inorganic nature is characteristic; it finds its parallel in many other instances, as, for example, in the limitation introduced by him into his statement (p. 282) of the thermodynamic axiom as to the impossibility of deriving mechanical energy *by means of inanimate material agency*, by cooling a body below the temperature of its surroundings. He regarded life, however certainly its operations were governed by chemical and dynamical laws, as essentially outside the range of physics. He utterly repudiated all idea of the generation of living matter by force or motion of dead matter alone. "That life proceeds from life,

[1] *Proc. Roy. Institution*, vol. iii. pp. 277-290.

and from nothing but life," was for him "true through all space and through all time" (p. 605). He declared[1] that whereas the fortuitous concourse of atoms was the sole philosophic foundation for the second law of thermodynamics, the fortuitous concourse of atoms was powerless to account for the directed operations of living matter. "The influence of animal or vegetable life on matter is," he declared, "infinitely beyond the range of any scientific inquiry hitherto entered on. Its power of directing the motions of moving particles, in the demonstrated daily miracle of our human free-will, and in the growth of generation after generation of plants from a single seed, are infinitely different from any possible result of the fortuitous concourse of atoms."

He had, in 1852 (p. 290), denied the probability that organized matter, either vegetable or animal, could reverse the thermodynamic dissipation of energy (a question which Helmholtz left open), and held this view unchanged in 1892. And again, in 1874, he had said: "If the materialistic hypothesis of life were true, living creatures could grow backwards with conscious knowledge of the future but no memory of the past, and would again become unborn. But the real phenomena of life infinitely transcend human science; and speculation regarding consequences of their ultimate reversal is utterly unprofitable."

"The only contribution of dynamics to theoretical biology is," he declared (*Popular Lectures*, vol. i.

[1] *Fortnightly Review*, March 1892; and *Popular Lectures*, vol. ii. p. 464.

p. 4.15), "absolute negation of automatic commencement or automatic maintenance."

But if he held that life was thus a thing apart from the physical forces which it controlled (and requiring in itself initially a creative act), his doctrine had little in common with the old crude vitalism which postulated a vital force as amongst the natural forces, so-called, which operate on matter. On the contrary, he maintained with the utmost strictness the rigorous insistence on physical forces to produce physical effects ; and his life-long struggle to reduce to terms of dynamics the other manifestations of energy rules out any such confusion. The human body, as a machine, was subject to the laws of mechanics ; as a digestive apparatus, to the laws of chemistry. No supposed vital force could maintain the animal heat : it was maintained by the consumption of food and the processes of respiration through metabolism. True, the body did not work as a thermodynamic engine, it more nearly resembled an electric motor ; but the doctrine of available energy might be extended hereafter to embrace the physiological workings of energy. But all this required only that some vital principle should direct or organize those transformations of energy which would otherwise be conducted as a purely inorganic, that is, unorganized process ; no vital principle could physically supply the energy needed in the process which it merely directed. But the real phenomena of life were infinitely beyond the range of all sound speculation in dynamics. So far as physics was concerned, free-will was a miracle. Emphatically, Science was not, for him, *eine entgötterte Natur*.

In 1897, at the request of Sir George Stokes, Lord Kelvin gave the Annual Address to the Victoria Institute, a Society whose object is the reconciliation of religion and science. The subject was the Age of the Earth; the scientific points of it are noted elsewhere. It ended with the statement that mathematics and dynamics fail us when we are confronted with the problem of the origin of life on the earth. "We must pause, face to face with the mystery and miracle of the creation of living creatures."

In the autumn of 1903, Sir Edward Fry, staying at Largs, wrote to a member of his family some personal impressions of visits to Netherhall, and he now kindly permits the publication of the following extracts:—

As we have seen a good deal of Lord and Lady Kelvin whilst at Largs, and as Lord Kelvin is a most remarkable man, and as I know you will like to hear about him, I will put down my recollections whilst they are fresh in my mind.

Lord Kelvin's house, Netherhall, is not a very large place—rather a seaside villa with grounds than a seat—and near him lives Dr. Watson, the Free Church minister of Largs, who married a sister of Lord Kelvin's first wife—and these two old people seem on terms of great intimacy with Lord Kelvin. He was till a few years ago a member of Dr. Watson's congregation, but the differences about Home Rule, I believe, led to Lord Kelvin frequenting the Episcopalian service; and one afternoon Canon Low came to consult Lord Kelvin as to the vacancy in the bishopric to which the Largs Episcopalians owe allegiance.

Lady Kelvin evidently devotes herself to the tender care of the precious piece of humanity in her keeping, and likes to talk about him. She showed us one of the series of books in which Lord Kelvin works. It is a quarto sized note-book which he always carries with him in a pocket made for the purpose, and in which he works away at his mathematical investigations when travelling by railway or at any time—but hardly ever alone, as conversation does not disturb him. Each entry begins with an exact date, day and hour, and seemed the most extraordinary web of mathematical formulæ, sometimes preceded by a statement of the question to be solved. He is now at work on the development of a set of lectures given by him ten years ago in Baltimore when he went to America under urgent pressure, and gave 21 lectures, generally of 1½ hours length, in 17 days, and got back in time for his Glasgow course. From that time till a little while ago he had not found time to develop the lectures he then delivered; but when he resigned his professorship Lady Kelvin says that he felt no lack in his life, but went steadily on with his work as if no change had occurred.

She told me that Lord Kelvin had been invited to deliver the Giffard lectures, and that he had considered the subject, but felt that whilst he could have given one lecture on the subject, he could not undertake to give two courses. He went back to Paley's *Natural Theology*, and studied it, and said that he could add nothing to what Paley had said, which rather surprised me, seeing how little Paley's argument is often thought of nowadays.

He is, she told me further, always willing to go anywhere for a definite purpose, and has a pleasure in scenery, but never really wishes to be away from home.

Lord Kelvin is much interested in the current questions of the day, strongly opposed to the behaviour of the motorists, and inclined to be very angry with Balfour for the transgressions of his driver ("He cannot get over it," interjected Lady Kelvin). He is rather smitten with the idea of a vast empire knit together for all purposes of mutual support and help,—feeling, I think, that electrical communication has made this possible which was impossible before, and much wishing that when we gave autonomy to Canada and Australasia we had reserved a right to free trade with them. He wanted to know how the Education Act was working in England. I told him a little about our or rather your experiences (he is a very good listener when you tell him what he likes to hear, which is, I think, almost anything). He took me into his study, where was a wonderful model which I cannot explain, to show the supposed structure of atoms and their elasticity, and besides a small globe. This set him off on a denunciation of the great omission in modern education in not teaching the use of the globes, for which he says Huxley is partly responsible. "No plane map," he said "could give any one a notion of the world; and with our Empire every English child," he said, "ought to be shown something of its extent." I said I would stir you up for our Failand school. "Tell her," he said, that I have a conscientious scruple against paying for education in geography without the use of the globes, and shall not pay the proportionate part of the rate."

Of course he was full of radium; indeed Dr. Watson said it was engrossing his thoughts, and was a real worry to him, as it seemed to him to imperil some of his conclusions as to matter. "The mystery of radium," Lord Kelvin said to me, "no doubt we shall solve it one day; but the freedom of the will, that is a mystery of another kind." He had two specimens of radium: one, impure, given him by Mons. Curie in 1900, in a small glass tube which he carries in his waistcoat pocket; the other, much purer, given him by Sir Wm. Crookes, which you observe through a magnifying glass. We went successively with him into his safe to see these specimens. It is a very dangerous thing, to judge by the stories we heard from him of Mons. Curie's fingers, and Dr. Waller's arm, on which a wound formed from inside outwards, a month after the application of a minute fragment of radium.

He talked a good deal about the expression "the fortuitous concourse of atoms," which I thought was to be found in Cicero's *De Natura Deorum*. He asked me a question I could not answer—whether Lucretius was edited by Cicero.

From fortuitous concourse of atoms he went on to speak of the whole frame of the universe, and propounded a view which I only inadequately grasped. "If," he said, "there were a mass of matter—a heap of stones at rest" (*i.e.* if I understood rightly, operated on by no force but what was inherent in it), "it would gather together in a system; and if there were a second mass it would do the same. If they were identical the two would gather together, but if they were in any respect unequal the effect of attraction would result in rotation. From these simple materials the whole starry universe might," he said, "have been evolved. I do not often mention it," he said, "for it sounds atheistic, and I am a firm believer in design."

He asked me whether I believed that other worlds than our own were inhabited. I naturally disclaimed expressing an opinion to him, and I found that he does not believe any other of the sun's planets to be inhabitable by life, but thinks there may be stellar planets in such a condition.

A course of lectures on "Christian Apologetics" was given at University College, London, in May 1903, the first of the course being on Present Day Rationalism, by the Rev. Professor Henslow. A vote of thanks was moved by Lord Kelvin in a short speech which attracted much attention. The following is *The Times*' report, as corrected by Lord Kelvin's own hand :—

Lord Kelvin, in moving a vote of thanks to the lecturer, said, "I wish to make a personal explanation with reference to Professor Henslow's mention of ether-granules. I had recently, at a meeting of the Royal Society of Edinburgh, occasion to make use of the expressions ether, atoms, electricity, and had been horrified to read in the Press that I had put forward a hypothesis of ether-atoms. Ether is absolutely non-atomic; it is absolutely structureless and homogeneous. I am in thorough sympathy with Professor Henslow in the fundamentals of his lecture. I do not say that, with regard to the origin of life, science neither affirms nor denies creative power. Science positively affirms creative power. Science makes every one feel a miracle in himself. It is not in dead matter that we live and move and have our being, but in the creating and directive Power which science compels us to accept as an article of belief. We cannot escape from that conclusion when we study the physics and dynamics of living and dead matter all around. Modern biologists are coming once more to a firm acceptance of something beyond mere gravitational, chemical, and physical forces; and that unknown thing is a vital principle. We have an unknown object put before us in science. In thinking of that object we are all agnostics. We only know God in His works, but we are absolutely forced by science to admit and to believe with absolute confidence in a Directive Power—in an influence other than physical, or dynamical, or electrical forces. Cicero, editor of Lucretius, denied that men and plants and animals could have come into existence by a fortuitous concourse of atoms. There is nothing between absolute scientific belief in Creative Power and the acceptance of the theory of a fortuitous concourse of atoms. Just think of a number of atoms falling together of their own accord and making a crystal, a sprig of moss, a microbe, a living animal. I admire throughout the healthy, breezy atmosphere of free-thought in Professor Henslow's lecture. Do not be afraid of being free-thinkers. If you think strongly enough you will be forced by science to the belief in God, which is the foundation of all Religion. You will find science not antagonistic, but helpful to Religion.

On the same day that this appeared Lord Kelvin sent to *The Times* the following amending letter :—

15 EATON PLACE, LONDON, S.W.
*May* 2 [1903].

SIR—In your report of a few words which I said in proposing a vote of thanks to Professor Henslow for his lecture "On Present Day Rationalism" yesterday evening in University College, I find the following :—"Was there anything so absurd as to believe that a number of atoms by falling together of their own accord could make a crystal, a sprig of moss, a microbe, a living animal?" I wish to delete "a crystal," though no doubt your report of what I said is correct. Exceedingly narrow limits of time prevented me from endeavouring to explain how different is the structure of a crystal from that of any portion, large or small, of an animal or plant, or the cellular formation of which the bodies of animals and plants are made; but I desired to point out that, while "fortuitous concourse of atoms" is not an inappropriate description of a crystal, it is utterly absurd in respect to the coming into existence, or the growth, or the continuation of the molecular combinations presented in the bodies of living things. Here scientific thought is compelled to accept the idea of Creative Power. Forty years ago I asked Liebig, walking somewhere in the country, if he believed that the grass and flowers which we saw around us grew by mere chemical forces. He answered, "NO, no more than I could believe that a book of botany describing them grew by mere chemical forces." Every action of human free-will is a miracle to physical and chemical and mathematical science.—Yours faithfully,

KELVIN.

A correspondence followed in the columns of *The Times*. Sir William Thiselton-Dyer queried whether Lord Kelvin was better equipped than any person of average intelligence for dogmatic utterance on biological questions, and complained that he wipes out by a stroke of the pen the whole position won for us by Darwin. Criticising Liebig's denial that flowers grew by mere chemical forces, he asked, "By what force do they grow?" And mistaking the new vitalism for the old, he added: "If growth is to be accounted for by a 'vital principle,' this must be capable of quantitative measurement like any other force. If it is physical energy in another form, Liebig's dictum is futile. If not, organisms are not subject to the principle of the conservation of energy." He overlooked the point that in the exercise of operations under the law of conservation of energy, the mere directing of a force—which may vastly affect the result—involves no necessary expenditure of energy. The old vitalism assumed that there was, and failed to account for it. Mr. W. H. Mallock inquired, "Does the evolution of organic life—does 'nature red in tooth and claw'—suggest to Lord Kelvin, what it failed to suggest to Tennyson, that the source of life is a Power which is not only creative, but is also wise, loving, and just in every comprehensible sense?" He further asked, "But is human free-will a fact? This is the great question which all the philosophers of the modern world have debated. The affirmative answer may be true; but Lord Kelvin merely assumes that it is and thus, so far as his recent letter goes, he seeks to reinforce our confidence in religion, not by meeting our difficulties, but by ignoring their existence." Sir John Burdon-Sanderson pointed out how Helmholtz had given the death-blow to the old vitalism. Another correspondent quoted Darwin's own words: "If ever it is found that life can originate on this world, the vital phenomena will come under some general law of nature. Whether the existence of a conscious God can be proved from the existence of the so-called laws of nature (*i.e.* fixed sequence of events) is a perplexing question, on which I have often thought, but cannot see my way clearly." Sir E. Ray Lankester denied the statement that any modern biologist showed signs of coming to a belief in the existence of a vital principle. On the other hand, some of the religious papers criticised Lord Kelvin's speech adversely, saying that unfortunately there was a great gulf between Lord Kelvin's affirmation of belief in a creative and directive power and the expression of Christian belief that is found in the propositions of the Apostles' Creed.

Lord Kelvin returned no rejoinder to his critics of either school of thought. He was invited by the editor of the *Hibbert Journal* to write out his speech in enlarged form for publication. A like request for the *Nineteenth Century* was made by Sir James Knowles, who wrote:—

To my thinking there is nothing more important in these whirling times than an anchorage for faith, such as common sense and science can approve, and your speech seems to me to indicate just that sort of anchorage. It brings back to my mind what Tennyson used so often to say to me about his personal faith.

Lord Kelvin consented, and sent to Sir James Knowles a version embodying the correction of his letter of May 2.

Eighteen months later Lord Kelvin returned to the question in an Address on presenting the prizes to students in the Medical School of St. George's Hospital, on October 23, 1904. The main thought is given in the following extract:—

Let it not be imagined that any hocus-pocus of electricity or viscous fluids will make a living cell. Splendid and interesting work has recently been done in what was formerly called inorganic chemistry, a great French chemist taking the lead. This is not the occasion for a lecture on the borderland between what is called organic and what is called inorganic; but it is interesting to know that materials belonging to the general class of food-stuffs, such as sugar, and what might also be called a food-stuff, alcohol, can be made out of the chemical elements. But let not youthful minds be dazzled by the imaginings of the daily newspapers, that because Berthelot and others have thus made food-stuffs they can make living things, or that there is any prospect of a process being found in any laboratory for making a living thing, whether the minutest germ of bacteriology or anything smaller or greater. There is an absolute distinction between crystals and cells. Anything that crystallizes may be made by the chemist. Nothing approaching to the cell of a living creature has ever yet been made. The general result of an enormous amount of exceedingly intricate and thorough-going investigation by Huxley and Hooker and others of the present age, and by some of their predecessors in both the nineteenth and eighteenth centuries, is that no artificial process whatever can make living matter out of dead. This is vastly beyond the subject of the chemical laboratory, vastly beyond my own subject of physics or of electricity—beyond it in depth of scientific significance and in human interest.

His own suggestion in 1871 of a possible introduction of life to this globe by meteoric sources, was often misunderstood or mis-stated. To a correspondent who wrote to him on this topic he replied in 1886:—

The "star germ theory" which I put forward as a possibility does not in the slightest degree involve or suggest the origination of life without creative power, and is not in any degree antagonistic to, or out of harmony with, Christian belief.

To another correspondent, in March 1887, he wrote:—

I think you will find nothing contrary to the Bible in the suggestion that some of the life at present on the earth may have come from seeds sown by meteoric stones. I have never thrown it out as more than a hypothesis that even so much was the case. But even if some of the living things on the earth did originate in that way so far as the earth is concerned, the origin of the species elsewhere in the universe cannot have come about through the functions of dead matter; and to our merely scientific judgment the origin of life anywhere in the universe seems absolutely to imply creative power. I believe that the more thoroughly science is studied the further does it take us from anything comparable to atheism.

Almost at the close of his life the following note was dictated by Lord Kelvin in reply to a correspondent in reference to a statement that "everybody knows Lord Kelvin's theory of life."

It is not quite true that everybody knows Lord Kelvin's theory, because certainly Lord Kelvin himself does not know it. He has put forward various suggestions at different times towards a Theory of Matter, but has never settled any in his own mind that could be called a new theory even of dead matter. The relations of matter and life are infinitely too complex for the human mind to understand. Science brings us face to face with creative power in the beginning of life on this earth and its continuance.

Lord Kelvin had a whole-hearted detestation of spiritualism and all that pertains to it; and would often go out of his way to denounce "that wretched superstition." He had seen its disastrous effects on his partner Varley. In the Address at the Midland Institute, Birmingham, in 1883 (p. 798), on the "Six Gateways of Knowledge," he spoke thus:—

Now I have hinted at a possible seventh sense—a magnetic sense—and though out of the line I propose to follow, and although time is precious, and does not permit much of digression, I wish just to remove the idea that I am in any way suggesting anything towards that wretched superstition of animal magnetism, and table-turning, and spiritualism, and mesmerism, and clairvoyance, and spirit-rapping, of which we have heard so much. There is no seventh sense of the mystic kind. Clairvoyance, and the like, are the result of bad observation chiefly; somewhat mixed up, however, with the effects of wilful imposture, acting on an innocent, trusting mind. But if there is not a distinct magnetic sense, I say it is a very great wonder that there is not.—*Popular Lectures*, vol. i. p. 258. 265.

To a correspondent, in March 1891, he wrote :—

You are perfectly right in your views about spiritualism and animal magnetism. There is no such thing as animal magnetism or any other influence, with which a table may be charged and in virtue of which it can be lifted, attracted, or repelled, or in any way influenced, without distinct mechanical action upon it. Faraday made an experimental examination of table-turning and found that the observed motions of the table were produced by the hands of the people touching it. The whole subject of spiritualism and animal magnetism is a tissue of superstition fostered by imposture.

Lord Kelvin was very fond of animal pets. His parrots have several times been mentioned. He had a horror of unnecessary slaughter of creatures, particularly of birds. He once seized the arm of a man who, while on board his yacht, was shooting a sea-gull, and he protested indignantly against such wanton cruelty.

On one occasion, while he was President of the Edinburgh Royal Society, a paper was read recounting certain experiments on living animals, —experiments which seemed to him to exceed any scientific necessity. The next day there appeared in the *Scotsman* the following letter of protest :—

UNIVERSITY, GLASGOW,
*March* 6, 1877.

SIR—In your print of this morning I see a report of Professor Rutherford's paper on "The Secretion of Bile," read at the meeting of the Royal Society yesterday evening, when, as president, I was in the chair. As chairman I did not feel that I had the right to express my opinion that experiments involving such torture to so large a number of sentient and intelligent animals are not justifiable by either the object proposed, or the results obtained, or obtainable, by such an investigation as that described by Professor Rutherford. I feel this opinion very strongly, after many years' serious consideration of the general question of the advisableness or justifiableness of experiments involving cruel treatment of the lower animals. I trust you will kindly give me this opportunity of expressing it, as my presence without protest yesterday evening might seem to imply that I approved of the experiments which were described.—I am, etc.,

WILLIAM THOMSON, LL.D., F.R.S.,
Professor of Natural Philosophy
in Glasgow University.

Though on this occasion he took an unusual course, he admitted that for adequate cause shown, vivisection was justifiable. His opposition was limited to such vivisection as was not justified by the acquisition of new knowledge. To the total suppression of vivisection he was also opposed, as the following letter to the Secretary of the Society for the Protection of Animals shows:—

*March* 21, 1885.

DEAR SIR—When applied to several years ago to join in a movement to obtain the absolute prohibition of vivisection, I refused to do so, and I still feel that I cannot approve of this movement. As I see the Society for the Protection of Animals liable to Vivisection has become united to the International Association for the Total Suppression of Vivisection, I feel that I cannot longer continue to be a vice-president, and I therefore request that you will take my name off the list of vice-presidents.—Yours truly, WILLIAM THOMSON.

Although nothing was more characteristic of Lord Kelvin than his intense passion for making scientific calculations, his incessant and laborious devotion to the deepest problems of physics did not prevent him from having other enjoyments and occupations. If he read little, he appreciated what he read. He was fond of sea-stories and of novels bearing on sea-faring life. To the sea, and all that pertains to seamanship, he was devoted, as we have seen; and sailing was his truest recreation. He was fond of sunshine. He hated to have the blinds in a room lowered, even partially, and from a boy had the habit of pulling them up to let in more sunlight.

Himself childless, he was fond of little children. He liked to take them on his knee and show them his repeater watch; and it was delightful to see him chatting with some small boy or little babbling maid, asking questions of them, and drawing out their childish answers.

When he became famous, as the world reckons fame, he had no desire to be lionized by society or courted by wealthy parvenus. If any of these self-seeking persons forced themselves on his company he was never discourteous, but shrunk into a seeming shyness. Exacting as he was in all pertaining to his work, and more exacting on himself and his own assistants than on any one else, he was essentially of a considerate and gentle turn of mind towards others.

Once the late Astronomer Royal had written an adverse criticism of one of his scientific works, in which his assistant was Ewing (afterwards Professor at Cambridge, now Director of Naval

Education), at that time a youth of twenty-one. The Astronomer Royal's condemnation rested upon a misapprehension, and Ewing, eager to correct it, wrote to his chief asking leave to publish a reply. "By all means answer," telegraphed Lord Kelvin, "but don't hit too hard. *Remember he is four times as old as you.*"

Lord Kelvin's aversion from controversy in matters of scientific import has been several times shown (see pp. 291, 383, 449). His rebuke to Edison in 1878 for accusing certain English electricians of piracy and bad faith in the invention of the microphone, is well known. He exerted himself on more than one occasion to avert public controversy as to priority in scientific discovery, but was most punctilious in acknowledging the priority of others wherever his own investigations came into question. Nothing could exceed the quiet courtesy and kindliness with which he treated younger men who brought to him the results of their own efforts in scientific research. His consideration in little things is illustrated by the following personal narrative communicated by Professor J. D. Cormack, now of University College, London:—

In 1884 the University of Heidelberg, at its tercentenary celebrations, desiring to confer an honorary degree upon Sir William Thomson, and finding that the only one at their disposal which he did not already possess was that of Doctor of Medicine, accordingly presented him with that diploma. On the very day on which it arrived his kindly eye noticed the indisposition of his lecture-room assistant, and thereupon he wrote out his first prescription in proper language and form. Nature was the remedy and a week the dose.

Vis medicatrix Naturae
viij dies R
W. T

From the possession of this diploma arose an amusing scene in 1893, when Lord Kelvin was giving evidence in the law courts in the notorious "electric belt" case. The learned counsel (now Lord Chief Justice of England) who was to cross-question him, opened fire by remarking that he presumed his Lordship's numerous qualifications did not include any medical ones. "I am M.D. of the University of Heidelberg," replied Lord Kelvin quietly; while a ripple of laughter went round the Court.

Once when a wild tale appeared in a Dundee paper about Lord Kelvin swearing at a student who had thrown a missile at his head, Professor Ewing wrote an indignant contradiction, and sent him a copy of his letter. Lord Kelvin in thanking him, said:—

The passage [in the newspaper] to which you particularly refer excited some surprise here, as there is not the slightest foundation for it. No pellet of paper ever struck my head in the Class Room during the 42 sessions I have been Professor. I never in all my life used the expression "By God," nor, I need scarcely say, the word "crucify," in such an application.

Lord Kelvin enjoyed a joke, and had a peculiar

humour of his own, often of a somewhat pedagogic flavour it must be confessed. Presiding in 1893 at the dinner to the physiologist Virchow, he announced at the close :—

> Gentlemen, the reduction of cellular tissue to free chemical molecules you may now commence—in other words, you may now smoke.

At the dinner of the Royal Literary Fund in the same year, in proposing the toast of the United Services, coupled with the health of Admiral Colomb, he said :—

> Admiral Colomb has done duty all over what the mathematicians would call the "great equipotential," or in plain language the *ingens aequor*.

With his intimate scientific friends there was an inexhaustible play of quiet fun underlying his severest studies. The pleasantries of his correspondence with Tait are but a faint reflexion of the incessant chaff and humour with which their intercourse was enlivened. In the *Memoirs* of Sir George Stokes, vol. i. p. 36, we read :—

> Lord Kelvin's visits were occasions of enjoyment to him, and great were the discussions between them, which anything served to begin; for instance, the eggs were always boiled in an egg-boiler on the table, and Lord Kelvin would wish to boil them by mathematical rule and economy of fuel, with preliminary measurement by the millimetre scale, and so on.

The love of music was a ruling passion with Lord Kelvin all his life, and, at seasons when sailing was impracticable, music formed almost his sole source of recreation. We have seen the part he took in the founding of the Cambridge University Musical Society. For his comrades Blow, Pollock, and Dykes he retained a life-long regard. When, in 1897, the life of Dykes was published, he wrote to Dykes's sister, Mrs. Cheape :—

The book you kindly sent me came one morning fifteen days ago, when I was just setting out for London. I was only then able to turn over two or three pages of it and see some interesting remembrances of my old friend John Dykes, and of our happy Cambridge musical times. I only returned home last night, and, after getting through my day's work, have been spending the evening in reading as much as I could of *The Life and Letters of Dr. Dykes*, with very great interest. I have delayed too long writing to thank you for the book, but I wished first to have at least made a beginning of reading it. It has been a great pleasure to me to read of your father and mother, and of your home in Wakefield, where I spent a week so happily 54 years ago. I shall always value the book also as a most interesting record of the life devoted to duty, which followed the bright cheerful days of your brother's youth.

In his last years he still loved, when in London, to go to the Opera, particularly if one of his old favourite masterpieces was to be performed. For the modern developments of music, particularly of German music, he had little liking. The Rev. H. F. Stewart, of St. John's College, has recorded in the *Cambridge Review* the following dialogue :—

A pianist, whom we will call X, has just played "the beautiful little *Traümerei* Op. 9 of Richard Strauss."

*Lord Kelvin* (approaching the piano, *loquitur*). The

piece is by Richard Strauss, a contemporary German composer?

*X.* Yes.

*Lord Kelvin.* Any friend of the Strauss family who wrote such excellent dance music?

*X.* None whatever.

*Lord Kelvin.* No, I should not have thought so. Has he written much?

*X.* Yes, quite a lot; but his early work has an entirely different character and style from his later compositions.

*Lord Kelvin.* Indeed! Do you know if there was any cause to account for this? We always look for cause with effect.

Thereupon X gives a short account of Strauss's abandonment of classical models and his development along the line of the symphonic poem created by Liszt.

*Lord Kelvin.* Very interesting. [*Exeunt.*]

Chatting in Lady Kelvin's drawing-room, in the spring of 1907, he asked me, on some chance turn in the conversation, if I was devoted to Wagner, and, without waiting for any reply, said that he preferred Weber. Did I know Weber's operas? Or the music of *Euryanthe*, *Oberon*, and *Freischütz*? I replied that the music was delightful, but that I thought the libretto of *Freischütz* extremely silly. "Not so silly as Wagner's operas," said Lord Kelvin. I protested that Wagner's operas—at least *Lohengrin*, *Parsifal*, and *The Ring*—were avowedly mythological and mystical. "What more mystical than the casting of the seventh bullet in *Freischütz*?" urged Lord Kelvin. I protested that I preferred the forging of the sword in *Siegfried*; at least it was not silly. He said he was going with

Lady Kelvin that night to see *The Gondoliers*, and hoped they would not have it under Wagnerian conditions with lowered lights. "I don't like," he said, "to look at the stage as from a prison or a dark cell. I like to be able to read my programme." Doubtless, Lord Kelvin would have joined in the views of the famous protest of Brahms and Joachim against the tendencies of modern music.

The French horn which he played at Cambridge used to be taken out once a year at Glasgow to illustrate his lectures in Acoustics; when, to the supreme joy of his students, he would attempt to draw out a few notes from it, playing it as he used to do with his hand inserted into the bell.

Lord Kelvin's lecture at Birmingham on the "Six Gateways of Knowledge" (1883) contains a reference to "the greatest master of sound, in the poetic and artistic sense of the word at all events, that ever lived—Beethoven." It also contains a wonderful account of the scientific implications of orchestral music.

. . . But now for what really to me seems a marvel of marvels: think what a complicated thing is the result of an orchestra playing—a hundred instruments—and two hundred voices singing in chorus accompanied by the orchestra. Think of the condition of the air, how it is lacerated sometimes in a complicated effect. Think of the smooth gradual increase and diminution of pressure—smooth and gradual though taking place several hundred times in a second—when a piece of beautiful harmony is heard! Whether, however, it be the single note of the most delicate sound of a flute, or the purest piece of harmony of two voices singing perfectly in tune; or whether it be the crash of an orchestra, and the high notes, sometimes even screechings and tearings of the air, which you may hear fluttering above the sound of the chorus—think of all that, and yet that is not too complicated to be represented by Professor Cayley, with a piece of chalk in his hand, drawing on the blackboard a single line. A single curve, drawn in the manner of the curve of prices of cotton, describes all that the ear can possibly hear, as the result of the most complicated musical performance. How is one sound more complicated than another? It is simply that in the complicated sound the variations of our one independent variable, pressure of air, are more abrupt, more sudden, less smooth, and less distinctly periodic, than they are in the softer, and purer, and simpler sound. But the superposition of the different effects is really a marvel of marvels; and to think that all the different effects of all the different instruments can be so represented! Think of it in this way. I suppose everybody present knows what a musical score is—you know, at all events, what the notes of a hymn tune look like, and can understand the like for a chorus of voices, and accompanying orchestra;—a "score" of a whole page, with a line for each instrument, and with perhaps four different lines for four-voice parts. Think of how much you have to put down on a page of manuscript or print, to show what the different performers are to do. Think, too, how much more there is to be done, than anything the composer can put on the page. Think of the expression which each player is able to give, and of the difference between a great player on the violin, and a person who simply grinds successfully through his part; think, too, of the difference in singing, and of all the expression put into a note or a sequence of notes in singing, that cannot be written down. There is, on the written or printed page, a little wedge showing a *diminuendo*, and a wedge turned the other way showing a *crescendo*, and that is all that the musician can put on paper to mark the difference of expression which is to be given. Well now, all that can be represented by a whole page or two pages of orchestral score, as the specification of the sound to be produced in, say, ten seconds of time, is shown to the eye with perfect clearness by a single curve on a riband of paper a hundred inches long. That to my mind is a wonderful proof of the potency of mathematics.—*Popular Lectures*, vol. i. pp. 274 277.

In lighter vein is the following letter to his niece, Miss May Crum, then in Berlin, studying under Joachim, himself a personal friend of Lord Kelvin, and a frequent visitor both at Netherhall and at Eaton Place:—

GLASGOW UNIVERSITY,
*Jan.* 18, '91.

DEAR MAY—Your father is with us for three days, and we are greatly delighted to have him.

I have just been telling him a little numerical experience, and he thinks you would like to hear it, as it may help, if only by an illustration, in some of your critical and scientific considerations.

I find that in Wagner's system (as in the Trilogy for instance) there is something perfectly natural in his invention of associating a certain sequence of notes with each chief character.

It has indeed been quite independently invented by one of the most natural, and at the same time most scientific of musicians. My brother-in-law, Raleigh Blandy, eleven years ago was well known to Dr. Redtail,[1] and used to highly appreciate his musical performances. One particular very definite sequence of notes somehow became associated with him in the composer's mind. Then he went away to Madeira, and only returned after nine years. The moment he came into the hall, before Dr. R. had even seen him, the latter sounded exactly the same cadence (I am not sure if this describes it correctly, perhaps I should simply say *sequence*)! This does prove an admirable naturalness in that musical idea or principle which Wagner has so splendidly developed.

I have been teaching Dr. Redtail the Pastoral Symphony, but I must confess it does get mixed up somewhat with the Raleigh Blandy strain. It looks, however, as if a wholly new idea may be evolved as the result.

It is very cold here just now, freezing; but not nearly so cold as London and the south of England, and I am afraid you are having it colder than anywhere on this island, but I hope you are all enjoying yourselves notwithstanding.

We are to have the violinist Ysaye on Tuesday night here. We are hoping Walter will come up to come with

[1] See p. 630 for a notice of this famous parrot.

us. Will you not take a place in Prof. von Helmholtz's new electric balloon and come with us also?

Aunt Fanny sends her love, and I am, yours very affec$^{y\cdot}$, WILLIAM THOMSON.

*P.S.*—You might come to us on the 19th of March for Joachim's concert.

Lord Kelvin often lamented that in the pressure of modern university and college education, and the too early specialization which is in fashion, students of science do not now receive training of the same breadth as used to be the case under the older curricula. Logic he held to be a study of almost vital importance to the scientific man; *zuerst Collegium logicum*, he would quote from *Faust*, not in jest but in earnest, recalling the logic lectures he had heard in his teens at Glasgow under Professor Buchanan. And though he could laugh at the mediæval jargon of *barbara celarent* and the rest of the doggerel hexameters which furnished mnemonics for the various orders of syllogism, he valued highly the aid which the study of logic gives to clear thinking. "More ships," he said, "have been lost by bad logic than by bad seamanship."

Greek, too, he put amongst the subjects in which the student should have some training, and advocated its retention in the universities. When Sir Richard Jebb asked him if his opinion in this matter might be published, he replied,[1] "Yes; I think, for the sake of mathematicians and science students, Cambridge and Oxford should keep Greek, of

[1] See Jebb in the *Cambridge Chronicle*, October 27, 1891.

which even a very moderate extent is of great value."

He had an instinctive distrust for the fallacies that arise from the loose usage of terms in different senses that are analogical only. He was emphatic (*Popular Lectures*, vol. i. p. 289) that "no relation exists between harmony of sound and harmony of colour." The fallacy lies, of course, in the use of the word harmony.

In his use of language he was precise—almost pedantically so at times; and when in his effort to make clear, in some lecture or communication to a learned society, the exact meaning of his expressions, he would half close his eyes and raise his chin, rapping out the words in a sort of staccato utterance that gave the impression of an exceedingly minute precision of definition. To some, however, the precision so imported savoured of Scottish caution. He hated ambiguities of language, and statements which mislead by looseness of phrasing. With painful effort he strove for clarity of expression, elaborating his phrases in a way that threatened at times to defeat the end intended. In "that hazy medium of words wherein we all drowse," he at least would attempt to observe the proprieties of language. As an example take this:—"Externally the sense of touch, other than heat, is the same in all cases—it is a sense of forces, and of places of application of forces, and of directions of forces." The sense of temperature he insisted on distinguishing from the sense of touch with which it is often confused.

Certain terms in current use he expressly avoided. The term *induction*, which Faraday had employed to denote the operation of inducing, he would never employ in the sense in which, unfortunately, Maxwell used it, in his *Treatise on Electricity and Magnetism*, as a synonym for the density of the magnetic flux. Induction [1] was the operation of inductive reception, not the quantity that resulted thereby.

He objected greatly to any physical phenomenon being described as *mysterious*. "When you call a thing mysterious, all that it means [2] is that you don't understand it."

Of the petty pedantries of text-book science he was scornful. Lecturing on the moments of forces and their equilibrium, as exemplified in the lever, he would refer to the empirical classification of levers in the text-books, and say: "There are said to be three kinds of levers; there are levers of the first order, and levers of the second order, and levers of the third order. I don't remember which is first, second, or third, and it doesn't matter; in all three kinds the lever turns about a fulcrum, and that is enough"—and then he proceeded to expound the theory of couples which supersedes the antiquated treatment of the subject.

Although daring in speculation as to the causes of physical phenomena, he was ever keen to subject any and every hypothesis to the test of

[1] "Induction or Electromotive Force," *Math. and Phys. Papers*, vol. ii. p. 133; see also *Electrostatics and Magnetism*, p. 494.

[2] An example occurs in what he said of Joule in 1848; see *Math. and Phys. Papers*, vol. i. p. 103, footnote.

numerical calculation. All hypotheses might be good, as guesses; but guesses they must remain until by finding appropriate mathematical forms for them they could be verified by actual computation, and be received as definite theories. He did not admit that the term theory could be validly applied to any unverified hypothesis. "Every theory," he wrote, in the *Athenæum* of October 4, 1856, "is merely a combination of established truths." Like Faraday, he hated uncertain knowledge and those half-truths which are often more misleading than downright errors.

He was particular as to the forms of certain words, using for instance the plurals "vacuums," "moduluses," "formulas," etc., spelling with letter *k* the words "disk" and "skulling," and insisting that the common noun "ampere" should have no accent. He would never sanction the vulgarism of "ammeter" for "amperemeter." He often used the word *aphasia* in an extended sense to denote the difficulty which the limitations of language present to the ready expression of scientific concepts: he also used the same word jocularly to describe the difficulty of a nervous student in answering a question in the class-room.

The words "physicist" and "scientist" he disliked, particularly the latter. Yet he used "physicist" once, when presenting Stokes with the Copley medal; and "scientists" occurs once[1] in his

[1] At the Royal Society Banquet of 1894, when the word *scientists* was again used by Lord Kelvin, a protest was raised against this interloper of doubtful etymology, and a newspaper correspondence arose. Huxley considered

*Popular Lectures*, vol. i. p. 405, where he refers with a touch of sarcasm to "scientists speaking, as now, each his own vernacular"!

As in all progressive researches, so in his own, he found new language necessary for the denotation of new ideas; and though in this respect he was surpassed by his brother James, he enriched the scientific vocabulary with many terms of his own coinage. To him we owe the term "kinetic" energy (the counterpart of Rankine's "potential" energy), the expressions "permeability" and "susceptibility" in the magnetization of iron, the adjectives "aeolotropic," "circuital," and "positional" ("a single adjective is used to avoid a sea of troubles here," *Thomson and Tait*, vol. i. p. 370), and the phrase "ohmic resistance" for one of the factors of energy-waste in a circuit traversed by an electric current. Old students of his class will recall how they listened year after year for the staccato enunciation of his definition[1] of the ideal magnet—"an infinitely long, infinitely thin, uniform and uniformly and longitudinally magnetized rod of steel." At an earlier time he had contented himself with a

"scientist" as degraded a word as "electrocution," and it was condemned by the Duke of Argyll, Lord Rayleigh, and Sir John Lubbock. A word seems needed, which is scarcely supplied by "philosopher," "naturalist," or "sçavant." A well-known Cambridge Professor was showing a distinguished foreigner around the place, when the foreigner remarked: "You do not appear to have any vat ve call *savans*." "Oh, yes," replied the Professor, "we have; but we call them p-p-p-prigs."

[1] There is a story told how his students looked forward to the coming of this definition, and punctuated it when it came by tapping on the floor with their feet, causing the Professor at the end to cry "Silence!"; and how, on one occasion, when by previous agreement they refrained from the usual running accompaniment of stamping, he still, by force of habit, cried "Silence!" at the end! It is at least *ben trovato*.

simpler definition. "A magnet is a substance which intrinsically possesses magnetic properties," a phrase irresistibly recalling Sydney Smith's famous definition of an archdeacon.

In an article in the *Glasgow Herald*, on the occasion of Lord Kelvin's elevation to the peerage, Dr. Hutchison, Headmaster of the Glasgow High School, a student of the university in the 'sixties, gave a number of lively reminiscences, including a graphic account of the great snow riot of 1864, when the Professor, as he then was, gave evidence in the police court in favour of certain students accused of assaulting the police in a great snowballing at the portals of the old college. In these reminiscences Dr. Hutchison says :—

Sir William in my time had three favourite subjects, on each of which he was sure to go off at a tangent whenever the smallest opening was presented. One of these was the Cambridge system of examinations, on the miserable nature of which Sir William with great warmth expatiated, and frequently instructed us, and in fact left us with very cheerful impressions of the value of our Scotch degrees. Equally severe he was on our "insular and barbarous" system of weights and measures,[1] which one

[1] For examples of his onslaughts, see pages 436, 794, 808, and 866. He was a keen supporter of the Bill for making the metric system compulsory. On this question he wrote, on February 7, 1895, to Herbert Spencer :—

"DEAR MR. SPENCER—It is the uniform simplicity of the French metric system which gives it its great advantage. This advantage it would still have even if the Arabs had counted by aid of their ten fingers and thumbs and two ideal digits, and had given us a duodecimal system of arithmetic. I quite agree with you as to the convenience of halves and quarters in coinage and in weights and measures. It is universally found convenient to take halves and quarters of the smallest unit in any particular measurement in practical use. Lengths on a road being measured in kilometres, all people would frequently use $\frac{1}{2}$ and $\frac{1}{4}$ kilometres for convenience : never $\frac{1}{8}$ or $\frac{1}{16}$. The $\frac{1}{8}$ and $\frac{1}{16}$ and $\frac{1}{32}$ are so

may be surprised to find surviving the vigour of his attacks.

But Sir William's anger on these two subjects[1] was as nothing compared with what it was on the third. This was — Hegel. If you wanted to see an illustration of pure white heat you should have seen Sir William castigating Hegel for the audacity of his assaults on the Newtonian philosophy. I remember on one occasion he sent over to the library for the learned volume containing Hegel's criticism of Newton, in order that we might hear the *ipsissima verba*, the downright nonsense, of this "arrant impostor." I remember the expression perfectly, both from its native force and the vehemence with which it was uttered. Had Home Rule been in the air one would have thought that Sir William was denouncing some prominent Gladstonian. Some, too, may remember the energy with which he assailed the professors of cheap materialism.

In my time nothing was more characteristic of Sir William than the thoroughness with which he thrashed out every subject that he took up. His students were

many curses to British mechanics, disabling them in comparison with mechanics of all other countries except England and America. Mechanics of other countries use ½ and ¼ centimetres when they find it convenient to do so. Miles, furlongs, perches, roods, yards, feet, inches, acres, square yards, square feet, square inches, cubic yards, cubic feet, cubic inches, cause enormous loss of efficiency to English engineers, and really involve a great national loss in useless labour, month after month, and year after year. We may extend the anathema to cwts., lbs. (avoirdupois), lbs. (troy), ounce (troy), dwt., scruples, grains, gunmakers' drams, apothecaries' drams, gallons, pints, quarts, gills.

"You say 'I see you have been tacitly urging,'—I have never tacitly when I have had an opportunity of speaking out.

"I wish you could be convinced, and give your powerful influence to a reform which is much needed, and from the want of which we in England, all of us, suffer every day of our lives.—Yours very truly, KELVIN.

"*P.S.*—Excuse haste, as I am writing, walking about in my laboratory among sufferers from our mischievous British system, or rather want of system, in weights and measures."

[1] Another topic of his diatribes was the incompetence of architects, and their ignorance of engineering principles; another was flying-machines and aerostation. He wrote in 1896 to Col. Baden-Powell:—"I have not the smallest molecule of faith in aerial navigation other than ballooning, or of expectation of good results from any of the trials we hear of. So you will understand that I would not care to be a member of the aëronautical society."

quite persuaded that he knew everything about a subject that could be known, and they felt that he was perfectly justified in his sometimes rather unsparing language about charlatans, "showmen," mere brilliant experimenters who wished to pose as philosophers. In fact, our faith in him went so far that I have heard some of the best students say that they were hardly satisfied with any philosophical theory till it had Sir William's *imprimatur*, and were almost content to accept it as all right if it got that.

The particular passage in Hegel which aroused Sir William to scorn was the attack[1] on Newton's theory of planetary motion around the sun:—

*The motion of the heavenly bodies is not a being pulled this way, or that (such as imagined by the Newtonians), but is free motion; they go along as the ancients said, as blessed gods. The celestial corporeity is not such a one as has the principle of rest or motion external to itself. Because stone is inert, and all the earth consists of stones, and the other heavenly bodies are of the same nature, is a conclusion which (wrongly) makes the properties of the whole the same as those of the part. Impulse, Pressure, Resistance, Friction, Attraction, and the like, are valid only for an existence of matter other than celestial.*

"Hear his words!" Sir William would say. "If,

[1] The passage is to be found in Hegel's *Encyklopädie* (second edition, 1827), part xi. p. 250, in § 269 in the section on Absolute Mechanics, of the part called *Naturphilosophie*. Its illogical silliness is best seen in the original German.

"Die Bewegung der Himmelskörper ist nicht ein solches Hin- und Hergezogen seyn, sondern die freie Bewegung; sie gehen, wie die Alten sagten, als selige Götter einher. Die himmlische Körperlichkeit ist nicht eine solche, welche das Princip der Ruhe oder Bewegung ausser ihr hatte. Weil der Stein träge ist, die ganze Erde aber aus Steinen besteht, und die andern himmlischen Körper eben dergleichen sind—ist ein Schluss, der die Eigenschaften des Ganzes denen des Theils gleichsetzt. Stoss, Druck, Widerstand, Reibung, Ziehen und dergleichen gelten nur von einer andern Existenz der Materie, als die himmlische Körperlichkeit. Das Gemeinschaftliche beider ist freilich die Materie, so wie ein guter Gedanke und ein schlechter beide Gedanken sind: aber der schlechte nicht darum gut, weil der gute eine Gedanke ist."

gentlemen, these be his physics, think what his metaphysics must be!"

It was perhaps this hatred of the school of thought which exalts wordy description above rigid demonstration, that gave Lord Kelvin more or less of a bias against all metaphysics, and to some extent against all non-symbolic philosophy. Herein he differed greatly from Helmholtz, whose habitual philosophical thought was of far wider sweep. He did not even appreciate the achievements of the greatest of contemporary English leaders in philosophy; for he wrote:[1]—

> I have never been of opinion that the philosophical writings of the late Mr. Herbert Spencer had the value or importance which has been attributed to them by many readers of high distinction. In my opinion a national memorial would be unsuitable.

Lord Kelvin's contempt for metaphysics[2] was often expressed in the interjected *obiter dicta* of his class-room. "Mathematics is the only true metaphysics" was another saying. He liked clear-cut

[1] *Nature*, vol. lxx. p. 521, Sept. 20, 1906.

[2] The following story (here a little softened from the vernacular) was narrated by Lord Kelvin himself when dining at Trinity Hall:—

A certain rough Highland lad at the university had done exceedingly well, and at the close of the session gained prizes both in mathematics and in metaphysics. His old father came up from the farm to see his son receive his prizes, and visited the College. Thomson was deputed to show him round the place. "Weel, Mr. Thomson," asked the old man, "and what may these mathematics be, for which my son has getten a prize?" I told him, replied Thomson, that mathematics meant reckoning with figures, and calculating. "Oo ay," said the old man, "he'll ha' getten that fra' me: I were ever a braw hand at the countin'." After a pause he resumed: "And what, Mr. Thomson, might these metapheesics be?" I endeavoured, replied Thomson, to explain how metaphysics was the attempt to express in language the indefinite. The old Highlander stood still and scratched his head. "Oo ay: maybe he'll ha' getten that fra' his mither. She were aye a bletherin' body."

conceptions in all matters of intellectual moment. "There are no paradoxes in science" was one of the *dicta* of the *Baltimore Lectures*. In a Royal Institution discourse in 1887, his phrase ran: "Paradoxes have no place in science. Their removal is the substitution of true for false statements and thoughts." Yet some of his own conclusions are intensely paradoxical; a striking one occurs in § 479 of *Thomson and Tait*, where the north edge of a crevasse running east and west is shown, in certain cases, to have a *lower* latitude than its southern edge!

When confronted with a new fact or discovery, Lord Kelvin's attitude of mind varied according to the circumstances of the case. Thus when Kerr in 1876 announced his discovery of electro-optic stress, Lord Kelvin was instantly and almost explosively excited; he had predicted this very effect thirty years before, and had written of it to Faraday, who had himself looked for it in vain at a still earlier date. When Röntgen's discovery of the X-rays was announced at the end of 1895, Lord Kelvin was entirely sceptical, and regarded the announcement as a hoax. On the other hand, when Crookes first showed him the radiometer, one evening in 1874, he sat down watching it in perfect silence for nearly an hour, gazing at it, shading the light from it at intervals with his hand, or moving it towards the lamp or from it, and thinking—thinking. Not even in 1906 was he satisfied that the true theory of the radiometer had ever been given.

Like Faraday, and the other great masters in

science, he was accustomed to let his thoughts become so filled with the facts on which his attention was concentrated that the relations subsisting between the various phenomena dawned upon him, and he *saw* them as if by some process of instinctive vision denied to others. It is the gift of the seer. Often he had to labour to devise explanations of that which had so come to him; and instances are known of his spending whole days upon trying to frame or recover a demonstration of something that had been previously obvious to him. Science has found no specific name for this real power of divination, the precious and incommunicable quality of her peculiar genius. Those who possess, or are possessed by it, are not always able to state the processes of thought that have guided them. Observation, experience, analysis, abstraction, imagination, all these are necessary—but are they all? Something seems yet wanting to account for what we call the intuition of the master mind. It is surely more akin to the innate faculty of the great artist than to the trained powers of the analyst or the logician. Nor is it without significance that in the sentence in which Lord Kelvin stated his inability to accept the Darwinian hypothesis of natural selection he should have said: "I have *felt* that this hypothesis does not contain the true theory of evolution."

Lord Kelvin had great faith in the benefits to be gained in science by co-operation in research. His assiduous attendance at meetings of the British

Association stimulated younger men into combined effort. He secured their service on Committees,[1] suggested often by himself, to investigate different subjects. He regarded the meetings of the British Association as peculiarly effective in advancing scientific investigation. In September 1888 he wrote to *The Times* :—

> No one not following the course of scientific progress, generally or in some particular department, can fully understand how much of practical impulse is owing to the British Association for the contributions made in the course of the year to the scientific societies and magazines, in which achieved results of scientific investigation are recorded and published.

Lord Kelvin possessed an extraordinarily exact memory of dates and events in matters which interested him. He could put his finger with almost unerring certainty upon paragraphs in his own earlier writings, and even upon the leaves of his notebooks of years before. His punctilious habit of dating everything he wrote doubtless helped him in such references. But at times his memory as to lesser matters suffered lapses. A letter from Tait to Lord Rayleigh touches on this :—

[1] From the year 1860, he appears to have served on no fewer than 41 such Committees, many of them being reappointed for several successive years. The Committee on the Investigation of Underground Temperatures sat for 18 years, that on Electrical Units for 7, and that on Electrical Standards for 10 years. The subjects of these Committees included such varied matters as the Printing of Mathematical Tables, Magnetic Surveys, Elasticity of Wires, Screw Gauges, Electrolysis, Endowment of Research, Patent Laws, Meteoric Dust, Lunar Disturbance of Gravity, Tides in the British Channel, Harmonic Analysis as Applied to Tides. He thus obtained data for tidal investigations and theories all too briefly mentioned on p. 729. Sir George Darwin says that the whole of his work on oceanic tides sprang from ideas initiated by Lord Kelvin, to whom in this sense he dedicated the volume of his papers relating to tidal theory.

12 *July* 1886.

. . . . . . . . . . . . .

I happen to have preserved Thomson's statements as made to you, pencilled (I suppose) *in your own house* on a note of mine which he returned. Please let me have it again, so that I may rub it well into him for his constant indulgence in prolepsis. I have no doubt that, when he pencilled the thing, he intended to do it, and therefore looked on it as done—then forgot it. . . .

The political views of Lord Kelvin are readily gathered from the preceding pages of this book. Briefly, he was, until the Home Rule split of 1886, a moderate Liberal, and an active supporter, on occasions, of the Glasgow University Liberal Association, though too much master of his own intellectual integrity to become a strong party politician. He once told me that he preferred Chamberlain's plan of Home Rule with four Irish Parliaments, one in each province. In 1871 he regarded war as a relic of barbarism probably destined to become as obsolete as duelling. He wrote to Tait, in 1879: "It must not be supposed that those who do not want to fight are devoid of patriotism." He did not care for Mr. John Bright, and he had as little regard for Mr. Gladstone's rhetoric as Mr. Gladstone had for science. More than once he expressed the wish[1] to see party government disappear, and to have a cabinet in which all parties should be represented, Lord Rosebery sitting beside Lord Salisbury. After the Home Rule disruption he became an active

[1] Another form of his wish was to see certain administrative questions removed from party politics, so as to restrict the range of party differences.

leader of the Liberal Unionists in the West of Scotland. In common with his Ulster kinsmen he regarded any concession to the Nationalist Irish party as tending to the dissolution of the bonds of empire: and he rarely spoke on the Unionist platform without proclaiming himself of Irish birth.

Once indeed, in 1881, at Birmingham, he alluded to the land of his birth in a non-political sense, and to the Irishman's claim to seven senses. "I presume," he said, "the Irishman's seventh sense was common sense; and I believe that the possession of that virtue by my countrymen—*I speak as an Irishman*—I say the large possession of the seventh sense, which I believe Irishmen have, and the exercise of it, will do more to alleviate the woes of Ireland than even the removal of the 'melancholy ocean' which surrounds its shores."

To Sir Guildford Molesworth, on June 25, 1902, he wrote regarding a proposed Imperial Customs Union League:—

I wish you had been born fifty years earlier and had suggested as an essential condition connected with the grant of self-government to our Colonies that there must be perfect inter-colonial free trade, and free trade between the mother-country and all her Colonies. Before 1840 we lost the opportunity of securing this. . . . I see great difficulties now, but I hope they may be partially, and perhaps at last wholly, overcome, through our Colonies joining us in free trade or in as near an approach to perfect free trade as possible. I cannot, however, feel that it would be wise for us in any case to return to protection.

Some contemporary notes on Lord Kelvin's later

opinions are afforded by the following extract from the diary of his niece, Miss Mary Hancock Thomson, when staying at Netherhall in 1904:—

*Jan.* 5, 1904.—Uncle William is looking very well. He conversed all through breakfast, and did not seem to suffer from face-ache. He spoke of the impending war between Russia and Japan as likely to take place, and much to be regretted. I spoke of Mr. Chamberlain's tariff proposals. Uncle William would like to see Free Trade all over the British Empire, Colonies, mother-country, and all; but thinks a small protective tariff against foreign goods would be beneficial, especially in the case of goods produced under conditions of long hours and too low wages, such as we wish to avoid in England. It is evidently the Imperial idea which attracts him. He says he sees no valid reason to exempt food and tax other necessities: it is, 1st dwelling-place, 2nd clothing, and only 3rd food, by which the poor feel the pinch of poverty.

I asked him would he approve of stopping the importation of ready-made doors and windows from Scandinavia by tariffs? He said not if it enabled English house joiners to make houses dearer than need be, by restriction of work. If tariffs on such things had a bad effect, and made building too costly, the tariffs could be taken off. He very strongly disapproves of trades unions restricting the number of bricks a bricklayer may lay in a day.

*Jan.* 11.—At breakfast Uncle William spoke of the Colonies, and blamed the "besotted Tories" of his early days for not wanting to keep the Colonies. The Tories then were worse than the Liberals. Uncle William says Benjamin Franklin was a staunch imperialist ten years before the war of Secession. Many Americans wished to maintain the union with England, but the home Government "kicked them away."

Speaking of party politics Uncle William says party politics and party animosities do such frightful harm that he would be glad to see a break-up of the system of two parties in Parliament. He says, as things are at present, members have to vote for what they do not approve of for the sake of their party, and have to vote against what they want if it is proposed by the other side, and he thinks this most deplorable. But as long as the present system exists it is often necessary for members of Parliament to vote in such a way as not to risk putting their party out of office, even to the voting of a thing they don't approve of, or against a thing they do approve of.

Though Lord Kelvin had had much experience in University work, the part he played in the development of University organization was limited. He served, though rather inconspicuously, on two University Commissions: that on the Scottish Universities in 1890 (see p. 896), and that on the University of London in 1889 (see p. 890). He warmly advocated (see pages 655 and 663) the creation of new Universities in England, and was the originator of the movement for granting a Charter to the Victoria University of Manchester—a fact not forgotten when, in 1895, the Victoria University conferred on him the Honorary Doctorate in Science. Several times he stated his view that the primary function of a University was to teach. He openly expressed his contempt for a University that occupied itself merely in holding examinations. He thought that there should be "some more suitable title" than the stamp of a University degree to mark the possession of knowledge by those who have not had the opportunity of being trained in any thoroughly equipped College or University. He had a great belief that one of the main uses of a University was to form character. Personal contact with the teacher,[1] if the teacher be a man of wide intellectual grasp and not a mere examination-grinder, was an element of the most formative potency.

[1] Lord Kelvin would assuredly have endorsed the words of von Helmholtz respecting the formative personal influence of his own master Johannes Müller, and equally true to-day of the influence of both Kelvin and Helmholtz on their respective students:—"Wer einmal mit einem oder einigen Männern ersten Ranges in Berührung gekommen ist, dessen geistiger Maasstab ist für das Leben verändert."

Lord Kelvin took part, as mentioned on p. 641, in the modification of the Cambridge Mathematical Tripos, which, originating mainly with Sir George Airy about 1848, took effect in 1873 when additional subjects (Heat, Electricity, and Magnetism) were included in the curriculum. He was additional examiner in 1874, and as member of the Mathematical Board of 1875 supported recommendations to the Senate for increasing the teaching staff of the University in Mathematics. In the later changes[1] in the Mathematical Tripos he took no active part, and was not in sympathy with them. A letter to Darwin of February 1900 will be found on p. 1151 *infra*. In the summer of 1906 he wrote to Routh :—

I hope the order of merit in the Mathematical Tripos will be maintained. The Senior Wranglership is an institution of Cambridge which ought not to be abandoned. I believe its maintenance has a good effect on the whole Tripos, and particularly on those who are at all near the running for the first place. It seems to me that the division into two parts does not work well, and that the very best reform that could be made would be to go back to the Tripos examination as it was in 1845, with only some modification in the way of physical mathematics.

He wrote his views more fully to Sir George Darwin :—

15 EATON PLACE, S.W.,
12*th Nov.* 1906.

DEAR DARWIN—I have received your letter of yesterday, and I am very sorry that you are seriously concerned

[1] These were : in 1882, when the Tripos was divided into Parts i., ii., and iii., and the Wranglerships were awarded on (the junior) Parts i., ii. ; and again in 1886, when the Tripos was reduced to Parts i. and ii., the Wranglerships being decided on (the junior) Part i. ; and in 1906-7, when Wranglerships (including the Senior) were abolished, and the Tripos was allowed to be taken by a candidate in his *first* year.

in respect to the reorganization of the Mathematical Tripos. I still feel as I did last June or July when I wrote to Routh the letter to which you refer. But the question of personal order of merit, though decided so as to abolish the Senior Wrangler, is less important than any reorganization which could diminish the importance and impair the prestige of the Mathematical Tripos. I think it would be a damage to the University of Cambridge and to the advancement of Science throughout the world, if any regulations should be adopted by which the efficiency of the Mathematical Tripos, and the number of undergraduates taking it, might be seriously diminished. The unique strength of Cambridge, as a place of experimental research, and as a leader in the advancement of Science generally, has depended greatly on the mathematical foundations given to a large proportion of all the undergraduates by the Mathematical Tripos, within the last hundred and fifty years. About a week ago I wrote to this effect to Dickson, with whom I had had some conversation about it when at Cambridge at the end of October.

I think it might be a good thing, tending to fertilize the mathematical studies of undergraduates taking the Mathematical Tripos, if it were made a condition for going in for its examination that every candidate should have attended lecture courses and gone through practical work, during at least two of his three years, in either the Cavendish Laboratory or the Chemical Department.

I do not understand the state of affairs in respect to placeting and non-placeting, and "appeal to the whole Senate" to which you refer. I feel that the whole affair must be settled by residents conversant with the whole working of University arrangements in respect to scientific studies, and I do not feel that my own opinions can be very helpful towards conclusions in the present circumstances.

I should have thanked you before for the copy of the three papers which you have sent me, and which I carefully note are all three to be returned to you. Your own two papers will be very useful to me in trying to think of the ancient history of the Moon, with or without radium to help !—Yours, KELVIN.

At an earlier date another question had agitated Cambridge, namely, the financial support which the Colleges were to be compelled to give to the University. A letter written by Sir William Thomson to the then Master of Peterhouse is of more than transient interest :—

18 *Nov.* 1880.

DEAR PORTER—I was very sorry indeed to receive the intimation this morning that I told you of in my telegram. As I shall be obliged to go up to London the week after next, to give evidence in compliance with a summons on the part of the Post Office, I am afraid I could not undertake to make the journey this week also; and therefore I am very sorry not to be able to attend the College meeting on Saturday.

As to the proposed Commissioners' Statutes, I don't feel so much averse to the Colleges being heavily taxed for University purposes as perhaps it is natural for Colleges and Fellows of Colleges to feel. It seems to me that if the money required by the University is made good use of, the diminution in the number of Fellowships in any or all of the Colleges which it will entail, will not be detrimental to the original object for which the College funds exist, and that it will not practically injure the Colleges themselves. If we at Peterhouse had only ten foundation Fellowships to give away instead of fourteen, but if, on the other hand, our undergraduates could look forward to University appointments (lectureships, readerships, with chances of ultimate promotion to professorships), etc., the present class of honour men would feel that they had really more to look forward to, and that of a more satisfactory character, than they have had hitherto. The prospect of obtaining one of the University appointments would be a great boon to those who really desire to remain doing the teaching or scientific work of the

University, and to such men would, I believe, be more than compensation for the shortened tenure of the Fellowships. Then the shortened term of the Fellowships will make a good and regular supply of vacancies; and so the inducements for a large number of undergraduates of the kind that look forward to Fellowships, etc., will be practically as good as it has been hitherto, or better.

I daresay, however, we are all pretty much agreed on this point; and that it will only be in respect to matters of proportion and adjustment of details that we really wish to protest, or to ask for changes in the ordinances. I should be much obliged by your letting me know if the matter is much discussed at the meeting of Saturday, and if it seems probable that it will be desirable to have further discussion with as many as possible of the Fellows present, about Xmas or any other time before May. . . . —Yours very truly, WILLIAM THOMSON.

The Revd. the Master,
Peterhouse Lodge, Cambridge.

From his earliest boyhood Lord Kelvin had been trained by his father to familiarity with not only the symbols but the methods of mathematical calculation, and, even before he went to Cambridge as an undergraduate of seventeen, he was familiar with the higher mathematics. Cambridge may have drilled him into the facile handling of analytical and symbolic processes, and systematized somewhat his impulsive genius, but happily she did not maim his own instinctive tendency toward thinking physically. His mathematics never degenerated into symbol-juggling, amid the most complicated of transformations the expressions still held[1] for him physical

[1] "Every equation has a physical meaning, and you should always try to realize it," he would keep saying.

significations. He used his mathematics—and used it with supreme ability—as a tool to his hand in the discovery of truth. He never allowed it to become his master, or permitted himself to be drawn by a mathematical argument into a non-physical inference. Again and again in reading even his most abstract writings one is struck by the tenacity with which physical ideas control in him the mathematical form in which he expressed them. An instance of this is afforded by a footnote which he appended to a passage in Vol. I. of his *Mathematical and Physical Papers* (p. 457), which is an example of a mathematical result that is, in his own words, "not instantly obvious from the analytical form of my solution, but which we immediately see must be the case by thinking of the physical meaning of the result." And again, in 1876 (*Mathematical and Physical Papers*, vol. iii. p. 320), he confessed to "a non-mathematical short-cut" which he had used in a paper fourteen years before. Not seldom did he, in his writings, set down some mathematical statement with the prefacing remark "it is obvious that," to the perplexity of mathematical readers, to whom the statement was anything but obvious from such mathematics as preceded it on the page. To him it was obvious for physical reasons that might not suggest themselves at all to the mathematician, however competent. So much was this so that during Lord Kelvin's lifetime purists in mathematical science have been known to carp at Lord Kelvin's "instinctive" mathematics. Yet, though few, if any—Clerk-Maxwell perhaps only excepted—ever possessed the same almost magical quality of physical insight, none could be more strict than Lord Kelvin in requiring demonstration freed from untenable assumptions or undemonstrable hypotheses. Daring as he was, at least in his earlier days, in the application of analytical methods to the phenomena of nature, he was in several ways very conservative. For example, he never would countenance the use in physics of the method of quaternions. At the British Association Meeting at Cambridge in 1845, he had met[1] Hamilton, who there read his first paper on Quaternions. One might have thought that the young enthusiast would have readily welcomed a new and ingenious method of symbolic analysis: but it was not so. He would not use quaternion notation or quaternion methods himself, nor did he admit the vector calculus into his works. His thirty-eight years' war with Tait[2] on this question is touched on in p. 452 *supra*. The following note furnishes an example :—

*July* 25, 1896.

DEAR TAIT—I see no possible objection to your now publishing the deferred "scrap" if you yourself approve of what is said in it in favour of quaternions.

I think you are right in your use of the word "Volapuk," but I don't think you should confine it to

[1] See *Popular Lectures and Addresses*, vol. ii. p. 139.

[2] Tait once urged the advantage of Quaternions on Cayley (who never used them), saying : "You know Quaternions are just like a pocket-map." "That may be," replied Cayley, "but you've got to take it out of your pocket, and unfold it, before it's of any use." And he dismissed the subject with a smile.

the vector part of quaternions. The whole affair has in respect to mathematics a value not inferior to that of "Volapuk" in respect to language. . . .—Yours, K.

Writing in April 27, 1889, to Oliver Heaviside, he says :—

I see I was wrong in attributing "curl" to Clifford. He gives a good many such words, but it was, as you say, Maxwell that first gave *curl*, as he in fact tells us himself in the first volume of his *E and M.* It is rather the symbolic system connected with it in your own and Maxwell's papers that I object to, than the word itself, and I cannot agree with any attack on Cartesian co-ordinates. All words that help us out of aphasia, provided they promote clearness instead of the reverse, are to be welcomed, so you will see I agree with a great deal of your letter, but not with all, and not with the attack on Cartesians in your last *Electrician* article.

All this, however, is a mere crust on the matter of electrodynamics. We want a thorough mechanical theory which shall include the undulatory theory of light with electrostatics, and electromagnetic force; and electro-magnetic induction; with the *mobility of the medium and all the bodies concerned*, which is part of the essential nature of the affair.

And again, writing in December 1892 to R. B. Hayward respecting his *Algebra of Coplanar Vectors and Trigonometry* :—

I do think, however, that you would find it would lose nothing by omitting the word "vector" throughout. It adds nothing to the clearness or simplicity of the geometry, whether of two dimensions or three dimensions. Quaternions came from Hamilton after his really good work had been done; and, though beautifully ingenious, have been an unmixed evil to those who have touched them in any way, including Clerk Maxwell.

About certain branches of mathematics he was

always enthusiastic. In praise of Fourier's "mathematical poem," the *Théorie analytique de la chaleur*, he never tired. He spoke of "Maclaurin's splendid theorem," "Lagrange's splendidly powerful method." He was immensely excited over harmonic analysis, and the adaptation of his brother James's mechanical integrator to perform harmonic analysis of the tides, and to solve differential equations.

"Do not imagine," he once said to an audience at the Midland Institute of Birmingham, "that mathematics is hard and crabbed, and repulsive to common sense. It is merely the etherealization of common sense." If the learning of mathematical processes had ever been bitter to him, he had early been "brought past the wearisome bitterness" of his learning. Was a thing hard? Then that was all the more reason why you should strive to master it. And so hard became easy[1] to him, and his mastery over difficult and intricate processes of thought grew into assurance. Hence arose the value which he set upon the study of analysis, for his own students[2] as well as for all students of

[1] Once when lecturing he used the word "mathematician," and then interrupting himself asked his class: "Do you know what a mathematician is?" Stepping to the blackboard he wrote upon it:—

$$\int_{-\infty}^{\infty} \epsilon^{-x^2}\, dx = \sqrt{\pi}.$$

Then, putting his finger on what he had written, he turned to his class and said: "A mathematician is one to whom *that* is as obvious as that twice two makes four is to you. Liouville was a mathematician." Then he resumed his lecture.

[2] The Glasgow University Calendar for the Session 1863-4, in the announcements as to the Senior Natural Philosophy Class, states: "Students who desire to undertake these higher parts of the business of the class ought to be well prepared on all the subjects of the Senior Mathematical Class." See also the footnote to p. 420 above.

physics. In the Preface to *Thomson and Tait* he spoke of those who "have the *privilege* which high mathematical attainments confer." He complained to Faraday (see page 204 *supra*) that a certain paper of Matteucci's, which he found some difficulty in understanding, was "not enlivened by any $x$'s or $y$'s." Yet he knew that an easy subject might be made difficult by being overlaid with too great an elaboration of mathematics, and declared (*Popular Lectures*, vol. ii. p. 329) that Laplace was so intensely mathematical that he often did not see the simplicity of the results attained by his complicated analysis. He could strive to simplify the results of analysis once obtained; witness his beautiful short statement of Green's theorem in *Electrostatics and Magnetism*, p. 463, and the passage in the lecture on "Capillarity" (*Popular Lectures*, vol. i. p. 19) where he says: "The dynamics of the subject" (the vibration of a dew-drop) "is absolutely comprised in the mathematics without symbols which I have put before you. Twenty pages covered with sextuple integrals could tell us no more." His deliberate attempt to write out for his Glasgow students in 1847-8 the chief theorems in the distribution of electricity in equilibrium (*Electrostatics and Magnetism*, p. 52, footnote) "without the explicit use of differential or integral calculus" is worthy of mention, if only for the next sentence stating that "the spirit, if not the notation, of the differential calculus must enter into any investigations with reference to Green's theory of the potential." The following sentence from the Preface (1872) to the smaller *Elements of Natural Philosophy* further illustrates the matter:—

It is particularly interesting to note how many theorems, even among those not ordinarily attacked without the help of the Differential Calculus, have here been found to yield easily to geometrical methods of the most elementary character.

Simplification of modes of proof is not merely an indication of advance in our knowledge of a subject, but is also the surest guarantee of readiness for further progress.

It is also interesting to note that he considered the reason why De Morgan's "great book" on the Differential and Integral Calculus (which he valued highly) was now less valued than formerly, was "because it is too good for examination purposes"! He himself sometimes seemed to regard as wasted the time he had to take from experimental research for working out mathematical solutions. There is an echo of this in his letter to Helmholtz on p. 433. He was particularly happy in the use he made in physics of the refined mathematical processes which are known by the names of the Principle of Least Action and the Principle of Varying Action. One result is the important general dynamical principle, discovered by him, that when a material system is set impulsively in motion, the state of motion instantaneously assumed by it is that one for which the kinetic energy is a minimum (see *Math. and Phys. Papers*, vol. ii. p. 109, and *Thomson and Tait*, §§ 312, 317, and 327).

More than once he pointed out, as constituting a difference between Stokes's mathematics and his own, that in those solutions of differential equations which contain in their expressions "imaginary" quantities, he himself would always transform them at the earliest stage so as to free the equations from the imaginary part, while Stokes would retain them in the more general form till the last. His passion for models by which any dynamical idea could be illustrated, and his often resort to geometrical illustration of analytical results, were part of the same frame of mind which sought for reality that it could perceive and grasp objectively. He had the engineering instinct to *feel* the magnitudes of quantities as well as calculate them. Strange as it may appear, he did not encourage the use of the slide-rule for calculations: he expected a much greater accuracy in numerical computation than one part in a thousand.

In a Memorial Address to the London Mathematical Society, in January 1908, its President, Professor A. E. H. Love, has spoken authoritatively, in sympathetic terms of Lord Kelvin's contributions to mathematics. From that address the following paragraphs are taken :—

It is fitting that on this occasion we should endeavour to appreciate Kelvin's greatness. Whether we have regard to the practical utility of his inventions, to the subtlety of his physical speculations, to the wide range of his contributions to natural knowledge, or to the power of his mathematical methods, we must hold him great, and we admire him not less for the generosity of his character.

Of his contributions to mathematics it is perhaps fair to say that they consisted rather in the development of new methods, leading to new results, than in the creation of new mathematical theories. He was an acute geometer, but geometry for its own sake did not satisfy him. The method of inversion became in his hands a means of investigating electrical distributions, the partitioning of space into equal and similar portions was a step in the theory of the molecular arrangement of crystals. He was a profound analyst, but his analysis always had an immediate application. In connexion with researches on the constitution of the earth, he incidentally revised the whole theory of those functions which previously had been known as Laplace's functions, and from that time forward became spherical harmonics. It may be that his most important contributions to mathematical analysis are contained in the short papers in which he gave examples of that method for the solution of partial differential equations which, unknown to him, had been initiated by Green. He showed how to build up the requisite solutions synthetically from simple special ones which involve the existence of singular points. The singular point in electrostatics is a point charge, in magnetism it is a magnetic particle. In the theory of the conduction of heat it is an instantaneous point source; and the memoir in which he developed the application of the corresponding special solution to obtain all the valuable solutions of the equation of diffusion is well known. In the theory of elasticity the corresponding singularity is a point at which a force is applied. The special solution for this singularity was given by him in a very early paper, and he pointed out its possible applications, though he did not work them out in detail. The corresponding solution for vibratory motion produced in an elastic solid by force applied at a point was given some years later by Stokes, and the paper which Kelvin read to our Society in June 1899 was occupied with a development of this solution. The subject was the advance of waves into a previously undisturbed elastic medium. In some of the papers which he published, within the last year or two, he was occupied with the very intractable question of the advance of surface waves into previously still water, and he there emphasized again the importance of a special solution of the equations.

It is perhaps too soon to attempt to estimate Kelvin's achievements, to place him, as it were, in a list in order of merit with Newton at the head; but it is not too soon to attempt to understand the special character of his genius, the qualities that distinguished him from others whom also we honour. I think perhaps we may take this special character to consist in a certain very rare, possibly unique, combination of qualities. He combined the ingenuity and enthusiasm of the inventor, the exact knowledge of the practised experimenter, the trained intellect of the mathematician, with a wonderful gift of imagination. His quality of imagination is nowhere better shown than in his early writings on electrostatics. In this theory he owed something to Poisson, who exerted a profound influence on the English school of mathematical physics; but where Poisson saw an effective formula, Kelvin discerned an electric image. We all know the brilliant mathematical investigations to which he was led by this simple intuition; what perhaps is not so well understood is that the concrete interpretation of the details of abstract formulæ, of which type of interpretation this was one of the first examples, has become the ideal of mathematical physics. According to the standard that Kelvin had set up it is not sufficient to obtain an analytical result, and to reduce it to numerical computation, every step in the process must be associated with some intuition, the whole argument must be capable of being conducted in concrete physical terms. Nothing illustrates this better than the interpretation in terms of circulation and vortex strength of the transformation of line integrals into surface integrals. But the most striking example of Kelvin's simultaneously concrete and imaginative mode of working is to be found in his theory of vortex motion and vortex atoms. Where Helmholtz had found interesting types of motions of air and water, depending upon his new integrals of the equations of hydrodynamics, Kelvin detected a possible interpretation of all nature, consisting in the permanence of vortices. His theory of vortex atoms became the type to which, as we now believe, a dynamical theory of ultimate physical reality must conform, inasmuch as it set forth in a realized example the doctrine that ether and atoms are one and the same stuff, the difference between matter and non-matter being kinematic.

He has focussed attention upon that which is in the end the fundamental problem of theoretical physics.

Truly we may be proud that such a man has been numbered among our Presidents.

The allusion to Newton in the foregoing address is a reminder of the fact that Lord Kelvin's mathematical method greatly resembled that of Newton in its close adherence to reality. Throughout the *Principia*, Newton's proofs are always closely related to that which is the essence of the thing to be proved, the physical problem in hand. In this Lord Kelvin belonged essentially to the school of Newton in which he had been trained. His reverence for Newton was not merely the veneration felt for an honoured name; it was a living devotion to the genius of the creator of the British school of Natural Philosophy.

Lord Kelvin gloried in the discovery that Newton—in the wonderful Scholium to the Third Law—was acquainted with the idea of Energy, and its Conservation; and he never failed, year after year, to tell his Glasgow students of the fact. On the occasion of one of his latest visits to Peterhouse he was astounded to hear that, in the schedules of the Elementary Examinations of the University of Cambridge, Newton's Laws are no longer named.

"And what do they put in their place?" he asked. "Nothing!"

With a smile of sadness and commiseration, he turned and gazed silently into the fire.

# CHAPTER XXVI

## THE CLOSING YEARS

RETIREMENT from the cares and duties of his Professorial Chair did not mean inactivity for Lord Kelvin. With the end of the London season he accompanied Lady Kelvin to Aix-les-Bains for her annual cure. While here he wrote letters to *Nature* on phenomena that were interesting him, the apparent "dark flashes" which are sometimes seen in thunderstorms, and on the "blue ray" of sunrise, the counterpart of the *rayon vert*[1] of Jules Verne seen at sunset, which he observed over Mont Blanc from Aix-les-Bains, on the morning of August 27, 1899.

He returned to the quiet of Netherhall to resume the task of completing the *Baltimore Lectures*, and wrote to me about phosphorescence.

NETHERHALL, LARGS,
*October* 10, 1899.

DEAR THOMPSON—I have looked in vain in Encyclopædias and text-books for something that every one

[1] When the sun sinks in a clear sky behind a distant horizon, just for a second or two, as the last trace of its disk disappears, the colour turns yellow, then green, owing to atmospheric dispersion. Lord Kelvin had written me of this in 1896: "As to the *green ray*, the first time I saw it, it passed quickly from white through green, to intense violet. The sun was very clear, with very little of the redness which we generally see at sunset."

doesn't know regarding the phosphorescence of luminous paint, Canton's phosphorus, etc. . . .

(1) Can you tell me what is known regarding the effect of temperature? I find with little copper plates and a glass plate painted with Balmain's luminous paint, that the warmth of my hand greatly increases the glow due to previous illumination; and that, if of two similar plates, equally dosed with light, I keep one for an hour or two warmer by 10° or 20° C. than the other, it glows more brightly than the other till it cools, and becomes darker than the other in a minute or two, when it is cold like the other. Hence it appears that the warmth causes the stored light to be given out faster. I suppose this is well known, but I haven't found it told anywhere that I can remember.

(2) Is there good information as to the excitement of ordinary phosphorescence by different parts of a homogeneous spectrum? I have heard it said that the phosphorescent light may be of either shorter or longer period than the originating light. In Stokes's fluorescence he found the fluorescent light always of longer period than the originating.

(3) Do you know Dewar's splendid phosphorescence of egg-shells and other ordinary solids at very low temperatures? Was it generated by incident light at the low temperature, and did it only appear brilliantly when the temperature was raised? I have been looking through the *Phil. Mag.* and can find nothing of it.

(4) Do you know what Edmond Becquerel did in respect of ultra-red radiation on phosphorescence? I remember him telling me of it or showing it to me a great many years ago, but I can't remember exactly what it was.

(5) Do you know anything of Stokes's experiments on the subject?—Yours very truly, KELVIN.

NETHERHALL, LARGS, AYRSHIRE,
*Oct.* 14, 1899.

DEAR THOMPSON— . . . I am most grateful to you for your letter of the 11th, and the copy of your Oxford

Lecture. I feel now that with your Oxford Lecture[1] and your "Light, Visible and Invisible," and the exceedingly interesting answers to my questions in your letter, I have all that is known on the subject, and as fair a view as possible towards the *omne scibile*.

In respect to Becquerel's effect of the extreme red, I thought it likely that the explanation would be what you tell me it has turned out to be.

As to Stokes, I hope to see him at Cambridge at the end of the month, and to extract all I can from him, which I believe will include something vitally important not yet published.

I hope to see yourself still sooner, as we are going to London on Monday next for the opening of Parliament. I would like to come and see you, if you will allow me, in Finsbury one of these days; and to see anything you could quite conveniently show me in the way of phosphorescence or the Phillips phenomenon, or any other of the splendid things you may chance to have at hand. Will you give me a line, addressed Fleming's Hotel, Hálf Moon St., Piccadilly (where we shall arrive on Monday evening), to say if there is any time on the forenoon of Wednesday, Thursday, or Friday (18th, 19th, or 20th) when it would be convenient for you that I should come?

As to the Electrical Engineers' Dinner on December 6, I am afraid I must not come. I expect to be settled here for the winter by that time. I am obliged to be very careful about exposure, and it too often happens that dining is almost an impossibility to me on account of a horrid demon of No. 5 nerve, who gives me frequent visits at most inconvenient times. If I could properly do anything to persuade Lord Salisbury or Mr. Balfour to come to the Dinner, I would be very glad to do so; but I am afraid it would not be right, considering the great tension and fatigues which fall upon them in the present anxious times. However, we can speak of this

[1] Luminescence: A Lecture delivered before the Oxford Junior Scientific Club, by Silvanus P. Thompson, May 26, 1896.

when we meet, as I hope we shall, next week.—Yours very truly, KELVIN.

There was a contest that autumn for the Lord Rectorship of Glasgow, Lord Cromer having been put forward by the Conservatives, and Lord Rosebery by the Liberals. Lord Cromer, however, withdrew, and Lord Kelvin was proposed as a Unionist. The election fell on Lord Rosebery.

At the end of the year 1899 an amusing public controversy arose in the press from a speech of Lord Kelvin's at a dinner given to him on December 21 by his fellow-members in the Imperial Union Club. Pulling out his watch he remarked that they were now within nine days and three hours of the twentieth century. The statement being received with cries of "No," he repeated it, admitting that there was a large number of estimable men who held that the twentieth century did not begin for another twelvemonth. For the next few days the newspapers were full of argumentative letters. Even *The Times* had a leading article on the dispute. Challenged to produce evidence that there ever was a year "0" in the calendar, Lord Kelvin maintained a discreet silence; but he would never admit that his reckoning was wrong.

NETHERHALL, LARGS, AYRSHIRE,
*Jan.* 2, 1900.

DEAR RAYLEIGH—Best wishes from us both, to you and Lady Rayleigh and all the boys, for the *New Century*. I hope you have good news from Arthur.

I had not noticed the last part of your October article on Capillary Att$^{rn.}$ (because of almost incessant absence from home during October) when I sent you my last P.C., which almost repeats the words of the last sentence of your last paragraph but one. I am very glad to have the separate copy, as also that of the latest blue sky, which I knew very well, having, because of it, taken the April No. of *Phil. Mag.* with me to Italy.

We are probably going to London on the 13th or 15th of the month. If you had been to be still at Terling we should probably have been offering a visit. Are you to be in London after the 15th? If so, I hope we shall meet, and settle a good many questions.—Yours,

K.

Lady Kelvin wrote:—

FLEMING'S HOTEL, HALF MOON ST., W.,
*Feb.* 6, 1900.

MY DEAR MR. DARWIN—Lord Kelvin asks me to write for him. He says he cannot but regret the old régime of Senior Wranglers and Mathematical Tripos generally, but he does not think the present system at all satisfactory. He does not know the circumstances sufficiently to have an opinion as to the proposal or proposals under consideration in respect to the future, or whether it would be practicable or desirable to return to something more like the system which has been followed previously to 1850 or 1860. He has never expressed any opinion on the subject. He is much obliged for the papers you send, which he will read with interest. . . .

We do not get into our own house, 15 Eaton Place, till April. We return to Largs on Thursday.

NETHERHALL, LARGS,
*February 9th*, 1900.

DEAR RAYLEIGH — À propos of the two Kinetic Theorems of Liouville for which I referred you to "isoperimetrical problems" at the end of Vol. II. of my *P.L. and A.*, I spent three or four hours at the R.S. and at Fleming's Hotel trying in vain to find them through more than 20 vols. of *Liouville's Journal* and by aid of the R.S. titles of all Liouville's papers.

To save you similar trouble, you may, if you care, look to *Nature*, vol. xlvi., 1892, pp. 386, 490, 541, where I found all I wanted to know on the subject, on our arrival home last night. Let me have a post-card by return to say where I can find your dis-archived paper of Waterston.—Yours,
KELVIN.

In April 1900 Lord Kelvin gave a Friday evening discourse at the Royal Institution. This was his ninth and last—the first having been given (see p. 312) in 1856. His title was: Nineteenth-Century Clouds over the Dynamical Theory of Heat and Light.[1] The clouds were these: (1) the old difficulty that the earth and heavenly bodies move freely through the ether, which yet acts in the propagation of light-waves as an elastic solid; (2) the difficulty about the proportion between the rotational and translational kinetic energies in a gas, which, according to the Boltzmann-Maxwell doctrine (see p. 900), is an equality. The former "cloud" he found no means of dissipating. The latter, in spite of the test-cases (see p. 947) which he had proposed, he regarded as yet unsettled, for the tests had failed to yield a decisive answer either way. But he could not rest with the mathematical verdict "not proven," and he now proposed new problems about impacts of particles during a large number of successive flights. He believed the conclusions to disprove the doctrine. Lord Rayleigh had recently found difficulty in applying the law of equal partition to the case of the newly discovered gases argon and helium, because the law of partition disregards

[1] Printed in the *Baltimore Lectures* (1904) as Appendix B, p. 486.

potential energy, and ought not to do so if potential energy involves constraint; and therefore he also sought for some escape from the destructive simplicity of the general conclusion. Lord Kelvin ended his lecture by saying that the simplest way of arriving at this desired result was to deny the conclusion.[1]

Lord Kelvin was now settled in London in his own house, No. 15 Eaton Place, whence on June 10 he wrote to M. Mascart:—

DEAR MASCART— . . . We think of going to Paris about the beginning of August, and to spend a few days there at that time. We are going to Aix-les-Bains, and might take time on our way back to see the Exhibition if that would be a better time than the beginning of August. I thought of the beginning of the month because of the "Congrès International de Physique" to be held from Aug. 6 to Aug. 12, at which I had promised to give a paper on the motion of ponderable matter. I hope to send the paper printed, in English, beforehand, to the Secretary; and I suppose the communication of it to the Congrès (as it is very mathematical, and not fit for *reading* to a meeting) will be quite a formal matter.

I would be greatly obliged if you would write me a line to say what would be the best time for us to choose, for about a week (we could scarcely take more), to see the Exhibition, and our good friends in Paris, if we can be so fortunate as to find any in August, or the early part of September. . . .—Always yours truly, KELVIN.

They went to Paris, and Lord Kelvin's paper, Sur le Mouvement d'un Solide élastique traversé par un

[1] An unexpected confirmation of the law of equi-partition in the case of the kinetic movements in liquids is afforded by the recent observations of Svedberg of Upsala, and of Perrin of Paris, on the Brownian motions.

corps agissant sur lui par Attraction ou Répulsion, was "read" at the Congress.

While in Paris a banquet was offered to him by the *Conférence Scientia*, organized by M. Louis Ollivier. Here he was hailed by MM. Mascart and Cornu as one of the world's greatest benefactors. In reply he alluded with emotion to his early studies under Regnault, by whom he was taught "a faultless technique, a love of precision in all things, and the highest virtue of the experimenter—patience." He also acknowledged his debt to the "inflexible logic and clear philosophy" of the French race.

An interesting feature of the year 1900 was the election of Lord Kelvin as Master of the Clothworkers' Company of the City of London for the period 1900-1901. Amongst the oldest of those ancient guilds which still hand on the mediæval traditions of mutual help and hospitality, the Clothworkers' Company has been conspicuous for the liberality with which it has applied vast sums from its resources to the promotion of education in the modern applications of science to the textile industries. It has also sought to widen its membership by bestowing its freedom upon men distinguished in statesmanship, science, and commerce. It had in 1891 presented the Freedom and Livery of the Company, *Honoris causa*, to Sir William Thomson, P.R.S., in recognition of his eminence in science. In 1896 he had been elected an Assistant of the Court. In July 1900 he was elected Master. He took a great interest in the work of the Company, and presided diligently at meetings of the Court and at the dinners of the Livery; and endeared himself to the members of the Company by his unassuming affability. At the close of his year of office the Court resolved to perpetuate the memory of his Mastership by causing a portrait of him to be painted and hung in Clothworkers' Hall. And on retiring from the Master's Chair, in June 1901, he presented to the Company a silver-gilt loving-cup of Elizabethan design, which, as "the Kelvin Cup," is treasured by the Company along with "the Pepys Cup" and other memorials of former Masters.

Reference has been made on pages 717 and 994 to Lord Kelvin's active part in the business of James White, optician, which had been taken over after White's death by Lord Kelvin, Mr. D. Reid, and Dr. J. T. Bottomley. In 1900 this business was formed into a limited company so that an interest in the ownership might be given to several of the employees who had helped in its development. At the same time, to prevent confusion with certain firms who had put on the market compasses and other instruments in imitation of Lord Kelvin's under names resembling those of White or Thomson, the style of the firm was changed to Kelvin and James White, Limited.

In November Lord Kelvin was busy in London, presiding at the Mathematical Society, where he read a paper on the Transmission of Force through a Solid; presenting prizes and speaking on technical education at the Birkbeck Institution; then running down to Glasgow to speak on Unionism to the Liberal Unionists, and on electricity supply systems to the local branch of the Electrical Engineers.

At the end of the year he wrote to Mascart:—

NETHERHALL, LARGS,
*December* 30, 1900.

DEAR MASCART—The medal in commemoration of Gramme and the dynamo reached me yesterday. It is most interesting and quite a beautiful work of art. I am exceedingly glad to have it, and I think M. Chaplain is to be congratulated on having produced so good a memorial of a great advance in the application of science for the use of man.

I hope you and Madame Mascart are well, and that the little grand-daughter had been quite freed from her illness and left with Louise[1] without anxiety. Lady Kelvin joins me in kindest regards and in wishing you all a happy New Year.—Yours always, KELVIN.

*P.S.*—I am very busy about a new form of electrostatic voltmeter, besides never-ending mathematical trials *re* Kinetic Theory of Gases, the Boltzmann-Maxwell Law, etc. etc.

The winter months of 1900-1901 slipped quietly by, with work at Netherhall on revising *Baltimore Lectures*, and visits to Edinburgh to preside or take part in the Royal Society meetings, or speaking in Glasgow on behalf of the Franco-Scottish Society (of which he was president), or of the South African War Relief Fund.

In his capacity as Master of the Clothworkers' Company Lord Kelvin visited the Yorkshire College

[1] Madame Louise Monnier, a daughter of M. Mascart, and a frequent correspondent with Lady Kelvin.

at Leeds, on May 10, 1901, to confer with the Textile Industries and Dyeing Committee, and to inspect the Textile Department, which was heavily subsidized by the Company. Replying to an Address presented to him by the students, he told them that he hoped all students in the technical departments, both engineering and textile, would take full advantage of the abstract sciences—mathematics, physics, and chemistry. They did not live merely for mechanics and the applications of engineering alone. There was a department of literature and philosophy; and he hoped the ancient classics would not be dropped or displaced by modern languages.

Lord Kelvin was this session appointed Chairman of a Royal Commission to inquire into arsenical poisoning which had been found in beer. They held numerous sittings in London and Manchester to take evidence, and presented their first report in July 1901.

He wrote to George Darwin:—

15 EATON PLACE, LONDON, S.W.,
*May* 20, 1901.

DEAR DARWIN—Was Dr. Erasmus Darwin your grandfather? or great-grandfather?

I have just been reading Muirhead's Life of James Watt with much interest, and learning, what I did not know before, that he and your ancestor were friends, with beautiful mutual sympathy.

I am in the sad straits of having to give an oration (oh, horror!) on James Watt on June 12 at the Glasgow University, 9th Jubilee celebration. . . .

Glasgow University celebrated on June 12, 1901,

the four hundred and fiftieth anniversary of its foundation. In the *Book of the Jubilee*,[1] produced by the students' Celebrations Committee, are found pictures of the old University, and a study by Dr. H. S. Carslaw of the student days of Lord Kelvin.

On this occasion, after the conferring of honorary degrees upon a large number of distinguished delegates, including the Marquis of Dufferin and Mr. Andrew Carnegie, Lord Kelvin delivered the James Watt Oration in the Bute Hall.

He gave a rapid sketch of the career of Watt, and in particular of his relations with the University, which had sheltered him, and for which he had worked as instrument maker. Lord Kelvin told of the old model Newcomen engine which belonged to the University :—

> Watt has told us that it was in the winter of 1763-4 that he was engaged repairing the model, and we see that his account for the work done was not given in till June 10, 1766; so we may fairly conclude that he had it in hand for more than two years, and made a great many experiments with it. In the course of these experiments he noticed with surprise the large quantity of water required to condense the steam, five or six times as much as the water primarily evaporated.
>
> In conference with James Black, lecturer on chemistry in Glasgow College, it was found that this was a splendid and previously unthought-of example of the doctrine of latent heat, then fresh from Black's original discovery of it. With very primitive and imperfect instrumental appliances Watt measured the amount of the latent heat of steam at different temperatures and pressures, and found for its variations a roughly approximate law. When, 81 years later, a student under Regnault in his laboratory in the Collège de France, I used to hear him speaking of "la loi de Watt," and telling us that it was the nearest approach to the truth which he found among the results of previous experimenters, I felt some pride in thinking that the

[1] Published by James MacLehose and Sons, Glasgow, 1901.

experiments on which it was founded had been made in Glasgow College.

Another passage from Lord Kelvin's oration evoked much laughter :—

In 1767 Watt was employed to make a survey for a small canal intended to unite the rivers Forth and Clyde. He attended Parliament on the part of the subscribers to this scheme, and it appears from some of his letters to Mrs. Watt that he was not much enamoured of the public life of which he thus obtained a glimpse ; "close-confined, attending this confounded Committee of Parliament," he says, "I think I shall not long to have anything to do with the House of Commons again : I never saw so many wrong-headed people on all sides gathered together."

He then referred to the association of Watt with his friends of the Lunar Society,[1] Erasmus Darwin, Matthew Boulton, Josiah Wedgwood, Thomas Day, Edgeworth, and Priestley. Lastly, he spoke of James Watt's foresight in founding, in 1808, the Watt prize[2]; of the foundation in Glasgow of the first University Laboratory of Chemistry, and of the first University Laboratory of Engineering ; and of the new James Watt Engineering Laboratory shortly to be opened.

Lord Lister and Sir Joseph D. Hooker were present at the celebration, and took part in the opening of the new Botanical Laboratory. Altogether the celebration was a notable one ;

[1] The Lunar Society, an association of kindred spirits who dined together at two o'clock on the day of the full moon, so as to have the benefit of its light in returning to their homes at night.

[2] Watt wrote to the Principal of the University : "Entertaining a due sense of the many favours conferred upon me by the University of Glasgow, I wish to leave them some memorial of my gratitude, and, at the same time, to excite a spirit of inquiry and exertion among the students of Natural Philosophy and Chemistry attending the College, which appears to me the more useful, as the very existence of Britain as a nation seems to me in great measure to depend upon her exertions in science and in the arts."

and the bringing together, as Lord Kelvin said, of representatives of "the intellects of all the countries of the world was a grand peace-promoting and peace-preserving influence."

For the supply of electric power "in bulk" from a central station, a large undertaking had been organized at Carville, near Newcastle-on-Tyne, to meet the needs of the Tyneside district. The formal inauguration by Lord Kelvin, on the invitation of Dr. J. Theodore Merz, one of the promoters of the enterprise, was fixed for June 17, 1901. He stayed at Jesmond Dene House with Sir Andrew Noble. Lord Kelvin's speech showed how greatly he was gratified with the work, the magnitude and simplicity of which he praised; he even praised the three-phase alternating-current motors.

The month of August was spent by Lord and Lady Kelvin at Aix-les-Bains.

An International Congress of Engineering was held at Glasgow early in September, and of this Lord Kelvin was honorary president. In this capacity, on September 3, he performed the ceremony of formally opening the James Watt Engineering Laboratory which had been organized by Professor Barr (the successor in 1889 of Professor James Thomson). Lord Kelvin spoke of the many engineering laboratories that had arisen since the days of James Watt's shop in the old College. He alluded to the creation in modern factories of testing laboratories, and to his own experiences in regard to the difficulties in 1857 of getting the manufacturers of copper wire to guarantee or even test the conductivity of the wire for the Atlantic cable. The University ought not to be speculating merely on the qualities of the ether, or on the possibility of explaining the electromagnetic theory of light by mechanical process, but it ought to be in touch with all knowledge of the properties of matter discovered by university men or scientific men, not merely for the abstract scientific value of the knowledge, but for the possibilities of its practical application.

A week later the British Association met in Glasgow under the presidency of Professor (now Sir Arthur) Rücker, whose inaugural address was a clear and broad summary of the accepted views of the atomic or particular nature of matter, of the dynamic nature of heat as consisting in movement among these particles, and of the existence of an all-pervading ether. In seconding the vote of thanks to the President, Lord Kelvin said :—

They had heard a most eloquent, convincing, and picturesque defence of atoms and ether—against what? He was afraid that rather jaw-breaking words must be used to describe what the defence was against. It was a most rude recrudescence of neo-pantheism, neo-Berkleyanism, neo-vitalism, and neo-nihilism, which had grown up in the last ten years of the nineteenth century, and grown up in a manner singularly inconsistent with the bright and clear teaching of realities and faith in realities with which the century commenced, and with which the century had in the main been conducted. From the time when Thomas Thomson, at the University of Glasgow, taught the atomic theory in his lectures, three years after he had learned it from Dalton, it was practically a reality, and it would be mere word-splitting to say that we were agnostics in respect of it. Agnostic meant not-knowing. Now we were all agnostic in most things in nature and religion, because we did not know them; but to be satisfied with agnosticism or to delight in it was absolutely unscientific.

There were mysteries far beyond the range of science, but when we found so-called scientific agnosticism, having doubts and difficulties regarding the influence of life on matter, leaping to the neo-vitalism in which Professor Haeckel, following Vogt, recently asserted that matter and ether were not dead, but were endowed with sense, that was neo-vitalism with a vengeance. Matter and ether, we were asked to believe, experienced a wish for condensation, and they struggled against strains! We were to picture, in fact, a lively struggle between the clay and the potter, the one with as much life as the other, but struggling together to make a pot! They had an eloquent, a picturesque, and an argumentative statement put before them by the President, against the idea of nihilism in science. He had declared for clearness and reality.

Lord Kelvin read a paper at the Association meeting, on the Clustering of Gravitational Matter in any Part of the Universe; he also took part in a discussion on the magnetic effect of electric convection. His paper was afterwards published as Appendix D of the *Baltimore Lectures*. In it he made an estimate of the amount of matter within the sensible universe, fixing the limit for purposes of calculation as a sphere of such a radius that a star on its surface would have a parallax of one-thousandth of a second of arc, that is $3 \times 10^{16}$ kilometres. From the darkness of the stellar sky, and the stellar velocities observed, he concluded that the total amount of matter in such a sphere would be about equal to 1000 million times the mass of our sun. So he imagined a beginning in which there should be 1000 million suns disseminated in minute particles or atoms uniformly throughout this gigantic sphere, and gradually falling together into nebulous masses. At first the density would be imperceptibly small, and as they shrank together it would take nearly

17 millions of years to reach one-sixth the density of water, when the velocity of the mutually attracting particles would be comparable to the velocity of light. Had the original amount of matter been greater the velocity generated would have been greater than any known stellar velocity. In those parts of such a universe where the density was greater than the average it would tend to increase, while in the less dense parts it would tend to become less; and so great void spaces would be found surrounding regions of greater density, in which the clustering matter would still further shrink together, forming nebulæ, solar systems, and worlds. Collisions would give rise to heat and light, which would be radiated away through the ether; and this loss of energy would reduce large condensing clusters to the condition of gas in equilibrium under the influence of its own gravity only, or rotating like our sun, or moving at moderate speeds as in the spiral nebulæ. The cooling of the condensing masses would give solid bodies, collisions between which would yield meteoric stones such as we see them.

This remarkable cosmic speculation he followed up by further calculations, in a paper on "The Problem of a Spherical Gaseous Nebula," which was left incomplete at his death, and published posthumously in the *Proc. Roy. Soc. Edin.*, xxviii. p. 259, March 9, 1908.

Professor Tait, who had retired, broken down in health, in the previous year, died on July 4, 1901.

At the meeting of the Royal Society of Edinburgh, of which he had for many years been the secretary and moving spirit, Lord Kelvin on December 2 read an appreciation of the life and work of his colleague. Extracts from this appreciation, relating to their conjoint work on the great treatise, have already been given on p. 478 *supra*. But Lord Kelvin's tribute to his friend contained much more, in particular an account of Tait's work for the Report of the *Challenger* expedition, and of his labours for the Society. "We all feel," he added, "that a great man has been removed; a man great in intellect, and in the power of using it, and in clearness of vision and purity of purpose, and therefore great in his influence, always for good, on his fellow-men: we feel we have lost a strong and true friend." "The cheerful brightness which I found on our first acquaintance forty-one years ago remained fresh during all these years, till first clouded when news came of the death in battle of his son Freddie, in South Africa, on the day of his return to duty after recovery from wounds received at Magersfontein."

On January 20, 1902, Lord Kelvin read two papers to the Royal Society of Edinburgh. A spectator has thus described the meeting:—

On the occasion under notice Lord Kelvin, the president, was the principal speaker. With mathematical exactness, his Lordship took the chair at eight o'clock, not a fraction of a second more or less. The meeting being duly constituted, Lord Kelvin vacated the chair in favour of Lord M'Laren, and at once dived into the heart of the subject. It was not a subject that people discuss at street corners. "The Specification of

Stress and Strain in the Mathematical Theory of Elasticity," is how he described it. It seems that hitherto people who are interested in such matters have been accustomed to use a notation which has this disadvantage, that the strain requires to be infinitesimally small. The method of notation which Lord Kelvin suggested—he modestly refrained from claiming it altogether as his own—was to take instead of a cube a tetrahedron, a figure with four corners, four planes, and six edges, and six ideal lines drawn in the directions of the six edges, any change in the shape or bulk of the tetrahedron being represented by the elongation of the six edges of the tetrahedrons. There now! The thing seemed so plain that it might be left to speak for itself. But, no. The aged seer, with voice and gesture, blackboard and model, shook the proposition inside out, turned it upside down, tried it by every test known to mathematicians, and finally he dragged the subject out of the maelstrom of weird words and expressions, arguments and proofs, laid it out flat, folded it neatly, and presented it to the audience as a thing which, although not greatly superior to that formerly in use, was at least worthy of their acceptance.

His other paper was on the Molecular Dynamics of a Crystal (see p. 1080).

In February and March Lord Kelvin was again busy with the meetings of the Royal Commission on Arsenical Poisoning; but he found time to attend meetings of the Civil and Electrical Engineers, and to take part in discussions. On March 19 he was present at the opening of the National Physical Laboratory, at Bushy, by the Prince of Wales. In seconding the vote of thanks to the Prince, Lord Kelvin spoke of the scientific importance of exceedingly minute and accurate measurements, which, though they might not strike the popular imagination, were the foundation of the most brilliant discoveries.

Early in April Lord and Lady Kelvin left England for a trip in the United States, landing at

New York, April 19. He had become Vice-Chairman of the Kodak Company of London, and one chief object of his journey was to view the Eastman Photographic Works at Rochester, N.Y., and to advise about supplying Rochester with electric power from the Genesee River. He arrived in time for the installation of President Butler as head of Columbia College, New York. On the evening of April 21 the American Institute of Electrical Engineers gave him a great reception at Columbia University, when he was welcomed by all the leading American electricians—Prof. Crocker, Prof. Elihu Thomson, Prof. A. G. Webster, Mr. T. A. Edison, Mr. Westinghouse, Mr. Nikola Tesla, Mr. C. F Scott, Mr. M. Pupin, and others. In responding to the speeches of welcome, he referred to the Atlantic cable days, to the telephone at the Centennial Exhibition, to the outburst of electric lighting in New York in 1884, and to wireless telegraphy on ships. Then he visited Washington[1] as the guest of Mr. Westinghouse, and attended a reception of the Academy of Science, and travelled thence *via* Jamestown to Niagara Falls, which were reached on April 26. He made a brief speech at Jamestown, eulogizing the memory of Cyrus Field, "who possessed an admirable and unapproachable quality, an attribute of heroes: he never knew when to give in." At Niagara he was greatly delighted to see

[1] While at Washington Lord Kelvin gave evidence before a House Committee on Coinage, Weights and Measures, on the advantage of the metric system. He stated that the English ladies in Madeira found the metre just as convenient for buying ribbons as the yard measure.

the enormous practical application of the power of the Falls to the development of electrical energy, —a project which, originating in Europe, had, in the hands of the Cataract Company and its engineers, achieved nothing short of an industrial revolution. While at Rochester, though in considerable pain from facial neuralgia, he visited the University and addressed the students in chapel. A sentence from his address claims notice :—

> It has come to be my belief that, as a man grows older, the pictures that he looks upon with the most pleasure by his fireside in the latter days of his life are those which bring before him again his college days. So then I urge upon you that your career here be so passed that when you become old men you may look back on these days with no regret or shame, that you may see no day wasted, no time spent in idleness, no duty neglected. Even in your days of vacation—and I hope that you may all achieve such places in the world that you can now and then enjoy periods of rest from your work—do something worth while the doing. Make your whole life full of pictures which are bright, and clear, and clean.

From Rochester Lord and Lady Kelvin went to Cornell University, where a reception for them was organized by President Schurman and Professors Nichols and Thurston. Thence, after two days, they travelled to New Haven, spending the week-end at Yale University, where the degree of LL.D. was conferred on Lord Kelvin. Here he spoke of research laboratories, and of the need of giving the professors the time and means to prosecute research, but not by cutting them off from their students, whom they should interest personally in the work. "A man comes away from his class-room with a new impulse to continue his work of research." From Yale Lord and Lady Kelvin returned to New York, where they were entertained by many friends; they also ran over to Philadelphia for two days as the guests of their old friend Mrs. W. P. Tatham.

Two of Lord Kelvin's dicta, preserved by the press, were—

> Beautiful as that wonderful work of nature [Niagara] is, it would be more beautiful still if those waters fell upon turbine wheels every one of which was turning the wheels of industry.
>
> The air-ship, on the plan of those built by Santos-Dumont, is a delusion and a snare. A gas balloon, paddled around by oars, is an old idea, and can never be of any practical use. Some day, no doubt, some one will invent a flying machine that one will be able to navigate without having to have a balloon attachment. But the day is a long way off when we shall see human beings soaring around like birds.

At Cornell he told the students:—

> A university is a place that fits some men for earning a livelihood, and that makes life better worth living for all men.

At Philadelphia, speaking of the Rhodes scholarships for Americans at Oxford, he spoke hopefully of this step towards unification of Anglo-Saxon peoples, as helping to correct "the political mishap" of the last half of the eighteenth century; adding:—

> We have shown that it is possible to get on well together under separate flags, but I wish we were all under one flag and under one government.

Bound for England, on board the *Lucania*, when off Nantucket, on May 12, the following farewell message was despatched to the *New York Herald*:—

By Marconi wireless telegraph through the ether, Lord and Lady Kelvin send warm thanks to kind friends, electric and not electric, who made their three weeks' visit to America most interesting and delightful.

The following letter from the Premier, the Marquis of Salisbury, to Lord Kelvin, is given by permission :—

*Private.*

20*th June* 1902,
DOWNING ST., S.W.

MY LORD—I have had the honour of receiving a command from the King to inform you that His Majesty has been pleased to direct that you should be sworn a member of His Majesty's Most Honourable Privy Council on the occasion of the approaching Coronation.—I have the honour to be, my Lord, your Lordship's obedient servant,

SALISBURY.

The Lord Kelvin, G.C.V.O., F.R.S.

His Majesty King Edward VII. decided to institute a new distinction, the Order of Merit, resembling the Prussian Order *Pour le Mérite*, but differing from it in admitting not only men most distinguished in Science, Literature, or Arts, but also naval and military men. The following letter was written by command of the King :—

BUCKINGHAM PALACE,
25 *June* 1902.

DEAR LORD KELVIN — I am commanded by the King to say that he hopes it will be agreeable to you to accept the "Order of Merit."

This order has just been instituted by him, and the number of members is limited to 24 ; twelve naval and military, and 12 civilians.—Believe me, yours very truly,

FRANCIS KNOLLYS.

15 EATON PLACE, S.W.,
*June* 26, 1902.

DEAR SIR FRANCIS KNOLLYS—The gracious message from the King which you give me in your letter of yesterday touches me deeply. I feel most grateful to His Majesty for his kindness in conferring on me the honour of the "Order of Merit" now instituted by him. I fervently pray that he may in due time be restored to health.

Will you, when it may be done, express my warmest gratitude to the King?

May I also through you express to the Queen my heartfelt sympathy, and my earnest hope that this great affliction may pass away happily, and that the sorrow of the subjects of the King in all parts of the world may soon be followed by joy in hearing of His Majesty's recovery.

Believe me, with hearty congratulations to yourself on the honour which the King has conferred on you, yours very truly, KELVIN.

By a happy coincidence Lord Kelvin received his honours on his birthday; he was seventy-eight on June 26th. The other civilian members of the Order of Merit were, Lord Rayleigh, Lord Lister, Sir William Huggins, the Rt. Hon. W. E. H. Lecky, the Rt. Hon. John Morley (now Lord Morley of Blackburn), and Mr. G. F. Watts, R.A.

The Athenæum Club entertained at dinner on July 25th all the twelve members of the Order of Merit, except Mr. Lecky and Lord Wolseley who were out of England, Lord Avebury presiding as trustee of the Club. Mr. Balfour described the company of about 120 members as one of "undiluted distinction."

The postponed Coronation kept Lord Kelvin in England somewhat later than usual; he and Lady Kelvin were present at that unique and splendid ceremonial on August 9, 1902, in Westminster Abbey. He was sworn in as a Privy Councillor on the 11th, and on the 12th they left for Aix-les-Bains.

On September 10th the British Association met in Belfast—the first time since 1874 in that city. Lord Kelvin sent a paper—on an Animal Thermostat—but he was not present, being away at Aix-les-Bains.

NETHERHALL, LARGS, AYRSHIRE,
*October* 11, 1902.

DEAR RAYLEIGH—Has anything been done to test with minute accuracy the angle of minimum reflected light due to incident light polarized perpendicular to the plane of incidence? Jamin's observations and yours are designed to give "*k*" with great accuracy; but I do not see in your investigation, and I suppose there is not in Jamin's, any indication of the minutely accurate angle of incidence which absolutely annuls the component of the reflected light polarized perpendicular to the plane of the incidence. This angle of incidence is very approximately, almost quite rigorously even for diamond, the exact angle of incidence for minimum intensity of the reflected light.

A few months ago, and particularly in the last few days, I have got a great lift in, and up out of, my 1888 (*Phil. Mag.* 2nd half-year) theory of slow condensational-rarefactional waves to account for Fresnel's "tangent law." My present difficulty is to explain deviations from it as large as those found by Jamin, and I take great comfort from the last three lines of your paper (*Scientific Papers*, vol. iii. p. 511). But do you really think your verdict "not guilty" of breaking of "Fresnel's law" can possibly acquit diamond? I wish it could.

My theory would allow comfortably, say in the case of diamond, one or two or three minutes' deviation from $\tan^{-1}\mu$, of the angle for minimum light, but not comfortably more than three minutes. But the residual component polarized perpendicular to the plane of incidence and reflection is enormously nearer nothing than, from the time of Airy, we have been led to believe is the truth, for diamond.

My amended theory makes the velocity of condensational waves practically infinite in pure ether; but in a transparent liquid or solid about $\frac{1}{100}$, or less than $\frac{1}{200}$ of the velocity of light.—Yours, KELVIN.

*P.S.*—If you are in Scotland you and Lady Rayleigh must give us a benefit here. If you are at Whittingehame give the Prime Minister my best wishes for a happy and not too long autumn session. The Education Bill seems to me sure to come through happily and work well when it becomes law.

On October 31, in a discussion at the Physical Society on Mr. Ridout's paper on the size of atoms, Lord Kelvin said:—

In dealing with the subject of atoms, it was necessary to consider the atoms of electricity; the atomic theory of electricity, now almost universally accepted, had been thought of by Faraday and Clerk Maxwell, and definitely proposed by Helmholtz. The atoms of electricity were very much smaller than the atoms of matter, and permeated freely through space not occupied by them. An atom of electricity in the interior of an atom of matter experienced electric force towards the centre of the atom. We were forced to conclude that every kind of matter had electricity in it, and Lorentz had named electricity as the moving thing in atomic vibrations. If the electrions, or atoms of electricity, succeeded in getting out of the atom of matter, they proceeded with velocities which may be as great as the velocity of light, or greater, or less; and the body would then be radioactive. It was therefore not surprising that some bodies showed radio-active properties, but rather surprising that such properties were not shown by all forms of matter.

M. Mascart sent to Lord Kelvin one of the Mascart medals which had been struck as a presentation to him by his friends, and received the following acknowledgment :—

NETHERHALL, LARGS,
*Jan.* 22, 1903.

DEAR MASCART—It was most kind of you to let one of the medals be sent to me, and to let me have the honour and pleasure of receiving the *first* of the twenty.

Lady Kelvin and I are delighted to have it, and we admire it very much as a work of art and as an admirable likeness of yourself. I hope Mme. Mascart is pleased with it in this respect. She must certainly be greatly pleased to have it as a perennial token to commemorate the admiration and gratitude which electrical engineers feel in return for the great benefits they have received from your scientific work in electricity.

I am just now feeling myself most grateful to you for your *Traité d'Optique*, a treasure from which I am incessantly drawing knowledge to help me in efforts to explain some of the outstanding difficulties of the dynamical theory of light. I almost believe that I have really succeeded at last in reconciling Fresnel's laws with definite real dynamics for transparent solids and liquids, and in explaining what we know by observation of metallic reflection. I hope with some confidence that a volume of my old *Baltimore Lectures* of 1884, amended, extended, and printed in the course of 19 years, will be published before summer, containing all I have been able to do, or tried to do, in Physical Optics.

Lady Kelvin thanks you for your letter, and joins in kindest regards to you and Mme. Mascart.—Yours,

KELVIN.

Stokes had died on February 1, 1903, and at the request of the editor of *Nature,* Lord Kelvin wrote the notice of his scientific work in that journal for February 12, 1903, vol. lxvii. p. 337. Lord Kelvin's views on the memorial to Stokes, which his friends were proposing, are succinctly expressed in the following extract from a letter he wrote to George Darwin on February 10, 1903:—

> As to the Stokes memorial, I have written to J. J. Thomson that the *very* best memorial, most useful practically and most suited to keep his memory ever fresh among workers in science, would be an earliest possible publication in collected form of *all* his scientific papers not in the three volumes already given by the University press, and including MSS., if found, intended for publication but not fully prepared for press. *That* would be a much more valuable and estimable memorial than any thing that could be done by a £3000 or £4000 or more collected by subscription.
>
> A strained effort to collect money would be a very painful thing in connexion with the memory of Stokes; and I am sure the idea of such an effort would have been most distasteful and painful to himself.

A short speech which Lord Kelvin made on May 1, 1903, on belief in creative power, was the cause of considerable public comment at the time. The matter has been dealt with, on p. 1091, in the summary of Lord Kelvin's religious opinions.

On May 11 Lord Kelvin went to Cambridge to the unveiling of a bust of the lamented Dr. John Hopkinson, in the Engineering Laboratory, which owed so much to him and his family. He spoke of Hopkinson's achievements, and expressed the hope that whatever new regulations were made for the classical tripos or for new triposes, they would never forget the old mathematical tripos which formed one sure foundation in the application of science, chemistry, physics, engineering, dynamics, and the arts in general.

On June 19, 1903, Lord Kelvin propounded to the Royal Society of Edinburgh the new views of the atom hinted at in his letter to Lord Rayleigh of October 11, 1902 and stated on p. 1081 *supra*.

On the 24th the reconstituted University of London for the first (and only) time conferred Honorary Degrees. The recipients were their Royal Highnesses the Prince and Princess of Wales, Lord Kelvin, and Lord Lister. There had been some opposition in the Senate on the ground that intellectual merit alone ought to be recognized by a University, not rank, nor political distinction, nor military glory, as a title for a degree. At the ceremony of presentation, which took place in the Royal Albert Hall, the capping of their Royal Highnesses was received with respectful and loyal cheers. But when Lord Kelvin and Lord Lister stepped on the dais to be presented to the Chancellor, the outburst of cheering was truly remarkable, and lasted many minutes. It was a spontaneous ovation to the princes of science. In presenting Lord Kelvin, Professor (now Sir William) Tilden, Dean of the Faculty of Science, pronounced the following oration :—

My Lord the Chancellor, I present to you William Thomson, Baron Kelvin of Largs, for the degree of Doctor of Science, *honoris causa*. The illustrious son of a family famous for mathematical talent, for more than half a century Lord Kelvin filled the office of Professor of Natural Philosophy in the ancient

University of Glasgow. Two generations have passed since he entered on his professorship, and the advances in physical science which have distinguished the nineteenth century from all preceding epochs have been largely due to the influence of Lord Kelvin in promoting true ideas concerning the conservation of energy, the laws of thermodynamics, and their application to the mechanics and physics of the universe. His untiring intellectual activity has led him also to inquire into problems interesting to the chemist and geologist as well as those which are important to the physicist and engineer. He has calculated the probable size of atoms; he has studied the structure of crystals; he has estimated the age of the earth. But the world knows him best as the man who has shown how practically to measure electric and magnetic quantities, and has made it possible to link together distant continents by the electric telegraph. It is he who has shown how to neutralize the effects of iron on the compasses of ships and how to predict the tides, and who has thus taught the mariner to steer safely over the surface of the ocean, and to sound, as he goes, its depths and shallows. A greater philosopher than Democritus, in him are united the qualities of Archimedes and Aristotle. Regarded with affectionate reverence by his contemporaries, it cannot be doubted that his name will shine brightly through long future generations. In offering a place of honour to such a man the University confers lustre on itself.

At a reunion at Peterhouse, on July 7th, 1903, of former members of that College, Lord Kelvin found himself the senior of the company. There was a reception by the Master in the garden, a Commemoration Service in the chapel, and a dinner. To the toast of *Floreat Domus Divi Petri*, Lord Kelvin responded, with a tale of the times of old, of Cookson, of Hopkins the renowned, of Fuller the kindly and inspiring, and of Routh who had picked up Hopkins's mantle; advocating the retention of some classics in the work of every student; and concluding with memories of the founding of the C.U.M.S. in 1843.

In the House of Lords on July 21, in committee on the Motor Car Bill, Lord Kelvin moved a provision to enable the Local Government Board to make regulations designed to prevent the use on public roads of racing cars of great power.

A week later he was in Paris, on his way to Aix-les-Bains, and was receiving a cordial welcome from the savants of the *Académie des Sciences* as of old.

Before he left Aix he wrote to Lord Rayleigh :—

AIX LES BAINS,
*Aug.* 24, 1903.

DEAR RAYLEIGH—Can you tell me roughly (or fairly approximately) how much heat is evolved per gramme of cordite fired? Quite a rough estimate will serve me. I want to compare it with the heat continually emitted by radium, found by Curie (and by Curie and Dewar at the temp$^{re}$ of liquid air) to be about 90 "calories" per gramme per hour, which I was assured meant 90 times the heat required to raise a gramme of water by 1° cent. This seems almost incredible: to raise to boiling-point its own weight of water in 1$^{h}$ 6$^{m}$! At this rate in 1000 hours it would emit many times as much heat as the greatest, so far as I know, ever found for the same weight, in the most intense chemical action or burning. It seems to me utterly out of the question to suppose, as Rutherford and others have done, that the heat emitted by radium is generated by a self-contained store of energy.

I have sent (told the printers to send) 24 copies of corrected revise of the enclosed to Lees; and I am answering him that I cannot be present at the B.A. discussion on radium, but that I intend to send a few words towards the discussion, in which I shall refer to this bit of my Sec. XX.

Suppose the radius of an atom of radium to be $\frac{1}{1000}$ that of O or N or H, etc. It would go through Robin's[1]

[1] Hon. R. J. Strutt.

centimetre-thick plate of lead with very little resistance. If each atom of radium takes an exceptionally great number of electrions in it to neutralize it, it will be exceptionally liable to have electrions knocked out of it by collisions with other atoms of radium, or by the thermal motions in bromide of radium. Thus the "$\alpha$ rays" would be atoms of radium with less than neutralizing quantum of electrions. These would be greatly obstructed ("very easily absorbed" according to Robin) by solids.

The "$\beta$ rays" are, in all probability, merely electrions or atoms of resinous electricity.

The "$\gamma$ rays" would be merely vapour of radium.

But Rutherford's experiments, and Ramsay's and Soddy's experiments (*Nature*, Aug. 13), seem to prove that "bromide of radium" is not merely bromine combined with a metallic atom, but has at least one other element combined with it which, when "radium bromide" is dissolved in water, leaves the solution and shows itself as a self-luminous gas!

I don't intend to do more for the discussion than to offer some suggestions for consideration.

We have been much grieved to hear of Lord Salisbury's death. I was very sorry when he retired from office, but hoped he would have many years of well-earned repose from public duty. His country and the world owe him gratitude for great benefits. What a sadly tragic time it has been, with the sudden illness and death of Lady Galloway, when Lord Salisbury's life was so near the end. Lady Kelvin joins in sympathy with you and Lady Rayleigh.—Yours ever, KELVIN

They left Aix for Netherhall, but took a few days at Southport to attend the meeting of the British Association, which opened September 9. Here Lord Kelvin propounded his electro-ethereal theory of light (see p. 1082), and took part in a discussion on the emanations from radium.

All his life Lord Kelvin had venerated the memory of John Pringle Nichol, the teacher who had awakened in him the love for experimental physics when he was a little lad. He had passed away in 1859, and his son, Professor John Nichol, had followed him in 1894. On October 24, 1903, a memorial tablet in the staircase of the University of Glasgow was unveiled[1] by Lord Kelvin, who feelingly recalled the days when he himself was a student under the elder Nichol.

An honorary Doctorate in Science was conferred on Lord Kelvin by the University of Wales, in Cardiff, on November 13, 1903. He was presented by the Dean of the Faculty of Science, Professor Claude M. Thompson, in a speech which laid emphasis on the knowledge-producing quality of his scientific work. *Domine illustrissime et eruditissime, Britanniarum omnium insigne decus, in doctoratum nostrum rite jam admissum, Cambria te salutat*, were the words of the deputy-Chancellor in admitting him to the degree.

On the same visit he was made an honorary member of the South Wales Institute of Engineers, and attended the laying of the foundation stone of the new University College.

Presiding on December 2 at the West of Scotland Liberal Unionist Association, Lord Kelvin declared that the duty of the Association remained as clear now as it was when it was founded. Though they might not agree on the fiscal question, their members

[1] The ceremony and speech are recorded in *Nature*, vol. lxviii. p. 623, October 29, 1903.

would be welcomed provided they were thoroughly sound in regarding it as a principle that nothing should be done to imperil the union between Great Britain and Ireland. A week later, apologising for absence from a meeting to establish a branch of the Tariff Reform League in Glasgow, he wrote that his adherence was based on the Colonial reason, but that in his opinion moderate and wisely chosen import duties would be beneficial to this country and be a convenient and proper mode of raising revenue.

Lord Kelvin had at last finished the rewriting of the *Baltimore Lectures*, and sent a copy to Mascart in February 1904. "I am taking advantage of my freedom," he wrote, "in having got the long-delayed volume off my hands, by going back to some old deferred work on Cauchy and Poisson's problem on water waves." He was now in London and spoke in the House of Lords in advocacy of the compulsory adoption of the metric system. In his speech on the Metric System Bill he told how his life had once been endangered by a mistake[1] between "drachms" and "drams." On March 16 he wrote again to Mascart that he had been for six weeks perfectly free from face-ache, but that within the last few days it had been troubling him again; so he and Lady Kelvin proposed to leave London for Mentone on March 25, "to enjoy ten degrees

[1] This was in his lecture-theatre, in the experiments which he showed once a year of measuring the speed of a rifle-bullet by firing it into a block of wood hung as a ballistic pendulum. His assistant would charge an old muzzle-loading rifle, Lord Kelvin would then, kneeling on one knee, take deliberate aim at the pendulum a few feet away, and fire; his students cheering frantically. On the occasion in question he stopped his assistant from loading with a charge that seemed dangerously large.

of more indulgent sky," and hoped to spend three or four days in Paris.

While Lord Kelvin was in the Riviera, the Earl of Stair, the Chancellor of the University of Glasgow, died. On April 6, 1904, the General Council of the University met to choose a successor. The Lord Provost moved and the Lord Advocate (Mr. C. Scott Dickson, K.C.) seconded the nomination of Lord Kelvin. There being no other nomination Principal Story declared him to have been unanimously elected. The installation was deferred till the autumn. "It will be a great happiness to me," Lord Kelvin wrote to Principal Story, "for the rest of my life to be Chancellor of my beloved University of Glasgow."

On his eightieth birthday, June 26, 1904, Lord Kelvin received many tributes of affectionate regard. To his niece, Miss May Crum, he wrote :—

15 EATON PLACE, S.W.,
*June* 26, 1904.

DEAR MAY—Your little note and the Scottish shamrock from the field of Bannockburn are delightful. Thank all the others too for thinking of me to-day. I had a telegram yesterday from six of my old students in Tokio. "Heartily congratulate your birthday, Masuda, Taniguchi, Watanabe, Mano, Goto, Tanakadate." Was it not good of them to remember me in the middle of their terrible war! Your Aunt Fanny joins me in thanks and love to you all.—Your affectionate uncle, KELVIN.

Miss Crum,
Auchenbowie, Bannockburn, N.B.

Lord Kelvin attended the Fourth of July dinner

of the Anglo-American Reunion in London, and proposed the health of Mr. Choate, the United States ambassador, whom he greeted as one who had worked for the public good, and as a resolute lover of peace.

On August 17 the British Association met in Cambridge under the presidency of the Rt. Hon. Arthur J. Balfour. His presidential address dealt with the modern doctrines of physics.

Here in Cambridge was the classic ground of physical dis covery. Nowhere in any corner of the world could be found a spot connected with so many men eminent as the originators of new and fruitful physical conceptions. He passed by Bacon and Darwin, being concerned rather with the illustrious line of physicists who had learned or taught in Cambridge,—a line stretching from Newton through Cavendish, Young, Stokes, Maxwell; through Kelvin, who embodies an epoch in himself, down to Rayleigh, Larmor, J. J. Thomson, and the scientific school centred in the Cavendish laboratory. The task which the physicists had set themselves to accomplish was not merely the discovery of laws connecting phenomena. The object of the physicist is physical reality. That such a reality exists, though philosophers have doubted, was the unalterable faith of science.

One of the tasks of physics had been to frame a conception of the physical universe in its inner reality. The conception framed in the end of the eighteenth century was that the universe consisted of various sorts of ponderable matter with action at a distance between its constituent masses, together with certain imponderables called heat, and electric fluids, and the corpuscles supposed to constitute light. About a hundred years ago Young had reopened the undulatory theory of light, and with its establishment came the acceptance of the notion of that new constituent, an all-pervading ether filling unbounded space.

But to-day the conception was different; for there were those who regard gross matter, the matter of everyday experience, as the mere appearance of which the physical basis is electricity; that the elementary atom of the chemist is but a connected system of sub-atoms which are electricity itself; that while in most cases these atomic systems may seem almost eternal they are not less obedient to the law of change than the everlasting heavens themselves. But what were these electrical sub-atoms or electrical monads? It might be, as Larmor has suggested, that they were mere modifications or knots in a universal ether. Surely here was a very extraordinary revolution. Two centuries ago electricity seemed but a scientific toy: now it was deemed to be the reality of which matter was but the sensible expression. Mass was not only explicable, it was actually explained; and so far from being unchangeable, it changes with its velocity. Further, each atom must now be regarded as a vast store of intrinsic energy: so that when a sun cooled down by radiating away its sensible heat, its energy was not (as used to be supposed) exhausted: the amount thus lost would be absolutely insignificant compared with that which remained stored up within its separate atoms. The new electric theory analyses matter into something which is not matter at all, so that matter is not merely explained, but explained away. Here was something of a paradox. We claimed to found all our scientific opinions on experience; yet the conclusions at which we arrive are to all appearance fundamentally opposed to it: our knowledge of reality is based on illusion.

"Arthur Balfour is just in the arms of metaphysics," was Lord Kelvin's comment, as he walked away from the Senate House on the arm of Sir Henry Roscoe. Probably no one would resent the dictum less than the distinguished statesman. And indeed if physics is defined as the science of dead matter; if the aim of natural philosophy were literally to reduce all the problems of the material universe to dynamics; if all else must be explained in terms of matter and motion only, then the twentieth-century doctrine which explains matter in terms of electricity and ether is truly metaphysical. Throughout the fifty-eight years of his long quest Lord Kelvin had retained with marvellous activity his formal powers of mathematical analysis and physical insight. But the very dominance of those dynamical conceptions which he had done so much to extend, militated against his acceptance of the new views which had been shaping themselves in the brains of the younger men who were intellectually his disciples.

Lord Kelvin read two papers at this meeting: one "On Insulation in a Vacuum," the other "On Models of Combinations of Atoms and Electrions to demonstrate the Properties of Polonium or Radium." He explained that he proposed as models those in which the equilibrium involved is purely statical, because he knew of no instance, in the whole of dynamics, of systems choosing successive stages of stable motion such as was proposed by the upholders of the most commonly-accepted form of the electronic theory of matter. He had in the previous year denied the possibility of that theory, and of the hypothesis of the disintegration of the atom to account for radio-activity, and had suggested that it was due to some external source of energy. His reason had been his alarm at the statements that the emission of energy from radium went on "for ever." Now that he saw that this was not meant, he frankly abandoned the suggestion of external source in favour of an intra-atomic source of energy.

On October 28 Lord Kelvin gave at St. George's Hospital School an address on the Living Cell, which is described on p. 1102.

At the installation of Lord Kelvin as Chancellor of the University of Glasgow on November 29, 1904, his first duty on assuming office was to confer the honorary degree of LL.D. upon H.R.H. the Princess Louise, Duchess of Argyll. Honorary degrees were also conferred by him upon Principal Lang of Aberdeen, the Hon. C. A. Parsons, Signor G. Marconi, and others. It was to him a real pleasure that his nephew, James T. Bottomley, who had for so many years been associated with him as deputy Professor, was also included amongst the recipients of the honorary degree. Then he pronounced his inaugural address. He spoke of himself as "a child of the University of Glasgow," in which he had lived for sixty-seven years (1832-1899). He alluded to his father, and to his father's experiences as student and afterwards as professor in the old Glasgow College; and added several touching reminiscences of his teachers in his own happy student days. He spoke of the pioneer work in laboratory teaching in anatomy, chemistry, natural philosophy, and engineering; of the recent reforms and extensions, and of the benefactions of recent date. Congratulating the city and the University on the splendid equipment and staff, he concluded by returning thanks for having been elected Chancellor, and by heartily wishing success and happiness to all professors, and lecturers, and teachers, and students in every department. One word, not in his printed address, he interposed, of regret that Greek should be dropped out of any University course; a word of the usefulness of teaching during early life the things that do not come naturally to the student; one little warning against the spirit of the age in too much following the bent of the mind, feeding it only with things convenient, and neglecting those things which yield increase only to toil and cultivation.

In the public press there were many notices of Lord Kelvin's succession to the Chancellorship. The *St. James's Gazette* spoke of the man himself:—

Wealthy and famous, and of prodigious learning, he is still one of the most modest and simple-minded of men, with an exquisitely gentle face framed in snow-white hair, and a smile which is like a benediction. It was a great day for him, but a still greater for Glasgow University.

He wrote to congratulate Lord Rayleigh:—

NETHERHALL, LARGS,
*Dec.* 23, 1904.

DEAR RAYLEIGH—. . . . It was only after I last wrote to you, and you told me you were starting for the Continent, but not *why*, that I heard the Nobel Prize had been awarded to you. I heartily congratulate you, and so does Lady Kelvin, who also joins in best wishes for Christmas and the New Year for you and Lady Rayleigh.

I hope you both enjoyed your trip to Stockholm, and are happily at home, all well, and caught no colds.

I am busy on canal ship-waves for R.S.E., January 23, and sounding machine (a very new plan) and compass, etc.—Yours, KELVIN.

[*P.S. by Lady Kelvin*].—The "demon" has been *very* wicked lately, and Lord Kelvin has had a very bad time.
F. A. K.

New Year 1905 found Lord Kelvin very far from well. On January 16 Lady Kelvin wrote to Mrs. Ramsay MacDonald:—

I am sorry to say he has been suffering terribly badly from the pain in his face. It has been quite the longest and worst attack that he has ever had, I think; and there is hardly any diminution yet. He has hardly been out since his installation on the 29th Nov$^{r}$. We went up to see a doctor in Glasgow last week, but doctors don't seem to be able to do anything for this particular ailment, and acknowledge themselves helpless. He was better for a day or two, but to-day he is very bad, owing partly, I think, to a roaring hard east wind, which is very cold and searching.

He was able to travel up to London a few days later to be present at the first meeting of the Admiralty Committee on Warships (see p. 733).

Lord Kelvin's medical advisers found it necessary that he should submit to a surgical operation, which was performed on March 29th. For several days there was great anxiety as to his state, and many were the inquirers at his house who called to see the bulletins of his medical attendants. The King telegraphed inquiries twice. By April 5th he was pronounced out of danger, but his recovery was necessarily slow. In July he was able to be removed to the quiet of Lady Winchelsea's house at Sleaford—"to keep him out of temptation" to exert himself in any way. Returning to London on July 18th, he wrote to his brother-in-law Mr. W. G. Crum :—

. . . . I have been greatly pressed by work which had accumulated during my illness, including a paper on Ship-Waves long promised for the last meeting of the R.S.E. session which was held yesterday. This is the *first* use I am making of my comparative freedom. I felt most deeply how kind and sympathetic you all were, and I am most happy to be able to tell you how exceeding well I have been feeling. I have had a good deal to do in the way of recovering strength, but this has all gone on well, and I sometimes forget my walking-stick. With it I feel pretty nearly all right, and can take fairly long walks (not really *long*!). I had a *very* wicked two days' and nights' attack of my No. 5 demon last week during our visit to Lady Winchelsea, from which we are now returning, but it ceased as suddenly as it began; and I am now *very* hopeful that I am never to be troubled much more by that enemy. Trials were going on on board H.M.S. *Mercury* last week of my new sounding machine, which have proved very satisfactory.

. . . We leave for Aix about July 28.

The British Association met that year in South Africa; and to it Lord Kelvin sent a paper on the equilibrium of ether in ponderable matter. He had returned from Aix to Netherhall, and wrote thence to Col. R. E. Crompton:—

NETHERHALL, LARGS, AYRSHIRE,
*Oct.* 7, 1905.

DEAR CROMPTON—I should long ago have thanked you for your letter of Sep. 8th. I hope young Archibald Smith will be found deserving of promotion from the ranks of the Volunteer Corps of Engineers.

I am much interested in all you tell me of the corps, and I shall be very glad if we can meet when I am next in London to talk over its affairs.

I shall be in London for a few days about the beginning of November. Would it be convenient to you to come and see me at 15 Eaton Place, on Wed[y]. Nov. 4, any time before 12?

Among other things I would like to hear more of the Coronation medals. I scarcely think it could be suggested to the King to give some additional medals for members of the corps who were in South Africa at the time; but you will tell me more about it.—Yours always truly, KELVIN.

Russell & Sons photo. Emery Walker Ph. sc.

F. A. Kelvin, Kelvin

[1906]

With rest and careful nursing Lord Kelvin recovered health, and was very well during the winter of 1905-6, and through the spring of 1906. His eyesight began, however, to trouble him: there was a threatening of detachment of the retina. He was able on May 17th to attend the annual dinner of the National Telephone Company, looking quite hale and hearty, and made a vigorous speech as a guest. A day or two later he was the guest at the South Africa dinner. On June 21st he spoke in the House of Lords on the disturbance caused to the Observatory of Greenwich by the adjacent electric lighting station. He suggested that the trouble from vibration might be practically annulled by the substitution of steam turbines for reciprocating engines; but if that should not prove effective, he trusted that both Houses of Parliament would unite in defending the Royal Observatory from the tremendous disaster that seemed to be impending. On June 24th he was welcomed at a banquet given by the Institution of Electrical Engineers to foreign electrical engineers then visiting England, and who were entertained in Glasgow on July 2nd. At a luncheon given to the visitors in that city, Lord Kelvin declared that electricity was a peace-maker for the whole world. Its present development would have seemed like a dream to the early pioneers; but there was another dream which he hoped would be realized — universal peace, universal alliance among the nations all over the world. Lord Kelvin had been elected on June 26 president of the

International Electrotechnical Commission for the unification of technical terms and formulæ.

An important controversy[1] about radium arose after the British Association meeting in August.

On August 5 Lord Kelvin wrote to *The Times* from Aix-les-Bains, to raise a warning against hasty generalization as to the transmutation of elements which had been suggested at the meeting at York as a deduction from the discovery by Sir W. Ramsay and Professor Soddy of the production of helium from radium. He also protested against the hypothesis that the heat of the sun or of the earth is due to radium; he held that it is mainly due to gravitation. To this Sir Oliver Lodge replied, regretting that physicists had been unable to carry their veteran leader with them in some recent developments, and the Hon. R. J. Strutt wrote to ask what becomes of the heat generated by the radium admitted to be present in the earth. Lord Kelvin, in *The Times* of August 20, rejoined that he thought he had spent more hours than any other person in reading Rutherford's *Radio-activity*. He contended that there was no valid experimental evidence to show that the heat of radium was sufficient to account for sun-heat or underground temperature. Several further letters appeared, including one from Professor Soddy, emphasizing the points of agreement with Lord Kelvin, and review-

[1] For an extended summary of this controversy see *Nature*, lxxiv. p. 539, September 20, 1906. Lord Kelvin's last contribution to the radium question was his paper in the *Philosophical Magazine* of March 1907, "An Attempt to explain the Radio-activity of Radium."

ing the evidence for the statement that radium is continually generating helium.

Another controversy arose a little later concern ing the international agreement as to wireless telegraphy, for which the Marconi Company was claiming a monopoly. Lord Kelvin interposed with a letter in *The Times* of October 16, describing how in June 1898 he had been taken by Lord Tennyson to the Marconi Station at Alum Bay, whence he had had the pleasure of sending messages through 15 miles of ether, and on by postal land telegraphs to Sir George Stokes and others. He believed that up to that time, or about then, "there had been no other practical advance upon Lodge's wireless telegraphy through one hundred and fifty yards in 1894."

An interesting ceremony, which linked together the names of Faraday and Kelvin, took place at Barnsbury on November 24, 1906, when Lord Kelvin unveiled a memorial tablet to Faraday in the room where Faraday, as an elder of the Sandemanian Church, used at times to preach, and which had now been transformed into a switchroom of the National Telephone Company. Lord Kelvin's words on this occasion were spoken with great emotion.

It is indeed a crowning privilege of my life to assist to-day in the inauguration of this monument to Faraday. Sixty-six years ago his was the inspiring influence of my early love for electricity. I did not then know him except in his published scientific papers. Ten years later I heard him lecture in the Royal Institution, and

I had the great honour of making his personal acquaintance. For many years after 1850 my chief attraction for making a journey from Glasgow to London was always a visit to the Royal Institution to see Faraday—at home in the midst of his work. Bright, lively, kind, he showed me what he was doing. He spoke to me of it, and he encouraged me to go on with any work I could take up, either in connection with electricity or any of the other interesting and important things in physical science in which he was working and making discoveries. His encouragement to me was most valuable. I treasure it now, and I can look back upon it as an inspiring influence throughout my life. Faraday had the gift of inspiration. . . . To-day we have inaugurated a memorial tablet[1] to Faraday, and here in the building in which it is placed we may well repeat the celebrated Latin words—*si monumentum requiris, circumspice*—which are inscribed in St. Paul's Cathedral over the tomb of Sir Christopher Wren. Look around. In the place itself you will see a monument of Faraday. These walls tell a story, not of a magnificent cathedral, but of the humble meeting-house of earnest Christian men. Here were carried on the religious services of the Sandemanians[2] in London, a very simple association devoted to faithful and earnest Christian work. Throughout his life Faraday adhered faithfully to this denomination as an officiating elder. . . . I well remember, at meetings of the British Association in Aberdeen and Glasgow, how he sought out the meetings of his denomination, and spent, as a preacher or worshipper there, the Sunday and any time he could spare from the work of the Association. How very interesting it is to think of Faraday's life-long faithfulness to his religious denomination. In another sense it is of very great interest to us to look around in this place and see the busy telephone operators doing their daily work in it. This also is a splendid monument to Faraday. Every electrician present, and all who are engaged in the telephonic business, will heartily agree with me in admiring and wondering at this beautiful and useful result of the vast discoveries in electromagnetism in which Faraday led the way. How much Faraday would have been delighted with the telephone some of us can imagine. How much, too, would he have delighted in the applications of electricity to the transmission of power! . . .

[1] The tablet, a large inscribed bronze plate, was subscribed for by the staff of the National Telephone Company, with whom the project of the memorial originated.

[2] A small sect which separated from the Scottish Presbyterian Church about 1730, and established in London about 1760 by Robert Sandeman. It has now practically died out.

In conclusion Lord Kelvin said he would ever remember the kindliness of Faraday, and his fertilizing influence throughout the scientific world, and the encouragements which he extended to any attempt to apply mathematical demonstration to the problems of science, which were the subjects of his daily work so long as health and life allowed him.

Lady Kelvin wrote to Miss King, saying, "Your uncle was the only person present who had ever seen Faraday, except some of Faraday's own family; and he made his little speech on his personal recollections. It was all quite a success."

He went down to Glasgow the next day to preside at a dinner of the Glasgow branch[1] of the Institution of Electrical Engineers, and to attend a Liberal Unionist meeting. He was back in London for the Royal Society anniversary on November 30.

On December 4, 1906, speaking in the House of Lords on the Trades Disputes Bill, Lord Kelvin asked the House to reject the Bill in the interests of those workmen who would lose their liberty if it passed, as it afforded no redress against a wrongful act committed by a trades union against a non-unionist. He had attended all the sittings of the House of Lords from October 24 to the end of

[1] At this dinner Lord Kelvin gave a long speech on the growth of electrical engineering, with reminiscences of Sir William Siemens, from whom he had in 1881 obtained a shunt-wound dynamo, to be driven by a Clerk gas engine in his laboratory. "He believed that the first house on this planet in which the whole lighting was done by electricity was his house in the University of Glasgow, getting the current from that Siemens' dynamo, and the Clerk gas engine of which he had spoken. In that house the rule given to the workmen (for making what began soon after to be called installation) was, 'Go through the whole house: wherever there is a gas burner put in an electric light.' There were 106 gas burners; and there were 106 electric lights in that house about the end of the year 1881."

the Committee and Report stages of the abortive Education Bill of that year.

The year 1907 found Lord Kelvin in fairly good health, and very active in mind. Christmas and the New Year had been spent at Netherhall; but Lord and Lady Kelvin came up to Glasgow on January 10 for the installation of Mr. Asquith as Lord Rector. Facial neuralgia had troubled him a little. Hardly had they returned to Netherhall before they heard of the death of Principal Story, whose funeral Lord Kelvin attended, as Chancellor, on the 16th. In the afternoon he spoke at a Women's Unionist Association meeting. The same afternoon they went off to Edinburgh, where Lord Kelvin presided at a dinner of the Royal Society of Edinburgh to the Prince of Monaco, the Oceanographer. He attended the anniversary James Watt dinner in Greenock on January 18, and spoke of the water-supply of Glasgow and its influence on public health. On the 21st he was in Edinburgh communicating to the Royal Society of Edinburgh two papers , one of which related to the motion of the earth through the ether. He sent a report of this to Lord Rayleigh :—

NETHERHALL, LARGS,
*Jan.* 23, 1907.

DEAR RAYLEIGH—Many thanks for your postcard. FitzGerald is very interesting and is clearly right. He suggested the electromagnetic idea, and in suggesting it proved that it was essentially nugatory. You will see by the last inch of the enclosed press-cutting, that I spoke of the subject at the R.S.E. on Monday evening. I referred to Michelson's celebrated experiment, seeming to prove that the relative motion between the earth and the ether at its surface could not be anything nearly as much as thirty kilometres per second. I referred very shortly to my proposed experiment with two insulated cylinders, and an uninsulated cylinder midway between them, to serve as an electrostatic screen, which I did not mention, perhaps I had not thought of it when I wrote to you and Silvanus Thompson a fortnight ago. The opposite electrifications induced on the two sides of it give an electromagnetic effect exactly equal and opposite to the electromagnetic effects of the two insulated cylinders. This is a fairly good illustration of the essential nugatoriness of the electromagnetic test, demonstrated by FitzGerald; and is somewhat interesting as showing a fixed electrostatic and electromagnetic screen instead of FitzGerald's movable one, which is the surface of the steel testing-needle.

The newspaper reports naturally reported only the wail of grief with which I ended my "paper."

Michelson's experiment is virtually a test for inequality in the velocity of light relatively to the earth, in the two opposite directions parallel to the tangent to the earth's orbit at any time. It might be imagined that Fizeau's method of measuring the velocity of light, in virtue of its greatly longer testing distance, might be more searching than Michelson's. But in reality Michelson's, with its comparatively short distance, is vastly more searching than the other, because of the vastly more delicate optical test, which his interferential method gives, than Fizeau's toothed wheel.—Yours truly,

KELVIN.

The Institution of Electrical Engineers desired to elect Lord Kelvin for the third time as its President, and entrusted Mr. W. M. Mordey with the duty of conveying the invitation. Lord Kelvin at first declined the proposal:—

NETHERHALL, *Feb.* 1, 1907.

DEAR MR. MORDEY.—I cannot tell you how much touched I am by your kindness, and the kindness of the Council of which you tell me in your yesterday's letter. After most mature thought, and daily experience of daily inevitable cares, I feel that it would be impossible for me to accept the honour of being President of the Institution, without neglecting arrears of work which I have promised from time to time during the last fifty years, and which I am now trying to get on with every day, with most disappointing experience of the smallness of the time that I find possibly available for doing so. Will you tell the Council how very sorry I am that I must deny myself the pleasure of being again President, and give my most cordial thanks to all for the great kindness with which they entertained your suggestion.

I am very sorry to hear that you have been laid up with influenza, but I hope you are now feeling perfectly well again.—Believe me, yours most truly,

KELVIN.

After further correspondence, and an assurance from the Council that everything possible would be done to lighten the cares of the office, he assented. The formal election took place in May.

Lord Kelvin invited Lord Rosebery to come to the Institution of Civil Engineers on the occasion of the reading of a paper on Modern Motor Vehicles by Col. R. E. Crompton. His reply was:—

DALMENY HOUSE, EDINBURGH,
*Feb.* 4, 1907.

MY DEAR LORD KELVIN.—I regard you as one of the glories of Scotland and your wishes as commands. Consequently, I am going to the Civil Engineers, which I had not intended to do.—Yours sincerely,

R.

On March 7 Lord Kelvin took part—it was for the last time—in a discussion at the Institution of Electrical Engineers on a paper by Mr. J. S. Highfield on the Electric Transmission of Energy by Direct Current on the Series System. The following are the principal points of Lord Kelvin's speech:—

For myself the subject is one of extreme interest. I have never swerved from the opinion that the right system for long-distance transmission of power by electricity is the direct-current system. I do not say a word in depreciation of the beauty of the polyphase idea, and of the development of that idea in the three-phase working.

Some of the most subtle problems of electricity, problems depending even on resonance, are more or less familiar to all electrical engineers at the present time. The perfect freedom of the direct-current system from complications of that kind is very remarkable.

You all understand something about square root of 2 being 1·41. Starting with that scientific fact, we have this practical application that 141 volts of direct current have no more tendency to produce sparks than 100 volts of alternating current. . . . Mr. Highfield finds, in respect to insulation, a much greater advantage of the direct current over the alternating current than has been hitherto known to us. I think I may safely say, as proved by the experiments which he has put before us, that 80,000 volts direct current are more easily worked than 40,000 volts alternating current.

I do not go so far as to say that the beautiful systems that are now worked out for distributing power over somewhat long distances would be better carried out by direct current than by alternating current. I do think, however, with regard to the ordinary power-supply stations, which are now with but few exceptions founded on the alternating-current system, it is just possible that, in future, there may be a change to direct-current.

Calling on Lord Kelvin at 15 Eaton Place on March 21, I was struck with his altered appearance. He looked feebler. He seemed bent, more lame than usual, and deafer. He told me that his sight was troubling him: something the matter with the central part of the retina of the left eye—all looked black—so that he saw things with the right eye only. He had been told, he said, that the trouble would not extend by sympathy to the other eye. "But I suppose it is *Anno Domini*," he added, rather sadly. He was busy, when I called, working at the compass corrector with his secretary Mr. George Green.

Lord Kelvin attended the anniversary dinner of the Chemical Society on March 22nd, and delivered a lively speech in reply to the toast of the Scientific Societies. He gently rallied the chemists on the difficulty of talking about the newly discovered phenomena of atomic disintegration. They had better not call "atom" a thing that could be broken. When was an atom not an atom? he asked. He had recently, when revisiting his old laboratory at the Collège de France, burned his fingers with phosphorus; for so old a person as himself a more appropriate day, he said, would have been the first of April!

At this period also he was much exercised in mind about the proposals of the Irish University Commission Bill. He strongly advocated that Trinity College, Dublin, should retain its entire independence.

On April 11 Lord Kelvin dined at the Constitutional Club to support Mr. Bonar Law, and made a vigorous speech in favour of Imperial reciprocity. On the 23rd he was presiding as Chancellor at the conferment of honorary degrees on the Prince and Princess of Wales in Glasgow.

He took the chair at the Royal Institution on May 28 for Professor Fleming's lecture on Wireless Telegraphy. It was a brilliant lecture, but Lord Kelvin seemed too weary to follow it with his usual keen interest. On June 19 was the annual conversazione of the Electrical Engineers at the Natural History Museum. Lord Kelvin came late, and alone; and his aged appearance was a matter of wide comment and regret. Five days later he spoke briefly, but vigorously, in the House of Lords on the insufficiency of existing laws to protect seamen in the merchant service from the dangers of deck loads.

He wrote to Lord Rayleigh :—

15 EATON PLACE, S.W.,
*July* 15, 1907.

DEAR RAYLEIGH.—§ 1. Do not write either a letter or a postcard in answer to this, but return the enclosed copy, with answers, or remarks, or nothing, written on the margin.

§ 2. Are you to be at Leicester at the B.A.? We are going, to remain from Wednesday till Saturday, and are to be guests of a Mr. Everard who lives about five miles away, but promises carriages or motors for everything we want. We shall not choose motor unless of necessity. We leave London for Aix-les-Bains on the 2nd or 3rd of August.

§ 3. I am designing a short communication to the B.A., showing numerous periods and corresponding spectrum lines in a glowing monatomic gas, even if there is only one electrion in each atom ; but much more numerous bright lines, or possibly hazy ones, if each atom has two or more electrions as its neutralizing quantum. You will save me much trouble if you will tell me something about the spectrums of argon and helium.

§ 4. It is possible that a diatomic gas such as hydrogen, may have some lines due to mutual vibration between its two atoms, or rotation of the lines joining their centres. But it seems to me more probable that all the bright lines of gases are due to vibrations of electrions.

§ 5. If we call by the name electron the ponderable atom deprived of all its electrions, the general statement (given first by Lorentz of Leyden) that the bright lines are due to vibrations of electrons would include ponderable atoms among the vibrators. But it seems to me more probable that it is only the equal and similar electrions that are the constituents of the vibrators, and that the enormously more passive "positive electrons" are not practically concerned in the vibrations, except as infinitely massive stands or supports. I am, however, quite ready to be convinced, if it is true, that the "positive electrons" are practical vibrators in the dynamics of light and radiant heat.

§ 6. It is possible (probable?) that part of the light and radiant heat of a monatomic gas may be Stokesian, due to non-vibratory collisions between pairs of atoms giving shocks to the ether, superimposed upon the trains of vibratory waves due to the electrions. Even if there were no electrions, we could, as you have shown, produce periodic light from the merely Stokesian light by a grating or by a prism.

At the British Association meeting at Leicester, on July 31, Lord Kelvin surprised his friends by the renewed vigour and activity which he displayed. In a neat little speech, when proposing a vote of thanks to the President Sir David Gill, he declared his delight in having been personally conducted on a magic tour through the universe, carried not on aeroplanes that would not fly, but on the stronger and surer wings of the scientific imagination. The next day Lord Kelvin read a paper on the motions of ether produced by collisions of atoms or molecules containing or not-containing electrions. He insisted strongly on the indivisibility of the atom of matter, whether it contained an atom of electricity or not. He laid down a number of dynamical propositions as to the atom, and its power to originate pulses in the ether when suddenly started or suddenly stopped. He inferred from the multiple bright line spectra of the monatomic gases that either the atom contained one electrion with many possible positions of stable equilibrium, or many electrions with one stable position for only one of them, or, thirdly, that there were many electrions with many stable positions for one of them. On August 2 one of the most instructive discussions ever known in the Association, upon the Constitution of the Atom, was opened by Professor Rutherford; the very title being a challenge to Lord Kelvin, who held that an atom as such could not have a constitution, but was an indivisible unit. After Sir Oliver Lodge, Sir William Ramsay, Professor Soddy, and Professor (now Sir Joseph) Larmor had spoken, Lord Kelvin took up the argument with all the keenness of a young debater.

He could not believe that the mere motion of electrions gave the widely different properties and degrees of stability which are recognised as belonging to the different kinds of matter called elements. Nor could he believe that radio-activity was a mere residue of the kinetic energy of the atom. He preferred rather to regard the atom as a gun loaded with a shell which again contained a charge; for when the atom threw off an electrion it did not cause the bursting of the atom any more than the firing of the gun effected its destruction. His gun—the atom—had probably been loaded before there were sun, stars, or any of the things that we now know, and it only required contact with matter to release the trigger and fire off the shell.

From Leicester, Lord and Lady Kelvin went straight to Aix-les-Bains for the usual month of rest and cure. Lady Kelvin wrote from the Hôtel Britannique on Sept. 5, that they would leave Aix on the 11th, reaching London on the 12th, and going on to Netherhall on the 14th. Lord Kelvin was due in Belfast on 20th to open the new laboratories of Queen's College.

Lady Kelvin, who had not been really well in the preceding months, found the journey to Scotland very trying, and on arriving at Netherhall she collapsed. In a few hours there came on a difficulty of speech and other symptoms. Complete rest was ordered her, and Lord Kelvin's engagement at Belfast was cancelled, to his intense disappointment. His address was, in his absence, read by his nephew James Thomson, of Newcastle-on-Tyne. In it he spoke of the opening of seven laboratories in one day as unique in University history. He also spoke of the benefactors of the University, of the uses of laboratories, of the discoveries to be made in them, and of his own recollections of the old Belfast Royal Academical Institution. He had written on the previous day to his nephew: "I am very glad that it has been arranged that you are to read my Address to-morrow. You will see that my statement that the Laboratories are open 'to-day' does not imply that the Opening Ceremony has been performed by myself. I wrote the whole Address carefully in the knowledge that I should not be present, but I expressed it in words as if I was personally addressing an audience. In the circumstances you must be regarded as absolutely representing me in all respects, except the actual opening of the Laboratories with a golden key, which Principal Hamilton told me had been prepared for the occasion." The Address was duly read, and the golden key brought back to Lord Kelvin at Netherhall.

Lady Kelvin's state remained critical for many days.

Lord Kelvin wrote to Sir George Darwin :—

NETHERHALL, LARGS,
*Sept.* 23, 1907.

DEAR DARWIN—Thank you very much for sending me the proof of your *preface.* I am much gratified that you are able to say what you say of me in it. I am glad you are publishing separately your papers, and I hope vol. ii. . . . . will soon follow.

Lady Kelvin's illness is very serious. It came on Sunday ev<sup>g.</sup> the 15th. Since then there has been *progress* toward recovery, tho' very slow. Since yesterday symptoms have been more favourable than any time before, and I am allowed to hope for continued progress to complete recovery. She would join me in love to you and Maud ; but the most absolute quiet and rest is enjoined as the first requisite for progress.—Yours very truly, KELVIN.

He wrote also to Madame Mascart, whose husband had been ill for some months :—

NETHERHALL, LARGS,
*October* 3, 1907.

DEAR MADAME MASCART—Lady Kelvin has been seriously ill since the 15th September. There has been some progress towards recovery, and some favourable symptoms during the last three days which have cheered me, but I am still in terrible anxiety. During her illness she has spoken of Mr. Mascart, feeling much concerned for you in your anxiety. Will you kindly tell me or ask Louise to tell me of his present condition? I earnestly hope you will be able to give a favourable report.

With our kindest regards to him, and Louise, and yourself, I remain, yours always truly, KELVIN.

He was himself unwell, suffering from indigestion and borne down by anxiety. His sadness was quite pathetic, he seemed unable to grasp the new conditions. A week later Lady Kelvin's symptoms showed distinct improvement, and she was regaining power. Lord Kelvin was himself better, and went up to Glasgow on October 10th for the day. He wrote again to Madame Mascart:—

NETHERHALL, LARGS,
*October* 13, 1907.

DEAR MME. MASCART—I thank you warmly for your most kind letter, and the sympathy with me of yourself and your husband which you express.

Lady Kelvin was very well at Aix les Bains, and felt that she was getting much benefit from the treatment there. On our arrival here on the evening of September 14 she was feeling very unwell, though she had felt perfectly well on our journey from Aix as far as Dover, and only began to feel ill on our railway journey to London where we remained a day. On the evening of Sept. 15 there came a terrible stroke, which has produced a loss of power. . . . Within the last seven days there has been a gradual, but very slow progress of recovery.

. . . She knows well what her illness is, and *hopes* for recovery. It is wonderful and most touching to see the cheerfulness with which she bears the calamity. I am allowed to be hopeful of a good recovery, but it must be very slow. To-day the doctor finds more of improvement since yesterday than on any previous day.

Absolute tranquillity has been ordered. Till to-day I have not been able to give her the kind message from you and Mr. Mascart, but I have given it to her to-day; and she was pleased and grateful; and she joined me in affectionate regards to you and him, and to Louise, and in best wishes for his recovery of health. I am glad you can tell me that he is now better, and I earnestly hope to hear soon that he is making good progress, and will soon be feeling really well.

Will you tell Louise that I hope to write to her in a few days, and give her good and better reports of Lady Kelvin.—Believe me, affectionately yours,

KELVIN.

The weeks dragged on with little change. Lord Kelvin was allowed occasionally to visit the bedside of the sufferer; any slight improvement seemed to buoy up his happiness, but he fretted at any symptom of relapse. The weather was cold and stormy, and he could not leave the house except for short walks. On November 8 he wrote to President Hamilton of Belfast:—

NETHERHALL, LARGS,
*Nov.* 8, 1907.

DEAR PRESIDENT HAMILTON—I am very glad to hear by your letter of yesterday that you strongly approve of the suggestion which I urged in my Address of September 20, that Queen's College, Belfast, should be made an independent National University. The more I think of it, the more it seems to me likely to gain general approval, and, if adopted, to be very beneficial, not only to Belfast, but to the whole of Ireland, and far beyond the limits of Ireland.

I remember strongly urging that Owens College, Manchester, should be made an independent University, and that Liverpool and Birmingham should as speedily as possible get Universities of their own; and pointing out that Scotland, a much smaller part of the United Kingdom than England, had four Universities, and not one too many for the good of its people, while England had only two. I remember the *Times* saying that Sir William Thomson was so fond of universities that he thought there could not be too many of them. At that time the tide was too strongly against me, and for a federal University; and Victoria University was founded. We know what the result is now.

To make Belfast an independent University now without delay would not hurt either the Royal University or Trinity College. Politically I believe it would be really popular, both among Nationalists and Unionists, and it could not possibly have any evil political influence of any kind.

Cork and Galway, it seems to me, might be made independent Universities with great advantage. They both I believe, have done really good work, and I owe a deep debt to Galway for being now in the excellent health in which I am, through the action of Mr. Frere, a thorough Irishman who had his medical education wholly and solely (I believe) in Galway. I cannot but think that the Roman Catholic difficulty might be satisfactorily got over somehow in Cork and Galway, if created independent Universities. But, feeling hearty good wishes for Ireland, and for all denominations of its people, I am afraid I am getting out of my depth and must not tax your patience with a longer letter.—Yours truly, KELVIN.

The Institution of Electrical Engineers was to open its session on November 14th. Any Presidential Address from Lord Kelvin was out of the question, but he wrote on November 9th to Dr. R. T. Glazebrook, the retiring President:—

NETHERHALL, LARGS,
*November* 9, 1907.

DEAR GLAZEBROOK—I wish I could be with you and our comrades of the Electrical Engineers on the 14th. Will you tell them how very sorry I am not to be present at the opening meeting of the Session. The work of Electrical Engineers throughout the world has grown marvellously from year to year, not only in practical importance, but also in profound scientific interest. From the time, in the middle of the nineteenth century, when their profession had its beginning with the electric telegraph over land and under sea, workers of all grades of Electrical Engineering have kept closely in touch with purely scientific investigation, which has indeed been largely advanced by their labours. The work of the present Session of the Institution will surely be carried out on the lines which have made its meetings so interesting and valuable ever since 1871, when it came into existence as the Society of Telegraph Engineers. Give the members of the Institution my best wishes for all their work, both in the Institution and in carrying out its practical objects, and tell them that I hope to be with them before the end of the present Session, and that in the meantime I thank them warmly for their great kindness in allowing me to be their President.—Yours very truly,

KELVIN.

Then he wrote Madame Monnier, daughter of M. Mascart:—

NETHERHALL, LARGS,
*November* 21, 1907.

MY DEAR LOUISE—It has been a great kindness to me your writing these last postcards and letters. I am very glad to hear by your letter of Nov. 18 that your father has been making such good progress, and that he is happy to be at home again, and that the doctors are much pleased by the quickness of his recovery. I hope your mother is now feeling quite rested again after the great fatigue and anxiety which she has had.

I am sending you by book post copies of two little papers on Atoms and Radium and Ether, which you might take to your father if you think that to turn over some of the pages of them might amuse him. But if you don't think he would care to see them, or if you think it might fatigue him for a moment, put them into the fire or the waste-paper basket.

I was glad to be able to tell Lady Kelvin to-day that I had a letter from you telling me that your father is much better. She would certainly join me in affectionate messages to your father and mother and yourself.

I hope your children are quite recovered from the influenza, and that you are well yourself.—Yours ever affectionately,

KELVIN.

Lord Kelvin felt acutely the anxious state of Lady Kelvin's health; but he never lost faith in her ultimate recovery, and resolutely set himself to his usual labours. Save for a passing cold he was very well, and until the middle of November he was able, not only to work at experiments, but to walk in his garden and take drives in the neighbourhood.

The season was very inclement; and he caught a chill, which on November 23 resulted in his being confined to bed with a severe attack of sickness, followed by other symptoms which happily passed away. Though he seemed better four days later, and even wrote a letter or two, he became rapidly worse again; and being unable to take much sustenance his strength began to fail. Afterwards symptoms of septic fever developed, under which he sank. On December 14th Lady Kelvin was able to visit his bedside, and subsequently two or three times, and he was able to greet her. On the evening of the 17th he passed peacefully away, at 10.15 p.m., in the presence of his physician, Dr. Sime, his nephews James Thomson Bottomley and James Thomson, and their wives, also his nephew Walter Ewing Crum; several of his faithful servants being also with him when he breathed his last.

By the universal expressions evoked by the news of Lord Kelvin's death, it became evident that the nation desired for him the homage of burial in Westminster Abbey. The funeral was fixed for Monday, December 23, at noon. A Memorial Service had been held on Sunday in the Bute Hall of the University of Glasgow. The coffin containing the remains had been removed quietly to St. Faith's Chapel in the Abbey. The spot selected for the grave was in the nave, next to the grave of Sir Isaac Newton. The funeral brought together one of the most wonderful congregations that has ever assembled in that historic building. The Royal Society was represented by all its principal officers. Glasgow sent twelve of the Professors of the University and the Lord Provost of the city. Cambridge was represented by the Vice-Chancellor, the Master of Trinity, Professors Sir George Darwin, Joseph Larmor, J. J. Thomson, and others; Oxford by its Vice-Chancellor and Professors Osler and Miers. Nearly every British University was represented, and various foreign Universities and Academies, and innumerable scientific and learned Societies at home. The Lord Mayor of London and the Lord Provost of Edinburgh were present. The King was represented by the Duke of Argyll, and the Prince of Wales by Sir Arthur Bigge. The Princess Louise, Duchess of Argyll, attended in person. The twelve pall-bearers were: Lord Rayleigh, President of the Royal Society; Mr. John Morley, Secretary of State for India; Sir Archibald Geikie, President of the Geological Society; Professor A. Crum Brown, of the Royal Society of Edinburgh; the Master of Peterhouse; Sir John Wolfe-Barry of the Institution of Civil Engineers; Sir Edward H. Seymour, Admiral of the Fleet; M. Gaston Darboux, Perpetual Secretary of the Académie des Sciences; Lord Strathcona, High Commissioner for Canada; Sir George Darwin, of the University of Cambridge; Dr. Donald M'Alister, Principal of Glasgow University; and Dr. R. T. Glazebrook, of the Institution of Electrical Engineers.

An unusually dreary December day added its darkness to the gloom. While the choir, unaccompanied, sang the verses of the hymn "Brief life is here our portion," the procession slowly moved from the little chapel, around the cloisters to the western end of the nave; and as it passed up the body of the cathedral, the effect on the vast assembly was most impressive. After the choir and the collegiate body of clergy came the Dean and Sub-Dean, followed by the coffin, the immediate mourners, and the delegates already enumerated. On the entrance of the body within the walls of the church, the voice of the Dean was heard beginning the Office for the Burial of the Dead: "I am the Resurrection and the Life." The choir took up the processional singing of these burial sentences in the music composed by Croft; and in the long minutes while the figures of the sad procession slowly filed into their places in the stalls, the organ played the marvellously beautiful funeral music of Purcell. The coffin rested immediately beneath the lantern; and the overhead darkness rendered more striking the lights that fell upon the scene from flickering candles and scattered lamps. Psalm xc., sung to Purcell's chant, was followed by the Lesson read by the Dean from the steps of the Sacrarium. The anthem was Sebastian Wesley's matchless "He will swallow up Death in Victory." Then the procession was re-formed, and the body, followed by the mourners, was borne down the choir, out into the nave, where the committal portion of the service was completed; and Goss's anthem "I heard a Voice from Heaven" was sung. After the concluding prayers, the whole congregation, with deepest feeling, sang "O God, our Help in Ages past." Once more they knelt while the Dean pronounced the Benediction; and then Sir Frederick Bridge played Handel's Dead March from the oratorio "Saul," while the clergy moved off in procession, and the assembly dispersed.

A simple flat slab marks Lord Kelvin's last resting-place.

For once, in the universal tribute rendered to the memory of Lord Kelvin, there seemed to be some revival of recognition of what the nation owes to science and to her great men. That which impressed Voltaire nearly two hundred years ago at the funeral of Newton was the public recognition which the England of that day accorded to the great representative of science. To-day the man of action looms larger in the world than the man of thought; and mankind which worships success is apt to heed little the thought and toil without which success is not achieved. In an age which has been pre-eminent over all that ever went before for the advances of science, the fashion of glorifying the warrior and the orator seems a grotesque anachronism. Mr. Gladstone's dictum, "that the present is by no means an age abounding in minds of the first order," did but reveal that he too shared the general blindness. The fact is that there never was an age so rich in minds of the first order in science. The nineteenth century has, intellectually, been the golden age, not of art or of poetry, not of drama or of adventure, but of science. It has been an epoch distinguished by a galaxy of men who made it great, and who, whether the world recognizes it or not, were great men. Though Lord Kelvin was not the last of these, he was assuredly the greatest; and his name will be revered and his memory cherished long after those who sat at his feet and listened to his voice shall have passed away. His words, his thoughts remain. And not his thoughts only; for though he was essentially a man of thought, he was also a man of effort to whom came the high privilege of achievement. That laborious humility for which he was conspicuous, that unceasing activity which drove him, as by an internal fire, from success to success, mark him as a man of purpose. In an age that threatens, now to fester into luxury, now to swell into the degenerate lust of bigness, now to drivel into sport, such a strenuous career as his, and such high ideals of intellectual endeavour as illuminated his whole life, are possessions not lightly to be lost.

> To know, to do, and on the tide of time
> Not to drift idly like the cockle-sailor,
> Whose pearly shallop dances on the blue,
> Fanned by soft airs and basking in brief sun;
> But to steer onward to some purposed haven,
> And make new waves with motion of our own,—
> That is to live.

# APPENDIX A

## List of Distinctions, Academic and Other, of the Right Honourable Lord Kelvin, O.M., etc.

### Universities

Bachelor of Arts, Cambridge University, 1845.
Master of Arts, Cambridge University, 1848.
Fellow of St. Peter's College, Cambridge, 1846-1852; re-elected for life, as Fellow, 1872.
Honorary Doctor of Laws, Dublin University, 1857.
Honorary Doctor of Civil Law, Oxford University, 1866.
Honorary Doctor of Laws, Cambridge University, 1866.
Honorary Doctor of Laws, Edinburgh University, 1869.
Honorary Doctor of Physical Science, Universitas Studiorum Ticinensis, 1878.
Honorary Doctor of Laws, M'Gill University, Montreal, 1884.
Honorary Doctor of Medicine, Heidelberg University, 1886.
Honorary Doctor of Laws, Columbia University, New York, 1887.
Honorary Doctor of Philosophy, Bologna University, 1888.
Honorary Doctor of Natural Philosophy, Padua University, 1892.
Honorary Doctor of Science, Victoria University, Manchester, 1895.
Honorary Doctor of Laws, Glasgow University, 1896.
Honorary Doctor of Laws, Princeton University, New Jersey, 1896.
Honorary Doctor of Philosophy, Royal Hungarian University, Budapest, 1896.
Honorary Doctor of Laws, Toronto University, 1897.
Honorary Doctor of Civil Law, Trinity College, Toronto, 1897.
Honorary Doctor of Mathematics, Christiania University, 1902.
Honorary Doctor of Science, University of Wales, 1902.
Honorary Doctor of Laws, Yale University, 1903.
Honorary Doctor of Science, London University, 1903.
Honorary Doctor of Science, Leeds University, 1904.
Chancellor of the University of Glasgow, 1904.

GOVERNMENTS AND MUNICIPALITIES

Knight Bachelor, 1866.
Freedom of the City of Glasgow, 1866.
Commander of the Imperial Order of the Rose (Brazil), 1873.
Legion of Honour (France), Commander, 1881.
" " " Grand Officer, 1889.
Knight of the Prussian Order *Pour le Mérite*, 1884.
Commander of the Order of Leopold (Belgium), 1890.
Freedom of the City of London, and Honorary Liveryman of the Clothworkers' Company, 1891.
Barony of Largs in the County of Ayr (U.K.), 1892.
Deputy Lieutenant of the County of the City of Glasgow, 1894.
Knight Grand Cross of the Royal Victorian Order, 1896.
Honorary Colonel of the Electrical Engineers (Volunteers), 1897.
Master of the Clothworkers' Company, 1900.
Order of the First Class of the Sacred Treasure of Japan, 1901.
Order of Merit, 1902.
Privy Councillor of Great Britain, 1902.

LEARNED SOCIETIES AND ACADEMIES

Cambridge Philosophical Society,
Member, 1845;
Hopkins Prize (for period 1874-1876), 1889.

British Association for the Advancement of Science,
Member, 1845;
President of Section A, 1852, 1867, 1876, 1881, 1884;
President, 1871.

Glasgow (Royal) Philosophical Society,
Member, 1846;
President, 1856, 1874;
Honorary Member, 1896.

Royal Society of Edinburgh,
Fellow, 1847;
Keith Medallist, 1864;
Gunning Victoria Jubilee Prize, 1887;
President, 1873-1878; 1886-1890; 1895-1907.

Royal Society of London,
Fellow, 1851;
Royal Medallist, 1856;
Copley Medallist, 1883;
President, 1890-1895.

Manchester Literary and Philosophical Society,
Honorary Member, 1851.

Regia Scientiarum Academia Svecica (Stockholm),
Foreign Member, 1851.

Regia Scientiarum Societas (Upsala),
Foreign Member, 1852.

Royal College of Preceptors,
Honorary Member, 1858.

Königliche Gesellschaft der Wissenschaften zu Göttingen,
Corresponding Member, 1859;
Foreign Member, 1864.

Academia Literarum et Scientiarum Regia Boica (München),
Corresponding Member, 1859;
Foreign Member, 1880.

Mathematical Society of London,
Member, 1867;
President, 1898-1899.

Royal Astronomical Society (London),
Fellow, 1868.

Regia Scientiarum Academia Borussica (Berlin),
Corresponding Member, 1871;
Helmholtz Medallist, 1892;
Foreign Member, 1900.

Institution of Electrical Engineers (London),
Foundation Member and Vice-President ("Society of Telegraph Engineers"), 1871;
Honorary Member, 1899;
President, 1874, 1889, 1907.

Glasgow Geological Society,
Honorary Member, 1872;
President, 1872-1893.

Institution of Naval Architects (London),
Honorary Member, 1873.

Hungarian Academy of Sciences,
Foreign Member, 1873.

Reale Accademia delle scienze di Bologna,
Corresponding Member, 1873.

American Academy of Arts and Sciences (Boston),
Foreign Honorary Member, 1873.

American Philosophical Society (Philadelphia),
Member, 1873.

Institution of Civil Engineers (London),
Member, 1874;
Honorary Member, 1889.

Institut de France,
Awarded the Prix Poncelet, 1874;
Foreign Associate, 1877;
Arago Medallist, 1896.

Académie royale des sciences (Amsterdam),
Foreign Member, 1874.

Physical Society of London,
Member, 1875;
President, 1880-1881.

Société française de physique,
Honorary Member, 1876.

Academy of Sciences (New York),
Honorary Member, 1876.

Société des arts de Genève,
Associé honoraire, 1876.

Società Italiana delle scienze (detta dei XL), Napoli,
Matteucci Prize, 1876;
Foreign Member, 1876.

Naturforschende Gesellschaft (Basel),
Honorary Member, 1876.

Societas Regia Scientiarum Hauniensis (Copenhagen),
Foreign Member, 1876.

American Electrical Society,
Honorary Member, 1876.

Academia Imperialis Scientiarum (St. Petersburg),
Corresponding Member, 1878;
Honorary Member, 1896.

Royal Irish Academy (Dublin),
Honorary Member, 1878.

Societas Physico-Medica Erlangensis,
Honorary Member, 1878.

Académie royale de Belgique (Bruxelles),
Associate, 1878.

Kaiserliche Akademie der Wissenschaften (Wien),
Foreign Corresponding Member, 1878;
Foreign Honorary Member, 1884.

Royal Society of Arts (London),
Albert Medallist, 1879;
Member, 1880.

Regia Taurinensis Academia Scientiarum,
Corresponding Member, 1881;
Member, 1882.

Reale Istituto Lombardo di scienze e lettere in Milano,
Corresponding Member, 1882.

Reale Accademia dei Lincei (Roma),
Corresponding Member, 1882.

New Zealand Institute,
Honorary Member, 1883.

National Academy of Sciences of the United States (Washington),
Foreign Associate, 1883.

Midland Institute (Birmingham),
President, 1883-1884.

Edinburgh Mathematical Society,
Honorary Member, 1883.

Société internationale des électriciens (Paris),
Member, 1884.

Reale Accademia Lucchese (Lucca),
Corresponding Member, 1884.

Société hollandaise des sciences (Haarlem),
Foreign Member, 1885.

Société impériale des amis des sciences naturelles d'anthropologie et d'ethnologie (Moscow),
Honorary Member, 1885.

Société batave de philosophie expérimentale (Rotterdam),
Corresponding Member, 1885.

Royal Institution of Great Britain,
Member, 1886.
Vice-President and Manager, 1892.

Academia Caesareae Leopoldino-Carolinae Germaniae Naturae Curiosorum (Halle),
Member, 1887.

Physikalischer Verein (Frankfurt am Main),
Honorary Member, 1887.

Canadian Society of Civil Engineers (Montreal),
Honorary Member, 1889.

Reale Istituto Veneto di scienze, lettere ed arti (Venezia),
Member, 1889.

National Electric Light Association of New York,
Honorary Member, 1890.

Association des ingénieurs-électriciens sortis de l'institut Montéfiore (Liége),
Honorary Member, 1890.

Institute of Marine Engineers (London),
Honorary Life Member, 1891.
President, 1892-1893.

Peterhouse Science Club (Cambridge),
President, 1892.

Sociedad cientifica "Antonio Alzate" (Mexico),
Honorary Member, 1892.

American Institute of Electrical Engineers,
Honorary Member, 1892.

Società reale di Napoli,
Foreign Member, 1893.

Academy of Natural Sciences (Philadelphia),
Corresponding Member, 1893.

Société d'encouragement pour l'industrie nationale (Paris),
Awarded the Ampère Medal, 1894.

Imperial University of Kasan,
Honorary Member, 1894.

Reale Accademia di scienze, lettere ed arti in Padova,
Honorary Member, 1894.

Junior Institution of Engineers (London),
Vice-President, 1895.

Reale Accademia di scienze, lettere ed arti in Modena,
Honorary Member, 1895.

Imperial University of St. Petersburg,
Honorary Member, 1896.

Société des sciences de Finlande,
Honorary Member, 1896.

Imperial Medical Military Academy of St. Petersburg,
Honorary Member, 1896.

Société de physique et d'histoire naturelle de Genève,
Honorary Member, 1896.

Royal Medical and Chirurgical Society (London),
Honorary Member, 1896.

Academy of Sciences, California,
Honorary Member, 1896.

British Institute of Preventive Medicine (London),
Honorary Fellow, 1896.

Imperial Society of Naturalists of the University of Moscow,
Honorary Member, 1896.

British Institute of Public Health (London),
Honorary Fellow, 1896.

Chester Society of Natural Science,
Honorary Member, 1897.

Scientific Society of Christiania,
Foreign Member, 1900.

Franco-Scottish Society,
President of the Scottish Branch, 1901.

National American Engineering Societies,
John Fritz Medallist, 1902.

Université impériale de Yourieff (Russia),
Honorary Member, 1903.

Society of Engineers of South Wales,
Honorary Member, 1903.

Royal Society of New South Wales (Sydney),
Honorary Member, 1903.

Verband deutscher Elektrotechniker (Berlin),
Honorary Member, 1904.

Optical Society (London),
Honorary Member, 1905.

Faraday Society (London),
President, 1905-1907.

Associazione elettrotecnica Italiana,
Honorary Member, 1906.

# APPENDIX B

## BIBLIOGRAPHY

### PART I.—PRINTED BOOKS

1. **Elements of Dynamics** [The "Glasgow Pamphlet"]. Edited, with permission, by John Ferguson, M.A., from notes of lectures delivered by William Thomson, LL.D. Glasgow, George Richardson, Printer to the University, 1863, pp. ii. + 81.
   2nd edition. Glasgow, 1869, pp. 84 + Appendix of vii.
2. **Atlantic Telegraph Cable.** Address delivered before the Royal Society of Edinburgh by Professor William Thomson, LL.D. . . , with other documents. 8vo, pp. 31. W. Brown and Co., London, 1866.
3. **Elementary Dynamics.** By Sir W. Thomson and P. G. Tait. 8vo, pp. ii. + 120 + 10. Oxford, 1867.
   Another edition of the same. 8vo, pp. ii. + 138. Oxford, 1868 [marked "Not Published"].
4. **Treatise on Natural Philosophy.** By Sir William Thomson, LL.D., D.C.L., F.R.S. . . . and Peter Guthrie Tait, M.A.
   Vol. I. 8vo, pp. xxiv. + 727. Oxford, 1867.
   New edition, Vol. I. Part I. 8vo, pp. xviii. + 508. Cambridge, 1879. Reprinted from stereotypes in 1886, 1888, 1890, 1896, 1903.
   Vol. I. Part II. 8vo, pp. xxviii. + 527. Cambridge, 1883. Reprinted from stereotypes in 1890, 1895, 1903.
5. **Handbuch der theoretischen Physik.** Von Sir William Thomson und Professor P. G. Tait. Autorisirte deutsche Uebersetzung von Dr. H. Helmholtz und G. Wertheim. Bd. I., Theile 1 und 2. 8vo. Braunschweig, 1871-1874.
6. Sir Isaac Newton's **Principia**, reprinted for Sir William Thomson, LL.D., late Fellow of St. Peter's College, Cambridge, and Hugh Blackburn, M.A., late Fellow of Trinity College, Cambridge, Professors of Natural Philosophy and Mathematics in the University of Glasgow. Glasgow. James MacLehose, Publisher to the University. Printed by Robert MacLehose. 4to, pp. xxxvi. (inc. 3 blanks) + 538 (2 blanks), 1871.

7. **Reprint of Papers on Electrostatics and Magnetism.** By Sir William Thomson, LL.D., D.C.L., F.R.S., F.R.S.E. 8vo, pp. xvi. + 592. London, 1872.
   2nd edition. London, 1884.
8. **Reports of the Committee on Electrical Standards** appointed by the British Association [1862-1869]. Revised by Sir W. Thomson, LL.D., F.R.S., Dr. J. P. Joule, LL.D., F.R.S., Professors J. Clerk Maxwell, M.A., F.R.S., and F. Jenkin, F.R.S., etc. 8vo, pp. 248 + iv. 7 pl. London, 1873.
9. **Elements of Natural Philosophy.** By Professors Sir William Thomson and P. G. Tait. Part I. 8vo, pp. viii. + 279. Oxford, 1873.
   2nd edition. 8vo, pp. viii. + 295. Cambridge, 1879.
10. **Tables for facilitating Sumner's Method at Sea.** Fol., pp. 22. London, 1876.
    2nd edition. London, 1886.
11. **Forms for use with Sir W. Thomson's Tables for facilitating Sumner's Method at Sea.** 2 Nos. 4to. London, 1876.
12. **Instructions for the Adjustment of Sir William Thomson's Patent Compass of 1879.** 8vo, 22 pp. Glasgow, 1880.
13. **Mathematical and Physical Papers.** By Sir William Thomson, LL.D., D.C.L., F.R.S.
    Vol. I. 8vo, pp. xiv. + 558. Cambridge, 1882.
    Vol. II. 8vo, pp. xii. + 407, xx. pl. Cambridge, 1884.
    Vol. III. Elasticity, Heat, Electro-Magnetism. 8vo, pp. xii. + 529. London, 1890.
14. **The Six Gateways of Knowledge.** An Address delivered in the Town Hall, Birmingham, on October 3, 1883, by Professor Sir William Thomson, LL.D., F.R.S., President [of the Birmingham and Midland Institute]. 8vo, 28 pp. Birmingham, 1883.
15. **Notes of Lectures on Molecular Dynamics and the Wave Theory of Light.** Delivered at the Johns Hopkins University, Baltimore, by Sir William Thomson, Professor in the University of Glasgow, Stenographically reported by A. S. Hathaway. [Papyrograph edition.] 4to, pp. x. + 328. Baltimore, Md., 1884.
16. Practical Guide for Compensation of Compasses without Bearings, by Lieut. Collet, with **Preface** by Sir W. Thomson. 8vo, 65 pp. Portsmouth, 1885.
17. **Popular Lectures and Addresses.**
    Vol. I. Constitution of Matter. *Nature Series*, 8vo. London, 1889. pp. xii. + 460. 2nd edition, London, 1894.
    Vol. II. Geology and General Physics. *Nature Series*, 8vo, pp. xii. + 599. London, 1894.
    Vol. III. Navigational Affairs. *Nature Series.* 8vo, pp. xii. + 511. London, 1891.
18. **Conférences scientifiques et allocutions de Lord Kelvin.**

Traduites et annotées par P. Lugol et M. Brillouin. 8vo, Paris, 1893.

19. Electric Waves, by Dr. Heinrich Hertz. Authorised English Translation by D. E. Jones, B.Sc., with a **Preface** [of 7 pp.], by Lord Kelvin, LL.D., D.C.L., President of the Royal Society. 8vo, pp. xviii. + 278. London, 1893.

20. **The Molecular Tactics of a Crystal,** being the Second Robert Boyle Lecture. Delivered before the Oxford University Junior Scientific Club on Tuesday, May 16, 1893. 8vo, 59 pp. Oxford, 1894.

21. **The Horsfall Destructor.** A Report by Lord Kelvin and Professor Archibald Barr, pp. 19 fol. Leeds, 1899.

22. **James Watt. An Oration** delivered in the University of Glasgow on the Commemoration of its Ninth Jubilee. 8vo, pp. 22. Glasgow, 1901.

23. **Baltimore Lectures on Molecular Dynamics and the Wave Theory of Light,** founded on Mr. A. S. Hathaway's Stenographic Report of Twenty Lectures delivered in Johns Hopkins University, Baltimore, in October 1884; followed by Twelve Appendices on Allied Subjects. By Lord Kelvin, O.M., G.C.V.O., P.C., F.R.S., etc., President of the Royal Society of Edinburgh, Fellow of St. Peter's College, Cambridge, and Emeritus Professor of Natural Philosophy in the University of Glasgow. 8vo, pp. xxii. + 694. London, 1904.

24. Hermann von Helmholtz, by Leo Königsberger. Translated by F. A. Welby, with a **Preface** [of 3 pages], by Lord Kelvin. 8vo, pp. xviii. + 440. Oxford, 1906.

25. Vorlesungen über Molekulardynamik und Theorie des Lichts [Baltimore Lectures]. Deutsch herausgegeben von B. Weinstein. 8vo, pp. 18 + 590. Leipzig, 1909.

## PART II.—SCIENTIFIC COMMUNICATIONS AND ADDRESSES

**1841.**

1. On Fourier's Expansions of Functions in Trigonometrical Series [signed P. Q. R.]. Camb. Math. Jl. II. pp. 258-262, May 1841; Math. and Phys. Papers I. art. i. pp. 1-6.

2. Note on a Passage in Fourier's *Heat* [signed P. Q. R.]. Camb. Math. Jl. III. pp. 25-27, Nov. 1841; Math. and Phys. Papers I. art. ii. pp. 7-9.

**1842.**

3. On the Uniform Motion of Heat in Homogeneous Solid Bodies, and its Connection with the Mathematical Theory of Electricity [signed P. Q. R.]. Camb. Math. Jl. III. pp. 71-84, Feb. 1842; Phil. Mag. VII. pp. 502-515 (Supplement), 1854; [with footnotes added in March 1854]; Math. and Phys. Papers I. art. iii. p. 9 [title only]; E. and M., art. i. pp. 1-14. [See No. 89.]

4. On the Linear Motion of Heat [signed N. N.]. Camb. Math. Jl. III. pp. 170-174, Nov. 1842; pp. 206-211, Feb. 1843; Math. and Phys. Papers I. art. iv. pp. 10-15, art. v. pp. 16-21.
5. Propositions in the Theory of Attraction [signed P. Q. R.]. Camb. Math. Jl. III. pp. 189-196, Nov. 1842; pp. 201-206, Feb. 1843; Math. and Phys. Papers I. art. vi. p. 21 [title only]; E. and M. art. xii. pp. 126-138.

**1843.**

6. On the Attractions of Conducting and Non-conducting Electrified Bodies [signed P. Q. R.]. Camb. Math. Jl. III. pp. 275-276, May 1843; Math. and Phys. Papers I. art. vii. p. 21 [title only]; E. and M. art. vii. pp. 98-99.
7. Notes on Orthogonal Isothermal Surfaces [signed P. Q. R.]. Camb. Math. Jl. III. pp. 286-288, May 1843; IV. pp. 179-182, Nov. 1844; Math. and Phys. Papers I. art. viii. pp. 22-24, art. xii. pp. 48-52.
8. On the Equations of the Motion of Heat referred to Curvilinear Co-ordinates [signed P. Q. R.]. Camb. Math. Jl. IV. pp. 33-42, Nov. 1843; Math. and Phys. Papers I. art. ix. pp. 25-35.

**1844.**

9. Elementary Demonstration of Dupin's Theorem [signed P. Q. R.]. Camb. Math. Jl. IV. pp. 62-64, Feb. 1844; Math. and Phys. Papers I. art. x. pp. 36-38.
10. Note on some Points in the Theory of Heat [signed P. Q. R.]. Camb. Math. Jl. IV. pp. 67-72, Feb. 1844; Math. and Phys. Papers I. art. xi. pp. 39-47.
11. Note on the Law of Gravity at the Surface of a Revolving Homogeneous Fluid [signed P. Q. R.]. Camb. Math. Jl. IV. pp. 191-192, Nov. 1844; Math. and Phys. Papers I. art. xiii. pp. 53-54.
12. Note sur la théorie de l'attraction. Liouville Journ. de Math. ix. pp. 239-244, 1844.

**1845.**

13. Demonstration of a Fundamental Proposition in the Mechanical Theory of Electricity. Camb. Math. Jl. IV. pp. 223-226, Feb. 1845; Math. and Phys. Papers I. art. xiv. p. 54 [title only]; E. and M. art. viii. pp. 100-103.
14. On the Reduction of the General Equation of Surfaces of the Second Order. Camb. Math. Jl. IV. pp. 227-233, Feb. 1845; Math. and Phys. Papers I. art. xv. pp. 55-62.
15. On the Lines of Curvature of Surfaces of the Second Order. Camb. Math. Jl. IV. pp. 279-286, May 1845; Math. and Phys. Papers I. art. xvi. pp. 63-71.
16. Démonstration d'un théorème d'analyse. Liouville Journ. de Math. X. pp. 137-147, 1845; Math. and Phys. Papers I. art. xvii. p. 71 [title only].
17. Note on the Elementary Laws of Statical Electricity. Brit. Assoc. Report (Sect.), pp. 11-12, 1845; Liouville Journ. de Math. X.

pp. 209-221, 1845; Phil. Mag. VIII. p. 42, July 1854. [See Nos. 20 and 89], with additions in brackets. Math. and Phys. Papers I. art. xviii. p. 71 [title only].

18. Extrait d'une lettre sur l'application du principe des images à la solution de quelques problèmes relatifs à la distribution d'électricité. Liouville Journ. de Math. X. pp. 364-367, 1845 [Oct. 8, 1845]; Math. and Phys. Papers I. art. xix. p. 71 [title only]; E. and M. art. xiv. pp. 144-146.

19. Note on Induced Magnetism in a Plate. Camb. and Dub. Math. Jl. I. pp. 34-37, Nov. 1845; Math. and Phys. Papers I. art. xx. p. 71 [title only]; E. and M. art. ix. pp. 104-107.

20. On the Mathematical Theory of Electricity in Equilibrium. Camb. and Dub. Math. Jl. I. pp. 75-95, Nov. 1845; Phil. Mag. VIII. pp. 42-62, July 1854 [with additions in brackets]. Math. and Phys. Papers XXI. p. 71 [title only]. [See Nos. 17 and 89]; E. and M. art. ii. pp. 15-37.

**1846.**

21. Note on the Rings and Brushes in the Spectra produced by Biaxal Crystals. Camb. and Dub. Math. Jl. I. pp. 124-126, 1846; Math. and Phys. Papers I. art. xxii. pp. 72-74.

22. On the Principal Axes of a Rigid Body. Camb. and Dub. Math. Jl. I. pp. 127-133, pp. 195-206, 1846; Math. and Phys. Papers I. art. xxiii. p. 75 [title only]; T and T', §§ 282-284.

23. Note on a Paper: "Sur une propriété de la couche électrique en équilibre à la surface d'un corps conducteur. Camb. and Dub. Math. Jl. I. pp. 281-282, Nov. 1846; Math. and Phys. Papers I. art. xxiv. p. 75 [title only]; E. and M. art. x. pp. 108-111.

**1847.**

24. On a Mechanical Representation of Electric, Magnetic, and Galvanic Forces. Camb. and Dub. Math. Jl. II. pp. 61-64, 1847; Math. and Phys. Papers I. art. xxvii. pp. 76-80.

25. On certain Definite Integrals suggested by Problems in the Theory of Electricity. Camb. and Dub. Math. Jl. II. pp. 109-122, Mar. 1847; Math. and Phys. Papers I. art. xxviii. p. 80 [title only]; E. and M. art. xi. pp. 112-125.

26. Notice of Stirling's Air-Engine. Glasg. Phil. Soc. Proc. II. pp. 169-170 [read Apr. 21, 1847].

27. On the Forces experienced by small Spheres under Magnetic Influence; and on some of the Phenomena presented by Diamagnetic Substances. Camb. and Dub. Math. Jl. II. pp. 230-235, 1847; Math. and Phys. Papers I. art. xxix. p. 80 [title only]; E. and M. art. xxxiii. pp. 493-499.

28. On a System of Magnetic Curves. Camb. and Dub. Math. Jl. II. p. 240, May 1847; Math. and Phys. Papers I. art. xxx. pp. 81-82.

29. Notes on Hydrodynamics. (1) On the Equation of Continuity. (2) On the Equation of the Bounding Surface. Camb. and

Dub. Math. Jl. II. pp. 282-286, 1847; III. pp. 89-93, Feb. 1848; Part (1) in T and T′, § 194; Math. and Phys. Papers I. art. xxxi. pp. 83-87.

30. Système nouveau de coordonnées orthogonales. Liouville Journ. de Math. xii. pp. 256-264, 1847; Math. and Phys. Papers I. art. xxxii. p. 87 [title only]; E. and M. art. xiv. pp. 146-154.

31. On Electrical Images. Brit. Assoc. Report, 1847, pt. ii. pp. 6-7; Math. and Phys. Papers I. art. xxv. p. 75 [title only].

32. On the Electric Currents by which the Phenomena of Terrestrial Magnetism may be produced. Brit. Assoc. Report, 1847, pt. ii. pp. 38-39; Math. and Phys. Papers I. art. xxvi. p. 75 [title only]; E. and M. art. xxix. pp. 462-465.

33. Note sur une équation aux différences partielles, qui se présente dans plusieurs questions de physique mathématique. Liouville Journ. de Math. XII. pp. 493-496, 1847; Math. and Phys. Papers I. art. xxxiii. p. 87 [title only]; E. and M. art. xiii. pp. 142-143.

**1848.**

34. Theorems with Reference to the Solution of certain Partial Differential Equations. Camb. and Dub. Math. Jl. III. pp. 84-87, Feb. 1848; Math. and Phys. Papers I. art. xxxvi. pp. 93-96; E. and M. art. xiii. pp. 139-141.

35. Note on the Integration of the Equations of Equilibrium of an Elastic Solid. Camb. and Dub. Math. Jl. III. pp. 87-89, Feb. 1848; Math. and Phys. Papers I. art. xxxvii. pp. 97-99.

36. On the Mathematical Theory of Electricity in Equilibrium. Camb. and Dub. Math. Jl. III. pp. 131-148, Mar., pp. 266-274, May, 1848; Math. and Phys. Papers I. art. xxxviii. p. 99 [title only]; E. and M. art. iv. pp. 42-51, art. v. pp. 52-68. [See Nos. 43, 45.]

37. On an Absolute Thermometric Scale, founded on Carnot's Theory of the Motive Power of Heat, and calculated from the Results of Regnault's Experiments on the Pressure and Latent Heat of Steam. Glasg. Phil. Soc. Proc. II. p. 262 [read Apr. 12, 1848, title only]; Camb. Phil. Soc. Proc. I. pp. 66-71, 1866 [read June 5, 1848]; Phil. Mag. xxxiii. pp. 313-317, Oct. 1848; Math. and Phys. Papers I. art. xxxix. pp. 100-106.

38. On the Equilibrium of Magnetic or Diamagnetic Bodies of any Form under the Influence of the Terrestrial Magnetic Force. Brit. Assoc. Report, 1848, pt. ii. pp. 8-9; Math. and Phys. Papers I. art. xxxiv. pp. 88-90.

39. On the Theory of Electromagnetic Induction. Brit. Assoc. Report, 1848, pt. ii. pp. 9-10; Math. and Phys. Papers I. art. xxxv. pp. 91-93.

**1849.**

40. An Account of Carnot's Theory of the Motive Power of Heat, with Numerical Results deduced from Regnault's Experiments on Steam. Edinb. Roy. Soc. Proc. II. pp. 198-204 [read

Jan. 2, 1849]; Edinb. Roy. Soc. Trans. XVI. pp. 541-574, 1849; Annal. de Chimie XXXV. pp. 248-255, 1852; Math. and Phys. Papers I. art. xli. pp. 113-164.

41. Notes on Hydrodynamics. On the Vis-viva of a Liquid in Motion. Camb. and Dub. Math. Jl. IV. pp. 90-94, Feb. 1849; Math. and Phys. Papers I. art. xl. pp. 107-112.

42. A Mathematical Theory of Magnetism. Roy. Soc. Proc. V. pp. 845-846 [read June 21, 1849]; Roy. Soc. Proc. V. pp. 975-978 [read June 20, 1850]; Phil. Trans. pp. 243-285, 1851; Phil. Mag. XXXV. p. 540, 1849; E. and M. art. xxiv. pp. 340-425.

43. Effects of Electrical Influence on Internal Spherical and on Plane Conducting Surfaces. Camb. and Dub. Math. Jl. IV. pp. 276-284, Nov. 1849; E. and M. art. v. pp. 68-77. [See Nos. 36, 45.]

44. On some Remarkable Effects of Lightning observed in a Farm-house near Moniemail, Cupar-Fife. Glasg. Phil. Soc. Proc. III. pp. 69-72 [read Dec. 5, 1849]. Phil. Mag. XXXVII. pp. 53-57, July 1850; E. and M. art. xvi. pp. 232-236.

**1850.**

45. On the Mathematical Theory of Electricity in Equilibrium. Camb. and Dub. Math. Jl. V. pp. 1-9, 1850; E. and M. art. v. pp. 77-85. [See Nos. 36 and 43.]

46. The Effect of Pressure in Lowering the Freezing Point of Water, experimentally demonstrated. Edinb. Roy. Soc. Proc. II. pp. 267-271, 1851 [read Jan. 21, 1850]; Phil. Mag. XXXVII. pp. 123-127, 1850; Annal. de Chimie. XXXV. pp. 381-383, 1852; Journ. de Pharm. XVIII. pp. 372-375, 1850; Poggend. Annal. LXXXI. pp. 163-168, 1850; Amer. Jl. of Science XI. (2nd Ser.) p. 115, 1851; Math. and Phys. Papers I. art. xlv. pp. 165-169.

47. Notes on a paper: "Problem respecting Polygons in a Plane" by Robert Moon. Camb. and Dub. Math. Jl. V. pp. 137-139, 1850; Math. and Phys. Papers I. art. xliii. p. 164 [title only].

48. On the Potential of a Closed Galvanic Circuit of any Form. Camb. and Dub. Math. Jl. V. pp. 142-148, 1850; Math. and Phys. Papers I. art. xliv. p. 164 [title only]; E. and. M. art. xxv. pp. 425-431.

49. On the Theory of Magnetic Induction in Crystalline Substances. Brit. Assoc. Report, 1850, pt. ii. p. 23; Phil. Mag. I. pp. 177-186, March 1851; E. and M. art. xxx. pp. 465-481; Math. and Phys. Papers I. art. xlii. p. 164 [title only]; art. lii. p. 471 [title only].

50. Remarks on the Forces experienced by Inductively-magnetised Ferromagnetic or Diamagnetic Non-crystalline Substances. Phil. Mag. XXXVII. pp. 241-253, Oct. 1850; Poggend. Annal. LXXXII. pp. 245-262, 1851; E. and M. art. xxxiv. pp. 500-513; Math. and Phys. Papers I. art. xlvi. p. 169 [title only].

51. On a Remarkable Property of Steam connected with the Theory of the Steam-Engine. [A Letter to J. P. Joule.] Phil. Mag. XXXVII. pp. 386-389, Nov. 1850; Poggend. Annal. LXXXI. pp. 477-480, 1850; Math. and Phys. Papers I. art. xlvii. pp. 170-173.

**1851.**

52. On the Dynamical Theory of Heat; with Numerical Results deduced from Mr. Joule's "Equivalent of a Thermal Unit" and M. Regnault's "Observations on Steam." Edinb. Roy. Soc. Proc. III. pp. 48-52 [read Mar. 17, 1851]; Edinb. Roy. Soc. Trans. XX. pp. 261-288, 1853; Liouville Journ. de Math. XVII. pp. 209-252, 1852; Phil. Mag. IV. pp. 8-21, pp. 105-117, pp. 168-176, 1852; Annal. de Chimie XXXVI. pp. 118-124, 1852; Math. and Phys. Papers I. art. xlviii. pp. 174-210.
53. On a Method of Discovering Experimentally the Relation between the Mechanical Work spent and the Heat produced by the Compression of a Gaseous Fluid. [Part IV. of Dynamical Theory of Heat.] Edinb. Roy. Soc. Proc. III. pp. 69-71 [read Apr. 21, 1851]. Edinb. Roy. Soc. Trans. XX. pp. 289-298, 1853; Annal. de Chimie LXIV. pp. 504-509, 1862; Phil. Mag. IV. pp. 424-434, Dec. 1852; Math. and Phys. Papers I. art. xlviii. pp. 210-222.
54. Note on the Effect of Fluid Friction in Drying Steam which issues from a High-Pressure Boiler through a small Orifice into the Open Air. Phil. Mag. I. p. 474, June 1851; II. pp. 273-274, Oct. 1851.
55. On the Dynamical Theory of Heat, Part V.:—On the Quantities of Mechanical Energy contained in a Fluid in different States as to Temperature and Density. Edinb. Roy. Soc. Proc. III. pp. 90-91 [read Dec. 15, 1851]; Edinb. Roy. Soc. Trans. XX. pp. 475-482, 1853; Phil. Mag. IX. pp. 523-531 (Supplement), 1855; Math. and. Phys. Papers I. art. xlviii. pp. 222-232.
56. On a Mechanical Theory of Thermo-electric Currents. [Part of the Dynamical Theory of Heat.] Edinb. Roy. Soc. Proc. III. pp. 91-98 [read Dec. 15, 1851]; Math. and Phys. Papers I. art. xlviii. pp. 316-323.
57. On the Mechanical Theory of Electrolysis. Phil. Mag. II. pp. 429-444, Dec. 1851; Math. and Phys. Papers I. art. liii. pp. 472-489.
58. Applications of the Principle of Mechanical Effect to the Measurement of Electromotive Forces and of Galvanic Resistances in Absolute Units. Phil. Mag. II. pp. 551-562, Dec. 1851; Math. and Phys. Papers I. art. liv. pp. 490-502.
59. Magnecrystallic Property of Calcareous Spar. Letter to Phil. Mag. II. p. 574, Dec. 1851.

**1852.**

60. On the Mechanical Action of Radiant Heat or Light. On the Power of Animated Creatures over Matter. On the Sources

available to Man for the Production of Mechanical Effect. Edinb. Roy. Soc. Proc. III. pp. 108-113, 1857 [read Feb. 2, 1852]; Phil. Mag. IV. pp. 256-260, Oct. 1852; Math. and Phys. Papers I. art. lviii. pp. 505-510.

61. Additional Note to J. P. Joule's paper on the Air-Engine. Phil. Trans. pp. 78-82, 1852; dated Glasgow College, Feb. 19, 1852, republished in Joule's Scientific Papers I. p. 350; Camb. and Dub. Math. Jl. VIII. pp. 250-256, 1853; Math. and Phys. Papers I. art. xlviii. pp. 326-332.

62. On a Universal Tendency in Nature to the Dissipation of Mechanical Energy. Edinb. Roy. Soc. Proc. III. pp. 139-142 [read Apr. 19, 1852]; Phil. Mag. IV. pp. 304-306, Oct. 1852. Math. and Phys. Papers I. art. lix. pp. 511-514.

63. Thomson and Joule.—On the Thermal Effects of Air rushing through Small Apertures. Brit. Assoc. Report, 1852, p. 16 [title only]; Phil. Mag. (ser. 4) IV. pp. 481-491, Dec. 1852; Math. and Phys. Papers I. art. xlix. pp. 333-345.

64. On the Sources of Heat generated by the Galvanic Battery. Brit. Assoc. Report, 1852, pt. ii. pp. 16-17; Math. and Phys. Papers I. art. lv. pp. 503-504.

65. On the Mutual Attraction and Repulsion between two Electrified Spherical Conductors. Brit. Assoc. Report, 1852, pp. 17-18; Phil. Mag. V. pp. 287-297, Apr. 1853, VI. pp. 114-115, Aug. 1853; Math. and Phys. Papers II. art. lxiv. p. 1 [title only]; E. and M. art. vi. pp. 86-97.

66. On certain Magnetic Curves; with Applications to Problems in the Theories of Heat, Electricity, and Fluid Motion. Brit. Assoc. Report, 1852, pt. ii. p. 18; Math. and Phys. Papers I. art. lvi. p. 504 [title only]; E. and M. art. xxxv. pp. 514-515.

67. On the Equilibrium of Elongated Masses of Ferromagnetic Substance in Uniform and Varied Fields of Force. Brit. Assoc. Report, 1852, pt. ii. pp. 18-20; Math. and Phys. Papers I. art. lvii. p. 504 [title only]; E. and M. art. xxxv. pp. 515-517.

68. On the Economy of the Heating or Cooling of Buildings by means of Currents of Air. Glasg. Phil. Soc. Proc. III. pp. 269-272 [read Dec. 1, 1852]; Phil. Mag. VII. pp. 138-142, Feb. 1854; Math. and Phys. Papers I. art. lx. pp. 515-520.

**1853.**

69. On the Mechanical Values of Distributions of Electricity, Magnetism, and Galvanism. Glasg. Phil. Soc. Proc. III. pp. 281-285 [read Jan. 19, 1853]; Phil. Mag. VII. pp. 192-197, March 1854; Math. and Phys. Papers I. art. lxi. pp. 521-533.

70. On Transient Electric Currents. Glasg. Phil. Soc. Proc. III. pp. 285-289 [read Jan. 19, 1853]; Phil. Mag. V. pp. 393-405, June 1853; Math. and Phys. Papers I. art. lxii. pp. 534-553.

71. On the Restoration of Mechanical Energy from an unequally heated Space. Phil. Mag. v. pp. 102-105, Feb. 1853; Math. and Phys. Papers I. art. lxiii. pp. 554-558.
72. Thomson and Joule.—On the Thermal Effects of Elastic Fluids. Roy. Soc. Proc. VI. pp. 331-332, 1850-1854 [read June 16, 1853]; Phil. Trans. pp. 357-365, 1853; Phil. Mag. VI. pp. 230-231, 1853; Annal. de Chimie LXV. pp. 244-254, 1862; Poggend. Annal. XCVII. pp. 576-589, 1856; Math. and Phys. Papers I. art. xlix. pp. 346-356, pp. 432-433.

**1854.**

73. Remarques sur les oscillations d'aiguilles non-cristallisées de faible pouvoir inductif paramagnétique ou diamagnétique et sur d'autres phénomènes magnétiques produits par des corps cristallisés ou non-cristallisés. Comp. Rend. XXXVIII. pp. 632-640, March 1854 [dated Glasgow, March 22, 1854]; Math. and Phys. Papers II. art. lxv. p. 1 [title only]; E. and M. art. xxxvi. pp. 518-526.
74. On the Mechanical Energies of the Solar System. Edinb. Roy. Soc. Proc. III. pp. 241-244 [read April 17, 1854]; Appendix on the Age of the Sun, added Aug. 15, 1854; Edinb. Roy. Soc. Trans. XXI. pp. 63-80, 1857; Comp. Rend. XXXIX. pp. 682-687, Oct. 1854; Phil. Mag. VIII. pp. 409-430, Dec. 1854; Math. and Phys. Papers II. art. lxvi. pp. 1-25.
75. Note on the Possible Density of the Luminiferous Medium and on the Mechanical Value of a Cubic Mile of Sunlight. Edinb. Roy. Soc. Proc. III. p. 253 [read May 1, 1854]; Edinb. Roy. Soc. Trans. XXI. pp. 57-61, 1857; Comp. Rend. XXXIX. pp. 529-534, Sept. 1854; Phil. Mag. IX. pp. 36-40, Jan. 1855; Math. and Phys. Papers II. art. lxvii. pp. 28-33.
76. Account of Experimental Investigations to answer Questions originating in the Mechanical Theory of Thermo-electric Currents. Edinb. Roy. Soc. Proc. III. p. 255 [read May 1, 1854].
77. On the Dynamical Theory of Heat, Part VI. (*continued*).—Thermo-electric Currents. Edinb. Roy. Soc. Proc. III. pp. 255-256, 1857 [read May 1, 1854]; Edinb. Roy. Soc. Trans. XXI. pp. 123-171, 1857; Phil. Mag. XI. pp. 214-225, Mar., pp. 281-297, Apr., pp. 379-388, May, pp. 433-446, June 1856; Math. and Phys. Papers I. art. xlviii. pp. 232-291, pp. 324-325.
78. Account of Researches in Thermo-electricity. Roy. Soc. Proc. VII. pp. 49-58, 1854-1855 [read May 4, 1854]; Phil. Mag. VIII. pp. 62-69, July 1854; Math. and Phys. Papers I. art. li. pp. 460-468.
79. On the Heat produced by an Electric Discharge. Letter to Phil. Mag. VII. pp. 347-348, May 1854.
80. Aperçu des recherches relatives aux effets des courants électriques dans des conducteurs inégalement échauffés, et

à d'autres points de la thermo-électricité. Comp. Rend. XXXVIII. pp. 828-829, May 8, 1854; Comp. Rend. XXXIX. pp. 116-119, July 10, 1854; Math. and Phys. Papers I. art. li. pp. 460-463, II. art. lxviii. p. 33.

81. Thomson and Joule.—On the Thermal Effects of Fluids in Motion [No. 2], Roy. Soc. Proc. VII. pp. 127-130 [read June 15, 1854]; Phil. Trans. pp. 321-364, 1854; Annal. de Chimie LXV. pp. 244-254, 1862; Math. and Phys. Papers I. art. xlix. pp. 357-400, pp. 433-436.

82. Note sur les effets de la pression et de la tension sur les propriétés thermo-électriques des métaux non-cristallisés. [Extrait d'une lettre de W. Thomson à M. Élie de Beaumont.] Comp. Rend. XXXIX. pp. 252-253, July 31, 1854.

83. Account of Experimental Researches in Thermo-electricity [Part IV.]. Brit. Assoc. Report, 1854, pt. ii. pp. 13-14; Math. and Phys. Papers I. art. li. pp. 469-471.

84. On the Mechanical Antecedents of Motion, Heat, and Light. Brit. Assoc. Report, 1854, pt. ii. pp. 59-63; Edinb. New Phil. Jl. I. pp. 90-97, 1855; Comp. Rend. XL. pp. 1197-1202, 1855 [abstract]; Math. and Phys. Papers II. art. lxix. pp. 34-40.

**1855.**

85. On Thermo-elastic and Thermomagnetic Properties of Matter, Part I. Quart. Jl. of Math. I. pp. 57-77, 1857 [dated from Glasgow College, Mar. 10, 1855]; Phil. Mag. V. pp. 4-27, 1878 [with additions]; Math. and Phys. Papers I. art. xlviii. pp. 291-316.

86. Elementary Demonstrations of Propositions in the Theory of Magnetic Force. Phil. Mag. IX. pp. 241-248, Apr. 1855; Math. and Phys. Papers II. art. lxx. p. 40 [title only]; E. and M. art. xxxvii. pp. 526-534.

87. On the "Magnetic Medium," and on the Effects of Compression (being a letter to Tyndall). Phil. Mag. IX. pp. 290-293, Apr. 1855; Math. and Phys. Papers II. art. lxxi. p. 41 [title only]; E. and M. art. xxxviii. pp. 535-538.

88. On the Theory of the Electric Telegraph. Roy. Soc. Proc. VII. pp. 382-399, 1854-1855 [read May 24, 1855]; Phil. Mag. XI. pp. 146-160, Feb. 1856; Math. and Phys. Papers II. art. lxxiii. pp. 61-76.

89. On the Electro-statical Capacity of a Leyden Phial and of a Telegraph Wire insulated in the Axis of a Cylindrical Conducting Sheath. Phil. Mag. IX. pp. 531-535 (Supplement), 1855 [an additional note to Nos. 3, 17, and 20]; Math. and Phys. Papers II. art. lxxiv. p. 76 [title only]; E. and M. art iii. pp. 38-41.

90. On the Effects of Mechanical Strain on the Thermo-electric Qualities of Metals. Brit. Assoc. Report, 1855, pt. ii. pp. 17-18; Math. and Phys. Papers II. art. lxxxvi. pp. 173-174.

91. On the Use of Observations of Terrestrial Temperature for the

Investigation of Absolute Dates in Geology. Brit. Assoc. Report, 1855, pt. ii. pp. 18-19; Math. and Phys. Papers II. art. lxxxvii. pp. 175-177.

92. On the Electric Qualities of Magnetized Iron. Brit. Assoc. Report, 1855, pt. ii. pp. 19-20; Poggend. Annal. XCIX. pp. 334-335, 1856; Math. and Phys. Papers II. art. lxxxviii. pp. 178-180.

93. On the Thermo-electric Position of Aluminium. Brit. Assoc. Report, 1855, pt. ii. pp. 20-21; Math. and Phys. Papers II. art. lxxxix. p. 181.

94. On New Instruments for measuring Electrical Potentials and Capacities. Brit. Assoc. Report, 1855, pt. ii. p. 22.

95. On Peristaltic Induction of Electric Currents in Submarine Telegraph Wires. Brit. Assoc. Report, 1855, pt. ii. p. 22; Math. and Phys. Papers II. art. lxxv. pp. 77-78.

96. On the Reciprocal Action of Diamagnetic Particles [Letter to Prof. Tyndall on Dec. 24, 1855]. Phil. Mag. XI. pp. 66-67, Jan. 1856; Math. and Phys. Papers II. art. lxxi. p. 41 [title only]; E. and M. art. xxxviii. pp. 540-542.

**1856.**

97. Thomson and Joule.—On the Thermal Effects of Fluids in Motion. Roy. Soc. Proc. VIII. pp. 41-42 [read Feb. 21, 1856]; Phil. Mag. XII. p. 466, Dec. 1856; Math. and Phys. Papers I. art. xlix. pp. 436-437.

98. On the Electrodynamic Qualities of Metals (Bakerian Lecture). Roy. Soc. Proc. VIII. pp. 50-55, 1856-1867 [delivered Feb. 28, 1856]; Phil. Trans. CXLVI. pp. 649-751, 1856; Annal. de Chimie LIV. pp. 105-120, 1858; Phil. Mag. XII. pp. 393-397, Nov. 1856; Math. and Phys. Papers II. art. xci. pp. 189-327, pp. 396-401.

99. On the Origin and Transformations of Motive Power. Roy. Instit. Proc. II. pp. 199-204, 1854-1858 [Feb. 29, 1856]; Chemist. III. pp. 607-612, 1856; Math. and Phys. Papers II. art. xc. pp. 182-188; Pop. Lect. II. pp. 418-432.

100. Elements of a Mathematical Theory of Elasticity. Roy. Soc. Proc. VIII. pp. 85-87 [abstract] [read Apr. 24, 1856]; Phil. Trans. CXLVI. pp. 481-498, 1856; Phil. Mag. XII. pp. 539-540 (Supplement), 1856; Math. and Phys. Papers III. art. xcii. pp. 84-112.

101. On Peristaltic Induction of Electric Currents. Roy. Soc. Proc. VIII. pp. 121-132, 1856-1857 [read May 22, 1856]; Phil. Mag. XIII. pp. 135-145, Feb. 1857; Math. and Phys. Papers II. art. lxxv. pp. 79-91.

102. Dynamical Illustrations of the Magnetic and the Helicoidal Rotatory Effects of Transparent Bodies on Polarized Light. Roy. Soc. Proc. VIII. pp. 150-158, 1856-1857 [read June 12, 1856]; Phil. Mag. XIII. pp. 138-204, Mar. 1857; Baltimore Lectures, Appendix F. pp. 569-583, 1904.

103. Thomson and Joule.—On the Thermal Effects of Fluids in Motion:—On the Temperature of Solids exposed to Currents of Air. Roy. Soc. Proc. VIII. pp. 178-185, 1856-1857 [read June 19, 1856]; Phil. Mag. XIII. pp. 286-291, Apr. 1857; Math. and Phys. Papers I. art xlix. pp. 437-444.

104. On the Discovery of the True Form of Carnot's Function. [Letter to the Phil. Mag., May 12, 1856]; Phil. Mag. XI. pp. 447-448, June 1856.

105. On Dellmann's Method of observing Atmospheric Electricity. Brit. Assoc. Report, 1856, pt. ii. pp. 17-18.

106. Sui fenomini magneto-cristallini. [Lettera al Prof. Matteucci.] Nuovo Cimento IV. pp. 192-198, 1856 [Aug. 27, 1856].

107. Discussion of J. P. Joule's Paper on "A Surface Condenser." Inst. Mech. Engrs. Proc. 1856, pp. 191-192, and p. 194. [Glasgow Meeting, Sept. 17 and 18, 1856.]

108. Telegraphs to America [Letters to the Athenæum of Oct. 4, 1856, and Nov. 1, 1856]; Math. and Phys. Papers II. art. lxxvi. pp. 92-102.

109. On Practical Methods for Rapid Signalling by the Electric Telegraph. Roy. Soc. Proc. VIII. pp. 299-307, 1856-1857; [two communications, read Dec. 11, 1856]; Phil. Mag. XIV. pp. 59-65, July 1857; Math. and Phys. Papers II. art. lxxvii. pp. 103-111.

**1857.**

110. Discussion of F. R. Window's Paper on "Submarine Telegraphs" [Jan. 13, 1857]; Inst. Civ. Engrs. Proc. XVI. pp. 210-211, 1857.

111. Discussion of R. Hunt's Paper on "Electromagnetism as a Motive Power." Inst. Civ. Engrs. Proc. XVI. pp. 400-401 [Apr. 21, 1857].

112. On the Electrodynamic Qualities of Metals:—Effects of Magnetization on the Electric Conductivity of Nickel and of Iron. Roy. Soc. Proc. VIII. pp. 546-550 [read June 15, 1857]; Phil. Mag. XV. pp. 469-472, June 1858; Math. and Phys. Papers II. art. xci. pp. 327-331.

113. On the Electrical Conductivity of Commercial Copper of various kinds. Roy. Soc. Proc. VIII. pp. 550-555, 1856-1857 [read June 15, 1857]; Phil. Mag. XV. pp. 472-476, June 1858; Math. and Phys. Papers II. art. lxxviii. pp. 112-117.

114. Thomson and Joule.—On the Thermal Effects of Fluids in Motion:—Temperature of a Body moving slowly through Air. Roy. Soc. Proc. VIII. pp. 556-564, 1856-1857 [read June 15, 1857]; Phil. Mag. XV. pp. 477-482, June 1858; Math. and Phys. Papers I. art. xlix. pp. 445-453.

115. On the Alterations of Temperature accompanying Changes of Pressure in Fluids. Roy. Soc. Proc. VIII. pp. 566-569, 1856-1857 [read June 15, 1857]; Phil. Mag. XV. pp. 540-542

(Supplement), 1858; Math. and Phys. Papers III. art. xcii. Appendix A. pp. 236-239.

116. On Mr. Whitehouse's Relay and Induction Coils in Action on Short Circuit. Brit. Assoc. Report, 1857, pt. ii. p. 21.

117. On the Effects of Induction in Long Submarine Lines of Telegraph. Brit. Assoc. Report, 1857, pt. ii. pp. 21-22.

118. On Machinery for laying Submarine Telegraph Cables. Brit. Assoc. Report, 1857, p. 199 [title only]; Engineer IV. p. 185, Sept. 11, 1857, and p. 280, Oct. 16, 1857.

119. Submarine Telegraphy [Lecture at the Brit. Assoc. Meeting in Dublin in 1857].

120. On the Calculation of Transcendents of the Form $\int_0^x e^{-x^2} fx\,dx$. Quart. Jl. of Math. I. pp. 316-320, 1857; Math. and Phys. Papers I. pp. 56-60.

121. Articles on "Thermomagnetism" and "Telegraph" (in Appendix). Nichol's Cyclopædia, 1857.

**1858.**

122. The Conductivity of Samples of Copper Wire [note of expts. shown]. Glasg. Phil. Soc. Proc. IV. pp. 185-186 [read Jan. 13, 1858].

123. Remarks on the Interior Melting of Ice. [In a letter to Prof. Stokes, Sec. R.S.] Roy. Soc. Proc. IX. pp. 141-143, 1857-1859 [read Jan. 25, 1858]; Phil. Mag. XVI. pp. 303-304, Oct. 1858.

124. Intorno ad alcune richerche di elettrostatica [letter to Prof. P. Volpicelli]. Atti dell' Accademia Pontificia de' Nuovi Lincei XI. pp. 177-185, 1857-1858 [read Mar. 7, 1858]; Nuovo Cimento VIII. pp. 115-123, 1858.

125. On the Stratification of Vesicular Ice by Pressure. [In a letter to Prof. Stokes, Sec. R.S.]. Roy. Soc. Proc. IX. pp. 209-213, 1857-1859 [read Apr. 22, 1858]; Phil. Mag. XVI. pp. 463-466, Dec. 1858.

126. On the Thermal Effect of drawing out a Film of Liquid [being an extract of two letters to J. P. Joule, F.R.S., dated Feb. 2 and 3, 1858]. Roy. Soc. Proc. IX. pp. 255-256, 1857-1859 [read June 10, 1858]; Phil. Mag. XVII. pp. 61-62, Jan. 1859.

127. On Phenomena of Submerged Atlantic Cable. [Letter to J. P. Joule, dated Sept. 25, 1858.] Manchester Phil. Soc. Proc. I. pp. 60-62 [Oct. 5, 1858].

**1859.**

128. On Atmospheric Electricity. [Letter to J. P. Joule.] Manchester Phil. Soc. Proc. I. pp. 108-109 [Mar. 8, 1859]; E. and M. art. xxi. pp. 311-312.

129. On Electrical "Frequency." Brit. Assoc. Report, 1859, pt. ii. p. 26; E. and M. art. xvi. pp. 226-227.

130. Remarks on the Discharge of a Coiled Electric Cable. Brit. Assoc. Report, 1859, pt. ii. pp. 26-27; Edinb. New Phil. Jl.

XI. pp. 112-113, 1860; Math. and Phys. Papers II. art. lxxx. pp. 129-130.

131. On the Necessity for Incessant Recording, and for Simultaneous Observations in Different Localities, to investigate Atmospheric Electricity. Brit. Assoc. Report, 1859, pt. ii. pp. 27-28; Edinb. New Phil. Jl. XI. pp. 108-110, 1860; E. and M. art. xvi. pp. 227-229.

132. On the Reduction of Periodical Variations of Underground Temperature, with Applications to the Edinburgh Observations. Brit. Assoc. Report, 1859, pt. ii. pp. 54-56; Math. and Phys. Papers III. art. xciii. pp. 291-294.

133. Apparatus for Atmospheric Electricity [Two Letters]. Manchester Phil. Soc. Proc. I. pp. 151-153 [Oct. 18, 1859], p. 156 [Nov. 1, 1859].

134. Recent Investigations of Le Verrier on the Motion of Mercury. Glas. Phil. Soc. Proc. IV. pp. 263-266 [read Dec. 14, 1859].

135. On Photographed Images of Electric Sparks. Glas. Phil. Soc. Proc. IV. pp. 266-267 [read Dec. 14, 1859].

**1860.**

136. On the Variation of the Periodic Times of the Earth and Inferior Planets, produced by Matter falling into the Sun. Glasg. Phil. Soc. Proc. IV. pp. 272-274 [read Jan. 4, 1860].

137. On Instruments and Methods for observing Atmospheric Electricity. Glasg. Phil. Soc. Proc. IV. pp. 274-280 [read Jan. 4, 1860].

138. Analytical and Synthetical Attempts to ascertain the Cause of the Differences of Electric Conductivity discovered in Wires of nearly Pure Copper. Roy. Soc. Proc. X. pp. 300-309, 1859-1860 [read Feb. 9, 1860]; Math. and Phys. Papers II. art. lxxix. pp. 118-128.

139. Measurement of the Electrostatic Force produced by a Daniell's Battery. Roy. Soc. Proc. X. pp. 319-326, 1859-1860 [read Feb. 23, 1860]; Phil. Mag. XX. pp. 233-239, Sept. 1860; E. and M. art. xviii. pp. 238-246.

140. Measurement of the Electromotive Force required to produce a Spark in Air between Parallel Metal Plates at Different Distances. Roy. Soc. Proc. X. pp. 326-338, 1859-1860 [read Feb. 23, 1860]; Phil. Mag. XX. pp. 316-326, Oct. 1860; E. and M. art. xix. pp. 247-259.

141. On the Reduction of Observations of Underground Temperature, with Application to Prof. Forbes's Edinburgh Observations and the continued Calton Hill Series. Edinb. Roy. Soc. Proc. IV. pp. 342-346 [read Apr. 30, 1860]; Edinb. Roy. Soc. Trans. XXII. pp. 405-427, 1861; Phil. Mag. XXII. pp. 23-34, July 1861, pp. 121-135, Aug. 1861; Math. and Phys. Papers III. art. xciii. pp. 261-290.

142. On Atmospheric Electricity. Roy. Instit. Proc. III. pp. 277-290

[May 18, 1860]; Annal. de Chimie VII. pp. 148-172, 1866; E. and M. art. xvi. pp. 208-226.

143. On the Importance of making Observations on Thermal Radiation during the coming Eclipse of the Sun [Abstract of Letter to the Editors dated June 5, 1860]. Astron. Soc. Monthly Notices XX. pp. 317-318, 1860.

144. Thomson and Joule.—On the Thermal Effects of Fluids in Motion. Roy. Soc. Proc. X. p. 502, 1859-1860 [read June 7, 1860]; Math. and Phys. Papers I. art. xlix. p. 453.

145. Thomson and Joule.—On the Thermal Effects of Fluids in Motion:—On Changes of Temperature experienced by Bodies moving through Air. Roy. Soc. Proc. X. p. 519 [read June 21, 1860]; Phil. Trans. pp. 325-336, 1860; Math. and Phys. Papers I. art. xlix. pp. 400-414, p. 454.

146. Report of Committee appointed to prepare a Self-recording Atmospheric Electrometer for Kew, and Portable Apparatus for observing Atmospheric Electricity. Brit. Assoc. Report 1860, pp. 44-45.

147. Notes on Atmospheric Electricity. Brit. Assoc. Report 1860, pt. ii. pp. 53-54; Phil. Mag. XX. pp. 360-363, Nov. 1860; E. and M. art. xxi. pp. 312-316.

148. Velocity of Electricity. Nichol's Cyclopædia [second edition], 1860; Math. and Phys. Papers II. art. lxxxi. pp. 131-137; Baltimore Lectures, Appendix L. pp. 688-694, 1904.

149. Telegraph. Nichol's Cyclopædia [second edition], 1860; Math. and Phys. Papers II. art. lxxxii. pp. 138-141.

150. Atmospheric Electricity. Nichol's Cyclopædia [second edition], 1860; E. and M. art. xvi. pp. 192-208; Annal. de Chimie XI. pp. 86-106, 1877; Soc. Telegr. Engrs. Jl. III. p. 13, 1874.

151. Telegraph. *Encyclopædia Britannica* (eighth edition), 1860.

**1861.**

152. On the Measurement of Electric Resistance. Roy. Soc. Proc. XI. pp. 313-328, 1860-1862 [read June 20, 1861]; Annal. de Chimie LXVII. pp. 501-506, 1863; Phil. Mag. xxiv. pp. 149-162, Aug. 1862.

153. Physical Considerations regarding the Possible Age of the Sun's Heat. Brit. Assoc. Report, 1861, pt. ii. pp. 27-28; Phil. Mag. XXIII. pp. 158-160, Feb. 1862.

154. Thomson and Joule.—On the Thermal Effects of Elastic Fluids. Brit. Assoc. Report, 1861, pt. ii. pp. 83-84.

155. Thomson and Jenkin.—On the True and False Discharge of a Coiled Electric Cable. Phil. Mag. xxii. pp. 202-211, Sept. 1861; Math. and Phys. Papers II. art. lxxxiii. pp. 142-152.

**1862.**

156. On the Convective Equilibrium of Temperature in the Atmosphere. Manchester Phil. Soc. Proc. II. pp. 170-176, 1860-1862 [Jan. 21, 1862]; Manchester Phil. Soc. Mem. II. pp.

125-131, 1865; Math. and Phys. Papers III. art. xcii. pp. 255-260.

157. New Proof of Contact Electricity [Extract of Letter to President of Manchester Lit. and Phil. Soc.]. Manchester Phil. Soc. Proc. II. pp. 176-178, Jan. 21, 1862; E. and M. art. xxii. pp. 317-318.

158. Observations on Atmospheric Electricity. Manchester Phil. Soc. Proc. II. pp. 204-207, 1860-1862 [March 4, 1862]; E. and M. art. xvi. pp. 230-232.

159. On the Age of the Sun's Heat. Macmillan's Magazine V. pp. 288-393, March 1862; Pop. Lect. I. pp. 349-368.

160. On a New Instrument for Measuring Electric Resistance in Absolute Units. Glasg. Phil. Soc. Proc. V. pp. 167-168, 1864 [read March 26, 1862].

161. On the Rigidity of the Earth. Glasg. Phil. Soc. Proc. V. pp. 169-170 [read Mar. 26, 1862]; Roy. Soc. Proc. XII. pp. 103-104, 1862-1863 [read May 15, 1862, abstract]; Phil. Trans. CLIII. pp. 573-582, 1863, appendix added Jan. 2, 1864; Math. and Phys. Papers III. art. xcv. pp. 312-336; Phil. Mag. XXV. pp. 149-151, Feb. 1863; T and T', pp. 689-704.

162. Note on Gravity and Cohesion. Edinb. Roy. Soc. Proc. IV. pp. 604-606 [read Apr. 21, 1862]; Edinb. New Phil. Jl. XVI. pp. 146-148, 1862; Pop. Lect. I. pp. 59-63.

163. On the Secular Cooling of the Earth. Edinb. Roy. Soc. Proc. IV. pp. 610-611 [read Apr. 28, 1862]; Edinb. Roy. Soc. Trans. XXIII. pp. 157-170, 1864; Phil. Mag. XXV. pp. 1-14, Jan. 1863; Edinb. New Phil. Jl. XVI. pp. 151-152, 1862; Math. and Phys. Papers III. art. xciv. pp. 295-311; T and T', Appendix D.

164. Discussion of H. C. Forde's Paper on "The Malta and Alexandria Submarine Telegraph Cable" [May 13 and 20, 1862]; Inst. Civ. Engrs. Proc. XXI. pp. 535-536.

165. Thomson and Joule.—On the Thermal Effects of Fluids in Motion. Roy. Soc. Proc. XII. pp. 202 [read June 19, 1862, abstract]; Phil. Trans. pp. 579-589, 1862; Math. and Phys. Papers I. art. xlix. pp. 415-431, pp. 454-455.

166. Thomson and Tait.—Energy. Good Words, 1862; pp. 601-607 [Sept.]

167. Dynamical Problems regarding Elastic Spheroidal Shells and Spheroids of Incompressible Liquid. Roy. Soc. Proc. XII. pp. 274-275 [read Nov. 27, 1862, abstract]; Phil. Trans. CLIII. pp. 583-616, 1863; Math. and Phys. Papers III. art. xcvi. pp. 351-394.

168. Note on the Electromotive Force induced in the Earth's Crust by Variations of Terrestrial Magnetism [being a Note on Mr. B. Stewart's Paper on "The Nature of the Forces concerned in producing the greater Magnetic Disturbances"]. Phil. Trans. CLII. pt. ii. pp. 637-638, 1862.

**1863.**

169. A Note relative to the Mathematical Theory of the Transmission of Signals through Submarine Cables [Appendix to F. Jenkin's Paper on "Experimental Researches on the Transmission of Electric Signals through Submarine Cables"]. Phil. Trans. CLII. pt. ii. pp. 1011-1017 [read Jan. 14, 1863].
170. On some Kinematical and Dynamical Theorems. Edinb. Roy. Soc. Proc. V. pp. 113-115, 1863 [read Apr. 6, 1863].
171. On Prof. Tyndall's "Remarks on the Dynamical Theory of Heat." [Letter to Phil. Mag. dated May 4, 1863]. Phil. Mag. XXV. p. 429, June 1863.
172. The Electric Telegraph: Three Lectures at the Royal Institution on May 30, June 6 and 13, 1863.
173. On the Result of Reductions of Curves obtained from the Self-recording Electrometer at Kew. Brit. Assoc. Report, 1863, p. 27.
174. On Sound emitted by Air when suddenly subjected to Electric Force [Letter to Prof. Tait]. Manchester Phil. Soc. Proc. III. pp. 159-160 [Oct. 10, 1863]; E. and M. art. xvii. pp. 236-238.
175. Sur le refroidissement séculaire du soleil. De la température actuelle du soleil. De l'origine et de la somme totale de la chaleur solaire. Les Mondes III. pp. 473-480, 1863.
176. Note on two paragraphs of Mr. C. Chamber's paper on "The Nature of the Sun's Magnetic Action on the Earth." Phil. Trans. CLIII. pp. 515-516, 1863 [recd. Oct. 28, 1863].
177. On the Theory of Jones's Method of controlling Clocks by Electricity. Glasg. Phil. Soc. Proc. IV. p. 334, Dec. 2, 1863, p. 335, Jan. 27, 1864. [Note only in Proc.].

**1864.**

178. On Centrobaric Bodies. Edinb. Roy. Soc. Proc. V. pp. 190-191, 1866 [read Mar. 7, 1864].
179. On the Elevation of the Earth's Surface Temperature produced by Underground Heat. Edinb. Roy. Soc. Proc. V. pp. 200-201, 1866 [read Mar. 21, 1864].
180. On the Protection of Vegetation from Destructive Cold every Night. Edinb. Roy. Soc. Proc. V. pp. 203-204, 1866 [read Apr. 4, 1864]; Pop. Lect. II. pp. 1-5.
181. On the Periodic Variations and the Secular Lowering of Terrestrial Temperature. Glasg. Phil. Soc. Proc. IV. p. 338 [Apr. 6, 1864]. [Note only in Proc.]
182. Réponse aux deux notes de M. Dupré "Sur la thermodynamique" insérées dans les Comptes Rendus, mars 21 et sept. 12, 1864; Comp. Rend. LIX. pp. 665-666, pp. 705-708, 1864.

**1865.**

183. Discussion of Joule's Paper on "A Self-acting Apparatus for

Steering Ships." Inst. of Engrs. in Scotland Trans. VIII. pp. 57 and 59 [Jan. 18, 1865].

184. On the Defects, etc., of the Ordinary Electrometer and on Heterostatic Electrometers. Glasg. Phil. Soc. Proc. V. p. 25, Jan. 25, 1865. [Note only in Proc.]

185. On the Secular Variations of Terrestrial and Atmospheric Temperature. Glasg. Phil. Soc. Proc. V. p. 27, Mar. 22, 1865. [Note only in Proc.]

186. On the Elasticity and Viscosity of Metals. Roy. Soc. Proc. XIV. pp. 289-297, 1865 [read May 18, 1865]; Phil. Mag. XXX. pp. 63-71, July 1865.

187. On the Forces concerned in the Laying and Lifting of Deep-Sea Cables [Address delivered at the request of the Council, Dec. 18, 1865]. Edinb. Roy. Soc. Proc. V. pp. 495-509, 1866; Math. and Phys. Papers II. art. lxxxiv. pp. 153-167; Pop. Lect. III. pp. 422-449.

188. On the Dynamical Theory of Heat. Edinb. Roy. Soc. Proc. V. pp. 510-512, 1866 [read Dec. 18, 1865].

189. The "Doctrine of Uniformity" in Geology briefly refuted. Edinb. Roy. Soc. Proc. V. pp. 512-513, 1866 [read Dec. 18, 1865]; Pop. Lect. II. pp. 6-9.

**1866.**

190. On Electrically impelled and Electrically controlled Clocks. Glasg. Phil. Soc. Proc. VI. pp. 61-64, 1868 [read Jan. 24, 1866].

191. On a Land Standard Electrometer and a Marine and Land Telegraphic Testing Electrometer. Glasg. Phil. Soc. Proc. VI. pp. 64-65, 1868 [read Jan. 24, 1866].

192. Note on Mr. Croll's Paper "On the Physical Cause of the Submergence and Emergence of Land during the Glacial Epoch." Phil. Mag. XXXI. pp. 305-306, Apr. 1866.

193. The Rede Lecture, Cambridge. On the Dissipation of Energy [May 23, 1866]. See Appendix (taken from this) republished in Pop. Lect. vol. ii. pp. 65-72. [See No. 191.]

194. On the Observations and Calculations required to find the Tidal Retardation of the Earth's Rotation [part of the Rede Lecture at Cambridge, May 23, 1866]. Phil. Mag. XXXI. pp. 533-537 (Supplement), 1866. Math. and Phys. Papers III. art. xcv. pp. 337-341. [See No. 190.]

**1867.**

195. The Atlantic Telegraph. Good Words, 1867, pp. 43-49 [Jan.].

196. On Vortex Atoms. Edinb. Roy. Soc. Proc. VI. pp. 94-105, 1869 [read Feb. 18, 1867]; Phil. Mag. XXXIV. pp. 15-24, July 1867; Glasg. Phil. Soc. Proc. VI. pp. 197-206, 1868 [read Mar. 6, 1867].

197. On the Rate of a Clock or Chronometer as influenced by the Mode of Suspension. Inst. of Engrs. in Scotland, Trans. X.

pp. 139-151, 1867 [Feb. 27, 1867]; Pop. Lect. II. pp. 360-386.

198. On a New Form of the Dynamic Method for Measuring the Magnetic Dip. Manchester Phil. Soc. Proc. VI. pp. 157-158, 1867 [read Apr. 16, 1867]; Manchester Phil. Soc. Mem. III. pp. 291-292, 1868.

199. On Vortex Motion. Edinb. Roy. Soc. Proc. VI. p. 167 [read Apr. 29, 1867 (title only)]; Edinb. Roy. Soc. Trans. XXV. pp. 217-260, 1869 [§§ 1-59 recast and augmented, Aug. 28 to Nov. 12, 1868].

200. On the Nature and Motions of Vortex Filaments [Abstract of Letter to Prof. Tait, May 17, 1867]. Phil. Mag. XXXIII. pp. 511-512 (Supplement), 1867.

201. On a Self-acting Apparatus for Multiplying and Maintaining Electric Charges, with Applications to illustrate the Voltaic Theory. Roy. Soc. Proc. XVI. pp. 67-72, 1862 [read June 20, 1867]; Phil. Mag. XXXIV. pp. 391-396, Nov. 1867; Annal. de Chimie XIII. pp. 446-448, 1868; E. and M. art. xxiii. pp. 319-325.

202. Report of Committee on Standards of Electrical Resistance:—Drawn up by Sir W. Thomson, Report on Electrometers and Electrostatic Measurements. Brit. Assoc. Report, 1867, pp. 489-512; E. and M. art. xx. pp. 260-286.

203. On a Series of Electrometers for Comparable Measurements through great range. Brit. Assoc. Report, 1867, p. 16.

204. On a Self-acting Electrostatic Accumulator. Brit. Assoc. Report, 1867, p. 16.

205. On a Uniform Electric Current Accumulator. Brit. Assoc. Report XXXVII., 1867, pt. ii. pp. 16-17; Phil. Mag. XXXV. pp. 62-64, Jan. 1868; Journ. de Phys. I. (2nd series), pp. 31-32, 1882; E. and M. art. xxiii. pp. 325-327.

206. On Volta-convection by Flame. Brit. Assoc. Report XXXVII., 1867, pt. ii. pp. 17-18; Phil. Mag. XXXV. pp. 64-66, Jan. 1868; Annal. de Chimie XIV. pp. 487-488, 1868; E. and M. art. xxiii. pp. 328-329.

207. On Electric Machines founded on Induction and Convection. Brit. Assoc. Report, 1867, pt. ii. pp. 18-19; Phil. Mag. XXXV. pp. 66-73, Jan. 1868; Annal. de Chimie XIV. pp. 488-493, 1868; E. and M. art. xxiii. pp. 330-337.

208. Plant Protection. The Gardeners' Chronicle and Agricultural Gazette, Nov. 30, 1867.

**1868.**

209. On Geological Time. Glasg. Geol. Soc. Trans. III. pp. 1-28, 1871 [read Feb. 27, 1868]; Pop. Lect. II. pp. 10-64.

210. On New Electrical Instruments for Delicate Testing Purposes, for Use at Sea. Glasg. Phil. Soc. Proc. V. p. 414 [Mar. 18, 1868]. [Note only in Proc.]

211. On Mr. C. F. Varley's Reciprocal Electrophorus. Phil. Mag.

xxxv. pp. 287-289, April 1868; E. and M. art. xxiii. pp. 337-339.

212. Sketch of Proposed Plan of Procedure in Tidal Observations and Analysis. Brit. Assoc. Report, 1868, pp. 490-505; Pop. Lect. III. pp. 209-223.

213. On a New Form of Centrifugal Governor. Inst. of Engrs. in Scotland, Trans. XII. pp. 67-69, and 70-71, 1869 [Nov. 25, 1868].

**1869.**

214. Determination of the Distribution of Electricity on a Circular Segment of Plane or Spherical Conducting Surface, under any given Influence. E. and M. art. xv. pp. 178-191, Jan. 1869.

215. Lecture at Greenock, "Elasticity viewed as a Mode of Motion," Jan. 1869.

216. On the Fracture of Brittle and Viscous Solids by "Shearing." Roy. Soc. Proc. XVII. pp. 312-313, 1869 [read Feb. 25, 1869]; Phil. Mag. XXXVIII. pp. 71-73, July 1869.

217. Note on the Meteoric Theory of the Sun's Heat [after Prof. Grant's paper on "The Physical Constitution of the Sun"]. Glasg. Phil. Soc. Proc. VII. pp. 111-112 [Mar. 24, 1869]; Glasg. Geol. Soc. Trans. III. pp. 239-240, 1871; Pop. Lect. II. pp. 127-131. [See No. 215.]

218. Geological Dynamics. Glasg. Geol. Soc. Trans. III. pp. 215-240, 1871 [read Apr. 5, 1869]; Geol. Mag. VI. pp. 472-476, 1869; Pop. Lect. II. pp. 73-127. [See No. 214].

219. On a New Astronomical Clock, and a Pendulum Governor for Uniform Motion. Roy. Soc. Proc. XVII. pp. 468-470, 1869 [read June 10, 1869]; Phil. Mag. XXXVIII. pp. 393-395, Nov. 1869; Brit. Assoc. Report, 1876, pt. ii. pp. 49-52 [with alterations and additions]; Nature XV. pp. 227-229, Jan. 11, 1877 [with alterations and additions]; Pop. Lect. II. pp. 387-394.

220. On the Effect of Changes of Temperature on the Specific Inductive Capacity of Dielectrics. Roy. Soc. Proc. XVII. p. 470 [read June 10, 1869, title only].

221. Discussion of J. Gaudard's Paper on "The Present State of Knowledge as to the Strength and Resistance of Materials." Inst. Civ. Engrs. Proc. XXIX. pp. 74-75 [Nov. 16, 23, 30, 1869].

**1870.**

222. Dr. Balfour Stewart's Meteorological Blockade. Nature I. p. 306, Jan. 20, 1870.

223. On the Forces experienced by Solids Immersed in a moving Liquid. Edinb. Roy. Soc. Proc. VII. pp. 60-63, 1872 [read Feb. 7, 1870]; E. and M. art. xli. pp. 567-571.

224. On the Equilibrium of Vapour at a Curved Surface of Liquid. Edinb. Roy. Soc. Proc. VII. pp. 63-68, 1872 [read Feb. 7,

1870]; Phil. Mag. XLII. pp. 448-452, December 1871; Pop. Lect. I. pp. 64-72.

225. On the Size of Molecules [Extract of a Letter to the Society dated Mar. 21, 1870]. Manchester Phil. Soc. Proc. IX. pp. 136-141 [Mar. 22, 1870]; Les Mondes XXII. pp. 701-708, 1870, XXVII. pp. 616-623, 1872; Annal. Chem. Pharm. CLVII. pp. 54-66, 1871; Amer. Jl. Sci. L. (2nd ser.) pp. 258-261, 1870; Nature II. pp. 56-57, May 19, 1870.

226. On the Size of Atoms. Nature I. pp. 551-553, Mar 31, 1870; Amer. Jl. Sci. L. pp. 38-44, 1870.

227. Reports of Committee for the Purpose of Promoting the Extension, Improvement, and Harmonic Analysis of Tidal Observations. Brit. Assoc. Report, 1870, pp. 120-151, and 1876, pp. 275-307.

228. On a New Absolute Electrometer. Brit. Assoc. Report, 1870, p. 26 [title only]; E. and M. art. xx. pp. 287-310.

229. On the Attractions and Repulsions due to Vibration observed by Guthrie and Schellbach. Glasg. Phil. Soc. Proc. VII. pp. 401-404, 1871 [read Dec. 14, 1870]; E. and M. art. xli. pp. 574-578.

**1871.**

230. Modification of Wheatstone's Bridge to find the Resistance of a Galvanometer-coil from a single Deflection of its own Needle. Roy. Soc. Proc. XIX. p. 253, 1871 [read Jan. 19, 1871]; Phil. Mag. XLI. pp. 537-538 (Supplement), 1871.

231. On a Constant Form of Daniells Battery. Roy. Soc. Proc. XIX. pp. 253-259, 1871 [read Jan. 19, 1871]; Phil. Mag. XLI. pp. 538-543 (Supplement), 1871; Nature III. pp. 350-351, Mar. 2, 1871.

232. On the Determination of a Ship's Place from Observations of Altitude. Roy. Soc. Proc. XIX. pp. 259-266, 1871 [read Jan. 19, 1871].

233. On Approach Caused by Vibration [a Letter from Sir W. T. to Prof. Guthrie, communicated by Sir W. T.]. Roy. Soc. Proc. XIX. pp. 271-273, 1871 [read Jan. 26, 1871]; Phil. Mag. XLI. pp. 423-429, June 1871; Annal. de Chimie XXV. pp. 205-206, 1872; E. and M. art. xli. pp. 571-574.

234. A Hint to Electricians. Nature III. pp. 248-249, Jan. 26, 1871.

235. On the Motion of Free Solids through a Liquid. Edin. Roy. Soc. Proc. VII. pp. 384-390, 1872 [read Feb. 20, 1871]. Baltimore Lectures, Appendix G. pp. 584-590, 1904.

236. Amended Rule for working out Sumner's Method of Finding a Ship's Place. Roy. Soc. Proc. XIX. pp. 524-526, 1871 [read June 15, 1871].

237. Address to the British Association at Edinburgh. Brit. Assoc. Report XLI., 1871, pp. lxxxiv.-cv.; Nature IV. pp. 262-270, Aug. 3, 1871; Amer. Jl. Sci. II. (3rd ser.), pp. 269-294,

1871; Math. and Phys. Papers II. art. lxvi. pp. 25-27; Pop. Lect. II. pp. 132-205.

238. On the General Canonical Form of a Spherical Harmonic of the $n^{th}$ Order. Brit. Assoc. Report XLI., 1871, pt. ii. pp. 25-26.

239. Ripples and Waves [Extract from a Letter to Mr. W. Froude]. Nature V. pp. 1-3, Nov. 2, 1871.

240. Hydrokinetic Solutions and Observations. Phil. Mag. XLII. pp. 362-377, Nov. 1871; Baltimore Lectures, Appendix G, pt. v. pp. 598-601, 1904.

241. Inverse Problems. E. and M. art. xxviii. pp. 451-462, Nov. 1871.

242. On Vortex Motion. Edinb. Roy. Soc. Proc. VII. pp. 576-577, 1872 [read Dec. 18, 1871].

243. On the Ultramundane Corpuscles of Le Sage, also on the Motion of Rigid Solids in a Liquid Circulating Irrotationally through Perforations in them or in a Fixed Solid. Edinb. Roy. Soc. Proc. VII. pp. 577-589, 1872 [read Dec. 18, 1871]; Phil. Mag. XLV. pp. 321-345, May 1873.

**1872.**

244. The Rigidity of the Earth. Letter to Nature V. pp. 223-224, Jan. 18, 1872.

245. On the Mechanical Values of Distributions of Matter, and of Magnets. E. and M. art. xxvi. pp. 432-443, Jan. 1872.

246. Hydrokinetic Analogy. E. and M. art. xxvii. pp. 444-450, Jan. 1872.

247. On the Internal Fluidity of the Earth. Nature V. pp. 257-259, Feb. 1, 1872.

248. Remarks on Contact-electricity. Edinb. Roy. Soc. Proc. VII. p. 648 [read Feb. 19, 1872, title only].

249. General Hydrokinetic Analogy for Induced Magnetism. E. and M. art. xlii. pp. 579-587, Feb. 1872.

250. On the Motion of Rigid Solids in a Liquid Circulating Irrotationally through Perforations in them or in any Fixed Solid. Edinb. Roy. Soc. Proc. VII. pp. 668-682, 1872 [read Mar. 4, 1872].

251. Magnetic Permeability, and Analogues in Electrostatic Induction, Conduction of Heat, and Fluid Motion. E. and M. art. xxxi. pp. 482-486, Mar. 1872.

252. General Problem of Magnetic Induction. E. and M. art. xl. pp. 544-566, Mar. 1872.

253. Inductive Susceptibility of a Polar Magnet. E. and M. art. xxxix. pp. 543-544, Mar. 1872.

254. Diagrams of Lines of Force, to Illustrate Magnetic Permeability. E. and M. art. xxxii. pp. 486-493, May 29, 1872.

255. On the Simultaneous Reduction of two Polynomial Quadratics to Sums of Squares. Lond. Math. Soc. Proc. IV. p. 120 [read June 13, 1872].

256. On the Identification of Lights at Sea. Brit. Assoc. Report, 1872, pt. ii. p. 251.
257. On the Use of Steel-Wire for Deep-Sea Soundings. Brit. Assoc. Report, 1872, pt. ii. p. 251.
258. On a New Form of Joule's Tangent Galvanometer [original communication]. Soc. Telegr. Engrs. Journ. I. pp. 392-394, 1872-1873.
259. On the Measurement of Electrostatic Capacity [original communication]. Soc. Telegr. Engrs. Journ. I. pp. 394-398, 1872-1873.
260. Tests of Battery [1873] [original communication]. Soc. Telegr. Engrs. Journ. I. pp. 399-403, 1872-1873.
261. Tray Battery for the Siphon Recorder [original communication]. Soc. Telegr. Engrs. Journ. I. pp. 403-407, 1872-1873.

**1873.**

262. Note on Homocheiral and Heterocheiral Similarity. Edinb. Roy. Soc. Proc. VIII. p. 70 [read Feb. 17, 1873, title only].
263. On Vortex Motion. Edinb. Roy. Soc. Proc. VIII. p. 80 [read Mar. 3, 1873, title only].
264. On Signalling through Submarine Cables, illustrated by Signals Transmitted through a Model Submarine Cable, exhibited by Mirror Galvanometer and by Siphon Recorder. Inst. of Engrs. in Scotland Trans. XVI. pp. 119-120, 1873 [Mar. 18, 1873]; Math. and Phys. Papers II. art. lxxxv. pp. 168-172.
265. On the Rope-Dynamometer, with Application to Deep-Sea Sounding by Steel Wire. Inst. of Engrs. in Scotland Trans. XVI. pp. 121-135, 1873 [Mar. 18, 1873].
266. Lighthouses of the Future. Good Words, 1873, pp. 217-224 [March]
267. A New Method of determining the Material and Thermal Diffusivities of Fluids. Edinb. Roy. Soc. Proc. VIII. p. 229 [read Dec. 22, 1873, title only].

**1874.**

268. Obituary Notice of Archibald Smith. Roy. Soc. Proc. XXII. pp. i.-xxiv. [dated Jan. 4, 1874]; Edinb. Roy. Soc. Proc. VIII. pp. 282-288 [read Feb. 2, 1874]. [Abridged (by direction of the author) from Proc. R.S.]
269. Inaugural Address to the Society of Telegraph Engineers. Soc. Telegr. Engrs. Journ. III. pp. 1-15 [Jan. 14, 1874]; Telegraphic Journal II. pp. 67-70, Jan. 15, 1874; Nature IX. pp. 269-271, Feb. 5, 1874; Pop. Lect. II. pp. 206-237.
270. The Mariner's Compass [Part I.] Good Words, 1874, pp. 69-72 [Jan.]; Pop. Lect. III. p. 228.
271. Influence of Geological Changes on the Earth's Rotation [Presidential Address]. Glasg. Geol. Soc. Proc. IV. pp. 311-313 [Feb. 12, 1874]; Nature IX. pp. 345-346, Mar. 5, 1874.
272. The Kinetic Theory of the Dissipation of Energy. Edinb. Roy. Soc. Proc. VIII. pp. 325-334, 1875 [read Feb. 16, 1874];

Nature IX. pp. 441-444, Apr. 9, 1874; Phil. Mag. XXXIII. pp. 291-299, Mar. 1892.

273. On a New Form of Mariner's Compass. Edinb. Roy. Soc. Proc. VIII. p. 363 [read Mar. 16, 1874, title only].

274. On Deep-Sea Sounding by Pianoforte Wire. Glasg. Phil. Soc. Proc. IX. pp. 111-117, 1875 [read Mar. 18, 1874].

275. On Deep-Sea Sounding by Pianoforte Wire. Soc. Telgr. Engrs. Journ. III. pp. 206-219, 1874 [Apr. 22, 1874], pp. 220-221 and pp. 222-224; Appendix, pp. 226-228; Van Nostrand's Engin. Mag. XI. pp. 6-13, 1874; Franklin Instit. Jl. LXVIII. pp. 1-4, July 1874; Pop. Lect. III. pp. 337-376.

276. On the Perturbations of the Compass produced by the Rolling of the Ship. Brit. Assoc. Report, 1874, pt. ii. p. 32 [title only]; Nature X. pp. 388-389, Sept. 10, 1874; Phil. Mag. XLVIII. pp. 363-369, Nov. 1874.

277. On Improvements in the Mariner's Compass [title only]. Brit. Assoc. Report, 1874, pt. ii. p. 231.

278. Note on Mr. Gore's Paper on Electrotorsion. Phil. Trans. CLXIV. pp. 560-562, 1874.

**1875.**

279. Exhibited and Described his Tide Calculating Machine, also his Improved Tide-Gauge; he also described certain Capillary Phenomena, with Experiments. Edinb. Roy. Soc. Proc. VIII. p. 445 [Jan. 4, 1875].

280. Lecture on "The Tides" (Glasgow Science Lectures Association) in Glasgow City Hall, Feb. 3, 1875. Reported in "Glasgow Herald," Feb. 4, 1875; Extract printed in Pop. Lect. III. pp. 191-202.

281. On the Oscillation of a System of Bodies with Rotating Portions. Edinb. Roy. Soc. Proc. VIII. p. 490 [read Feb. 15, 1875, title only].

282. Diagrams in Illustration of the Capillary Surfaces of Revolution. Edinb. Roy. Soc. Proc. VIII. p. 500 [exhibited Mar. 1, 1875].

283. Vibrations and Waves in a Stretched Uniform Chain of Symmetrical Gyrostats. London Math. Soc. Proc. VI. pp. 190-194, 1874-1875 [read Apr. 8, 1875].

284. On Tides. Roy. Instit. Proc. VII. pp. 447-448, 1875 [Apr. 9, 1875].

285. Thomson, Sir W., and Perry, J.—On the Capillary Surface of Revolution. Edinb. Roy. Soc. Proc. VIII. p. 520 [read Apr. 19, 1875, title only].

286. On the Oscillation of a System of Bodies with Rotating Portions. Part II. Vibrations of a Stretched String of Gyrostats (Dynamics of Faraday's Magneto-optic Discovery) with experimental Illustrations. Edinb. Roy. Soc. Proc. VIII. p. 521 [read Apr. 19, 1875, title only].

287. On the Theory of the Spinning-Top, with experimental

Illustrations. Edinb. Roy. Soc. Proc. VIII. p. 521 [read Apr. 19, 1875, title only].

288. Electrodynamic Qualities of Metals (Continued from Phil. Trans. for June 15, 1857). Part vi. Effects of Stress on Magnetization; Part vii. Effects of Stress on the Magnetization of Iron, Nickel, and Cobalt. Roy. Soc. Proc. XXIII. pp. 445-446, 1875 [read May 27, 1875]; XXVII. pp. 439-443, 1878 [read May 23, 1878, abstract]; Phil. Trans. CLXVI. pp. 693-713, 1877; CLXX. pp. 55-85, 1880; Math. and Phys. Papers II. art. xci. pp. 332-353, pp. 401-403; II. art. xci. pp. 358-395, pp. 403-407.

289. Electrolytic Conduction in Solids. First Example, Hot Glass. Roy. Soc. Proc. XXIII. pp. 463-464, 1875 [read June 10, 1875].

290. Effects of Stress on Inductive Magnetism in Soft Iron (Preliminary Notice). Roy. Soc. Proc. XXIII. pp. 473-476, 1875 [read June 10, 1875]; Math. and Phys. Papers II. art. xci. pp. 353-357.

291. General Integration of Laplace's Differential Equation of the Tides. Brit. Assoc. Report, 1875, pt. ii. p. 23 [title only]; Phil. Mag. L. (series 4), pp. 388-402, Nov. 1875.

292. On Laplace's Process for Determining an Arbitrary Constant in the Integration of his Differential Equation for the Semi-Diurnal Tide. Brit. Assoc. Report, 1875, pt. ii. p. 23 [title only].

293. On the Integration of Linear Differential Equations with Rational Coefficients. Brit. Assoc. Report, 1875, pt. ii. p. 23 [title only].

294. On some Effects of Laplace's Theory of Tides. Brit. Assoc. Report, 1875, pt. ii. p. 23 [title only].

295. On the Effect of Stress on the Magnetism of Soft Iron. Brit. Assoc. Report, 1875, pt. ii. p. 29 [title only].

296. On a Machine for the Calculation of the Tides. Brit. Assoc. Report, 1875, pt. ii. p. 253 [title only].

297. Thomson, Sir W., and Hopkinson, J.—On Methods of giving Distinctive Characters to Lighthouses. Brit. Assoc. Report, 1875, pt. ii. p. 253 [title only].

298. On an Alleged Error in Laplace's Theory of the Tides. Phil. Mag. L. pp. 227-242, Sept. 1875.

299. Note on the "Oscillations of the First Species" in Laplace's Theory of the Tides. Phil. Mag. L. pp. 279-284, Oct. 1875.

300. On Navigation, Lecture in City Hall, Glasgow, Nov. 11, 1875. Pop. Lect. III. pp. 1-138; Nature XV. pp. 403-404, Mar. 8, 1877.

301. Vortex Statics. Edinb. Roy. Soc. Proc. IX. pp. 59-73, 1878 [Dec. 20, 1875, abstract]; Phil. Mag. X. pp. 97-109, Aug. 1880.

302. Instructions for the Observation of Atmospheric Electricity for

the use of the Arctic Expedition of 1875. In the Manual of the Natural History, Geology, and Physics of Greenland, prepared for the use of the Arctic Expedition of 1875, under direction of the Arctic Committee of the Royal Society. London, 1875, 8vo. pp. 20-24.

**1876.**

303. On Two-dimensional Motion of Mutually Influencing Vortex Columns, and on Two-dimensional Approximately Circular Motion of a Liquid. Edinb. Roy. Soc. Proc. IX. p. 98 [read Jan. 3, 1876, title only].

304. On an Instrument for Calculating ($\int\phi(x)\psi(x)dx$), the Integral of the Product of two given Functions. Roy. Soc. Proc. XXIV. pp. 266-268, 1876 [read Feb. 3, 1876].

305. Mechanical Integration of the Linear Differential Equations of the Second Order with Variable Coefficients. Roy. Soc. Proc. XXIV. pp. 269-271, 1876 [read Feb. 3, 1876].

306. Mechanical Integration of the General Linear Differential Equation of any Order with Variable Coefficients. Roy. Soc. Proc. XXIV. pp. 271-275, 1876 [read Feb. 3, 1876].

307. An Application of Prof. James Thomson's Integrator to Harmonic Analyses of Meteorological, Tidal, and other Phenomena, and to the Integration of Differential Equations. Edinb. Roy. Soc. Proc. IX. p. 138 [read Mar. 6, 1876, title only].

308. On the Vortex Theory of Gases, of the Condensation of Gases on Solids, and of the Continuity between the Gaseous and Liquid State of Matter. Edinb. Roy. Soc. Proc. IX. p. 144 [read Apr. 3, 1876, title only].

309. On Thermodynamic Motivity. Edinb. Roy. Soc. Proc. IX. p. 144 [read Apr. 3, 1876, title only]; Edinb. Roy. Soc. Trans. XXVIII. pp. 741-744, 1879; Phil. Mag. VII. pp. 348-352, May 1879; Journ. de Phys. VIII. p. 316, 1879.

310. Electrical Measurement. [Lecture before the Section of Mechanics at the Conferences held in connection with the Special Loan Collection of Scientific Apparatus at the South Kensington Museum, May 17, 1876]. Pop. Lect. I. pp. 423-454.

311. Presidential Address to the Mathematical and Physical Science Section of the British Association at Glasgow, Sept. 7:—Review of Evidence Regarding the Physical Condition of the Earth; its Internal Temperature; the Fluidity or Solidity of its Interior Substance; the Rigidity, Elasticity, Plasticity of its External Figure; and the Permanence or Variability of its Period and Axis of Rotation. Brit. Assoc. Report, 1876, pt. ii. pp. 1-12; Nature XIV. pp. 426-431, Sept. 14, 1876; Archives Sci. Phys. Nat. LVII. pp. 138-161, 1876; Amer. Jl. Sci. XII. pp. 336-354, 1876; Math. and Phys. Papers III. art. xcv. pp. 320-335; Pop. Lect. II. pp. 238-272.

312. On the Precessional Motion of a Liquid. Brit. Assoc. Report, 1876, pt. ii. pp. 33-35; Nature XV. pp. 297-298, Feb. 1, 1877.
313. Secular Illustration of the Laws of the Diffusion of Liquids. Brit. Assoc. Report, 1876, pt. ii. p. 35 [title only].
314. On a New Case of Instability of Steady Motion. Brit. Assoc. Report, 1876, pt. ii. p. 35 [title only].
315. On the Nutation of a Solid Shell containing Liquid. Brit. Assoc. Report, 1876, pt. ii. p. 35 [title only].
316. On Compass Correction in Iron Ships. Brit. Assoc. Report, 1876, pt. ii. p. 45 [title only].
317. Effects of Stress on the Magnetization of Iron. Brit. Assoc. Report, 1876, pt. ii. p. 45 [title only].
318. On Contact Electricity. Brit. Assoc. Report, 1876, pt. ii. p. 45 [title only].
319. On a Practical Method of Tuning a Major Third. Brit. Assoc. Report, 1876, pt. ii. p. 48 [title only]. [See No. 342.]
320. On a New Form of Astronomical Clock with Free Pendulum and Independently Governed Uniform Motion for Escapement-wheel. Brit. Assoc. Report, 1876, pt. ii. pp. 49-52.
321. Physical Explanation of the Mackerel Sky. Brit. Assoc. Report, 1876, pt. ii. p. 54 [title only]; Symon's Meteor. Mag. XI. p. 131, 1876; Année Sc. Industr. XXI. pp. 53-54, 1877.
322. On Navigational Deep-Sea Soundings in a Ship moving at High Speed. Brit. Assoc. Report, 1876, pt. ii. p. 54 [title only].
323. On Naval Signalling. Brit. Assoc. Report, 1876, pt. ii. p. 233 [title only].

**1877.**

324. Geological Climate. Glasg. Geol. Soc. Trans. V. pp. 238-250 [Feb. 22, 1877]; Pop. Lect. II. pp. 273-298.
325. On Compass Adjustment on the Clyde. Inst. of Engrs. and Shipbuilders in Scotland Trans. XX. pp. 143-144 [Apr. 24, 1877].
326. Remarks on the Glasgow Philosophical Society's Memorials and Petitions of January 1877, to the Lord Chancellor and to the two Houses of Parliament, regarding Patent Law Amendment. Glasg. Phil. Soc. Proc. X. pp. 269-273, 1877.
327. On the Effect of Transverse Stress on the Magnetic Susceptibility of Iron. Brit. Assoc. Report, 1877, pt. ii. p. 37 [title only].
328. Thomson, Sir W., and Evans, Capt.—On the Tides of Port Louis, Mauritius, and Freemantle, Australia. Brit. Assoc. Report, 1877, pt. ii. p. 40 [title only].
329. Solutions of Laplace's Tidal Equation for certain special Types of Oscillation. Brit. Assoc. Report, 1877, pt. ii. p. 43 [title only].
330. Diurnal and Semi-diurnal Harmonic Constituents of the

Variation of Barometric Pressure. Brit. Assoc. Report, 1877, pt. ii. p. 43 [title only].
331. On a Marine Azimuth Mirror and its Adjustments. Brit. Assoc. Report, 1877, pt. ii. p. 43 [title only].
332. On the Possibility of Life on a Meteoric Stone falling on the Earth. Brit. Assoc. Report, 1877, pt. ii. p. 43 [title only].
333. On a Navigation Sounding Machine for Use at Full Speed. Brit. Assoc. Report, 1877, pt. ii. p. 218 [title only].
334. On an Improved Method of Recording the Depth in Flying Soundings. Brit. Assoc. Report, 1877, pt. ii. p. 218 [title only].
335. On the Importance of giving a Distinctive Character to the Needles Light. Brit. Assoc. Report, 1877, pt. ii. p. 218 [title only].
336. On the Mariner's Compass, with Correctors for Iron Ships. Brit. Assoc. Report, 1877, pt. ii. p. 218 [title only].
337. Rapports sur les Machines Magnéto-électriques Gramme [présentées à l'Exposition de Philadelphie]. Journ. de Phys. VI. pp. 240-242, Aug. 1877.
338. On the Resonance of Cavities. Glasg. Phil. Soc. Proc. XI. pp. 1-2 [read Nov. 21, 1877, abstract].

**1878.**

339. Discussion of Higgs' and Brittle's Paper on "Some Recent Improvements in Dynamo-Electric Apparatus." Inst. Civ. Engrs. Proc. LII. pp. 81-83, 1878 [Jan. 22, 1878].
340. On Compass Adjustment in Iron Ships, and on a New Sounding Apparatus. Nature XVII. pp. 331-334, Feb. 21; pp. 352-354, Feb. 28; pp. 387-388, March 14, 1878; United Service Instit. Journ. XXII. pp. 91-114, 1879 [Feb. 4, 1878, with Appendix]; Pop. Lect. III. p. 228, p. 322, p. 329, p. 377.
341. The Internal Condition of the Earth; as to Temperature, Fluidity, and Rigidity. Glasg. Geol. Soc. Trans. VI. pp. 38-49, 1882 [read Feb. 14, 1878]; Pop. Lect. II. pp. 299-318.
342. On Beats of Imperfect Harmonies. Edinb. Roy. Soc. Proc. IX. pp. 602-612 [read Apr. 1, 1878]; Pop. Lect. II. pp. 395-417. [See No. 319.]
343. On Vortex Vibrations, and on Instability of Vortex Motions. Edinb. Roy. Soc. Proc. IX. p. 613 [read Apr. 15, 1878, title only].
344. Floating Magnets. Nature XVIII. pp. 13-14, May 2, 1878.
345. Harmonic Analyzer. Roy. Soc. Proc. XXVII. pp. 371-373, 1878 [May 9, 1878].
346. The Effects of Stress on the Magnetization of Iron, Cobalt, and Nickel. Roy. Instit. Proc. VIII. pp. 591-593, 1879 [May 10, 1878].
347. On the Effect of Stress on the Magnetization of Iron, Nickel, and Cobalt. Phys. Soc. Proc. III. p. 3 (proc.) [May 11, 1878, title only]; Nature XVIII. p. 215, June 10, 1878.

348. A Mechanical Illustration of the Vibrations of a Triad of Columnar Vortices. Edinb. Roy. Soc. Proc. IX. p. 660 [read May 20, 1878, title only].
349. Problems Relating to Underground Temperature. A Fragment. [Written eighteen years before—kept back for time to solve two parts which have not been done.] Phil. Mag. V. pp. 370-374, May 1878; Journ. de Phys. VII. pp. 397-402, 1878; Amer. Jl. of Science XVI. pp. 132-135, 1878.
350. On the Effect of Torsion on the Electrical Conductivity of Brass. Phys. Soc. Proc. III. p. 4 (proc.) [May 25, 1878, title only]; Nature XVIII. pp. 180-181, June 13, 1878.
351. The Microphone. Letter to Nature XVIII. pp. 355-356, Aug. 1, 1878.
352. Thomson, Sir W., and Evans, Capt.—On the Tides of the Southern Hemisphere, and of the Mediterranean. Brit. Assoc. Report, 1878, pp. 477-481; Nature XVIII. pp. 670-672, Oct. 24, 1878; Pop. Lect. III. pp. 204-208.
353. On Gaussin's Warning regarding the Sluggishness of Ships' Magnetism. Brit. Assoc. Report, 1878, pp. 496-497; Nature XIX. pp. 127-128, Dec. 12, 1878.
354. Influence of the Straits of Dover on the Tides of the British Channel and North Sea. Brit. Assoc. Report, 1878, pp. 639-640; Nature XIX. pp. 152-154, Dec. 19, 1878; Pop. Lect. III. pp. 201-204.
355. The Distinction of Lighthouses [Lecture to the Shipmasters' Society]. Times, Oct. 26, 1878.
356. On a Machine for the Solution of Simultaneous Linear Equations. Roy. Soc. Proc. XXVIII. pp. 111-113, 1879 [read Dec. 5, 1878].
357. Elasticity, article in "Encyclopædia Britannica," 1878; Math. and Phys. Papers III. art. xcii. (pt. i.) pp. 1-112. [See No. 523.]

**1879.**

358. The Sorting Demon of Maxwell. Roy. Instit. Proc. IX. pp. 113-114 [Feb. 28, 1879]; Nature XX. p. 126, June 5, 1879; Pop. Lect. I. pp. 137-141.
359. On Gravitational Oscillation of Rotating Water. Edinb. Roy. Soc. Proc. X. pp. 92-100 [read Mar. 17, 1879, abstract]; Phil. Mag. X. pp. 109-116, Aug. 1880.
360. Note on the preceding letter ["On the Dissipation of Energy," by Prof. P. G. Tait]. Phil. Mag. VII. pp. 346-348, May 1879; Journ. de Phys. VIII. pp. 236-237, 1879; Math. and Phys. Papers I. art. l. pp. 456-459.
361. Evidence before the Select Committee on Electric Light [Abstract]. Nature XX. pp. 110-111, May 29, 1879.
362. Terrestrial Magnetism and the Mariner's Compass. Good Words, 1879, pp. 383-390 [June]; pp. 445-453 [July]; Pop. Lect. III. p. 228.

363. Distinguishing Lights for Lighthouses [Abstract from a Letter to *The Times*]. Nature XXI. pp. 109-110, Dec 4, 1879.

**1880.**

364. On Steam-Pressure Thermometers of Sulphurous Acid, Water, and Mercury. Edinb. Roy. Soc. Proc. X. pp. 432-441 [read Mar. 1, 1880].

365. On a Sulphurous Acid Cryophorus. Edinb. Roy. Soc. Proc. X. pp. 442-443 [read Mar. 1, 1880 (Abstract)].

366. Vibrations of a Columnar Vortex. Edinb. Roy. Soc. Proc. X. pp. 443-456 [read Mar. 1, 1800]; Phil. Mag. X. pp. 155-168, Sept. 1880.

367. On a Realised Sulphurous Acid Steam-Pressure Thermometer, and on a Sulphurous Acid Steam-Pressure Differential Thermometer; also a Note on Steam-Pressure Thermometers. Edinb. Roy. Soc. Proc. X. pp. 532-536 [read Apr. 19, 1880].

368. On a Differential Thermoscope founded on Change of Viscosity of Water with Change of Temperature. Edinb. Roy. Soc. Proc. X. p. 537 [read Apr. 19, 1880].

369. On a Thermomagnetic Thermoscope. Edinb. Roy. Soc. Proc. X. pp. 538-539 [read Apr. 19, 1880].

370. On a Constant Pressure Gas Thermometer. Edinb. Roy. Soc. Proc. X. pp. 539-545 [read Apr. 19, 1880].

371. On the Elimination of Air from Water. Phys. Soc. Proc. IV. p. 4 [May 8, 1880, title only].

372. On Steam-Pressure Thermometers. Phys. Soc. Proc. IV. p. 4 [May 8, 1880, title only].

373. On the Radiation of Water-Steam Pressure Thermometers. Phys. Soc. Proc. IV. p. 4 [May 8, 1880, title only].

374. Recent Improvements in the Compass, with Correctors for Iron Ships. United Service Instit. Journ. XXIV. pp. 404-410, 1881 [May 10, 1880]; Pop. Lect. III. p. 228.

375. On the Maximum and Minimum Energy in Vortex Motion. Brit. Assoc. Report, 1880, pp. 473-476. Nature XXII. pp. 618-620; Oct. 28, 1880; with corrections and additions, Phil. Mag. XXIII. pp. 529-539, 1887.

376. On a Septum Permeable to Water and Impermeable to Air, with Practical Applications to a Navigational Depth-Gauge. Brit. Assoc. Report, 1880, pp. 488-489; Nature XXII. pp. 548-549, Oct. 7, 1880.

377. On an Experimental Illustration of Minimum Energy in Vortex Motion. Brit. Assoc. Report, 1880, pp. 491-492; Nature XXIII. pp. 69-70, Nov. 18, 1880.

378. On a Disturbing Infinity in Lord Rayleigh's Solution for Waves in a Plane Vortex Stratum. Brit. Assoc. Report, 1880, pp. 492-493; Nature XXIII. pp. 45-46, Nov. 11, 1880; p. 70. Nov. 18, 1880.

379. On a Method of Measuring Contact Electricity. Brit. Assoc.

Report, 1880, pp. 494-496; Nature XXIII. pp. 567-568, Apr. 14, 1881.

380. On a Method of Determining without Mechanism the Limiting Steam-Liquid Temperature of a Fluid. Brit. Assoc. Report, 1880, pp. 496-497; Nature XXIII. pp. 87-88, Nov. 25, 1880; Journ. de Phys. X. p. 414, 1881.

381. On an Improved Sounding Machine. Brit. Assoc. Report, 1880, p. 703 [title only].

382. Heat, article on, in "Encyclopædia Britannica," 1880. Math. and Phys. Papers II. art. lxxii. pp. 41-55; III. art. xcii. (pt. ii.) pp. 113-235.

**1881.**

383. On his Siphon Recorder [Letter to La Lumière Électrique, dated Jan. 25, 1881]. Lum. Élec. III. p. 126, Feb. 5, 1881.

384. On Lighthouse Characteristics. [Paper read at Naval and Marine Exhibition, Glasgow, Feb. 11, 1881.] Pop. Lect. III. pp. 389-421.

385. On the Effect of Moistening with Water the Opposed Metallic Surfaces in a Volta-condenser, and of Substituting a Water Arc for a Metallic Arc in the Determining Contact. Edin. Roy. Soc. Proc. XI. p. 135 [read Feb. 21. 1881, title only].

386. On Vortex Sponge. Edin. Roy. Soc. Proc. XI. p. 135 [read Feb. 21, 1881, title only].

387. Discussion of M. am Ende's Paper "On the Weight and Limiting Dimensions of Girder Bridges." Inst. Civil Engrs. Proc. LXIV. pp. 277-278, 1881 [Feb. 22, 1881].

388. The Tide-Gauge, Tidal Harmonic Analyzer, and Tide Predicter. Inst. Civil Engrs. Proc. LXV. pp. 2-25, 1881 [Mar. 1, 8, and 15, 1881]; Discussion and Reply, pp. 26-31, pp. 58-64.

389. On his New Navigational Sounding Machine and Depth-Gauge. United Service Instit. Journ. XXV. pp. 374-381, 1882 [Mar. 4, 1881].

390. Elasticity viewed as Possibly a Mode of Motion. Roy. Instit. Proc. IX. pp. 520-521, 1882 [Mar. 4, 1881]; Pop. Lect. I. pp. 142-146.

391. The Tide Predicter. Letter to Nature XXIII. p. 482, Mar. 24, 1881; p. 578, Apr. 21, 1881.

392. On the Average Pressure due to Impulse of Vortex Rings on a Solid. Edinb. Roy. Soc. Proc. XI. p. 204 [read Apr. 18, 1881, title only].

393. The Storage of Electric Energy. Nature XXIV. pp. 137, 156, 157, June 16, 1881.

394. Presidential Address to the Mathematical and Physical Science Section of the British Association at York, Sept. 1.—On the Sources of Energy in Nature available to Man for the Production of Mechanical Effect. Brit. Assoc. Report, 1881, pp. 513-518; Nature XXIV. pp. 433-436, Sept. 8, 1881,

Franklin Inst. Journ. LXXXII. pp. 376-385, Nov. 1881; Pop. Lect. II. pp. 433-450.

395. On some Uses of Faure's Accumulator in Connection with Lighting by Electricity. Brit. Assoc. Report, 1881, p. 526.

396. On the Economy of Metal in Conductors of Electricity. Brit. Assoc. Report, 1881, pp. 526-528; Lum. Élec. V. pp. 65-66, Oct. 12, 1881.

397. On the Proper Proportions of Resistance in the Working Coils, the Electromagnets and the External Circuits of Dynamos. Brit. Assoc. Report, 1881, pp. 528-531; Comp. Rend. XCIII. pp. 474-479, 1881; Nature XXIV. pp. 526-527, Sept. 29, 1881; Lum. Élec. IV. pp. 385-387, Sept. 24, 1881.

398. On an Electro-ergometer. Brit. Assoc. Report, 1881, p. 554 [title only].

399. Thomson (Sir W.) and Bottomley (J. T.).—On the Illuminating Powers of Incandescent Vacuum Lamps with Measured Potentials and Measured Currents. Brit. Assoc. Report, 1881, pp. 559-561.

400. On Photometry, with Experiments. Brit. Assoc. Report, 1881, p. 561 [title only].

401. Accélération Thermodynamique du Mouvement de Rotation de la Terre. Paris Soc. Phys. Séances, pp. 200-210, 1881 [Sept. 23, 1881]; Edinb. Roy. Soc. Proc. XI. pp. 396-405 [read Jan. 16, 1882]; Journ. de Phys. I. pp. 61-70, 1882; Nuovo Cimento XI. pp. 240-243, 1882; Math. and Phys. Papers III. pp. 341-350; L'Astronomie IV. pp. 230-231, June 1885.

**1882.**

402. Solar Observations. Letter to Nature XXV. pp. 316-317, Feb. 2, 1882.

403. Discussion of J. J. Coleman's Paper on "Air Refrigerating Machinery and its Applications," Inst. Civ. Engrs. Proc. LXVIII. pp. 210-211, 1882 [Feb. 14, 1882].

404. On the Figures of Equilibrium of a Rotating Mass of Fluid. Edinb. Roy. Soc. Proc. XI. pp. 610-613 [read Apr. 3, 1882].

405. Discussion of D. Clerk's paper "On the Theory of the Gas Engine." Inst. Civ. Engrs. Proc. LXIX. pp. 278-279, 1882 [Apr. 4, 1882].

406. On a New Form of Galvanometer for Measuring Currents and Potentials in Absolute Units. Brit. Assoc. Report, 1882, p. 464 [title only].

407. On the Transmission of Force through an Elastic Solid. Brit. Assoc. Report, 1882, p. 474 [title only].

408. On a Method of Investigating Magnetic Susceptibility. Brit. Assoc. Report, 1882, p. 474 [title only].

409. On the Tides. [Lecture to B.A. at Southampton, Aug. 25, 1882]; Pop. Lect. III. pp. 139-190.

410. Approximate Photometric Measurements of Sun, Moon, Cloudy Sky, and Electric and other Artificial Lights. Glasg. Phil.

Soc. Proc. XIV. pp. 80-85, 1883 [read Nov. 29, 1882]; Nature XXVII. pp. 277-279, Jan. 18, 1883; Journ. de Phys. III. (2nd Ser.) p. 50, 1884; Lum. Élec. VII. pp. 638-640, Dec. 30, 1882; Math. and Phys. Papers III. art. xcii. pp. 249-255.

**1883.**

411. The Size of Atoms. Roy. Instit. Proc. X. pp. 185-213, 1884 [Feb. 2, 1883]; Nature XXVIII. pp. 203-205, June 28, 1883, pp. 250-254, July 12, 1886, pp. 274-278, July 19, 1883; Pop. Lect. I. pp. 147-217.

412. On the Dynamical Theory of Dispersion. Edinb. Roy. Soc. Proc. XII. p. 128 [read Mar. 5, 1883, title only].

413. On Gyrostatics. Edinb. Roy. Soc. Proc. XII. p. 128 [read Mar. 5, 1883, title only].

414. Oscillations and Waves in an Adynamic Gyrostatic System. Edinb. Roy. Soc. Proc. XII. p. 128 [read Mar. 5, 1883, title only]. [See No. 499.]

415. Electrical Units of Measurement. Inst. Civ. Engrs. [abstract of lecture at the Inst. Civ. Engrs., May 3, 1883]; Nature XXVIII. pp. 91-92, May 24, 1883; Chemical News XLVII. pp. 242-243, 1883; Pop. Lect. I. pp. 73-136.

416. On a Model illustrating Helicoidal Asymmetry, and particularly the Formation of Right- and Left-handed Helicoidal Crystals from a Non-Helicoidal Solution. Brit. Assoc. Report, 1883, p. 405 [title only].

417. Gyrostatic Determination of the North and South Line and the Latitude of any Place. Brit. Assoc. Report, 1883, p. 405 [title only].

418. The Six Gateways of Knowledge [Address at the Midland Institute, Birmingham, Oct. 3, 1883]. Nature XXIX. pp. 438-440, Mar. 6, 1884, pp. 462-465, Mar. 13, 1884; Pop. Lect. I. pp. 253-299.

419. Obituary Notice of Sir C. W. Siemens. Nature XXIX. pp. 97-99, Nov. 29, 1883.

**1884.**

420. On the Measurement of Electric Currents and Potentials. Glasg. Phil. Soc. Proc. XV. 1884, pp. 96-101 [read Jan. 9, 1884].

421. On Measurements made by Mr. Abdank and himself in the Vienna Electric Exhibition on the Current yielded by the New Voltaic Element of Lalande. Glasg. Phil. Soc. Proc. XV. p. 378 [read Jan. 9, 1884, title only].

422. On the Efficiency of Clothing for Maintaining Temperature. Edinb. Roy. Soc. Proc. XII. p. 563 [Mar. 3, 1884, title only]; Nature XXIX. p. 567, Apr. 10, 1884.

423. On a Modification of Gauss's Method for Determining the Horizontal Component of Terrestrial Magnetic Force, and the Magnetic Moments of Bar Magnets in Absolute Measure.

Edinb. Roy. Soc. Proc. XII. p. 578 [Mar. 3, 1884, title only].

424. Discussion of W. H. Preece's paper on "The Electrical Congresses of Paris." Soc. Telegr. Engrs. Proc. XIII. pp. 381-388 [May 22, 1884].

425. Steps towards a Kinetic Theory of Matter. [Presidential Address to the Mathematical and Physical Science Section of the British Association, Montreal, Aug. 28, 1884.] Brit. Assoc. Report, 1884, pp. 613-622; Nature XXXI. pp. 461-463. Mar. 19, 1885: Pop. Lect. I. pp. 218-252.

426. On a Gyrostatic Working Model of the Magnetic Compass. Brit. Assoc. Report, 1884, pp. 625-628.

427. On Safety Fuses for Electric Circuits. Brit. Assoc. Report, 1884, p. 632 [title only].

428. On the Distribution of Potential in Conductors experiencing the Electromagnetic Effects discovered by Hall. American Association for the Advancement of Science. Paper read in Section B, Philadelphia meeting, Sept. 1884, vol. xxxiii. 137 [title only, given in Proc.].

429. The Wave Theory of Light. [Lecture delivered in the Academy of Music, Philadelphia, under the auspices of the Franklin Institute]; Jnl. Franklin Instit. LXXXVIII. pp. 321-341, Nov. 1884 [delivered Sept. 29, 1884]; Nature XXXI. p. 62, Nov. 20, 1884, pp. 91-94, Nov. 27, 1884, p. 115, Dec. 4, 1884, pp. 508-510, Apr. 2, 1885; Pop. Lect. I. pp. 300-348.

430. On the Ether and Gravitational Matter through Infinite Space. [Amplification of Baltimore Lect. XVI., Oct. 15, 1884. Phil. Mag. II. pp. 161-177, Aug. 1901; Amer. Jnl. Sci. XII. (4th Ser.) pp. 390-391, 1901.

431. On the Weights of Atoms. [Lecture XVII. of Baltimore Lectures extended.] Phil. Mag. IV. pp. 177-198, Aug. 1902, pp. 281-301, Sept. 1902.

432. On the Dynamics of Reflection and Refraction in the Wave Theory of Light. Edinb. Roy. Soc. Proc. XIII. p. 2 [Dec. 1, 1884, title only].

433. On the Distribution of Energy between colliding Groups of Molecules. Edinb. Roy. Soc. Proc. XIII. p. 2 [Dec. 1, 1884, title only].

434. On Kerr's Discovery Regarding the Reflection of Light from a Magnetic Pole. Edinb. Roy. Soc. Proc. XIII. p. 2 [Dec. 1, 1884, title only].

**1885.**

435. The Bangor Laboratories [Address at Univ. Coll. Bangor, Feb. 2, 1885]; Nature XXXI. pp. 409-413, Mar. 5, 1885; Pop. Lect. II. p. 475-501.

436. On Energy in Vortex Motion. Edinb. Roy. Soc. Proc. XIII. p. 114 [Feb. 16, 1885, title only].

437. Letter giving Corrections of the Papyrograph Report of Baltimore Lectures. Nature XXXI. p. 407, Mar. 5, 1885.
438. On Constant Gravitational Instruments for Measuring Electric Currents and Potentials. Brit. Assoc. Report, 1885, pp. 905-906; Nature XXXII. p. 535, Oct. 1, 1885.
439. On a Method of Multiplying Potential from a Hundred to several Thousand Volts. Brit. Assoc. Report, 1885, p. 907.
440. Discussion on "Electrolysis" at the Brit. Assoc. opened by Prof. Lodge. Brit. Assoc. Report, 1885, pp. 723-772; Nature XXXIII. p. 20, Nov. 5, 1885.
441. On the Motion of a Liquid within an Ellipsoidal Hollow. Edinb. Roy. Soc. Proc. XIII. pp. 370-378 [Dec. 7, 1885]. [See No. 611.]
442. Obituary Notice of Fleeming Jenkin. Roy. Soc. Proc. XXXIX. pp. 1-3 [1885].

**1886.**

443. Capillary Attraction [1886]. Roy. Instit. Proc. XI. 1887, pp. 483-507 [Jan. 29, 1886]; Pop. Lect. I. pp. 1-55. [See No. 451.]
444. On the Magnitude of the Mutual Attraction between Two Pieces of Matter at a distance of less than Ten Micro-millimetres. Edinb. Roy. Soc. Proc. XIII. p. 625 [Mar. 1, 1886, title only].
445. On a New Form of Portable Spring Balance for the Measurement of Terrestrial Gravity. Edinb. Roy. Soc. Proc. XIII. pp. 683-686 [Apr. 19, 1886].
446. Description of a Differential Gravity Meter Founded on the Flexure of a Spring. Brit. Assoc. Report, 1886, pp. 534-535.
447. On Stationary Waves in Flowing Water. Brit. Assoc. Report, 1886, p. 546; Phil. Mag. XXII. pp. 353-357, 445-452, 517-530, 1886; XXIII. pp. 52-58, 1887; Nature XXXIV. pp. 507-508, Sept. 23, 1886.
448. Artificial Production and Maintenance of a Standing Bore. Brit. Assoc. Report, 1886, p. 547 [title only].
449. Velocity of Advance of a Natural Bore. Brit. Assoc. Report, 1886, p. 547 [title only].
450. Graphical Illustrations of Deep-Sea Wave Groups. Brit. Assoc. Report, 1886, p. 547 [title only].
451. Capillary Attraction. Nature XXXIV. 1886, pp. 270-272, July 22, 1886; 290-294, July 29, 1886; 366-369, Aug. 19, 1886. [See No. 443.]
452. On the Waves Produced by a Ship Advancing Uniformly into Smooth Water. Edinb. Roy. Soc. Proc. XIV. p. 37 [Dec. 20, 1886, title only]; Phil. Mag. XXIII. pp. 52-58, Jan. 1887.
453. On the Ring-Waves Produced by Throwing a Stone into Water.

Edinb. Roy. Soc. Proc. XIV. p. 37 [Dec. 20, 1886, title only]; Phil. Mag. XXIII. pp. 52-58, Jan. 1887.

**1887.**

454. On the Front and Rear of a Free Procession of Waves in Deep Water. Edinb. Roy. Soc. Proc. XIV. pp. 38-46 [Jan. 7, 1887]; Phil. Mag XXIII. pp. 113-120, Feb. 1887.

455. The Sun's Heat. Roy. Instit. Proc. XII. 1889, pp. 1-21 [Jan. 21, 1887]; Nature XXXV. pp. 297-300, Jan. 27, 1887; Ciel et Terre III. pp. 79-80, pp. 281-288, 1887-1888; Pop. Lect. I. pp. 369-422.

456. On the Waves Produced by a Single Impulse in Water of any Depth, or in a Dispersive Medium. Roy. Soc. Proc. XLII. pp. 80-83, 1887 [read Feb. 3, 1887]; Phil. Mag. XXIII. pp. 252-255, Mar. 1887.

457. On the Formation of Coreless Vortices by the Motion of a Solid through an Inviscid Incompressible Fluid. Roy. Soc. Proc. XLII. pp. 83-85, 1887 [Feb. 3, 1887]; Phil. Mag. XXIII. pp. 255-257, Mar. 1887.

458. On the Equilibrium of a Gas under its own Gravitation only. Edinb. Roy. Soc. Proc. XIV. pp. 111-118 [read Feb. 21, 1887], pt. iii. p. 118 [title only]; Phil. Mag. XXII. pp. 287-292, Mar. 1887.

459. On Laplace's Nebular Theory, considered in Relation to Thermodynamics. Edin. Roy. Soc. Proc. XIV. p. 121 [Mar. 7, 1887, title only].

460. The Sun's Heat. Good Words, 1887, pp. 149-153 [Mar.], pp. 262-269 [Apr.].

461. On Ship-Waves. Edinb. Roy. Soc. Proc. XIV. p. 194 [Apr. 18, 1887, title only].

462. On the Instability of Fluid Motion. Edinb. Roy. Soc. Proc. XIV. p. 194 [Apr. 18, 1887, title only]; Phil. Mag. XXIII. pp. 459-464, May 1887.

463. On a Double Chain of Electrical Measuring Instruments to Measure Currents from the Millionth of a Milliampere to a Thousand Amperes and to Measure Potentials up to Forty Thousand Volts. Glasg. Phil. Soc. Proc. XVIII. pp. 249-256 [Apr. 20, 1887]; La Lumière Électrique XXIV. pp. 476-479, June 4, 1887.

464. Nouveaux Appareils électriques de mesure. La Lumière électrique XXIV. pp. 501-506, June 11, 1887.

465. Stability of Fluid Motion—Rectilineal Motion of Viscous Fluid between Two Parallel Planes. Edinb. Roy. Soc. Proc. XIV. pp. 359-368 [read July 15, 1887]; Phil. Mag. XXIV. pp. 188-196, Aug. 1887.

466. Ship-Waves [a Lecture in connection with the Institution of Mechanical Engineers' Conference in Edinburgh. (E. H. Carbutt, President.) Marquis of Tweeddale in the Chair,

Aug. 3, 1887]. Inst. Mech. Engrs. Proc. 1887, pp. 409-433, Pop. Lect. III. p. 450.

467. On the Vortex Theory of the Luminiferous Æther. (On the Propagation of Laminar Motion through a Turbulently Moving Inviscid Liquid.) Brit. Assoc. Report, 1887, pp. 486-495; Phil. Mag. XXIV. pp. 342-353, Oct. 1887; Nature XXXVI. pp. 550-551, Oct. 6, 1887.

468. On the Turbulent Motion of Water between Two Planes. Brit. Assoc. Report, 1887, p. 581 [title only]; Nature XXXVI. p. 523, Sept. 29. 1887.

469. New Electric Balances. Brit. Assoc. Report, 1887, pp. 582-583 [title only]; Electrician XX. pp. 130-131, Dec. 16, 1887; Nature XXXVI. p. 522, Sept. 29, 1887.

470. On the Application of the Centi-ampere or the Deci-ampere Balance for the Measurement of the E.M.F. of a Single Cell. Brit. Assoc. Report, 1887, pp. 610-611; Electrician XX. pp. 130-131, 272, 1888; Nature XXXVI. p. 522, Sept. 29, 1887; with additions, Phil. Mag. XXIV. pp. 514-516, Dec. 1887, and XXV. p. 164, Feb. 1888.

471. Stability of Motion. Broad River flowing down an Inclined Plane Bed. Phil. Mag. XXIV., Sept. 1887, pp. 272-278.

472. Note on the Contributions of Fleeming Jenkin to Electrical and Engineering Science. Papers of Fleeming Jenkin, Memoir, Appendix I. pp. 155-159, pub. 1887.

473. On Cauchy's and Green's Doctrine of Extraneous Force to Explain Dynamically Fresnel's Kinematics of Double Refraction. Edinb. Roy. Soc. Proc. XV. pp. 21-33 [read Dec. 5, 1887]; Phil. Mag. XXV. pp. 116-128, Feb. 1888.

474. On Models of the Minimal Tetrakaidekahedron. [Note of the Exhibition of]; Edinb. Roy. Soc. Proc. XV. p. 33 [Dec. 5, 1887].

475. On the Division of Space with Minimum Partitional Area. Phil. Mag. XXIV. pp. 503-514, Dec. 1887; Acta. Math. XI. pp. 121-134, 1887-1888.

**1888.**

476. On a New Composite Electric Balance. Glasg. Phil. Soc. Proc. XIX. pp. 273-274 [read Feb. 4, 1888].

477. Note on the Determination of Diffusivity in Absolute Measure from Mr. Coleman's Experiments. Edinb. Roy. Soc. Proc. XV. p. 256 [Mar. 19, 1888, title only].

478. Polar Ice-caps and their Influence in Changing Sea-levels. Glasg. Geol. Soc. VIII. pp. 322-340 [Feb. 16, 1888]; Pop. Lect. II. pp. 319-359.

479. On his New Standard and Inspectional Electric Measuring Instruments. Soc. Teleg. Engrs. Jl. XVII. p. 540, pp. 560 and 563 [May 24, 1888].

480. A Simple Hypothesis for Electromagnetic Induction of Incomplete Circuits, with consequent Equations of Electric Motion

in Fixed Homogeneous or Heterogeneous Solid Matter. Brit. Assoc. Report, 1888, pp. 567-570; Nature XXXVIII. pp. 569-571, Oct. 11, 1888.

481. On the Transference of Electricity within a Homogeneous Solid Conductor. Brit. Assoc. Report, 1888, pp. 570-571; Nature XXXVIII. p. 571, Oct. 11, 1888.

482. Five Applications of Fourier's Law of Diffusion, illustrated by a Diagram of Curves with Absolute Numerical Values. Brit. Assoc. Report, 1888, pp. 571-574; Nature XXXVIII. pp. 571-573, Oct. 11, 1888; Lum. Élec. XXX. pp. 80-83, Oct. 13, 1888; Math. and Phys. Papers III. art. xcviii. pp. 428-435.

483. Discussion on Lightning Conductors at the British Association. Brit. Assoc. Report, 1888, pp. 603-606; Nature XXXVIII. p. 546, Oct. 4, 1888.

484. Thomson (Sir William), Ayrton (W. E.), and Perry (J.).—Electrometric Determination of "*v*." Brit. Assoc. Report, 1888, pp. 616 [title only]; Electrician XXI., 1888, p. 681; Lum. Elec. XXX. p. 79, Oct. 13, 1888.

485. Note of a Letter to *The Times* Correcting an Error in the Report of the B.A. Meeting, and Criticising Remarks on the British Association. Nature XXXVIII. p. 500, Sept. 20, 1888.

486. On the Reflexion and Refraction of Light. Phil. Mag. XXVI. 1888, pp. 414-425, Nov. 1888, and 500-501, Dec. 1888.

**1889.**

487. Ether, Electricity, and Ponderable Matter. Inst. Elec. Engrs. Jl. XVIII. 1890, pp. 4-37 (Inaug. Address, Jan. 10, 1889); Correction XVIII. 1890, p. 128; Correction vol. XXIV. 1895, inserted between pp. 396 and 397; Math. and Phys. Papers III. art. cii. pp. 484-515.

488. Electrostatic Measurement. Roy. Inst. Proc. XII., 1889, pp. 561-562 [Feb. 8, 1889]; Nature XXXIX. pp. 465-466, Mar. 14, 1889.

489. Exhibition of a Gyrostatic Model of a Medium capable of Transmitting Waves of Transverse Vibration. Edinb. Roy. Soc. Proc. XVI. p. 811; Nature XXXIX. p. 527, Mar. 28, 1889.

490. Gyrostatic Experiments. Belfast Nat. Hist. Soc. Rep. and Proc. (1888-89), pp. 89-91 [Apr. 16, 1889].

491. Discussion of Dr. O. J. Lodge's paper on "Lightning, Lightning Conductors, and Lightning Protector," Inst. Elec. Engrs. Jl. XVIII. pp. 516-520 [May 16, 1889].

492. On the Security against Disturbance of Ships' Compasses by Electric Lighting Appliances. Inst. Elec. Engrs. Jl. XVIII. 1890, pp. 567-571 [May 23, 1889]; p. 579; Lum. Élec. XXXIII. pp. 290-291, Aug. 10, 1889.

493. Discussion of W. M. Mordey's Paper on "Alternate Current Working." Inst. Elec. Engrs. Jl. XVIII. pp. 667-668 [May 30, 1889].

494. Molecular Constitution of Matter. Edinb. Roy. Soc. Proc. XVI. pp. 693-724 [read July 1, 1889]; XXIV. p. 583 [Dec. 2, 1901]; Math. and Phys. Papers III. art. xcvii. pp. 395-427. [See Nos. 496 and 508.]
495. Electrification of Air by Flame. Edinb. Roy. Soc. Proc. XVI. pp. 262-263 [read July 15, 1889].
496. Sur la tactique moléculaire de la macle artificielle du spath d'Islande produite par Baumhauer au moyen d'un couteau. Comp. Rend. CIX. pp. 333-337 [Aug. 26, 1889]; Edinb. Roy. Soc. Proc. XVI. p. 707 [see No. 494].
497. Sur l'équilibre des atomes et sur l'élasticité des solides, dans la théorie Boscovichienne de la matière. Comp. Rend. CIX. pp. 337-341 [Aug. 26, 1889].
498. On Boscovich's Theory. Brit. Assoc. Report, 1889, pp. 494-496; Nature XL. pp. 545-547, Oct. 3, 1889.
499. Sur une constitution gyrostatique adynamique pour l'Éther. Comp. Rend. CIX. pp. 453-455 [Sept. 16, 1889]; Math. and Phys. Papers III. art. c. pp. 466-467 [see No. 414].
500. Note on the Direction of the Induced Longitudinal Current in Iron and Nickel Wires by Twist when under Longitudinal Magnetizing Force. Phil. Mag. XXIX. pp. 132-133, Jan. 1890 [dated Dec. 21, 1889].
501. On the Stability and Small Oscillation of a Perfect Liquid Full of Nearly Straight Coreless Vortices. Irish Acad. Proc. I. 1889-91, pp. 340-342 [1889].

**1890.**

502. Remarks on Retiring from the Presidential Chair, and Reply to Vote of Thanks. Inst. Elec. Engrs. Jl. XIX. pp. 2-6, [Jan. 9, 1890].
503. Remarks on the President's (Dr. J. Hopkinson) Inaugural Address (Magnetism). Inst. Elec. Engrs. Jl. XIX. p. 36 [Jan. 9, 1890].
504. On Electrostatic Stress. Edinb. Roy. Soc. Proc. XVII. p. 412 [Jan. 20, 1890, title only]; Nature XLI. p. 358, Feb. 13, 1890.
505. Rainbow due to Sunlight Reflected from the Sea. [Letter publishing a Letter from William Scouller.] Nature XLII. p. 271, Jan. 23, 1890.
506. Eight Rainbows seen at the same Time. [Letter publishing a Letter from Mr. P. Frost.] Nature XLI. pp. 316-317, Feb. 6, 1890.
507. Communication in reference to Mr. Preece's Remarks on Submarine Cables in the Discussion on Mr. Addenbrooke's Paper on "Electrical Engineering in America." Inst. Elec. Engrs. Jl. XVIII. p. 863 [Feb. 11, 1890].
508. On the Moduluses of Elasticity in an Elastic Solid according to Boscovich's Theory. Edinb. Math. Soc., Feb. 1890, *not printed*, but substance of the paper is in Edinb. Roy. Soc.

Proc. XVI. pp. 693-724, and Math and Phys. Papers III. art. xcvii. pp. 395-498 [see No. 494].

509. On a Mechanism for the Constitution of Ether. Edinb. Roy. Soc. Proc. XVII. pp. 127-132 [read Mar. 17, 1890]; Math. and Phys. Papers III. art. c. pp. 467-472.

510. On an Accidental Illustration of the Effective Ohmic Resistance to a Transient Electric Current through a Steel Bar. Edinb. Roy. Soc. Proc. XVII. pp. 157-167 [read Mar. 17, 1890]; Math. and Phys. Papers III. art. xci. pp. 473-483.

511. On the Time-integral of a Transient Electromagnetically Induced Current. Phil. Mag. XXIX. pp. 276-280, March 1890.

512. L'Éclairage électrique et la securité publique. Lum. élec. XXXVI. pp. 40-43, Apr. 5, 1890.

513. On the Researches of Hertz and Lodge, being a Short Speech after a Paper by Lodge on "Electrical Oscillations." Glasg. Phil. Soc. Proc. XXI. p. 224 [Apr. 16, 1890].

514. Motion of a Viscous Liquid; Equilibrium or Motion of an Elastic Solid; Equilibrium or Motion of an Ideal Substance called for Brevity *Ether*; Mechanical Representation of Magnetic Force. Math. and Phys. Papers III. art. xcix. pp. 436-465, May 1890.

515. Address on Presenting the Gunning Victoria Jubilee Prize for 1887-90 to Professor Tait. Edinb. Roy. Soc. Proc. XVII. pp. 417-418 [July 7, 1890].

516. On the Submarine Cable Problem, with Electromagnetic Induction. Edin. Roy. Soc. Proc. XVII. p. 420 [title only]; Nature XLII. pp. 287-288, July 17, 1890.

517. On an Illustration of Contact Electricity presented by the Multicellular Electrometer. Brit. Assoc. Report, 1890, p. 728; Nature XLII. p. 577, Oct. 9, 1890; Lum. Élec. XXXVIII., Dec. 6, 1890.

518. On Alternate Currents in Parallel Conductors of Homogeneous or Heterogeneous Substance. Brit. Assoc. Report, 1890, pp. 732-736; Nature XLII. p. 577, Oct. 9, 1890.

519. On Anti-effective Copper in Parallel Conductors or in Coiled Conductors for Alternate Currents. Brit. Assoc. Report, 1890, pp. 736-740; Nature XLII. pp. 577-578, Oct. 9, 1890; Lum. Elec. XXXVIII. pp. 317-321.

520. On a Method of Determining in Absolute Measure the Magnetic Susceptibility of Diamagnetic and Feebly Magnetic Solids. Brit. Assoc. Report, 1890, pp. 745-746; Nature XLII. p. 578, Oct. 9, 1890; Lum. Élec. XXXVIII. p. 611, Dec. 27, 1890.

521. A New Electric Meter. The Multicellular Voltmeter. An Engine-room Voltmeter. An Ampere Gauge. A New Form of Voltapile, useful in Standardising Operations. Brit. Assoc. Report, 1890, p. 956 [title only]; Nature XLII. p. 534, Sept. 25, 1890.

522. Prof. Tait's Experimental Results regarding the Compressibility of Fresh Water and Sea Water at Different Temperatures; Compressibilities at Single Temperatures of Mercury, of Glass, and of Water with Different Proportions of Common Salt in Solution; being an Extract from his Contribution to the Report on the Scientific Results of the Voyage of H.M.S. *Challenger*, vol. ii. Physics and Chemistry, Part iv. (published in 1888). Math. and Phys. Papers III. art. ciii. pp. 516-518.
523. Velocities of Waves of Different Character, and Corresponding Moduluses in Cases of Waves due to Elasticity; being Appendix to art. xcii. § 51. [See No. 357.] Math. and Phys. Papers III. art. civ. pp. 519-522.

**1891.**

524. Remarks in proposing a Vote of Thanks to the President (Sir W. Crookes) for his Address. Inst. Elec. Engrs. Jl. XX. p. 49, Jan. 15, 1891.
525. On Electrostatic Screening by Gratings, Nets, or Perforated Sheets of Conducting Material. Roy. Soc. Proc. XLIX., 1891, pp. 405-418 [Apr. 9, 1891].
526. On Variational Electric and Magnetic Screening. Roy. Soc. Proc. XLIX., 1891, pp. 418-423 [Apr. 9, 1891]; Lum. Élec. XL. pp. 229-232, May 2, 1891; Baltimore Lectures, Appendix K, pp. 681-687, 1904.
527. Electric and Magnetic Screening. Roy. Inst. Proc. XIII. 1893, pp. 345-355 [Apr. 10, 1891].
528. Discussion of Dr. J. A. Fleming's Paper on "Effects of Alternating Current Flow in Circuits having Capacity and Self-Induction," and W. H. Preece's Paper on "Some Points Connected with Mains for Electric Lighting." Inst. Elec. Engrs. Jl. XX. pp. 466-469 [May 21, 1891].
529. On some Test Cases for the Maxwell-Boltzmann Doctrine regarding Distribution of Energy. Roy. Soc. Proc. L., 1892, pp. 79-88 [June 11, 1891]; Nature XLIV. pp. 355-358, Aug. 13, 1891.
530. On Periodic Motion of a Finite Conservative System. Brit. Assoc. Report, 1891, p. 566 [title only]; Phil. Mag. XXXII. pp. 375-383, Oct. 1891; pp. 555-560, Dec. 1891 [read to R.S.]; [§§ 1-10 and 17-22, read at Brit. Assoc.]. [See No. 532.]
531. On a Theorem in Plane Kinetic Trigonometry suggested by Gauss's Theorem of *Curvatura Integra*. Phil. Mag. XXXII. pp. 471-473, Nov. 1891.
532. On Instability of Periodic Motion. Roy. Soc. Proc. L., 1892, pp. 194-200 [Nov. 26, 1891]; Phil. Mag. XXXII. pp. 555-560, Dec. 1891 [see No. 530].
533. Presidential Address. Roy. Soc. Proc. L., 1892, pp. 219-229 [Nov. 30, 1891]; Nature XLV. pp. 110-112; Pop. Lect. II. pp. 502-508.

**1892.**

534. On the Dissipation of Energy. Fortnightly Review, March 1892; Pop. Lect. II. pp. 451-474.

535. Discussion of J. Swinburne's Paper on "Electrical Measuring Instruments." Inst. Civ. Engrs. Proc. CX. pp. 52-54 [Apr. 26, 1892].

536. On a Decisive Test-Case Disproving the Maxwell-Boltzmann Doctrine regarding Distribution of Kinetic Energy. Roy. Soc. Proc. LI., 1892, pp. 397-399 [Apr. 28, 1892]; Phil. Mag. XXXIII. pp. 466-467, May 1892; Nature XLVI. pp. 21-22, May 5, 1892.

537. On a New Form of Air Leyden, with Application to the Measurement of Small Electrostatic Capacities. Roy. Soc. Proc. LII., 1893, pp. 6-10 [June 2, 1892]; Nature XLVI. pp. 212-213, June 30, 1892; Lum. Élec. XLV. pp. 139-141, July 16, 1892.

538. Inaugural Address as President of the Institute of Marine Engineers. June 2, 1892. Rep. of Proc. at the Second Dinner of Inst. Marine Engineers, 1892, pp. 7-9.

539. On the Stability of Periodic Motion. Brit. Assoc. Report, 1892, p. 638 (title only); Nature XLVI. p. 384, Aug. 18, 1892.

540. On Graphic Solution of Dynamical Problems. Brit. Assoc. Report, 1892, pp. 648-652; Nature XLVI. pp. 385-386, Aug. 18, 1892; Phil. Mag. XXXIV. pp. 443-448, Nov. 1892.

541. Reduction of every Problem of Two Freedoms in Conservative Dynamics to the Drawing of Geodetic Lines on a Surface of given Specific Curvature. Brit. Assoc. Report, 1892, pp. 652-653; Nature XLVI. p. 386, Aug. 18, 1892.

542. Discussion at the Brit. Assoc. on a National Physical Laboratory; Brit. Assoc. Report, 1892, p. 648 [title only]; Nature XLVI. pp. 383, Aug. 18, 1892.

543. Generalisation of "Mercator's" Projection performed by Aid of Electrical Instruments. Nature XLVI. pp. 490-491, Sept. 22, 1892.

544. To Draw a Mercator Chart on one Sheet representing the whole of any Complexly Continuous Closed Surface. Nature XLVI. pp. 541-542, Oct. 6, 1892.

545. Comment le soleil a commencé à brûler. L'Astronomie XI. pp. 361-367, Oct. 1892; Nature XLVI. p. 597, Oct. 20, 1892.

546. Presidential Address. Roy. Soc. Proc. LII. 1893, pp. 300-310 [Nov. 30, 1892]; Nature XLVII. pp. 107-111, Dec. 1, 1892; Pop. Lect. II. pp. 508-527.

547. On the Velocity of Crookes' Cathode Stream. Roy. Soc. Proc. LII. 1893, pp. 331-332 [Dec. 8, 1892]; Nature XLVII. pp. 164-165, Dec. 15, 1892.

**1893.**

548. Isoperimetrical Problems. Roy. Instit. Proc. XIV., 1896, pp. 111-119 [May 12, 1893]; Nature XLIX. pp. 515-518, Mar. 29, 1894; Pop. Lect. II. pp. 571-592.
549. On the Molecular Tactics of a Crystal. [The Robert Boyle Lecture, delivered before the Oxford University Junior Scientific Club, May 16, 1893.] Baltimore Lectures, Appendix H, pp. 602-642, 1904.
550. On the Elasticity of a Crystal According to Boscovich. Roy. Soc. Proc. LIV., 1894, pp. 59-75 [June 15, 1893]; Phil. Mag. XXXVI. pp. 414-430, Nov. 1893; Baltimore Lectures, Appendix I. pp. 643-661, 1904.
551. Quartz, Piezo-electric Property of Quartz. Brit. Assoc. Report, 1893, p. 691 [title only]; Phil. Mag. XXXVI, 1893, pp. 331-342, Oct. 1893, and p. 384, Oct. 1893; Lum. Élec. L. pp. 37-41, Oct. 7, 1893, L. pp. 236-237, Nov. 4, 1893.
552. On a Piezo-electric Pile. Brit. Assoc. Report, 1893, pp. 691-692; Phil. Mag. XXXVI. pp. 342-343, Oct. 1893; Electrician XXXI. p. 664, Oct. 20, 1893; Lum. Élec. L. pp. 41-42, Oct. 7, 1893.
553. Sur la théorie de la pyro-électricité et de la piézo-électricité. Comp. Rend. CXVII. pp. 463-472 [Oct. 9, 1893]; Lum. Élec. L. pp. 238-242, Nov. 4, 1893; Phil. Mag. XXXVI. pp. 453-459, Nov. 1893.
554. Presidential Address. Roy. Soc. Proc. LIV., 1894, pp. 377-389 [Nov. 30, 1893]; Nature XLIX. pp. 135-138, Dec. 7, 1893; Écl. Élec. VI. pp. 423-426, Feb. 29, 1896; Pop. Lect. II. pp. 529-557.
555. The Work of Joule: an Address on the Occasion of the Unveiling of Joule's Statue in Manchester Town Hall [Dec. 7, 1893]. Nature XLIX. pp. 163-165, Dec. 14, 1893; Pop. Lect. II. pp. 558-570.

**1894.**

556. On Homogeneous Division of Space. Roy. Soc. Proc. LV. 1894, pp. 1-16 [Jan. 18, 1894]; Nature XLIX. pp. 445-448, Mar. 8, 1894, pp. 469-471, Mar. 15, 1894.
557. Kelvin (Lord) and Maclean (Magnus).—On the Electrification of Air. Roy. Soc. Proc. LVI., 1894, pp. 84-94 [May 31, 1894]; Errata LVII., 1895, p. xxxii.; Nature L. pp. 280-283, July 19, 1884; Phil. Mag. XXXVIII. pp. 225-235, Aug. 1894; Écl. Élec. L. pp. 568-570, Dec. 1, 1894.
558. Kelvin (Lord), Maclean (Magnus), and Galt (Alexander).—Preliminary Experiments to find if Subtraction of Water from Air electrifies it. Brit. Assoc. Report, 1894, pp. 554-555; Écl. Élec. II. p. 92, Jan. 12, 1895.
559. Kelvin (Lord) and Galt (Alexander).—Preliminary Experiments for Comparing the Discharge of a Leyden Jar through Different Branches of a Divided Channel. Brit. Assoc.

Report, 1894, pp. 555-556; Écl. Élec. II. pp. 35-37, Jan. 5, 1895.
560. On the Resistance Experienced by Solids in Moving through Fluids. Brit. Assoc. Report, 1894, p. 557 [title only].
561. Towards the Efficiency of Sails, Windmills, Screw-propellers in Water and Air, and Aeroplanes. Letter dated Aug. 17, to Nature L., 1894, pp. 425-426, Aug. 30, 1894.
562. On the Doctrine of Discontinuity of Fluid Motion, in Connection with the Resistance against a Solid Moving through a Fluid. Nature L., 1894, pp. 524-525, Sept. 27, p. 549, Oct. 4, pp. 573-575, Oct. 11, pp. 597-598, Oct. 18, 1894.
563. On the Resistance of a Fluid to a Plane kept Moving Uniformly in a Direction Inclined to it at a Small Angle. Phil. Mag. XXXVIII. pp. 409-413, Oct. 1894.
564. Presidential Address. Roy. Soc. Proc. LVII., 1895, pp. 37-54 [Nov. 30, 1894].

**1895.**

565. Kelvin (Lord), Maclean (Magnus), and Galt (Alexander).—Electrification of Air and other Gases by Bubbling through Water and other Liquids. Roy. Soc. Proc. LVII., 1895, pp. 335-346 [Feb. 21, 1895]; Nature LI. pp. 495-499, Mar. 21, 1895; Écl. Élec. III. pp. 521-524, June 15, 1895.
566. The Age of the Earth. Nature LI. pp. 438-440, Mar. 7, 1895.
567. Kelvin (Lord), Maclean (Magnus), and Galt (Alexander).—On the Diselectrification of Air. Roy. Soc. Proc. LVII., 1895, pp. 436-439 [Mar. 21, 1895]; Nature LI. pp. 573-574, Apr. 11, 1895.
568. Kelvin (Lord) and Murray (Erskine).—On the Temperature Variation of the Thermal Conductivity of Rocks. Glas. Phil. Soc. Proc. XXVI. pp. 227-232 [Mar. 27, 1895]; Roy. Soc. Proc. LVIII. pp. 162-167 [May 30, 1895]; Nature LII. p. 70, May 16, pp. 182-184, June 20, 1895; Amer. Jl. Sci. L. (3rd Ser.) pp. 419-423, 1895.
569. On the Electrification of Air. Glas. Phil. Soc. Proc. XXVI. pp. 233-243 [Mar. 27, 1895]; Nature LII. pp. 67-70, May 16, 1895.
570. On the Translational and Vibrational Energies of Vibrators after Impacts on Fixed Walls. Brit. Assoc. Report, 1895, p. 612 [title only]; Nature LII., 1895, p. 533, Sept. 26, 1895.
571. Kelvin (Lord), Maclean (Magnus), and Galt (Alexander).—On the Electrification and Diselectrification of Air and Other Gases. Brit. Assoc. Report, 1895, pp. 630-633; Nature LII. pp. 608-610, Oct. 17, 1895.
572. Discours prononcé, en réponse à l'Allocution du Président de l'Institut, à l'occasion du centenaire. Comptes Rendus CXXI. pp. 581-582 [Oct. 28, 1895].
573. Presidential Address. Roy. Soc. Proc. LIX., 1896, pp. 107-

124 [Nov. 30, 1895]; Nature LIII. pp. 110-115, Dec. 5, 1895.

**1896**

574. Velocity of Propagation of Electrostatic Force [Letters]; Nature LIII. p. 316, Feb. 6, 1896; Nature LIII. pp. 364-365, Feb. 20, 1896; Ecl. Élec. VI. pp. 494-495, Mar. 14, 1896.

575. On the Generation of Longitudinal Waves in Ether. Roy. Soc. Proc. LIX., 1896, pp. 270-273 [Feb. 13, 1896]; Nature LIII. pp. 450-451, Mar. 12, 1896; Écl. Élec. VI. pp. 493-494, Mar. 14, 1896.

576. Note on Lord Blythswood's paper ["On the Reflection of Roentgen Light from Polished Speculum Metal Mirrors"]; Roy. Soc. Proc. LIX., 1896, pp. 332-333 [read Mar. 19, 1896].

577. On the Motion of a Heterogeneous Liquid, commencing from Rest with a given Motion of its Boundary. Edinb. Roy. Soc. Proc. XXI., 1897, pp. 119-122 [read Apr. 6, 1896]; Nature LIV. pp. 250-251, July 16, 1896.

578. On Lippman's Colour Photography with Obliquely Incident Light. Nature LIV., 1896, pp. 12-13, May 7, 1896.

579. Kelvin (Lord), Bottomley (J. T.), and Maclean (Magnus).—On Measurements of Electric Currents through Air at Different Densities down to One Five-millionth of the Density of Ordinary Air. Brit. Assoc. Report, 1896, pp. 710-711.

580. Kelvin (Lord), Maclean (Magnus), and Galt (Alexander).—On the Communication of Electricity from Electrified Steam to Air. Brit. Assoc. Report, 1896, p. 721 [title only]; Electrician XXXVIII., 1897, p. 115; Nature LIV. pp. 622-623, Oct. 29, 1896.

581. On the Molecular Dynamics of Hydrogen Gas, Oxygen Gas, Ozone, Peroxide of Hydrogen, Vapour of Water, Liquid Water, Ice, and Quartz Crystal. Brit Assoc. Report, 1896, pp. 721-724.

582. Kelvin (Lord), Beattie (J. C.), and Smolan (M. S. de).—Experiments on the Electrical Phenomena Produced in Gases by Roentgen Rays, by Ultra-violet Light, and by Uranium. Edinb. Roy. Soc. Proc. XXI. pp. 393-428, 1897. Part I. Read Dec. 21, 1896; Nature LV. pp. 199-200, Dec. 31, 1896. Part II. Read Feb. 15, 1897; Nature LV. pp. 472-474, Mar. 18, 1897; Electrician XXXVIII. p. 401, Jan. 22, 1897; Part III. Read Mar. 1, 1897; Nature LV. pp. 498-499, Mar. 25, 1897; Phil. Mag. XLV. pp. 277-278, Mar. 1898; Part IV. Read Feb. 1, 1897; Nature LV. pp. 343-347, Feb. 11, 1897. Part V. Read Apr. 4, 1897; Nature LVI. p. 20, May 6, 1897.

583. On Atomic Configurations in Molecules of Gases according to Boscovich. Edinb. Roy. Soc. Proc. XXI. p. 511 [read Dec. 21, 1896, title only]; Nature LV. p. 238, Jan. 7, 1897.

**1897**

584. On Osmotic Pressure against an Ideal Semi-permeable Membrane. Edinb. Roy. Soc. Proc. XXI. pp. 323-325 [read Jan. 18, 1897]; Nature LV. pp. 272-273, Jan. 21, 1897.
585. On a Differential Method of Measuring Differences of Vapour Pressure, of Liquids at one Temperature and at Different Temperatures. Edinb. Roy. Soc. Proc. XXI. pp. 429-432 [read Jan. 18, 1897]; Nature LV. pp. 273-274, Jan. 21, 1897.
586. Osmotic Pressure [Letter]; Nature LV. p. 272, Jan. 21, 1897.
587. Note on a Method suggested for Measuring Vapour Pressures. Nature LV., 1896-97, pp. 295-296, Jan. 28, 1897.
588. Crystallization according to Rule. Edinb. Roy. Soc. Proc. XXI. p. 513 [read Feb. 1, 1897, title only].
589. On Configurations of Minimum Potential Energy in Clusters of Homogeneous Molecules with Application to the Theory of Crystalline forms. Edinb. Roy. Soc. Proc. p. 513 [read Feb. 15, 1897, title only].
590. Kelvin (Lord), Beattie (J. Carruthers), and Smolan (M. S. de). —Electric Equilibrium between Uranium and an Insulated Metal in its Neighbourhood. Edinb. Roy. Soc. Proc. XXII. pp. 131-133 [read Mar. 1, 1897]; Nature LV. pp. 447-448, Mar. 11, 1897; Écl. Élec. XI. pp. 119-120, Apr. 10, 1897.
591. On Models illustrating the Dynamical Theory of Hemihedral Crystals. Edinb. Roy. Soc. Proc. XXI. p. 514 [Mar. 1, 1897].
592. Kelvin (Lord) and Maclean (Magnus).—On Electrical Properties of Fumes proceeding from Flames and Burning Charcoal. Edinb. Roy. Soc. Proc. XXI. pp. 313-322 [read Apr. 5, 1897]; Nature LV. pp. 592-595, Apr. 22, 1897.
593. Contact Electricity of Metals; Roy. Instit. Proc. XV., 1899, pp. 521-554 [May 21, amplified Feb. 1898]; Phil. Mag. XLVI. pp. 82-120, July 1898.
594. Discussion of Mr. Mordey's Paper on "Dynamoes." Inst. Elec. Engrs. Jl. XXVI. p. 610 [May 27, 1897].
595. Contact Electricity and Electrolysis according to Father Boscovich. Nature LVI. pp. 84-85, May 27, 1897.
596. Note to "The Electrification of Air by Uranium and its Compounds," by J. Carruthers Beattie. Edinb. Roy. Soc. Proc. XXI. pp. 471-472 [read June 7, 1897]; Phil. Mag. XLIV. pp. 107-108, July 1897.
597. Kelvin (Lord), Maclean (Magnus), and Galt (Alexander).—Electrification of Air, of Vapour of Water, and of other Gases. Roy. Soc. Proc. LXI. pp. 483-485 [read June 17, 1897]; Phil. Trans. (A), CXCI., 1898, pp. 187-228.
598. Kelvin (Lord) and Maclean (Magnus).—Leakage from Electrified Metal Plates and Points placed above and below Uninsulated Flames. Edinb. Roy. Soc. Proc. XXII. pp.

38-46 [read July 5, 1897]; Nature LVI. pp. 233-235, July 8, 1897.

599. Address on Presenting the Keith Prize, the Makdougall-Brisbane Prize, and the Neill Prize. Edinb. Roy. Soc. Proc. XXI. pp. 518-522 [July 19, 1897].

600. On the Fuel-Supply and the Air-Supply of the Earth. Brit. Assoc. Report, 1897, pp. 553-554.

601. Chairman's Opening Address. Edinb. Roy. Soc. Proc. XXII. pp. 2-10 [read Dec. 6, 1897].

602. On the Food, Fuel, and Air of the World. Edinb. Roy. Soc. Proc. XXII. p. 701 [read Dec. 6, 1897, title only].

603. The Age of the Earth as an Abode fitted for Life. [Address 1897.] Journ. Victoria Institute, London, XXXI. pp. 11-35, 1899; Ann. Rep. Smithsn. Inst. 1897, pp. 337-357, 1898; Phil. Mag. XLVII. pp. 66-90, Jan. 1899 [with additions].

**1898.**

604. On Thermodynamics of Volta-contact Electricity. Edinb. Roy. Soc. Proc. XXII. pp. 118-125 [read Feb. 21, 1898].

605. On Thermodynamics founded on Motivity and Energy. Edinb. Roy. Soc. Proc. XXII. pp. 126-130 [read Mar. 21, 1898].

606. The Dynamical Theory of Refraction, Dispersion, and Anomalous Dispersion. Brit. Assoc. Report, 1898, pp. 782-783; Nature LVIII. pp. 546-547, Oct. 6, 1898.

607. Continuity in Undulatory Theory of Condensational-rarefactional Waves in Gases, Liquids, and Solids, of Distortional Waves in Solids, of Electric Waves in all Substances capable of Transmitting them, and of Radiant Heat, Visible Light, Ultra-violet Light. Brit. Assoc. Report, 1898, pp. 783-787; Nature LIX. pp. 56-57, Nov. 17, 1898; Phil. Mag. XLVI. pp. 494-500, Nov. 1898.

608. On Graphic Representations of the Two Simplest Cases of a Single Wave; (*a*) Condensational-rarefractional, (*b*) Distortional. Brit. Assoc. Report, 1898, p. 792.

609. On the Reflection and Refraction of Solitary Plane Waves at a Plane Interface between Two Isotropic Elastic Mediums—Fluid, Solid, or Ether. Edinb. Roy. Soc. Proc. XXII. pp. 366-378 [read Dec. 19, 1898]; Phil. Mag. XLVII. pp. 179-191, Feb. 1899.

**1899.**

610. Application of Sellmeier's Dynamical Theory to the Dark Lines $D_1$, $D_2$ produced by Sodium-vapour. Edinb. Roy. Soc. Proc. XXII. pp. 523-531 [read Feb. 6, 1899]; Phil. Mag. XLVII. pp. 302-308, Mar. 1899.

611. On the Motion of a Liquid in an Ellipsoidal Hollow. [A continuation of a paper in Proc., Dec. 7, 1885]. (See No. 441.) Edinb. Roy. Soc. Proc. XXII. p. 724 [read Feb. 6, 1899, title only].

612. On the Application of Force within a Limited Space, required to produce Spherical Solitary Waves, or Trains of Periodic Waves, of both Species, Equivoluminal and Irrotational, in an Elastic Solid. Edinb. Roy. Soc. Proc. XXII. p. 726 [read May 1, 1899, title only]; Phil. Mag. XLVII. pp. 480-493, May 1899; XLVIII. pp. 227-236, Aug. 1899, pp. 388-393, Oct. 1899; also read before London Math. Soc. XXXI. p. 147 [June 8, 1899, title only].

613. Magnetism and Molecular Rotation. [Correspondence.] Electrician XLIII. p. 411, July 14, 1899, p. 532, Aug. 4, 1899, p. 574, Aug. 11, 1899.

614. Magnetism and Molecular Rotation. Edinb. Roy. Soc. Proc. XXII. pp. 631-635 [read July 17, 1899]; Phil. Mag. XLVIII. pp. 236-239, Aug. 1899.

615. Apparent Dark Lightning Flashes. Nature LX., 1899, p. 341, Aug. 10, 1899.

616. Blue Ray of Sunrise over Mont Blanc. Letter to Nature LX. pp. 411 and 442, Aug. 31, 1899 [letter dated Aug. 27].

617. Chairman's Opening Address. Edinb. Roy. Soc. Proc. XXIII. pp. 2-11 [delivered Dec. 4, 1899]; Nature LXI. p. 166, Dec. 14, 1899.

**1900.**

618. Nineteenth-Century Clouds over the Dynamical Theory of Heat and Light. Roy. Inst. Proc. XVI., 1902, pp. 363-397 [Apr. 27, 1900]; Phil. Mag. II. pp. 1-40, July 1901; Amer. Jl. Sci. XII. p. 391, 1901; Baltimore Lectures, Appendix B, pp. 486-527, 1904.

619. On the Motion produced in an Infinite Elastic Solid by the Motion through the Space occupied by it of a Body acting on it only by Attraction or Repulsion. Edinb. Roy. Soc. Proc. XXIII. pp. 218-235 [read July 16, 1900]; Phil. Mag. L. pp. 181-198, Aug. 1900; Congrès Internationale de Physique à l'Exposition de 1900 II. pp. 1-22; Baltimore Lectures. Appendix A, pp. 468-485, 1904.

620. On the Number of Molecules in a Cubic Centimetre of a Gas. Edinb. Roy. Soc. Proc. XXIII. p. 435 [read July 16, 1900, title only].

621. On the Duties of Ether for Electricity and Magnetism. Phil. Mag. L. pp. 305-307, Sept. 1900.

622. On Transmission of Force through a Solid. [Address.] London, Math. Soc. Proc. XXXIII. p. 2 [Nov. 8, 1900, title only].

623. The Transmission of Force. Edinb. Roy. Soc. Proc. XXIII. p. 442 [read Dec. 17, 1900, title only].

**1901.**

624. One-dimensional Illustrations of the Kinetic Theory of Gases. Edinb. Roy. Soc. Proc. XXIII. p. 443 [read Jan. 21, 1901, title only].

625. Discussion on the Magnetic Effects of Electric Convection opened by Dr. V. Crémieu. Brit. Assoc. Report, 1901, p. 531 [title only]; Nature LXIV. p. 586, Oct. 10, 1901.
626. On the Clustering of Gravitational Matter in any part of the Universe. Brit. Assoc. Report, 1901, pp. 563-568; Nature LXIV. pp. 626-629, Oct. 24, 1901; Phil. Mag. III. pp. 1-9, Jan. 1902; Baltimore Lectures, Appendix D, pp. 532-540, 1904.
627. Obituary Notice of Professor Tait. Edinb. Roy. Soc. Proc. XXIII. pp. 498-504 [read Dec. 2, 1901].

**1902.**

628. A New Specifying Method for Stress and Strain in an Elastic Solid. Edinb. Roy. Soc. Proc. XXIV. pp. 97-101 [read Jan. 20, 1902]; Nature LXV. p. 407, Feb. 27, 1902; Phil. Mag. III. pp. 95-97, Jan. 1902, pp. 444-448, Apr. 1902.
629. Molecular Dynamics of a Crystal. Edinb. Roy. Soc. Proc. XXIV. pp. 205-224 [read Jan. 20, 1902]; Phil. Mag. IV. pp. 139-156, July 1902; Nature LXV. p. 407, Feb. 27, 1902; Baltimore Lectures, Appendix J, pp. 662-680, 1904.
630. Discussion of W. M. Mordey and B. M. Jenkin's paper on "Electrical Traction on Railways." Inst. Civ. Engrs. Proc. CXLIX. pp. 87-89 [Feb. 18, 1902].
631. Æpinus Atomized. [From the Jubilee Volume presented to Prof. Bosscha in Nov. 1901.] Phil. Mag. III. pp. 257-283, Mar. 1902; Baltimore Lectures, Appendix E, pp. 541-568, 1904.
632. Animal Thermostat. Brit. Assoc. Report, 1902, pp. 543-546; Nature LXVII. pp. 401-402, Feb. 26, 1903; Phil. Mag. V. pp. 198-202, Feb. 1903.
633. Becquerel Rays and Radio-activity. [Letter correcting the Report of the Physical Soc. Meeting on Oct. 31.] Nature LXVII. p. 103, Dec. 4, 1902.

**1903.**

634. Reflection and Refraction of Light. Edinb. Roy. Soc. Proc. XXIV. p. 606 [read Jan. 19, 1903, title only].
635. The Scientific Work of Sir George Stokes. Nature LXVII. pp. 337-338, Feb. 12, 1903.
636. Creative Power. Speech at University College. Nineteenth Century and After, June 1903.
637. On the Electro-ethereal Theory of the Velocity of Light in Gases, Liquids, and Solids. Brit. Assoc. Report, 1903, p. 535; Phil. Mag. VI. pp. 437-442, Oct. 1903; Baltimore Lectures XX. pp. 463-467.
638. Discussion on the Nature of the Emanations from Radium [opened by Prof. E. Rutherford]. Brit. Assoc. Report, 1903, pp. 535-537; Nature LXVIII. p. 609, Oct. 22, 1903; Phil. Mag. VII. pp. 220-222, Feb. 1904.
639. Recollections of Professor J. P. Nichol (Speech by Lord

Kelvin when unveiling a stained-glass window in the Bute Hall in Memory of J. P. Nichol). Nature LXVIII. pp. 623 and 624, Oct. 29, 1903.

**1904.**

640. On Deep-Water Two-Dimensional Waves produced by any given Initiating Disturbance. Edinb. Roy. Soc. Proc. XXV(i). pp. 185-196 [read Feb. 1, 1904]; Phil. Mag. VII. pp. 608-620, June 1904. [See No. 642.]
641. On the Destruction of Cambric by Radium Emanations. Phil. Mag. VII. p. 233, Feb. 1904.
642. On the Front and Rear of a Free Procession of Waves in Deep Water. Edinb. Roy. Soc. Proc. XXV(i). pp. 311-327 [read June 20, 1904]; Phil. Mag. VIII. pp. 454-470, Oct. 1904; Nature LXX. p. 263, July 4, 1904. [See Nos. 640 and 647.]
643. Electrical Insulation in Vacuum. Brit. Assoc. Report, 1904, p. 472 (title only); Phil. Mag. VIII., 1904, pp. 534-538, Oct. 1904.
644. Plan of a Combination of Atoms having the Properties of Polonium or Radium. Brit. Assoc. Report, 1904, p. 472 (title only); Phil. Mag. VIII. pp. 528-534, Oct. 1904.
645. Models of Radium Atoms to give out $\alpha$ and $\beta$ Rays respectively. Nature LXX. p. 516, Sep. 22, 1904.
646. On the Living Cell. [Speech at St. George's Hospital Medical School.] Nature LXXI. p. 13, Nov. 3, 1904.

**1905.**

647. Deep-Water Ship-Waves. Edinb. Roy. Soc. Proc. XXV(i). pp. 562-587 [read Jan. 23, 1905]; Phil. Mag. IX. pp. 733-757, June 1905. [See Nos. 642 and 648.]
648. On Deep-Sea Ship-Waves. Edinb. Roy. Soc. Proc. XXV(ii). pp. 1060-1084 [read July 17, 1905]; Phil. Mag. XI. pp. 1-25, Jan. 1906; Nature LXXI. p. 382, Feb. 16, 1905. [See Nos. 647 and 651.]
649. On the Kinetic and Statistical Equilibrium of Ether in Ponderable Matter at any Temperature. Brit. Assoc. Report, 1905, pp. 346-347; Phil. Mag. X. pp. 285-290, Sept. 1905; Nature LXXII. p. 641, Oct. 26, 1905.
650. Plan of an Atom to be capable of Storing an Electron with Enormous Energy for Radio-activity. Phil. Mag. X. pp. 695-698, Dec. 1905.

**1906.**

651. Initiation of Deep-Sea Waves of Three Classes: (1) From a Single Displacement; (2) From a Group of Equal and Similar Displacements; (3) By a Periodically Varying Surface-pressure. Edinb. Roy. Soc. Proc. XXVI. pp. 399-436 [read Jan. 22, 1906]; Phil. Mag. XIII. pp. 1-36, Jan. 1907. [See No. 648.]
652. Letter from Mr. Barber Starkey describing Remarkable Results

of a Discharge of Lightning upon an Oak Tree; with a Note by Lord Kelvin. Phil. Mag. XII. pp. 62-63, July 1906.

653. The Recent Radium Controversy. [Letter.] Nature LXXIV. p. 539, Sept. 27, 1906.

**1907.**

654. On the Means of Testing Experimentally the Motion of the Earth Relatively to the Ether. Edinb. Roy. Soc. Proc. XXVII. p. 375 [read Jan. 21, 1907, title only].

655. On Homer Lane's Problem of a Spherical Gaseous Nebula. Edinb. Roy. Soc. Proc. XXVII. p. 375 [read Jan. 21, 1907, title only]; Nature LXXV. pp. 368-369, Feb. 1907.

656. Discussion of Mr. Highfield's Paper on "Transmission of Energy by Direct Current." Inst. Elec. Engrs. Jl. XXXVIII. p. 502 [Mar. 7, 1907].

657. An Attempt to Explain the Radio-activity of Radium. Phil. Mag. XIII. pp. 313-316, Mar. 1907.

658. Discussion on the Constitution of the Atom, opened by Prof. E. Rutherford. Brit. Assoc. Report, 1907, p. 439 (title only); Nature LXXVI. p. 458, Aug. 29, 1907.

659. On the Motions of Ether produced by Collisions of Atoms or Molecules, containing or not containing Electrions. Brit. Assoc. Report, 1907, p. 439 (title only); Electrician LIX. pp. 714-716, Aug. 16, 1907; Phil. Mag. XIV. pp. 317-324, Sept. 1907; Nature LXXVI. p. 457, Aug. 29, 1907; Écl. Élec. LII. pp. 415-417, Sept. 21, 1907; LIII. pp. 14-16, Oct. 5, 1907.

**1908.**

660. The Problem of a Spherical Gaseous Nebula, with a Note by G. Green. Edinb. Roy. Soc. Proc. XXVIII. pp. 259-302 [MS. read Mar. 9, 1908], issued separately May 6, 1908; Phil. Mag. XV. pp. 687-711, June 1908; XVI. pp. 1-23, July 1908.

661. On the Formation of Concrete Matter from Atomic Origins. [Communicated by Dr. J. T. Bottomley.] Phil. Mag. XV. pp. 397-413, Apr. 1908.

---

Lord Kelvin was also one of the editors of the *Philosophical Magazine* from 1871 till his death in 1907.

# APPENDIX C

## LIST OF PATENTS

1854. 2547.—Thomson (W.), Rankine (W. J. M'Q.), and Thomson (J.). Improvements in electrical conductors for telegraphic communication.

1858. 329.—Improvements in testing and working electric telegraphs. (Disclaimer added in 1871.)

437.—Improvements in apparatus for applying and measuring resistance to the motion of rotating wheels, shafts, or other rotating bodies.

1860. 2047.—Thomson (W.) and Jenkin (F.). Improvements in the means of telegraphic communication.

1865. 1784.—Thomson (W.) and Varley (C. F.). Improvements in electric telegraphs.

1867. 2147.—Improvements in receiving or recording instruments for electric telegraphs.

1870. 3069.—Improvements in electric telegraph transmitting, receiving, and recording instruments, and in clocks.

1871. 252.—Improvements in transmitting, receiving, and recording instruments for electric telegraphs.

810.—Improvements in clocks and apparatus for giving uniform motion.

1873. 2086.—Thomson (Sir W.) and Jenkin (F.). Improvements in telegraphic apparatus.

1876. 1095.—Improvements in telegraphic apparatus.

1339.—Improvements in the mariner's compass and in the means of ascertaining and correcting its errors. (Disclaimer and memorandum of alteration—1877.)

3452.—Improvements in apparatus for navigational deep-sea soundings.

4876.—Improvements in the mariner's compass and in appliances for ascertaining and correcting its errors.

1879. 679.—Improvements in the mariner's compass and in appliances for correcting its errors.

1880. 781.—Improvements in navigational sounding apparatus.

1881. 3032.—Improvements in regulating electric currents, and in the apparatus or means employed therein.

1881. 5668.—Improvements in dynamo-electric machinery, and apparatus connected therewith.

1883. 2028.—Improvements in apparatus and processes for generating, regulating, and measuring electric currents.

4617.—Apparatus for generating, règulating, measuring, recording, and integrating electric currents.

5675.—Improved navigational sounding apparatus.

5676.—Improvements in the mariner's compass and in the means for ascertaining and correcting its errors.

1884. 4655.—New or improved suspensions for electrical incandescent lamps.

5335.—Thomson (Sir W.) and Ferranti (S. Z. de). Improvements in dynamo-electric machinery.

6410. — Improvements in breaking electric contact to prevent over-heating by imperfect contact.

10,530.—Thomson (Sir W.) and Bottomley (J. T). Safety fuses for electric circuits.

11,106.—Improvements in apparatus for measuring electric currents.

1885. 12,240.—An improved navigational sounding machine.

1886. 9016.—Improved apparatus for measuring the efficiency of an electric circuit (amended, Oct. 4, 1897).

1888. 18,035.—Electrostatic apparatus for measuring potentials.

18,035*a*.—An improved ampere-gauge and connections.

18,035*b*.—Improved apparatus for continually measuring potentials or currents.

1889. 4923.—Improvements of compass cards for use on board ship (amended, July 1, 1890).

5471.—Improvements in valves for water, steam, or other liquids or gases.

15,769.—Apparatus for measuring and recording electric currents.

1890. 8959.—Improvements in the mariner's compass.

1891. 1004.—An improved indicator for electric potentials.

3864.—Improvements in valves for water, steam, or other liquids or gases.

18,436.—Improved apparatus for measuring and recording electric currents.

1892. 10,230.—An improved electric condenser.

1893. 2198.—Improvements in balances.

2199.—An instrument for measuring electric currents.

5733.—Improved arrangement for reading the deflections of electric instruments.

24,471.—Improvements in electric supply meters.

24,841.—Improvements in the mariner's compass.

24,979.—Improvements in instruments for measuring and recording electric pressures and currents.

1894. 15,034.—Improvements in instruments for measuring electric currents.

1895. 22,661.—Improvements in apparatus for indicating and recording electric supply.

24,868.—Improvements in recording instruments for telegraphic and other purposes.

1896. 4207.—Improvements in instruments for measuring electric currents.

1897. 18,438.—Improved coil for electric instruments.

25,178.—Improvements in navigational sounding machines.

1898. 18,522.—An improved electrolytic apparatus for the production of alkali.

21,716.—Improvements in electric measuring instruments.

1900. 3937.—Apparatus for indicating and recording electric pressure and current.

1902. 17,267.—Improved electrostatic high-pressure voltmeter.

22,030*.—Improvements in and connected with navigational sounding machines.

22,031*.—Improvements in the mariner's compass.

22,032*.—Improvements in the mariner's compass.

28,588*.—Improved magnetic brake and damper of vibrations.

1903. 24,526*.—Improved sounding machine.

1904. 22,695*.—Improvements in the mariner's compass.

1905. 14,897.—Kelvin (Lord), Blakiston (R.), Hope (W.), and Richards (G. B.). Improvements in liquid meters.

20,813*.—Improvements in and connected with navigational sounding machines, especially for flying soundings.

20,813*a**.—Improvements in and connected with navigational sounding machines, especially for flying soundings.

26,759*.—An improved apparatus for adjusting the compass at sea without sights.

26,759*a**.—An improved apparatus for adjusting the compass at sea without sights.

1906. 26,132*.—An improvement in connection with navigational sounding machines, especially for flying soundings.

1907. 12,783*.—A combined level and shadow pin for use at sea in connection with the mariner's compass.

23,570*.—An improved compass suspension for use on board ship.

*N.B.*—Those patents marked with an asterisk in the above list were taken in the name of Lord Kelvin and of the firm Kelvin and James White, Ltd.

# INDEX

THE END

*Printed by* R. & R. CLARK, LIMITED, *Edinburgh.*

Printed in the United States
By Bookmasters